Sexual Biology and Reproduction in Crustaceans

Sexual Biology and Reproduction in Crustaceans

Thanumalaya Subramoniam, Ph.D, D.Sc, (Madras Univ); FNA, FNASc, FAAS
Senior Scientist,
Centre for Climate Change Studies,
Sathyabama University,
Rajiv Gandhi Salai, Chennai,
Tamil Nadu, India

ELSEVIER

AMSTERDAM • BOSTON • HEIDELBERG • LONDON
NEW YORK • OXFORD • PARIS • SAN DIEGO
SAN FRANCISCO • SINGAPORE • SYDNEY • TOKYO

Academic Press is an imprint of Elsevier

Academic Press is an imprint of Elsevier
125 London Wall, London EC2Y 5AS, United Kingdom
525 B Street, Suite 1800, San Diego, CA 92101-4495, United States
50 Hampshire Street, 5th Floor, Cambridge, MA 02139, United States
The Boulevard, Langford Lane, Kidlington, Oxford OX5 1GB, United Kingdom

Library of Congress Cataloging-in-Publication Data
A catalog record for this book is available from the Library of Congress

British Library Cataloguing-in-Publication Data
A catalogue record for this book is available from the British Library

ISBN: 978-0-12-809337-5

For information on all Academic Press publications visit our website at https://www.elsevier.com/

Publisher: Sara Tenney
Acquisition Editor: Kristi Gomez
Editorial Project Manager: Pat Gonzalez
Production Project Manager: Edward Taylor
Designer: Victoria Pearson

Dedication

This book is dedicated to the memory of my mother, Chellammal, and my wife, Lakshmi Bai.

Contents

Preface

Gametes development and their final fusion during fertilization assume center stage in the reproductive biology of animals. In Crustacea, gamete production and the several modes of fertilization are accomplished, by evolving different sexual systems, together with elaborate mating behaviors. Crustaceans express unique ways of sex determination that form the bases for sexuality, meaning acquisition of different sex characteristics, which are normally developed in both sexes. Nevertheless, significant numbers of crustacean species exhibit dual sexuality by which production of both gametes rests with a single individual. In accordance with the display of many sexual systems, the mechanistic modes of sex determination are likewise highly diverse. Furthermore, genetic and environmental determinants could bring about changes in the gender of many crustaceans. Despite variability in sex determination pattern, sex differentiation in malacostracan crustaceans is under the control of a male hormone, secreted by the androgenic gland, an equivalent of which has never been reported in any other invertebrate. Apart from many modes of sexual reproduction in crustaceans, there are instances where certain primitive crustaceans practice asexual reproduction by means of parthenogenesis.

Inasmuch as the mechanism of fertilization shows variability among different species, the crustaceans have evolved a variety of fully operational sperm-delivery systems that protect the sperm's viability as well as their safe transfer on to the female. Concomitant with the development of diversified sexual systems and sperm transfer mechanisms, the origin of different mating systems and the associated male–female behaviors play pivotal roles in successful fertilization in Crustacea. Besides, many coral reef-inhabiting caridean shrimps, living in symbiotic relationship with other marine invertebrates such as sponges, have reached the pinnacle of complexity in sexuality and peculiar mating behaviors, resulting in communal living and establishing advanced social systems, such as eusociality. The elaborate mating systems, as occurring in several crustaceans, provide illustrative examples for sexual selection among males, with respect to acquiring dominance over other members of the same sex as well as in attracting females for mating. Similarly, mate selection by the females has also evolved among some decapod species. Thus, crustaceans have been used as model systems to investigate current issues related to sexual selection such as sexual conflict, mate choice, sperm limitation, and sperm allocation. Undoubtedly, evolution of complex sexual systems and the associated physiological and behavioral adaptations has reference to crustacean species diversity in aquatic as well as terrestrial environments.

Crustacean reproductive endocrinology is at the crossroads of its potential application to the development of aquaculture of commercially significant species such as shrimps, lobsters, crayfish, and crabs. In this respect, gamete biology has attracted utmost attention to control egg production in captivity. There is a spurt of research activities on the control mechanisms in yolk formation in an attempt to augment egg production by manipulating endocrine systems in the aquaculture-important species. A simple eyestalk extirpation in a grass shrimp resulted in accelerated ovarian activity and the discovery of a gonad inhibitory hormone in the eyestalk ganglia of crustaceans. As a result, the idea of eyestalk ablation to induce oocyte maturation and spawning found universal acceptance for the hatchery production of commercially important crustacean species. Further work revealed the existence of many proximate endocrine factors to counteract the inhibitory eyestalk hormones, besides promoting gonadal maturation. Equally interesting is the crustacean spermatology. Most crustacean spermatozoa are aflagellate and nonmotile. There is also incredible diversity of sizes and shapes among the spermatozoa, even within a single genus, providing important phylogenetic clues. The secondary loss of motility in crustacean sperm and their subsequent encasement within spermatophores, in which form sperm are transported to females, could be viewed as a novelty in the evolution of fertilization mechanisms among marine invertebrates.

Crustaceans originated about 500 million years ago during Precambrian period and have since then undergone dynamic species radiation occupying diverse niches in aquatic ecosystems, as well as a few species venturing into terrestrial environment. Both morphological and recent molecular data have uncovered the phylogenetic closeness existing between Crustacea and Insecta, placing them under a separate clad Pancrustacea within Arthropoda. Obviously, many important reproductive processes show similarity between crustaceans and insects. Furthermore, crustaceans have emerged as the most important, economically useful invertebrate, by virtue of their aquaculture potentials in augmenting production of seafood in both developing and developed countries. However, no comprehensive treatment of crustacean reproductive biology is currently available. Information is scattered only in research papers and reviews and book chapters. Hence, a book on crustacean reproductive biology is not only timely but also useful to students of crustacean and invertebrate biology, researchers, and aqua culturists all over the globe.

My interest in the reproductive biology of crustaceans began when studying the role of fat body in cockroach vitellogenesis during my postdoctoral research in the University of Madras. With inspiration derived from Prof. Michael Locke, Case Western Reserve University, Cleveland, Ohio, who was then the Sir C.V. Raman visiting professor in our University, I switched from my doctoral study on millipede fat body to insect reproduction. When I started my research career as Lecturer in the Zoological Laboratory, Madras University, located right in front of the world-famous Marina Beach, my attention shifted to a host of marine invertebrates, which were available to me, as experimental model animals. One such animal was the mole crab, *Emerita asiatica*, and I just stumbled on it. At this time, Academic Press, New York, published the first volume of a multivolume treatise on marine invertebrate reproduction in the year 1974. Professors A.C. Giese and J.S. Pearse meticulously delineated the scope and content of marine invertebrate reproduction in their introductory chapter. There was no looking back for me from that point. I began my work on crustacean reproduction, using *Emerita* as the model organism, and my students investigated crustacean reproduction on a comparative basis by selecting a wide variety of crustaceans that abound on the Madras (now Chennai) Coast. As we progressed in this endeavor for over three decades, the idea of writing a monograph on crustacean reproduction blossomed in my mind.

Essentially, this book gives a comprehensive understanding on the basic phenomena governing reproductive processes in crustaceans. It is fondly hoped that this book will serve the need for students in crustacean biology, invertebrate reproduction, and comparative reproductive endocrinologists. Biology teachers will find this book useful in teaching comparative reproduction and endocrinology. And for aquaculturists, this book will serve as a manual to obtain information on aspects of reproduction in the edible crustacean species. Over and above all, this book will serve as a framework from which continued research on crustacean reproduction will proceed.

Thanumalaya Subramoniam

Acknowledgments

Several chapters of this book were read and reviewed by experts in the respective fields. Dr. Murali C. Pillai of Sonoma State University, USA, reviewed the chapter on fertilization, Dr. G. Anil Kumar (VIT, Vellore) and Dr. Vidya Jeyasankar (CMFRI, Chennai) reviewed chapters on accessory reproductive glands and spermatogenesis, respectively. Dr. E.Vivekanandan (CMFRI, Chennai) reviewed the chapter on yolk utilization. I thank them all for their valuable suggestions. My colleagues in the Centre for Climate Change Studies, Sathyabama University, Dr. Vinitha Ebinazer, Dr. Prakash Sanjeevi, Dr. Suja Srinivasan, and Dr. Vinu Siva extended unstinting support in the final preparation of the manuscript. I extend my sincere thanks to my students Persia Jothy, Carlton Ranjith, and Umar for their valuable assistance in the preparation of the references. Dr. Nandhini (Mother Theresa University), Dr. Jeyalakshmi (Madurai Kamaraj University) and Dr. Beema of Sathyabama University helped me with line diagrams. I thank all of them wholeheartedly. I also thank the Department of Science and Technology (Government of India) for financial support. My grateful thanks are due to Dr. Jeppiaar, the Chancellor and the Dean, Dr. T. Sasipraba, of Sathyabama University for their encouragement and support. Finally, I thank Pat Gonzalez, Editorial Project Manager, Kristi Gomez, Senior Acquisitions Editor, and Edward Taylor, Production Project Manager, of Elsevier, for their help and advice on finishing the book in time.

Chapter 1

Sex Determination

1.1 INTRODUCTION

Sex determination in animals is an integral part of reproduction. In general, sex determination describes the genetic and environmental processes that influence sex differentiation, whereas sex differentiation is the physical realization of these events in terms of testicular or ovarian development (Mittwoch, 1996). In other words, sex determination is concerned primarily with the determination of gonadal sex and the associated physiological processes that support gonadal development and function. Sex determination systems use different genes and regulatory mechanisms to establish activities in males and females to control a shared gene switch that regulates sexual development (Zarkower, 2001). The most highly evolved systems for sex determination in animals feature a single segregating pair of sex chromosomes that determine the sex, viz., XX/XY, as in placental mammals. Sex-determining genes, located in sex chromosomes, determine the cytodifferentiation of the indeterminate gametes to bring about sexual dimorphism into male and female gametes (see Bull, 1983).

In Crustacea, sex determination is a complex process, involving a large network of interactions among genes as well as between environment and genes. The genes so far discovered to be involved in gonad differentiation pathway are predominantly transcription factors. For example, *DMRT-1*, *DSX-1*, and *SOX-9*, responsible for the induction and regulation of gene expression, emphasize the importance of regulatory factors in development and differentiation of the gonad (Farazmand et al., 2010; Zarkower, 2001; Kato et al., 2010). In Crustacea, target genes for these transcription factors are not known, but could well be responsible for gonad differentiation. Similarly, the genes and the target cells involved in the temperature-dependent sex determination also remain to be established in Crustacea. On the contrary, in vertebrates, these genes are related to temperature-dependent modulation of aromatase activity (Kettlewell et al., 2000).

Crustaceans exhibit diversified mechanisms of sex determination but conform to the common genetically controlled sex determination pattern, found in other animals. Interestingly, genetic determination of sex in Crustacea varies vastly ranging from the most primitive, weak polygenic system to strong chromosomal sex determination. Sex determination can even differ markedly within a species and between closely related species in Crustacea. Significantly,

Sexual Biology and Reproduction in Crustaceans. http://dx.doi.org/10.1016/B978-0-12-809337-5.00001-0

different epigenetic factors (temperature and photoperiod) are known to exercise a strong influence on genetic sex-determining processes, yielding a wide variety of sexualization in these arthropods. Cytoplasmic sex determination is yet another system, found exclusively in crustaceans, with the possible exception of certain insects, like aphids (Legrand et al., 1987).

1.2 MODES OF SEX DETERMINATION

Sex determination in crustaceans could be categorized under the following four major types:

1. Genetic sex determination (GSD) with male heterogamety XY or Xo (male); XX (female) and female heterogamety ZZ (male); ZW (female)
2. Polygenic or polyfactorial sex determination
3. Environmental sex determination (ESD)
4. Cytoplasmic sex determination (CSD)

1.2.1 Genetic Sex Determination

Genetic sex determination systems are those in which the development of one sex or the other is triggered by the presence or absence of one or more critical genetic factors. Crustaceans exhibit different modes of genetic sex determination such as male and female heterogamety. Different types of male heterogamety are known in Crustacea, with Xo present in branchiopods, isopods, and copepods, XY males in copepods such as *Tortanus gracilis*, several decapods and in the isopod *Anisogammarus anandalei*. X_1X_2O is known in ostracods and X_1X_2Y is found in the decapod *Cervimunida princeps* (Legrand et al., 1987). On the other hand, female heterogamety is represented by WZ chromosomal pattern with examples of the anostracan *Artemia salina* (WZ), and the isopodan superspecies *Jaera albifrons* (W_1W_2Z). Table 1.1 summarizes the heterogametic sex determination prevalent in various crustacean species. It is seen from Fig. 1.1 that in the genetic system existing in crustaceans, the mechanisms involved in sex determination range from purely polygenic controls, to those with dominant sex-determining factors along with autosomal controls, and to highly evolved sex chromosomes with heterogametic (XY) males or heterogametic (ZW) females. This may indicate the possible occurrence of evolutionary pathways for different sex determination systems all within the group.

In general, heteromorphic chromosomes are difficult to discern by virtue of their large numbers and smaller size in crustaceans. However, crossing of neomale, which results from experimental sex reversal of amphigenous (genetic) female, with the normal thelygenous (producing all female progeny) female can be used to determine which sex is heterogametic. In the terrestrial isopods *Helleria brevicornis* and *Porcellio dilatatus*, the perfect thelygeny is observed when neomales were mated to normal females, also demonstrating female heterogamety (Juchault and Legrand, 1964).

TABLE 1.1 Heterogametic Sex Determination in Crustacea

Chromosomal Type	Species	Mechanism	References
Male Heterogamety			
♂ XO	**Branchiopoda**	Chromosomal-based sex determination	Baker and Rosof (1928)
	Branchipus vernalis		
	Chirocephalus nankinensis		
	Copepoda	Chromosomal-based sex determination	
	Acartia centura		Pai (1949)
	Acartia spinicauda		Goswami and Goswami (1974)
	Acartia negligens		Heberer (1932)
	Centropages typicus		Beermann (1954)
	Ectocyclops strenzkei		
♂ X1X2O	**Ostracoda**	Chromosomal-based sex determination	Dietz (1954)
	Cyprinotus incongruens		
	Notodromas monacha		
♂ XnO	**Ostracoda**	Chromosomal-based sex determination	
	Physocypria kliei		Dietz (1958)
	Platycypris baueri		
	Scottia browniana		
	Notodromas monacha		Dietz (1954)

Continued

TABLE 1.1 Heterogametic Sex Determination in Crustacea—cont'd

Chromosomal Type	Species	Mechanism	References
♂ XnY	**Ostracoda**	Chromosomal-based sex determination	
	Cyclocypris ovum		Dietz (1958)
	Cypria exculpta		
	Cypria opthalmica		
	Cypris compacta		
	Cypris dietzei		
	Cypris fodiens		
	Cypris whitei		
	Heterocypris incongruens		Bauer (1939)
♂ XY	**Ostracoda**	Chromosomal-based sex determination	Dietz (1958)
	Cyclocypris laevis		
	Copepoda	Chromosomal-based sex determination	Goswami and Goswami (1974)
	Tortanus gracilis		
	Tortanus barbatus		
	Tortanus forcipatus		
	Acartiella keralensis		
	Acartiella gravely		

♂ XO	**Decapoda**	Chromosomal-based sex determination	Niiyama (1941)
	Ovalipes punctatus		
♂ XY	**Decapoda**	Chromosomal-based sex determination	
	Eriocheir japonicas		Niiyama (1941)
	Plagusia dentipes		
	Hemigraspus sanguine		
	Hemigraspus pencillatus		
	Gaetice depressus		
	Pachygrapsus crassipes		
♂ X1X2Y	**Decapoda**	Chromosomal-based sex determination	Niiyama (1941)
	Cervimunida princeps		
♂ XY	**Isopoda**	Chromosomal-based sex determination	Niiyama (1950)
	Anisogammarus anandalei		
♂ XX	**Isopoda**	Chromosomal-based sex determination	Juchault and Legrand (1979)
	Armadillidium nasatum		
♂ XO	**Isopoda**	Chromosomal-based sex determination	Niiyama (1956)
	Tecticeps japonicas		
♂ XY	**Amphipoda**	Chromosomal-based sex determination	Niiyama (1950)
	Anisogammarus anandalei		

Continued

TABLE 1.1 Heterogametic Sex Determination in Crustacea—cont'd

Chromosomal Type	Species	Mechanism	References
♂ XY,XX	*Helleria brevicornis*	Chromosomal-based sex determination	Juchault and Legrand (1964)
♂XY,XX	*Porcellio dilatatus*	Chromosomal-based sex determination	Juchault and Legrand (1964)
♂ XX	*Orchestia cavimana*	Chromosomal-based sex determination	Ginsburger-Vogel (1972)
	Orchestia gammarellus		
Female Heterogamety			
♀ WZ	**Branchiopoda**	Chromosomal involvement in sex determination	Stefani (1963)
	Artemia salina		
	Copepoda	Chromosomal involvement in sex determination	
	Ectocyclops strenzkei		Beermann (1954)
	Megacyclops viridus		Rusch (1958)
	Acanthocyclops vernalis		Rusch (1960)
	Eucyclops serrulatus		
	Mytilicola intestinalis		Orlando (1973)
♀ WX	**Isopoda**	Chromosomal involvement in sex determination	Mittal and Pahwa (1980, 1981)
	Porcellio laevis		
	P. rathkei		
♀ WZ	**Amphipoda**	Chromosomal involvement in sex determination	Legrand et al. (1987)
	Porcellio dilatatus		

♀ZW1W2	**Isopoda**	Chromosomal involvement in sex determination	Staiger and Bocquet (1954) and Lecher (1963)
	Jaera albifrons		
	Jaera (a.)ischiosetosa		
	Jaera(a.) praehirsuta		
	Jaera(a.) syei		
♀ZW	**Decapoda**	Chromosomal involvement in sex determination	Malecha et al. (1992)
	Macrobrachium rosenbergii		
Polygenic or Polyfactorial			
Copepods, amphipods, and isopods	**Isopod**	Interaction between the major factors such as the strong W factor and multiple autosomal factors	Legrand et al. (1987)
	Idotea balthica		
	Porcellio scaber	Presence of a dominant autosomal gene Ma	
	Amphipod *Gammarus pulex subterraneus*	Presence of series of multiple alleles (R square, r^+, r) responsible for body color	De Lattin (1951) and Anders (1957)
Environmental Sex Determination			
Influence of parasites	**Isopod**	First-settled parasitic female releases a masculinizing substance to induce male differentiation Release of pheromone secreted by Bellonci organ found in the head	Chaigneau (1972)
	Ione thoracica		
	Anilocra frontalis		
	Decapod		
	Callianassa laticaudata		
	Copepod		
	Pachypygus gibber		

Continued

TABLE 1.1 Heterogametic Sex Determination in Crustacea—cont'd

Chromosomal Type	Species	Mechanism	References
Influence of temperature	**Isopod**	Sex change determining the adult sex ratio	Chaigneau (1972) and Dharani and Atlaff (2002)
	Anilocra frontalis		
	Calanoid		
	Sinodiaptomus indicus		
Food quality/nutritional status	*Acrocalamus gracilis*	Food quality/quantity provides cues to males for sex change	Gusmão and McKinnon (2009)
		Nutritional status: gut infection by dinoflagellates	
Cytoplasmic Sex Determination			
Bacterial (*Wolbachia pipientis*)	**Isopods**	Feminizing action of a protobacterium *Wolbachia* and intracytoplasmic factor	Vitigliano et al. (1996)
	Asellus aquaticus		
	Armadillidium vulgare		
Microsporidian protozoan	**Amphipods**	Absence of f-factor and feminizing action of protozoans	Legrand et al. (1987)

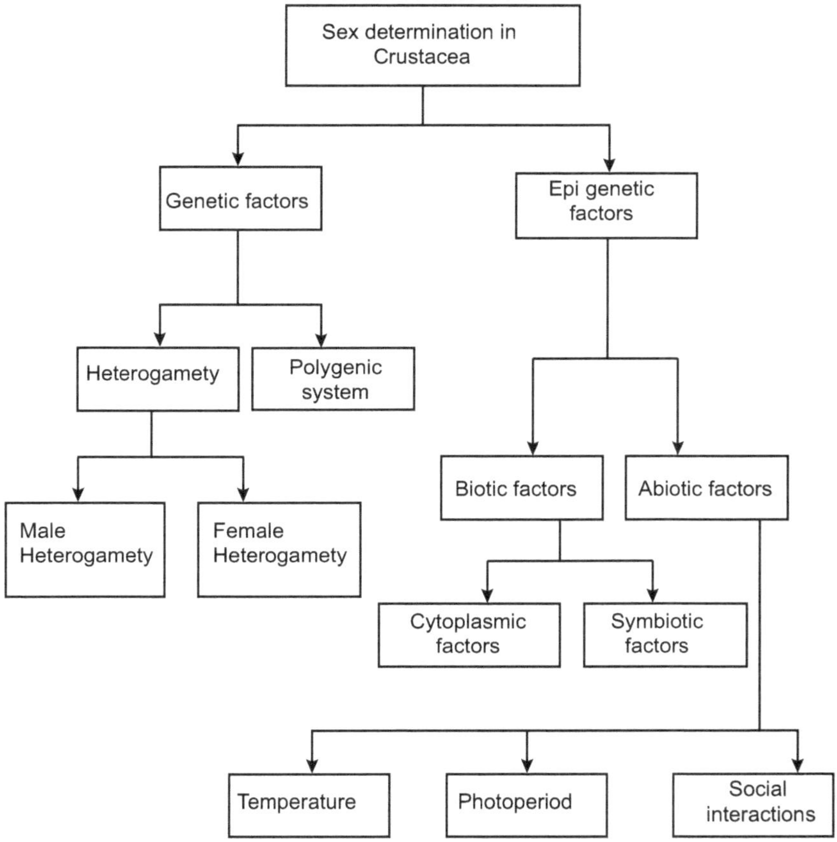

FIGURE 1.1 Classification of sex-determining systems in Crustacea.

In addition, a cross between intersex neofemale and normal male produced a sex ratio 0.75 in the progeny, confirming male heterogamety. Accordingly, the following Mendelian ratio is obtained in the F_2 generation of the above crosses.

F_1: XY (♂)×XY (intersex♀)→XX(♀)+2XY(♂♂)+YY (♂) (sex ratio 0.75)
F_2: XY (♂)×XX(♀)→amphigenous progeny (sex ratio 0.5);
and: YY (♂)×XX(♀)→XY(♂♂) (all-male progeny) (Legrand and Juchault, 1972).

In another experiment to demonstrate female heterogamety, Legrand et al. (1987) used the land isopod subspecies *P. dilatatus petiti*. The following result of crossing neomales (genetic females) to normal female is obtained.

F_1: WZ(neo ♂)×WZ(normal ♀)→ZZ (♂)+2WZ (♀♀)+WW(♀).

In this, the observed F_1 sex ratio is 0.25 suggesting female heterogamety. In the F_2 generation, the existence of the WW female has been confirmed by their all-female progenies. The above-described sex ratio studies employing sex-reversed individuals suggest that sex determination in these peracarid crustaceans is under the influence of a two-factor system comprising X/Y or W/Z chromosomes. Implicitly, the major sex factors are situated on these heterochromosomes. However, the analysis of successive generations of these crosses revealed several abnormalities in the sex ratios indicating that major sex factors are not the only ones to determine sex, but the sex determination is also under the control of a polygenic system (see below).

The sex determination studies in isopods have been further extended by Katakura (1961) and Hasegawa and Katakura (1983), who used sex-reversed individuals from androgenic implantation, in mating experiments to determine the genetic basis of sex determination. They produced viable all-female progeny from the mating of masculinized (neomales) and normal female isopods.

1.2.2 Polygenic Sex Determination

In a polygenic system, many genes, each with a small effect, have either male- or female-determining effect (Bull, 1983). If the expression of the determining genes for one sex is collectively stronger, then the zygote differentiates as that sex. Furthermore, when polygenic sex determination occurs, there has always been an interaction between the major sex factors present in the sex chromosomes (such as W and Z in isopods) and several minor sex factors present in autosomes. These minor factors themselves interact in a combinatorial way to interact with major sex factors to bring about sex determination. In other words, the sum total influence of all genetic factors involved in sex determination in the genome brings about the genetic effects in a polygenic sex determination system. Polygenic sex determination could be viewed as a primitive type which could ultimately be replaced by single gene sex determination. The genetic determination may then evolve to chromosomal sex determination.

Polygenic or polyfactorial sex determination is usually regarded as a mechanism with sex determined by many factors, so that no few of them have a major influence (Bull, 1983). Some authors consider multiple-factor systems as examples of polyfactorial sex determination. However, no practical method is available for ascertaining the number of sex factors in these polyfactorial systems. In this system, the family sex ratio also varies according to the genotype of both the father and mother, implying paternal and maternal effects on the family sex ratio. Furthermore, the polygenic system quite often incorporates the environmental effects on sex determination. Hence, the magnitude of the environmental effects within a particular population may be ascertained from the analysis of family sex ratio.

Prevalence of polygenic mechanism of sex determination was understood in Crustacea from the genetic studies in copepods, amphipods, and isopods, where the shift in the sex ratio of the progeny is the result of this kind of polygenic sex determination. Besides, sex inversion due to genetic influences also points to the existence of polygenic systems of sex determination. In several strains of terrestrial isopods, where sex inversion has a polygenic basis, an autosomal dominant M factor inhibits the major W sex factor of a genetic female. Crossing experiments of subspecies of the marine isopod *Idotea balthica* has revealed the occurrence of both major factor and polygenic mechanism of sex determination in this species (Legrand-Hamelin and Legrand, 1982). The interaction between the major factors such as the strong W factor and multiple autosomal (and sex-linked) factors are responsible for bringing about sex inversions in WZ females. Furthermore, W chromosome bears a major sex factor which in many individuals act to oppose sex reversal. The efficiency of this act increases when two copies of this major sex factor are present in the W chromosome. In a similar way, numerous minor sex factors are capable of inhibiting the masculinizing Z major sex factor. Both masculinizing and feminizing major factors are resident in W and Z chromosomes, respectively.

The sex ratio bias is also under the influence of autosomal color genes with additive effects on sex determination. In some species of terrestrial isopods, sex ratio is correlated to color phenotypes. For example, in *Porcellio scaber*, a dominant autosomal gene Ma is responsible for a mottled aspect of the body. When Ma/Ma strains and Ma/ma heterozygotes were crossed, Ma showed a strong feminizing influence, altering the sex ratio in the progeny significantly (Legrand et al., 1987).

Similarly, in the amphipod *Gammarus pulex subterraneus*, a series of multiple alleles are responsible for body color. They are R^2, r^+, r, responsible for brown, olive (wild type), or red body color, respectively. Unlike the isopods, R^2 and r exert a strong masculinizing influence on the progeny. The sex ratio in r^+/r^+ strain is 0.24; it reaches 0.68 in R^2 strain and 0.96 in r/r strains. In the reciprocal crosses between r/r and r^+/r^+ strains, the masculinizing effect of r is chiefly manifested, when the female is r/r (De Lattin, 1951; Anders, 1957).

These color genes are considered to be the sex realizers, as they are unable to induce a complete monogeny. They can only tilt the sex ratio of the progeny in favor of either male or female as the case may be. Furthermore, it has been assumed that other autosomal genes also interact with the color genes in accomplishing the polygenic sex realization in the amphipods and isopods.

The crossing experiments described above reveal that in a polygenic system autosomal genes play a major role in sex inversion. These small-effect autosomal minor sex genes, also called "modifier genes," seem to be able to repress the major sex genes found in the sex chromosomes. Some of these modifier genes have an additive effect in masculinization by repressing the W major sex gene of a WZ female and, perhaps the pair of the major sex genes in a WW female. Evidently, the masculinizing-gene complex, which inhibits the W major sex factor, simultaneously stimulates the Z major sex factor. On the contrary, other modifier genes have a feminizing effect from their ability to repress the

Z major sex genes in a ZZ male. Legrand et al. (1987) consider the polygenic system as a practice switch mechanism, as it allows the differentiation of normal male or female phenotype, without producing any intersexes. This condition is to be contrasted with the epigenetic mechanisms of sex determination involving environmental or cytoplasmic factors, which often induce intersexual phenotypes, irrespective of the mechanism of sex determination.

The action mechanism involved in the genetic or polygenic sex determination is revealed from the study on the experimental sex inversion in isopods and amphipods, which are amenable for sex inversion by manipulation of androgenic gland. A female transformed into a neomale acquires all the male characters, including functional androgenic glands. A female has therefore all the genes required for male differentiation, except the major sex factor that allows the natural development of the androgenic gland. Under the influence of the genetic modifier complexes, morphogenesis and functioning of androgenic gland begin in a WZ individual and are inhibited in a ZZ one; thus these complexes act as "trigger" of the sex determination, as do major sex genes.

1.3 SEX DETERMINATION IN CLAMP SHRIMP

One of the best examples for polygenic sex determination is found in a primitive branchiopod *Eulimnadia texana*, commonly termed as the clamp shrimp (Table 1.2). These shrimps characteristically produce mixtures of males and hermaphrodites, a reproductive trait known as androdioecy (Week et al., 2006). Androdioecy is a reproductive system found in species composed of a male population and distinct self-compatible hermaphrodites in the natural population. In the concostracans, hermaphrodites are of two types: amphigenic (producing both male and hermaphroditic offspring) and monogenic (producing only hermaphroditic offspring). In *E. texana*, the mode of sex determination involves a single gene or genetic element with two allelic states, S and s, with s recessive to S. Under this model, males are genetically ss, amphigenic hermaphrodites are Ss, and monogenic hermaphrodites are SS, respectively. The genetic analysis to explain this polymorphism shows that males, amphigenics and monogenics can be interpreted as three alternative phenotypes of a one-locus system of sex determination (Sassaman and Weeks, 1993). Pedigree analysis of *E. texana* also supports this interpretation. Regulation of sex differences using a single locus is unique to crustaceans, as this genetic system is prevalent only in plants.

Subsequent studies have, however, indicated that sex-determining mechanism may be a set of linked genes (or possibly an entire chromosome with reduced crossing over) and that there may be numerous genes within this linkage group that encode sexual dimorphism (Weeks et al., 2000). Androdioecy is exceptionally rare in animals, but in the genus *Eulimnadia* it is the dominant mode of reproduction.

Similarly, in another concostracan species, a tadpole shrimp, *Triops newberryi*, sex appears to be controlled by a single locus, with a recessive allele coding

TABLE 1.2 Occurrence of Androdioecy in Crustaceans

Class	Order	Family	Genus	Species	References
Branchiopoda	Notostraca	Triopsidae	*Triops*	*newberryi*	Weeks et al. (2006)
Branchiopoda	Spinicaudata	Limnadiidae	*Eulimnadia*	*texana*	Hollenbeck et al. (2002)
Branchiopoda	Spinicaudata	Limnadiidae	*Eulimnadia*	*africana*	Weeks et al. (2006)
Branchiopoda	Spinicaudata	Limnadiidae	*Eulimnadia*	*agassizii*	Weeks et al. (2006)
Branchiopoda	Spinicaudata	Limnadiidae	*Eulimnadia*	*antlei*	Weeks et al. (2006)
Branchiopoda	Spinicaudata	Limnadiidae	*Eulimnadia*	*braueriana*	Weeks et al. (2006)
Branchiopoda	Spinicaudata	Limnadiidae	*Eulimnadia*	*brasiliensis*	Weeks et al. (2006)
Branchiopoda	Spinicaudata	Limnadiidae	*Eulimnadia*	*colombiensis*	Weeks et al. (2006)
Branchiopoda	Spinicaudata	Limnadiidae	*Eulimnadia*	*cylindrova*	Weeks et al. (2006)
Branchiopoda	Spinicaudata	Limnadiidae	*Eulimnadia*	*diversa*	Weeks et al. (2006)
Branchiopoda	Spinicaudata	Limnadiidae	*Eulimnadia*	*feriensis*	Weeks et al. (2006)
Branchiopoda	Spinicaudata	Limnadiidae	*Eulimnadia*	*follisimilis*	Weeks et al. (2006)
Branchiopoda	Spinicaudata	Limnadiidae	*Eulimnadia*	sp. A	Weeks et al. (2006)
Branchiopoda	Spinicaudata	Limnadiidae	*Eulimnadia*	sp. B	Weeks et al. (2006
Thecostraca	Thoracica	Heteralepadida	*Paralepas*	*klepalae*	Weeks et al. (2006)
Thecostraca	Thoracica	Koleolepadidae	*Koleolepas*	*avis*	Weeks et al. (2006)

Continued

TABLE 1.2 Occurrence of Androdioecy in Crustaceans—cont'd

Class	Order	Family	Genus	Species	References
Thecostraca	Thoracica	Iblidae	*Ibla*	*quadrivalvis*	Weeks et al. (2006)
Thecostraca	Thoracica	Iblidae	*Ibla*	*cumingii*	Weeks et al. (2006)
Thecostraca	Thoracica	Iblidae	*Ibla*	*idiotica*	Batham (1945)
Thecostraca	Thoracica	Iblidae	*Ibla*	*pygmaea*	Weeks et al. (2006)
Thecostraca	Thoracica	Scalpellidae	*Scalpellum*	*vulgare*	Weeks et al. (2006)
Thecostraca	Thoracica	Scalpellidae	*Scalpellum*	*peronii*	Weeks et al. (2006)
Thecostraca	Thoracica	Scalpellidae	*Scalpellum*	*scalpellum*	Weeks et al. (2006)
Thecostraca	Thoracica	Scalpellidae	*Scillaelepas*	*arnaud*	Weeks et al. (2006)
Thecostraca	Thoracica	Chelonibiidae	*Chelonibia*	*patula*	Crisp (1983)
Thecostraca	Thoracica	Chelonibiidae	*Chelonibia*	*testudinaria*	Rawson et al. (2003)
Thecostraca	Thoracica	Bathylasmatidae	*Bathylasma*	*alearum*	Foster (1983)
Thecostraca	Thoracica	Balanidae	*Balanus*	*galeatus*	Gomez (1975)
Thecostraca	Thoracica	Balanidae	*Balanus*	*calceolus*	Weeks et al. (2006)
Thecostraca	Thoracica	Balanidae	*Balanus*	*merrilli*	Weeks et al. (2006)
Thecostraca	Thoracica	Balanidae	*Balanus*	*masignotus*	Weeks et al. (2006)

for males and a dominant allele coding for hermaphrodites/females (Sassaman, 1991). Interestingly, there occur a low number of intersex (mixed sex) individuals in *Eulimnadia texana* and other concostracan species, probably produced from limited crossing over between the sex chromosomes in the heterogametic sex of these species. In *E. texana*, the intersexuality is of two types: (1) a morphological intersex, possessing secondary male characteristics (eg, claspers) and an egg-producing gonad, and (2) a gonadal intersex, possessing primarily male traits (eg, male secondary sexual characters and male gamete production) but also producing low levels of abortive eggs. While low frequencies of crossing over between the sex-determining sex chromosomes result in the array of observed mixed sexual phenotypes, the preponderance of intersexuality as well as their habitation in ephemeral ponds may predispose these concostracans to the evolution of androdioecy. In the branchiopod species inhabiting temporary ponds, the mixed production of males, hermaphrodites, and females has been termed "trioecy," which are highly unstable and usually breaks down to either androdioecy or gynodioecy (females and hermaphrodites) (Sassaman, 1991).

1.4 SEX-LINKED GENES AND SEX DETERMINATION IN *ARTEMIA SALINA*

A. salina is the brine shrimp of worldwide distribution. Some of the geographical populations are cross-fertile, while others are reproductively isolated from each other (Clark and Bowen, 1976). Cytochemical studies have demonstrated female heterogamety (ZW) in this branchiopod species. Stefani (1963) has shown that the sex chromosomes of *Artemia* are of unequal length, with the male Z chromosome slightly longer than the W. Many crossbreeding studies using different populations of *A. salina*, together with karyotypic examinations have yielded information on their sex determination pattern. Studies of Mendelian genetics in *A. salina* have indicated the presence of sex-linked genes involved in sex determination. Bowen (1963) demonstrated female heterogamety (ZW) in this branchiopod species. This author observed one white-eyed male and the white locus was found to be partially sex linked. Studying the mode of inheritance of white eye in successive generations, the genes responsible for sex determination in *A. salina* is deduced. The mode of inheritance of white eyes in *A. salina* resembles that of the white eye in *Drosophila*. The mutant gene w which determines white eyes is recessive to its wild type allele W. The females are heterogametic. The chromosomal constitution of female will be represented as WZ, and the males will be ZZ. The white locus is partially sex linked. Because it is on the homologous segment of the sex chromosomes, both males and females may be WW, Ww, or ww. The first white female arose as a result of a crossover between the white locus and the sex locus. Both the sex chromosomes are of same length, but consist of two segments, namely differential segment and homologous segment. The white locus and the sex locus are found in the homologous segment. Crossing over may occur between the white locus and the sex locus.

1.5 EPIGENETIC FACTORS ON SEX DETERMINATION

In addition to the genetic system of sex determination, epigenetic factors are known to play a significant role in the sex determination of several crustaceans, particularly isopods, amphipods, copepods, and branchiopods. Two main types of epigenetic factors have been distinguished in these groups of crustaceans (Legrand et al., 1987). They are environmental sex determination (ESD) and cytoplasmic sex determination (CSD) systems. The ESD includes both abiotic (temperature, photoperiod, salinity, pH, food) and (2) biotic (substances released from the same species or by the host of a parasite, pheromone, excrements etc.) factors. In organisms with environmental sex determination, the epigenetic factors act on the zygote by modifying the expression of the genetic sex factors. In other cases, the epigenetic factors decide the initial orientation toward the male or female sex by acting on gametogenesis of a heterogametic female or of a parthenogenetic female. Systems with environmental sex determination generate extrabinomial variation for the primary sex ratio, if clutches of offspring encounter different environmental conditions during development. In these systems, the population sex ratio is driven by the sensitivity of the sex-determining mechanisms to the environmental factors and by the range of environmental variation (Bulmer and Bull, 1982).

In cytoplasmic sex determination, sex is determined by intracytoplasmic microorganisms such as proteobacteria or protozoan symbionts present in the germline of isopods and amphipods (Legrand et al., 1987). CSD also includes cytoplasmic sex factors like the f factor present in the isopods. These maternally inherited cytoplasmic feminizers manipulate the sex determination of their hosts to increase their transmission to the next generation. They also cause extrabinomial variation in the sex ratio. Thus, variation in the primary sex ratio is intimately connected with the sex-determining mechanisms. Genetic variation for the primary sex ratio is documented mainly in animal taxa with polygenic or environmental mechanisms of sex determination.

1.5.1 Influence of Environmental Factors on Sex Determination

In this type of sex determination, sex is determined after conception in response to the individual's immediate environment. That means environmental factors bring about alterations in the already existing genetic sex determination. Submammalian vertebrates such as reptiles provide the best example for the environmental influence of sex determination. In them, although sex is normally determined by heterogametic sex determination, certain environmental extremes override these genetic effects. Thus both male and female heterogamety have the potential to evolve ESD and this evolution simply requires selecting specific environmental sensitivity (Moreno-Mendoza et al., 1999). Crustaceans are even more amenable for such environmental sex determination inasmuch as their genetic sex determination is more labile than those of vertebrates.

In the bopyrid isopod, *Ione thoracica*, a gill parasite on the decapod *Callianassa laticaudata*, the planktonic larvae begin parasitism in a sexually indifferent state (Reverberi and Pittoti, 1942). The first larva to settle on a host becomes a female. Subsequent larvae to settle on the same host become dwarf males. Presumably, the first-settled parasitic female releases a masculinizing substance to induce male differentiation in the other larvae. However, the chemical nature as well as its mechanism of action is not known. In an isopod *Anilocra frontalis*, parasitic on fish, the larva is at first male, possessing testes, androgenic glands, and rudimentary ovaries and oviducts. However, when this male remains alone on a fish, it changes sex to female at a size of 14 mm, with concomitant disappearance of male gonopods and androgenic glands. The functional female state is reached at 20 mm size. If a male settles close to a female, it remains male and may grow to 20 mm size, until the female dies or is experimentally removed. The female probably influences the sexual differentiation of the younger individual through the release of a pheromone secreted by Bellonci organ found in the head. This pheromone is expected to maintain the androgenic gland activity of the male (Chaigneau, 1972).

1.5.1.1 Influence of Temperature, Nutrition, and Parasitism

Intersexuality and sex change in copepods, in which sex is determined late in development, provide excellent examples for environmental sex determination in crustaceans (Ginsburger-Vogel and Charniaux-Cotton, 1982). Several environmental conditions/factors such as temperature, food quality/quantity, and parasitization are ascribed to cause sex change and intersexuality in copepods (Michaud et al., 2004). Sex determination is temperature-dependent in some species of copepods. In *Cyclops viridis*, the female is heterogametic. At 2°C, 60% are females. When the rearing temperature of females and broods is raised to 13°C, the sex ratio increases, and at 23°C, it is 0.60% (Legrand et al., 1987). Voordouw and Anholt (2002) investigated the effect of temperature-dependent sex determination on the primary sex ratio of the harpacticoid copepod *Tigriopus californicus*. At higher temperatures (15°C and 22°C), the primary sex ratio is always biased in favor of males. Higher temperatures induce masculinization and the change in sex ratio is not caused by differential mortality of the sexes. However, the mechanism or the adaptive significance of the temperature-dependent sex determination in Crustacea is not known. In reptiles, the temperature of incubation during a critical period preceding sexual differentiation determines the future sex of the embryo, by affecting the temperature-dependent regulatory factors like the *Dmrt1* gene (Kettlewell et al., 2000). Extreme environmental fluctuations can result in highly biased sex ratio that may predispose the population toward extinction (Bulmer and Bull, 1982).

In planktonic calanoid copepods, food environment is linked to sex determination and the possible occurrence of intersexuality and sex change (Miller et al., 2005). From the microcosm incubation experiments using *Acrocalamus gracilis*, it has been shown that food quality/quantity may provide cue to males for change of sex: animals growing slow would experience the poor food environment longer than animals growing faster, thus promoting more animals to change sex (Gusmão and McKinnon, 2009). Interestingly sex change in copepods differs from that of sequential hermaphrodites, in which sex can change after reproduction has begun. But in copepods, sex change occurs at the last point in development in which the physiological condition of both sexes is similar. That means the copepod gonad will not contain ovotestis any time during sex change. The intersexes will invariably possess ovary along with fully or partially developed male secondary sex characters. Intersexuality is caused by changes in sex differentiation mechanisms. In copepods, food limitation could disrupt endocrine signaling of androgenic hormone to promote sex change from male to female.

Although nutritional status is considered to be an important environmental factor controlling sex determination in copepods, food limitation itself is linked to parasitization in some species. In paracalanoid copepods, gut infection by the dinoflagellates of the genus *Blastodinium* makes the host undernourished. This in turn leads to a physiological trigger for sex change and the appearance of intersexes in the population, probably through degeneration of androgenic gland. It has to be mentioned here that not all parasitized individuals in the population are intersexuals and that intersexuality is also found in nonparasitized animals (Gusmão and McKinnon, 2009). When it occurs, parasitization "boosts" the intersexuality-inducing trigger that is naturally found in noninfected populations. In planktonic copepods, sex change from male to female has obvious advantages in terms of increased reproductive output. Thus, a higher proportion of males would mature during rich food conditions, and conversely, when the trophic conditions become limiting, males would switch sex and mature as phenotypic females and have access to the males, matured during the previous favorable period.

The genetic basis of sex determination has not been understood fully in copepods, for the lack of chromosomal studies. However, laboratory rearing and crossing of *Tisbe reticulata* have resulted in changes in sex ratio in the generations. These results have been explained in terms of the existence of a polyfactorial mechanism of sex determination under the influence of numerous dominant feminizing genes and recessive masculinizing genes situated on different chromosomes (see Ginsburger-Vogel and Charniaux-Cotton, 1982).

1.5.1.2 Influence of Photoperiod

The amphipods are good examples for environmental sex determination, influenced by factors such as photoperiod. In the brackish water amphipod *Gammarus duebeni*, Bulnheim (1967, 1969, 1978a,b) found that the sex ratio varies according to photoperiod during the posthatching, sexually indifferent stage. Further, within the progeny from a single pair of parents, sex ratio is significantly higher with a long rather than a short photoperiod. In the laboratory, under laboratory conditions, L:D intervals of 8:16 h have induced sex ratio less than 0.25 and L:D 16:8 produced excess of males, thus revealing that day length rather than light intensity was responsible for the sex ratio response (Bulnheim, 1978a,b). However, not all the species of *Gammarus* is responsive to photoperiodic sex determination; *G. locusta* is one example to be unresponsive to photoperiodic sex determination.

1.5.2 Influence of Cytoplasmic Sex Factors on Sex Determination

In genetic sex determination, it is assumed that the sex-determining sex factors are typical nuclear genes, transmitted to zygotes according to Mendelian, sexual populations (Bull, 1983). On the contrary, there are animal species, in which sex is determined partly by factors inherited in the cytoplasm (Engelstädter and Hurst, 2009). The term "cytoplasmic sex factor" commonly refers to self-replicating sex factors transmitted from mother to daughter, but not transmitted through males. In certain groups of crustaceans, these factors are reported to cause sex ratio distortion toward the female sex. Cytoplasmic sex determination is unique to crustaceans and not found in any other animal species (Cordaux et al., 2011). In the isopods, for instance, these sex ratio distorters (SRDs) are ascribed to the feminizing action of an obligatory intracellular and maternally inherited proteobacterium called *Wolbachia pipientis* (Rigaud et al., 1977). These parasitic sex factors (F) also induce the occurrence of another genetic factor (f factor) originating from a sequence of the F bacterial DNA, unstably integrated into the host genome (Legrand and Juchault, 1984). The f factor may also act as a transposable element and therefore, unlike F, it can be inherited paternally, although partly.

One of the ways in which cytoplasmic factors influence sex determination is by controlling sexual development in the embryo or juvenile, overriding the sex tendencies of the nuclear genes (Bull, 1983). Under male heterogamety, cytoplasmic sex determination would produce XX males or XY females, by affecting the segregation of sex chromosomes from the parents. This results in a change in the frequency of XX and XY zygotes, thereby altering the sex ratio of the progeny. It follows further that there would not be any effect on sexual development after conception (contrast it with temperature dependant sex determination in copepods). Investigations on the skewed sex ratio, monogeny, and maternal sex determination in the terrestrial isopod *Armadillidium vulgare* belonging to different populations and strains have yielded significant insight into the mechanism of cytoplasmic sex determination in crustaceans (reviewed in Legrand et al., 1987). Cytoplasmic sex determination is prevalent in the peracarids, isopods, and amphipods.

1.6 EVOLUTION OF SEX-DETERMINING MECHANISM IN ISOPODS

1.6.1 Genetic Sex Determination

Early cytological investigations and breeding experiments using experimentally sex-reversed individuals showed that most isopod species have heterochromosomal sex determination (Vandel, 1941; Johnson, 1961). Interestingly, both heterogametic systems (male XY/female XX and male ZZ/female WZ) occur in isopods, sometimes within the same genus and occasionally within the same species, such as *P. dilatatus*. However, several studies have shown that both sex determination and sex differentiation is very labile in these crustaceans. Males can be readily changed to females and the females into males by simple experimental manipulation of androgenic gland, indicating that both sexes possess all the genetic programs necessary for expression of the opposite sex (Ginsburger-Vogel and Charniaux-Cotton, 1982).

In isopods, the sex chromosomes contain large homologous segments, allowing frequent crossing over between sex chromosomes, thereby creating viable and fertile WW females or YY males in some species. The absence of heteromorphy as well as the presence of large homologous segments shared by sex chromosomes is all generally considered to be signs of a primitive stage of sex chromosome differentiation. Thus, Rigaud et al. (1997) proposed that males and females of the woodlouse *A. vulgare* share common genotypes, except that the female chromosome (W chromosome) is thought to be Z chromosome, carrying an additional "female gene" that inhibits the activity of the "male gene" located on the Z chromosome. Hence, both male and female sex chromosomes differ only by the presence or absence of a single sex factor to impart differences in sex determination. This sex factor is only a "trigger" gene that induces a series of reactions, which are controlled by realizer genes present in the genotypes of both sexes.

In malacostracan crustaceans, androgenic gland controls male differentiation, and its development in *A. vulgare* is under the control of male sex-determining genes (*tm*, for "trigger male") carried by the Z chromosomes (Rigaud et al., 1997). The W and Z chromosomes probably share large homologous segments that can recombine, except in the region of the female sex factor on W (*tf* gene). This female gene may inhibit the activity of *tm*, preventing differentiation of the androgenic gland. In the absence of the male hormone synthesized by this gland, the undifferentiated gonad spontaneously differentiates into ovaries and the sexual characters evolved toward the female phenotype. On the contrary, a masculinizing autosomal gene (M gene) inhibits the feminizing effect of the *tf* gene which allows the differentiation of male phenotype in the presence of female genotype.

Based on this, the following sequence of sex determination of male and female *A. vulgaris* is proposed (Rigaurd et al., 1997).

For maleness: tm —(+)→ Male realizing gene → AGH synthesized → Male phenotype

For femaleness: tf —(−)→ tm in Z chromosome → Male realizing gene inactivated → No AG differentiated → Female phenotype

The generalization is that in Crustacea, the cue for sex determination and differentiation seems to be the "male gene," which controls the development of the androgenic gland that produces the male hormone. In female heterogamety, the female gene inhibits the activity of the male gene, allowing the female sex to develop.

In addition to heterogametic sex determination, a multiple-factor system of sex determination is operative in some isopod species. For example, in the marine isopod *Idotea balthica*, heterochromosomic sex determination can be overridden by a polyfactorial system of "minor sex factors" (Tinturier-Hamelin, 1963). These minor sex factors are invariably linked to body color genes and have an additive effect. Some of them are masculinizing (they change the heterochromosomic WZ females into males), while others are feminizing (they change homochromosomic ZZ males into females). Thus, the terrestrial isopod *P. scaber* has a dominant autosomal gene that is responsible for a mottled body appearance and possessing a strong feminizing influence. On the other hand, *A. vulgaris* has an autosomal dominant gene (M gene) that possesses a masculinizing influence. Males carrying this gene are often heterozygous (ZZ*Mm*). This gene strongly inhibits the W sex factor, since WZ*Mm* and even WW*Mm* females are changed into functional males.

1.6.2 Cytoplasmic Sex Determination in Isopods

Another peculiarity found in Isopoda is that more than 40% of the terrestrial woodlouse species have been reported to display sex ratio distortion toward the female sex; in contrast, most of their aquatic counterparts have a sex ratio close to 1:1. Sex ratio distortion and intersexuality in isopods have been ascribed to the influence of an endocytoplasmic bacterium (called F bacterium) that lives in the cells of all tissues of females, but not males (Bouchon et al., 2008). They cause female-biased sex ratios in the thelygenic females. In the isopod *A. vulgare*, the female:male ratio in a *Wolbachia*-infested population frequently exceeds 10:1. This bacterium belongs to the genus *Wolbachia* and is a Proteobacteria member. Incidentally, these microorganisms are concentrated in the oocytes and are maternally inherited. Therefore, all zygotes inheriting these bacteria will develop a female phenotype, regardless of their sexual genotypes. Notably, the ZZ males are changed to phenotypically functional females, which in turn produce female-biased broods. According to Juchault and Legrand (1981), some population of *Armadillium* were entirely ZZ, with females produced only by such "epigenetic" factors (Fig. 1.2). Antibiotic elimination of these bacteria, however, suppresses this feminizing effect on the host. *Wolbachia* bacteria are also thermosensitive, temperature above 30°C destroy many of the symbionts in the host tissues. Usually symbiont transmission is 90% efficient and hence the sex ratio in broods of infected females is strongly female biased. However, intersexual phenotypes possessing features of both sexes, and sometimes total sterility do appear in the progeny of certain *Wolbachia*-affected individuals, as a result of incomplete

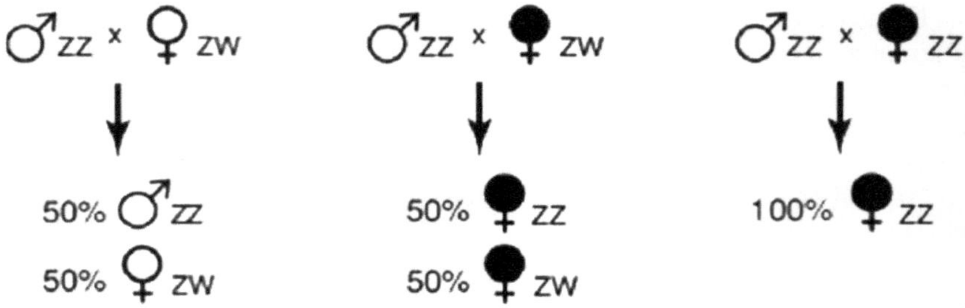

FIGURE 1.2 Endosymbiont-mediated host sex determination: feminization of genetic males by *Wolbachia* infection in *A. vulgare*, characterized by sex ratio tilting toward females through conversion of genetic ZZ males into phenotypic ZZ females. *Adapted from Cordaux, R., Bouchon, D., Greve, P., 2011. The impact of endosymbionts on the evolution of host sex-determination mechanisms. Trends Genet. 27, 332–341.*

feminization (Legrand et al., 1987). Evidently, *Wolbachia* does not alter the genotype of the infected isopods and therefore, they have their effect on the postembryonic sexual differentiation, such as the androgenic gland formation and functioning. Alternatively, the parasitic sex factors inhibit the expression of the "male gene" carried by the Z heterochromosome, thereby preventing the growth of androgenic gland, which would induce the development of male gonads through secretion of the androgenic hormone. The feminizing factors can be considered to be selfish genetic elements because they bias their host's sex ratio to increase their own transmission. Conversion of males into females by such intracellular symbionts has obvious genetic advantage to *Wolbachia*, as they are primarily transmitted down the maternal line. Hence, a male host represents an evolutionary "dead end" for the bacterium; their ability to convert male to female host confers a strong selective advantage.

Female-biased sex ratios have also been reported in certain strains of *A. vulgare*, which do not harbor *Wolbachia*. The feminizing agent responsible for the abnormal sex ratios in these strains is known to be the "f factor." Legrand and Juchault (1984) first indicated a link between the f factor and the *Wolbachia* bacteria in *A. vulgaris*. When WZ females were inoculated with *Wolbachia* and maintained in inbred lines, an f lineage appeared spontaneously, following a lack of *Wolbachia* transmission in the descendants of one female. This experiment revealed the possibility that the f factor could be a virus associated with *Wolbachia* or a part of the unstable DNA element of the infected *Wolbachia*, which becomes incorporated into the nuclear genome of the isopod. As no experimental transfection of the f element is possible by inoculation, f could be transposon-like, carrying bacterial feminizing information. Crossing experiments have indicated that this stable binding of f factor occurs to a male chromosome (Z chromosome), thereby creating a W-like chromosome, resulting once more in purely chromosomal sex determination. This adduces further evidence to the fact that the only difference between female (Z) and male (W)

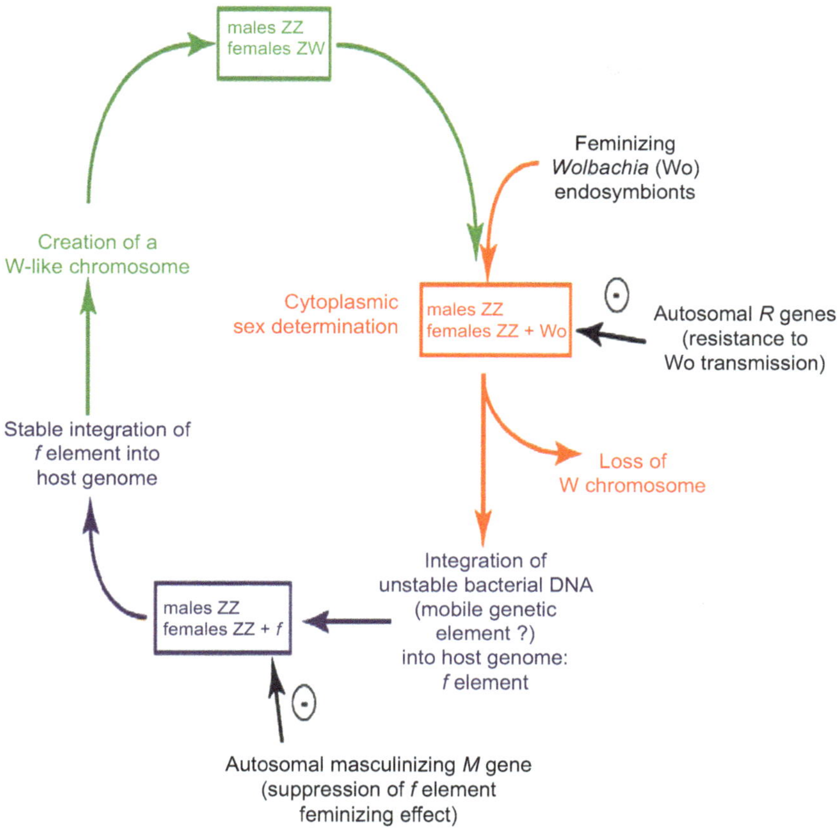

FIGURE 1.3 Diagrammatic representation of chromosomal sex differentiation affected by *Wolbachia* infection. Infection of feminizing *Wolbachia* endosymbionts in an *A. vulgare* population leads to a shift from chromosomal to cytoplasmic sex determination (red—*light gray in print version*). *Wolbachia* gives rise to the f element, a non-Mendelian feminizing sex factor (blue—*dark gray in print version*). The f element ultimately becomes stably integrated into a Z chromosome, hence generating a W-like chromosome and restoring chromosomal sex determination with female heterogamety (green—*gray in print version*). Genetic conflicts between host genome and the feminizing *Wolbachia* or the f element result in selection of nuclear resistance genes (black). *Adapted from Cordaux, R., Bouchon, D., Greve, P., 2011. The impact of endosymbionts on the evolution of host sex-determination mechanisms. Trends Genet. 27, 332–341.*

seems to be the presence or absence of the female determinant, as the rest of the chromosome is identical. In this way, the evolution of a heterochromosome has been suggested to have occurred from a sex-determining homologous pair, by cytoplasmic sex-determining factors from infection of an unstable element of the bacterial DNA into the nuclear genome of the host (Fig. 1.3). This phenomenon also illustrates the appearance of a new sex-determining mechanism in isopods through cytoplasmic factors. It is well known that in animals, the original sex-determining mechanism is nuclear, but in Crustacea, cytoplasmic effects on the evolution of the sex-determining mechanism are evidenced in isopods. This is similar to the reported conditions in other animals and plants, where nuclear–cytoplasmic interactions might have favored evolution from hermaphroditism to separate sexes (Avise, 2011).

1.6.3 Intragenomic Conflicts in Isopods

Isopods as well as amphipods exhibit a polyfactorial or polygenic system of sex determination, probably to compensate the female-biased sex ratio and to restore a balance in the sex ratio of the population. These interactions induce the so-called "genetic conflict" between nuclear autosomal and cytoplasmic genetic elements, contributed by the egg parasites as well as the f factor described above. According to Werren (1987), cytoplasmic sex ratio distorter genes are in conflict with the autosomal genes of the host, because Fisherian selection acting on the autosomal genes of the host favors an overall sex ratio of 1:1. This can result in selection for a male-biased sex ratio in lineages uninfected by the symbionts to compensate the excessive production of females from the infected population. In other words, the female bias caused by a cytoplasmic factor in turn selects nuclear factors that produce males, and the population evolves toward a state in which all uninfected zygotes become male. More importantly, this conflict between cytoplasmic and autosomal genes leads to the selection of host autosomal genes that resist the sex ratio distorter. These polygenic autosomal "resistance" genes (*R* genes) show their effect to limit the transmission of bacteria to offspring, rather than overriding the feminizing activity of the symbionts. Such infected mothers no longer produce a female-skewed F1; on the contrary, a few of them are overwhelmingly male-biased, a phenomenon referred to as ARF (ARhenogenous Female) trait. Such an intervention of these *R* genes in modifying the pattern of *Wolbachia* transmission may even be considered as a new type of sex-determining gene in isopods. Similarly, the presence of the f factor also leads to intragenomic conflict, by involving the autosomal masculinizing gene *M*, which can restore a male phenotype in the presence of f factor. The frequency of *M* increases with the frequency of the f factor, suggesting that it is selected to repress f factor activity. This instance of autosomal *M* gene acting as a repressor of cytoplasmic f factor is an example of multifactor or polygenic sex determination in isopods, as this *M* gene is also previously shown to have exerted an overriding effect on the female sex factor, resident in the W chromosome. It is henceforth inferred that the presence of multiple-sex factors, as found in *A. vulgare*, would have had major consequences for the evolution of sex chromosomes in crustaceans (Juchault and Rigaud, 1995).

Genetic resistance to the feminizing influence of the cytoplasmic factor is thought to enhance the maintenance of infected and uninfected females in the *Armadillium* population. Nuclear genes that abolish the feminizing, cytoplasmic influence occur at higher frequencies in males than females (Bull, 1983).

These resistance genes therefore become sex factors, since they provide part of the inherited basis of sex. The resistance locus can even become almost the entire inherited basis of sex determination. Evidently, there is a joint evolution of cytoplasmic factor and these Mendelian resistance factors in the population which invariably evolved to a sex ratio of ½ (Bull, 1983).

In some of *Wolbachia*-infected population, the feminization of genetic males is found to be incomplete, resulting in sterile intersexual animals possessing features of both sexes. The origin of these intersexuals is also thought to be due to the action of male restorer gene (M) against the feminizing action of f factor. In general, sex ratio distorters should increase in frequency in host populations previously uninfected, and that uninfected genetic females should be replaced by infected neofemales in a few generations. However, their frequency is often lower than those predicted. This is mainly due to the fact that males prefer noninfected females for mating than the infected females. Similarly, the infected neomales have lower mating rates and received less sperm relative to uninfected females. The infected females quite often exhibited behaviors that stop the mating sequence. Since these neofemales are genetically males, they lack the W chromosome, thereby lacking the genes necessary for the regulation of female mating behavior.

The original sex-determining mechanism is nuclear in crustacean species, influenced by feminizing factors, as in isopods. Apparently, factors external to the host genotype, well represented by the bacteria *Wolbachia* and the cytoplasmic f factor, provide a driving force of the complex genetic sex-determining system evolved within crustaceans. The existence and operation of multiple sex factors, as found in terrestrial isopods, have an ultimate consequence on the evolution of sex chromosomes itself in higher animals.

1.7 AMPHIPODS

The influence of cytoplasmic sex factors on sex determination has also been reported in another peracarid, Amphipoda, as revealed from the anomalous sex ratios found among natural populations. In addition, various intersex types have also been described in some species such as *Orchestia gammarellus* (Charniaux-Cotton, 1962) and *G. duebeni* (Bulnheim, 1978b). The intersexuality is characterized by the modified external sexual characters alone or the whole range of sexual characters, including a hermaphroditic gonad in the form of ovotestis. In *G. duebeni*, Bulnheim and Vavra (1968) interpreted the existence of thelygeny as well as intersexuality as resulting from an inversion of the sexual phenotype of genetic males under the influence of epigenetic feminizing factors such as two protozoan microsporidia namely *Octosporea effeminans* and *Thelohania hereditaria*. Interestingly, the *O. effeminans*-infected females produced both female offspring as well as intersexuals containing ovotestis, whereas *T. hereditaria*-infected *G. duebeni* produced only females. However, amphipods lack the cytoplasmic "f factor" for sex change, as in the case of isopods. The Microsporidia are transmitted from the infected mother to the daughters through the egg cytoplasm. Accordingly, females experimentally infected with these microsporidians became thelygenic and hence males could arise only from the noninfected eggs. Obviously, Microsporidia do not influence the genetic composition of the zygote, but seem to control sex determination by overriding any male tendency of the nuclear genes. Thelygeny and intersexuality caused by these protists are not irreversible. However, high salinity (above 27‰) inhibited the infection of the eggs by *Octosporea* and its feminizing influence during postembryonic development. Similarly, in most of the strains, exposure to 22°C or above, either during oogenesis or during embryonic and early postembryonic development can totally abolish both thelygeny and intersexuality (Legrand et al., 1987).

The amphiod *G. duebeni* is also infected by another feminizing microsporidian parasite, *Nosema granulosis*, to convert males into functional females. The parasite manipulates host sex by preventing androgenic gland differentiation, androgenic gland hormone production, and consequently male differentiation (Rodgers-Gray et al., 2004). However, compared to other *Microsporidium* sp., the intensity of feminization by *N. granulosis* is less (Jahnke et al., 2013).

1.8 DECAPODS

Mechanism of sex determination in crustaceans is well understood in Peracarida (Ginsburger-Vogel and Charniaux-Cotton, 1982). However, the large-bodied decapods have received only scanty attention in this regard. The caridian freshwater prawn *Macrobrachium rosenbergii* is the only decapods wherein successful sex inversion using the androgenic gland has been demonstrated. Studies on freshwater prawn *M. rosenbergii* have shown that the neomales produced by androgenic gland implantation, when crossed with normal females, produced a progeny with sex ratio skewed toward females (Malecha et al., 1992). Complete reversal of sexual phenotype and function by the surgical implantation of androgenic gland tissues from adult males into small juvenile females forms the basis of such crossing studies. The observed production of males and females in the sex ratio of 1:1.29 from most normal mating suggest that sex in prawns is probably controlled by a chromosomal mechanism. The presence of males in the progeny of the neomales crossed with normal females has, however, suggested a female heterogametic sex determination with ZW chromosomes in this decapod crustacean. Similarly XX neofemales have also been produced from androectomized males. When these neofemales are crossed with normal males a sex ratio of 108:1 and 458:0 has been obtained in the all male progeny confirming the occurrence of ZZ/ZW chromosomal sex-determining mechanism in *M. rosenbergii*. All this type of sex determination reveals the occurrence of genetic and endocrinological control of sex determination in the Malacostraca. Each individual crustacean is endowed with both male- and female-determining genes. When a female-determining gene is to be expressed in either sex, the male-determining factor is repressed and vice versa.

1.9 HORMONAL REGULATION OF SEX DETERMINATION IN CRUSTACEA

In crustaceans, sex differentiation is mediated by the unique androgenic gland hormone, under which male sexual differentiation occurs. In addition, two other hormones, viz., ecdysteroids and a terpenoid hormone, methyl farnesoate (MF), are shown to have morphogenetic control over embryogenesis and early development (Subramoniam, 2000). In addition, Olmstead and LeBlanc (2002) proposed a sex-determining role for MF in the water flea, *Daphnia* spp. Specimens exposed to elevated concentrations of MF or its synthetic analogs resulted in all male broods, which are highly unusual in these otherwise parthenogenetic species. This sex-determining effect of the hormone is restricted to the later stages of ovarian oocyte maturation. Once the eggs are transferred to the brood chamber, sex is fixed. In addition, increased temperature synergized with these crustacean juvenoids to cause elevated intersex progeny in *Daphnia magna* (Mitchell, 2001). Similar effects were seen in four other species of cladocerans (Kim et al., 2006), suggesting that the role of MF in sex determination may be a general phenomenon for cladocerans.

Manar et al. (2009) reported that chemical exposure could interfere with normal sex differentiation in *D. magna*. Environmental contaminants like the insecticide toxaphene influenced the timing of sex determination in this primitive crustacean. In response to the exposure of this chemical compound, the sex is determined in the immature oocytes in the ovary, before they are extruded into the brood chamber (Ignace et al., 2011). Toxaphene has juvenile hormone-like effects in disrupting the sex ratio by increasing the male production (Flaherty and Dodson, 2005).

Furthermore, MF that transduces environmental signals is capable of programming daphnid embryos to develop into males and is likely the endocrine factor responsible for initiating the sexual reproductive phase of this cladoceran. Both MF and its synthetic analogs induced male offspring production and elevation of hemoglobin, suggesting that these two physiological activities are coregulated by this hormone (Rider et al., 2005).

MF control of sex determination in cladoceran can be considered as a downstream step in the environmental sex determination in crustaceans. Rider et al. (2005) has proposed a terpenoid signaling pathway for these branchiopodan crustaceans, which live in ephemeral freshwater pools, where environmental fluctuations have a telling effect on the life cycle of these ephemeral organisms. Male sex determination required only a pulse of MF during a critical period of ovarian oocyte maturation (Olmstead and LeBlanc, 2002). In daphnids, MF is a major hormonal factor that is responsible for transducing environmental signals. Conceivably, environmental factors that stimulate a sustained elevation in hormonal (MF) levels would be likely to impact sex determination in the cladocerans. For instance, elevated temperature may prove to be an environmental signal by adversely impacting the uptake or assimilation of nutrients resulting in male production via the terpenoid signaling pathway (Rider et al., 2005). The terpenoid signaling pathway may include the hormone MF, its receptor (MF receptor), and the response elements on the responsive genes for sex determination such as *sex*-I (as in *C. elegans*), *dsx* in *Drosophila*, or *csd* in the honey bee. The recent discovery of the sex-determining gene *Dsx* in *D. magna* lends supporting evidence toward the possible involvement of endocrine pathway in the determination of sex in cladocerans (Kato et al., 2011). During embryogenesis of *D. magna*, expression of *DapmaDsx1* was increased only in males and its transcripts were primarily detected in male-specific structures. Knock-down of *DapmaDsx1* in male embryos resulted in the production of female traits including ovarian maturation, whereas ectopic expression of *DapmaDsx1* in female embryos resulted in the development of male-like phenotypes. Taken together, it may imply that there is a strong link between genetic and environmental sex determination in crustaceans, as demonstrated in cladoceran sex determination. The environmental signal may be perceived by MF, which by binding to its receptor in the target cells may trigger sex gene (*DapmaDsx1* and *2*) expression to induce male sex determination.

Moreover, the exposure of adult daphnids to juvenile hormones and their analogs induces parthenogenetically reproducing *D. magna* to produce male neonates (Oda et al., 2005; Tatarazako et al., 2003). Females less than 24 h old were exposed to MF, JH III, methoprene, and the insecticides pyriproxyfen and fenoxycarb and were observed for 24 days. These chemicals invariably triggered the appearance of male daphnids, whose percentage in the population was significantly greater than females under ordinary environmental conditions. These results again confirm the mediation of MF in the environmental signal transduction to induce male sex determination in *Daphnia*, as proposed by Olmstead and LeBlanc (2002).

Chapter 2

Sex Differentiation

2.1 INTRODUCTION

In Crustacea, the mode of sex determination ranges from heterogametic to polyfactorial and environmental regulatory mechanisms (Bull, 1983). Further, it is assumed that sex-determining genes or factors, present in sex chromosomes and/or autosomes, could be present in both male and female crustaceans for sexual development. In addition, experimental studies imply that the major regulatory genes of sex development are susceptible to epigenetic influences (see chapter on Sex Determination). Sex differentiation in crustaceans differs from that of other arthropods in that the male sexual differentiation and maintenance of sexual characteristics are regulated by the unique androgenic gland (AG) present in the male, with the concurrent inhibition of female differentiation. Differentiation of female gonad thus occurs in the absence of AG by default. Ovarian differentiation in the isopod *Armadillidium vulgare* seems to be spontaneous in the absence of androgenic hormone, regardless of chromosome constitution (Katakura, 1982). However, Charniaux-Cotton (1952) showed that the female secondary sex characteristics such as oostegites develop under the control of ovary in the amphipod *Orchestia gammarellus*. Recent identification of AG-specific insulin-like factors in decapod crustaceans is a step forward in our understanding of sexual differentiation mechanism in crustaceans. Indeed, with the discovery of AG, the endocrine regulation of sex differentiation has been established in Crustacea.

2.2 GONAD DIFFERENTIATION

Sex determination is a series of events whereby the sexually indeterminate embryo progressively acquires male or female characteristics in the gonads, genital tract, and external genitalia. The formation of male and female gonad is, however, different from those of somatic organs in the sense that the somatic gonads, after their embryonic origin, are populated by the migrating primordial germ cells from extraembryonic membranous regions to impart sex specificity to the undifferentiated gonad (Josso et al., 2001). Primordial germ cells or simply germ cells play a unique role in gamete production. There are fundamental distinctions between germ cells and somatic cells in their origin, differentiation, and maintenance throughout development. However, segregation of germ cells

Sexual Biology and Reproduction in Crustaceans. http://dx.doi.org/10.1016/B978-0-12-809337-5.00002-2

from other embryonic cells during cleavage and gastrulation of the embryo is crucial to developmental and reproductive biology of any animal. Basically, there are two distinct modes of germ cell line segregation in animals (Extavour and Akam, 2003). In some animals, during early embryogenesis, germ cells are identified with localization of maternally inherited determinants before or immediately following fertilization. This mode of germ cell segregation is termed as "preformation" and the best example is found in *Drosophila melanogaster*. In the other mode, viz., "epigenesis," the germ cells are not observed until later in development and arise as a result of the inductive signal from the surrounding embryonic cells or tissues.

In Crustacea, these two modes of germ cell segregation namely epigenesis and preformation are known to exist. However, progenitor germ cells (PGCs) of most crustaceans appear to be formed late in development from the mesodermal cells of the coelomic cavity, although early segregation by preformation do occur in some copepods and cladocerans (Amma, 1911; Kühn, 1913). All copepods that have been studied segregate germ cells by preformation. In *Cyclops fuscus*, embryonic cleavages are holoblastic and equal. During the first cleavage, germplasm, found in the form of dense granular material, associates with only one of the centromeres and hence inherited by only one of the two daughter cells. The granular material continues to segregate asymmetrically into a single blastomere in subsequent cleavages. Finally two cells representing PGCs are getting located at the tip of the invaginating archenteron (Amma, 1911). On the contrary, in all higher decapod crustaceans, the germline develops during larval stages, or continuously throughout adult development from the mesodermally derived cells of gonadal epithelium (Extavour and Akam, 2003).

The formation of genital apparatus involves segregation of the germ cells, elaboration of genital anlage, and colonization of anlage by germ cells, and differentiation of the gonoducts. In Malacostraca, sexual differentiation is acquired after embryonic or larval stage. The sexual differentiation might last for several molts. The male and female sexual characteristics appear in a well-defined order and develop during successive molt stages. The internal genital apparatus may develop before or after the appearance of external secondary sexual characteristics, according to the species.

Gonadal differentiation is well studied in the amphipod, *O. gammarellus*, in which sexual differentiation of the genital apparatus begins long before the appearance of external characteristics (Charniaux-Cotton and Payen, 1985). Soon after hatching, the young is sexually undifferentiated, with no sign of external or internal sexual characteristics. At the second postembryonic intermolt, the genital apparatus makes its appearance in the form of two strands of mesodermal cells extending from the second to the seventh thoracic segment on both sides of the digestive track. Each strand is then directed ventrally toward the sternite to end in the mesenchymatic tissue of the coxopodite of the last thoracic segment. Some of the mesenchymatic cells near the terminal region of the genital strand forms the presumptive cells of the AG (Fig. 2.1). Whereas the

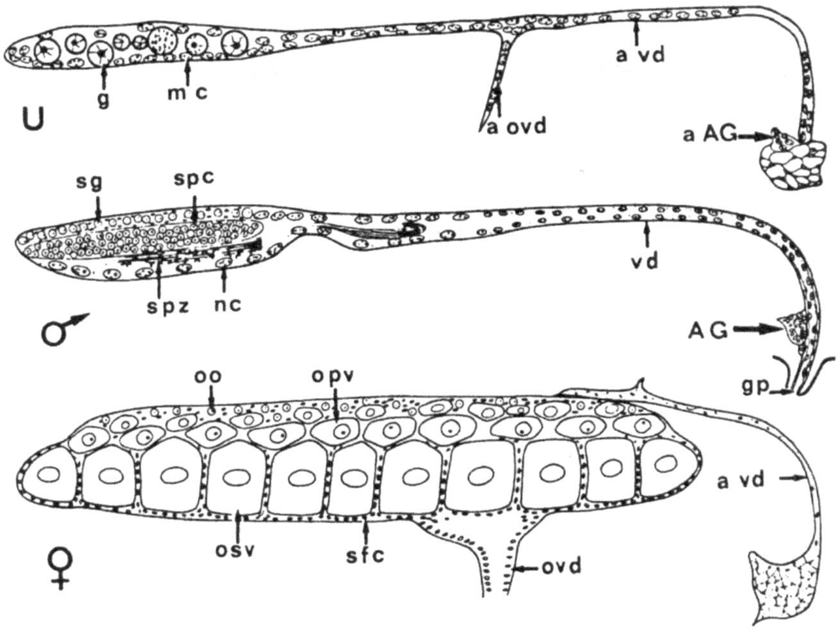

FIGURE 2.1 Organogenesis of the genital apparatus in *O. gammarellus*. U, undifferentiated stage; ♂, juvenile male; ♀, female in secondary vitellogenesis. The scale for U is larger than that for ♂ and ♀. *AG*, androgenic gland; *a AG*, anlage of AG; *a ovd*, anlage of oviduct; *a vd*, anlage of vas deferens; *g*, protogonia; *gp*, genital papilla; *mc*, nucleus of mesodermal cell; *nc*, nucleus of nurse cell; *oo*, oogonia (embedded in stroma of mesodermal cells); *opv*, oocyte in primary vitellogenesis; *osv*, oocyte in secondary vitellogenesis; *ovd*, oviduct; *sfc*, secondary follicular cell; *sg*, spermatogonia (embedded in stroma of mesodermal cells); *spc*, spermatocyte; *spz*, spermatozoa; *vd*, vas deferens. *Reproduced from Charniaux-Cotton, H., Payen, G., 1985. Sexual differentiation. In: Bliss, D.E., Mantel, L.H. (Eds.), The Biology of Crustacea, vol. 9. Academic Press, New York, pp. 217–299, Fig. 2, p. 224.*

anterior part of each cellular strand contains gonia, the posterior region constitutes the anlage of the sperm duct. As shown in the figure, a separate thin strand gives origin to the formation of the oviduct that opens at the coxae of the fifth thoracic segment. As for the external sexual characteristics, in *O. gammarellus*, the genital papillae in the male and the oostegites in females appear during the fifth molt. In the males, after this molt, the relative growth of the second gnathopod becomes more rapid than in the females (Charniaux-Cotton, 1957).

2.3 FEMALE DIFFERENTIATION IN *ORCHESTIA GAMMARELLUS*

Early histological observations on the amphipod *O. gammarellus* by Charniaux-Cotton and her associates have provided important information on the process of gonad morphogenesis facilitating a basic understanding of the mechanism of gonadal sex differentiation in Crustacea. In the development of ovary, mesodermal cells play a major role. However, the germinal zone starts developing in the undifferentiated gonad during the second intermolt, as a result of the migration of gonia in the anterior part of genital tract up to the sixth segment. In Crustacea, the germinal zone is present throughout the adult life of the female. Mitoses of the gonial cells take place only within the germinal zone. During the third intermolt, the oogonia leave the germinal zone and enter into prophase of meiosis. At this stage, the juvenile ovary consists of a longitudinal germinal zone and primary follicles between 20 and 160 μm in diameter. Mesodermal tissue forms around each oocyte to form a multilayered follicular epithelium. The oocytes in the primary follicles remain in the previtellogenic stage (see Fig. 2.1). It is around the 10th intermolt, the secondary follicular tissue differentiates at the periphery of the ovary. In the secondary follicles, each oocyte is surrounded by a single layer of follicle cells, the process being called secondary folliculogenesis. The oocytes of these follicles undergo secondary vitellogenesis. Thus, the differentiation of secondary follicular tissue corresponds to the attainment of puberty. The oviduct becomes functional and opens at the coxae of the fifth thoracic leg. Interestingly, rudiments of the sperm ducts persist in females, but lose their continuity with the ovaries after the fifth molt (Zerbib, 1964).

2.3.1 Secondary Sexual Characteristics

In *O. gammarellus*, secondary sexual characteristics make their appearance with the development of oostegites at the fifth postembryonic molt, coinciding with the differentiation and maturation of the ovary, suggesting that the control of female secondary sex characters in amphipods is under an ovarian hormone (Charniaux-Cotton and Payen, 1985). The oostegites develop as small lamellar outgrowths on the inner side of coxae of the second gnathopods as well as periopods 3, 4, and 5. They continue to develop further at each molt and form the brood pouch or marsupium in puberal females. The oostegites are permanent secondary sexual characters, but the ovigerous setae that makes matrix of the brood chamber are temporary sex characteristics, associated with the incubation of embryos. However, the allometric growths of female appendages are similar to those of sexually undifferentiated amphipods. Similarly, the appearance of vitellogenin in the hemolymph could be considered as a temporary female characteristic related to ovarian activity in amphipods.

2.4 MALE SEX DIFFERENTIATION IN *ORCHESTIA GAMMARELLUS*

The differentiation of the male genital apparatus starts with the formation of androgenic anlage. During the postembryonic intermolt, differentiation begins at the extremities of the sperm ducts with the multiplication of mesodermal cells that represent rudiments of the AG. The anterior region of the tract containing the germ cells will form the germinal zone of the testis. The spermatogenesis starts during the third intermolt, similar to oogenesis. With the formation of a testicular cavity, the bordering mesodermal cells enlarge and their nuclei become polyploidy. The longitudinal germinal zone of the tubular testis contains mitotically dividing spermatogonia. After leaving this zone, the spermatogonia form a longitudinal zone of contiguous secondary spermatogonia. The mesodermal cells of this zone envelop the secondary spermatogonia, which become spermatocytes rather synchronously. At this time, new spermatogonia leave the germinal zone (Meusy, 1963). Spermiogenesis occurs within the testicular cavity bordered by the polyploid mesodermal cells. These mesodermal cells act as nurse cells to secrete mucopolysaccharide substances that form the extracellular tubules surrounding bundles of elongating spermatids and spermatozoa.

2.4.1 External Male Characteristics

In the male *O. gammarellus*, several appendages such as the second antennae, first and second gnathopods, and last pereopods are longer and robust than in females. The propodus and carpus of the gnathopod in the male start growing after the fifth molt, the molt at which the genital papillae also appear. There are three phases in the growth of gnathopod. In the immature phase, the gnathopod looks similar to an undifferentiated appendage. In the next phase, the second gnathopod effects the transition from the immature to the definitive male form in three molts. The relative growth of the carpus of the second gnathopod also presents three phases. The male amphipods are capable of copulating only when the second gnathopod has acquired the puberal morphology. The second gnathopods have important functions in holding the female during copulation.

2.5 SEXUAL DIFFERENTIATION IN ISOPODS

In isopods, the gonads are tubular in both sexes, but in males, each genital tract is formed by three testicular tubules, which release their contents into a seminal vesicle. In addition, the AGs also differ in their location: in most isopods such as *Porcellio dilatatus* and *A. vulgare*, the AGs are found attached to the extremities of the testicular utricles (Legrand and Juchault, 1960; Katakura, 1982). However, in *Helleria brevicornis*, they are found along the subterminal region of the vas deferens (Juchault, 1967). In the sexually undifferentiated phase, the genital apparatus is formed by two cellular strands containing germ cells in the anterior region. In *P. dilatatus*, there are three pairs of thin cellular strands emerging from the external side of the anterior region, each one found in the second, third, and fourth thoracic segments (Legrand and Vandel, 1948). Sexual differentiation of the genital apparatus starts after the third molt. Just like the amphipods, external sexual differentiation begins after the differentiation of the testis. In *Ligia italica*, the appendix masculina, however, appears in the first posthatching molt, when the genital apparatus also begins to differentiate (Berreur-Bonnenfant and Inagaki, 1973).

Katakura (1984) described the sex differentiation in the isopod *A. vulgare*. In the first stage manca, the primordium of the reproductive organs contains, at its central portion, a cyst of primordial germ cells as well as rudiments of both oviducts and spermiducts. In the third stage, dimorphic development of the reproductive organs is recognizable. In animals larger than 3.5 mm in body length, the testis and ovary attain full growth. Female differentiation begins with the progression of germ cells toward the posterior region of the genital tract. The germinal zone runs along the external border on the side of the oviduct (see Fig. 2.2). Oviduct originates at the mesal side of the ovary and opens into the oviductal aperture at the base of the fifth pereopod. As the female approaches the pubescent stage, the proximal funnel-like oviduct differentiates into the seminal receptacle. The middle region of the oviduct also enlarges at the moment of copulation into a temporary copulatory pocket due to the presence of a muscular layer surrounding it. The distal oviduct is lined internally with a chitinous layer that is continuous with the external cuticle. In the fully developed males, the anterior part of each testis is connected by a threadlike AG.

The external female characteristics of isopods include the appearance of oostegites in young females. They grow gradually with each successive molt and become large and concave in shape to attain the functional form after the molt following egg laying. Oostegites are, however, temporary characteristics, related to the incubation of the embryos; they disappear at the time of genital rest.

2.6 SEXUAL DIFFERENTIATION IN DECAPODA

The genital apparatus in decapods also has a sexually undifferentiated phase in the young crabs and crayfishes. However, some of the characteristics that differ from those of peracarids are (1) there is only one rudiment for the gonoducts, with the possible exception that the genital apparatus of the young hermit crabs has two rudiments of genital ducts for both sexes; (2) no rudiment of an AG is visible in their genital apparatus and hence; (3) male differentiation begins before a discrete AG is visible. Whereas the sexually undifferentiated phase lasts for the first two postembryonic intermolts in the crayfishes (Payen, 1973), it persists for the entire larval period in the hermit crab, and up to the second crab stage in the brachyuran crabs (Payen, 1974).

In the zoea and megalopa stages of the crabs, the genital rudiments extend forward, while they anastomose posteriorly. Protogonia residing in the genital rudiment give rise to gonial cells by mitotic multiplication. The mesodermal cells also multiply in number in the megalopa stage. The mesodermal cells of the genital tract, beyond the point of posterior fusion, multiply in number and form two separate, short threadlike strands, representing the rudiment of the genital ducts. The genital apparatus assumes H-shape during the first two crab stages. The gonad is still filled with undifferentiated germ cells, and the pair of genital ducts extends in a posterior direction. The actual sexual differentiation of the genital apparatus begins in the third crab stage with the orientation of the

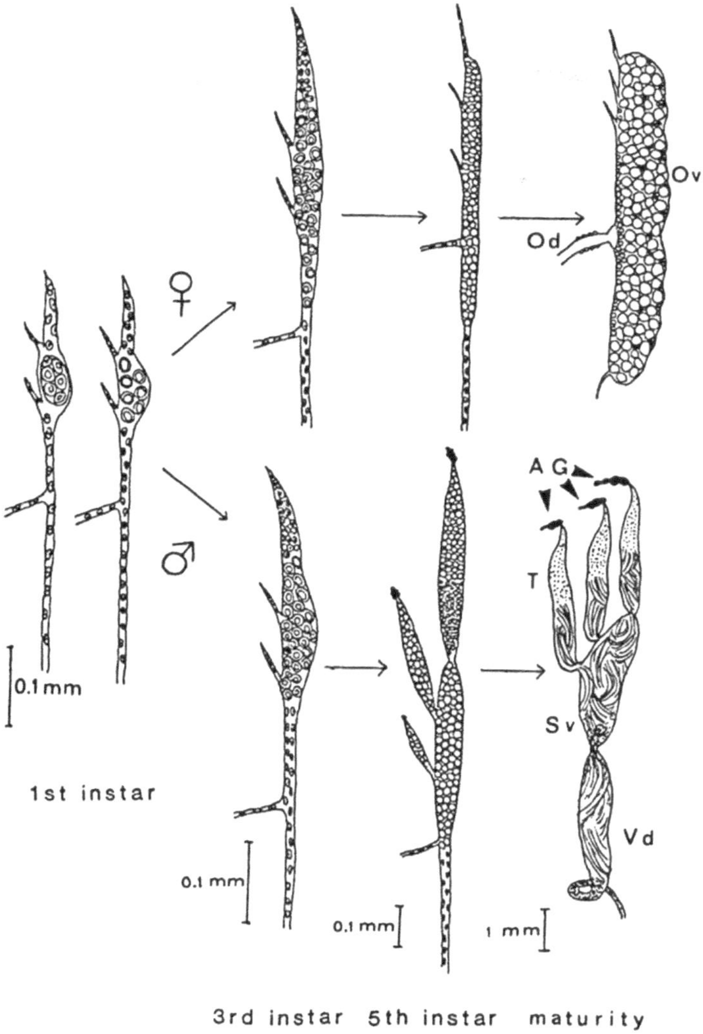

FIGURE 2.2 Postembryonic development of left internal reproductive organs of *A. vulgare*. *AG*, androgenic glands; *Od*, oviduct; *Ov*, ovary; *Sv*, seminal vesicle; *T*, testis; *Vd*, vas deferens. *Reproduced from Charniaux-Cotton, H., Payen, G., 1985. Sexual differentiation. In: Bliss, D.E., Mantel, L.H. (Eds.), The Biology of Crustacea, vol. 9. Academic Press, New York, pp. 217–299, Fig. 11, p. 238.*

single pair of gonoducts toward the sixth thoracic sternite in the female and the last thoracic sternite in males. In the females, the ovary also extends posteriorly (Charniaux-Cotton, 1985).

In the crayfishes, during the first postembryonic molt stage, the genital rudiment, containing cellular masses of protogonia, lengthens posteriorly and fuses partially to form a paired anterior half and an unpaired posterior half. This structure extends from the second to the sixth thoracic segment. The rudiments of the paired genital ducts become distinguishable at the median region of the genital apparatus. During the second postembryonic molt stage, proliferation of mesodermal cells in the genital rudiments leads to the formation of the testis, whereas such a proliferation is not evident in the future ovaries. However, the sexual dimorphism in the genital apparatus occurs only in the third embryonic stage. The crayfishes have a paired vas deferens; the orientation of them makes the identification of the sex possible. The onset of meiosis starts in the ovary during the third embryonic stage, whereas in the males the meiosis of the spermatogonia starts only in the seventh stage (Payen, 1973).

Several studies on the organogenesis of penaeid shrimps have revealed that the differentiation of the AG occurs before or at the onset of spermatogenesis by the end of the second month of the postlarval stage (Garza-Torres et al., 2009). Male sex differentiation, however, takes place in these shrimps within the first two weeks of the postlarval stage by the formation of vas deferens from the still undifferentiated testis. In various penaeid shrimps, the gonad makes its appearance long after hatching of the larva. For example, in *Marsupenaeus japonicus*, the gonad appears first as two masses of germ cells associated with somatic cells in the postlarva (PL) 8 (Laubier et al., 1983). In *Litopenaeus vannamei*, the bilateral anterior lobe of the gonad was histologically recognizable only by PL12. However, Garza-Torres et al. (2009) identified the female oviduct and the filliform male vas deferens in PL16. The external sexual characteristics for both males (the gonopores) and females (thelycum) appear only during PL50. Fig. 2.3 shows the timeline of events of internal and external differentiation of female and male *L. vannamei*. The stage of sexualization in the males corresponds to the histological identification of AG on the exterior of distal vas deferens. The primordial AG cells probably originate from the mesodermal cells found among the connective tissue that interconnects the vas deferens with the testicular lobes. Mesodermal cells play an important role in the formation of gonadal lobes, with the gonial cells gathering around their lumen, forming into the germinal zone. On the other hand, the oviducts and vas deferens differentiate from the connective tissue surrounding the gonad. Interestingly, the ovary and testis could be visualized only in PL72, although the germinal epithelium in the ovary is distinguishable at PL52. Early developing oocytes indicating the ovarian differentiation are also noticeable at PL72. In the males, the spermatogonia are arranged at the testicular lobe, and the lumen is filled with differentiating secondary spermatocytes and spermatids. Here again, in the testicular differentiation, mesodermal supporting epithelial cells, called sustentacular cells play a major role in the differentiation of the spermatocytes from spermatogonia, as well as transferring the genital products into the lumen of the seminiferous tubules by their contractile activity on the lumen facilitating movement of the spermatids down the proximal vas deferens. In the penaeid shrimps, as in other decapods, spermatogenesis continues in the vas deferens lumen, before being packaged into spermatophores.

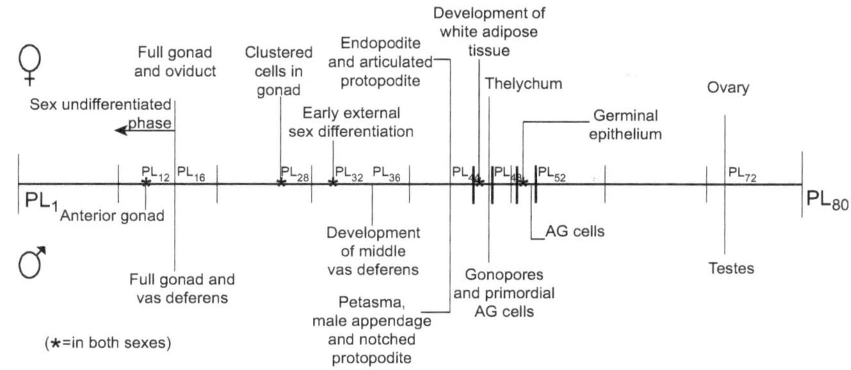

FIGURE 2.3 Timeline events of internal and external differentiation of female and male Pacific white shrimp *P. vannamei. Reproduced from Garza-Torres, R., Campos-Ramos, R., Maeda-Martínez, A.M., 2009. Organogenesis and subsequent development of the genital organs in female and male Pacific white shrimp* Penaeus (Litopenaeus) vannamei. *Aquaculture 296, 136–142, Elsevier, Fig. 1, p. 137 with modification.*

In the shrimps, gonadal differentiation starts soon after the larval transformation into postlarval stages. But the AG differentiation occurs almost at the same time of gonadal maturation at around PL72 stage. Nevertheless, AG could exert its influence on testis maturation, as spermatogenesis begins after AG is formed. Concurrently, AG could negatively regulate oogenesis in the undifferentiated gonad by some unknown inhibitory mechanism. However, external sex characteristics in shrimps such as *L. vannamei* develop around 15–20 days earlier than gonad differentiation in both sexes (Garza-Torres et al., 2009).

2.7 DISCOVERY OF ANDROGENIC GLAND

Historically, sex differentiation in Crustacea has been explained in terms of classical parasitic castration caused by the parasite *Saculina* on the male crabs (Charniaux-Cotton and Payen, 1988). In the sacculinized *Carcinus maenas*, Courrier (1921) was the first to suggest that a parasite caused the atrophy of testis resulting in the feminization of the host, thereby implying that the testis could be the endocrine source to control the male secondary sex characteristics in the crabs. While describing the anatomy and histology of the swimming crab, *Callinectes sapidus*, Cronin (1947) noted the presence of a ductless accessory gland attached to the posterior vas deferens. Neither the endocrine nature nor its functional role in male reproduction was postulated at this time. A clear understanding on the sex differentiation in crustaceans was obtained only with the discovery of a male sex hormone secreting gland namely the AG in the male amphipod *O. gammarellus* by Charniaux-Cotton (1955). However, AG is not forming a part of testis but is found attached to the terminal part of the vas deferens. Only in the isopod *A. vulgare*, AG is located at the cephalic end of each of the three pairs of testes and was observed as a fine continuation of the testes proper (Katakura, 1959). After the discovery of AG in the amphipods, several reports appeared on its occurrence in almost all malacostracan crustaceans (Table 2.1), suggesting that this could be the endocrine gland to produce the male sex hormone. However, AG has not been detected in any of the lower nonmalacostracan crustacean orders. Several ultrastructural studies on AG have revealed certain common characteristics such as large amounts of endoplasmic reticulum, Golgi apparatus, mitochondria, and multivesicular bodies with variable contents (King, 1964; Charniaux-Cotton et al., 1966; Radu and Cracium, 1976). Similar to other reproductive organs, AG activity is shown to be negatively regulated by X organ/sinus gland neuropeptides (Khalaila et al., 2001).

TABLE 2.1 On the Occurrence of Androgenic Gland and Experimental Studies in Malacostracan Crustaceans

Species	Order	Method of Detection	Nature of Hormone and Mode of Action
Orchestia gammarellus	Amphipoda	AG ablation from males	Reduction in size of male appendages, cessation of spermatogenesis, and development of oocytes from primary germ cells
O. gammarellus	Amphipoda	Implantation of AG in genetically females	Secondary gonia in the ovary changed to form an all-sperm-cell lineage
Lysmata seticaudata	Decapoda	AG ablation from males and implantation into females	Development of external male secondary characteristics
Pandalus borealis	Decapoda	AG ablation from males and implantation into females	Development of external male secondary characteristics
Carcinus maenus	Decapoda	AG ablation from males and implantation into females	Occasional development of external male secondary characteristics
Rhithropanopeus harrisii	Decapoda	AG ablation from males and implantation into females	Occasional development of external male secondary characteristics
Armadillidium vulgare	Isopoda	Injection of AG extracts into immature females	Appearance of secondary male characteristics

TABLE 2.1 On the Occurrence of Androgenic Gland and Experimental Studies in Malacostracan Crustaceans—Cont'd

Species	Order	Method of Detection	Nature of Hormone and Mode of Action
A. vulgare	Isopoda	Implantation of ablated AG into immature females	Appearance of secondary male characteristics
A. vulgare	Isopoda	AG ablation in males	The elongation of endopod that serves as copulatory organ disappeared, and Vg titers rose to levels exceeding those of normal female when vitellogenesis was commenced
A. vulgare	Isopoda	AG ablation in male larvae	Complete and fully functional sex reversal was obtained
Macrobrachium rosenbergii	Decapoda	AG implanted into early stage females	Conversion of genotypic females into phenotypic males
M. rosenbergii	Decapoda	AG ablation from males	Conversion of genotypic males into phenotypic females
M. rosenbergii	Decapoda	Implantation of AG in female	Reduction in the width of second pleuron, generation of the appendix masculine, and masculinization of the chelipeds
M. rosenbergii	Decapoda	Implantation of AG in the early stage of female	Ovotestis were observed in females
M. rosenbergii	Decapoda	AG implantation in female before ovarian differentiation	Full and functional sex reversal was accomplished and neomales were obtained
Cherax destructor	Decapoda	Injection of AG extract into immature females	Development of male pores on 5th walking legs. Inhibition of oostegite development

Continued

TABLE 2.1 On the Occurrence of Androgenic Gland and Experimental Studies in Malacostracan Crustaceans—Cont'd

Species	Order	Method of Detection	Nature of Hormone and Mode of Action
Procambarus clarkii	Decapoda	AG implantation in females	The first pair of pleopods of female which are flexible and slender were modified into gonopod-like, stiff robust appendages as in males
P. clarkii	Decapoda	AG implanted in vitellogenic female	Vitellogenesis was inhibited due to a lack of yolk protein accumulation in the oocyte
P. clarkii	Decapoda	AG ablation from males and implantation into females	Development of secondary male characteristics in females
Cherax quadricarinatus	Decapoda	Ablation of AG from intersex males	Ovarian development in ovary which was permanently arrested in non-AG ablated individuals
C. quadricarinatus	Decapoda	Implantation of AG into immature females	>90% development of secondary male characteristics. Vitellogenesis suppressed but not entirely eliminated
C. quadricarinatus	Decapoda	AG ablation in intersex males	Vitellogenin-specific genes expressed after ablation from intersex male crayfish. No expression in control intersex males
C. quadricarinatus	Decapoda	Implantation of AG in females	Male courtship and false copulation when encountered with intact females

TABLE 2.1 On the Occurrence of Androgenic Gland and Experimental Studies in Malacostracan Crustaceans—Cont'd

Species	Order	Method of Detection	Nature of Hormone and Mode of Action
C. quadricarinatus	Decapoda	AG implantation in females	The distinct female characters such as ovigerous setae lining the endopod of pleopods and the wide abdomen on which the fertilized eggs are carried were reduced
C. quadricarinatus	Decapoda	Functional male crossed with normal female	Offsprings obtained were in the ratio 1:3 (male:female) which shows that AG^{+} males are genetically females
C. quadricarinatus	Decapoda	Functional female crossed with normal male	All the offsprings obtained were males suggesting that AG^{-} females are genetically males
Eriocheir japonicus	Decapoda	AG implantation	Female crabs develop male-like appendages
M. rosenbergii	Decapoda	RNAi silencing of IAG encoding gene (*Mr IAG*)	*Mr IAG* silencing impeded spermatogenesis and male secondary sexual characteristics, and also inhibits growth
			Full and functional sex reversal of male
M. rosenbergii	Decapoda	RNAi silencing of insulin-like receptor gene (*Mr IR*)	Hypertrophy of AG and unusual abundance of immature sperm cells. No growth impact after silencing
			No sexual shift suggesting that the involvement more than a single *Mr-IAG* receptor

2.8 ROLE OF ANDROGENIC GLAND IN MALE SEX DIFFERENTIATION

AG's role in the sex differentiation in Crustacea was demonstrated largely by classical experimental studies involving removal and implantation of this endocrine gland. Charniaux-Cotton (1955) demonstrated for the first time that the surgical removal of this gland transforms testis into ovary with the concurrent disappearance of male secondary characters and the appearance of female secondary characters such as the egg-bearing oostegites in *O. gammarellus*. This experimental study on the amphipods provided first-hand information on the role of AG in the control of sex differentiation for both male and female crustaceans. Later work by Suzuki and Yamasaki (1997) revealed the bipotential nature of the gonad, which could be manipulated to successful sex reversal by removal and grafting of the AG in the terrestrial isopod, *A. vulgare*. Schematic representation of gonadal masculinization two months after implantation of AG into stage 5–9 females is shown in Fig. 2.4. It is seen from the figure that only in early stage 5 complete transformation of ovary into testis along with seminal vesicle and vas deferens has taken place. Three pairs of AGs were also observed in the cephalic end of each testis. Furthermore, the masculinized females did not develop their rudimental oviducts. On the other hand, there is no effect of AG implantation on gonadal transformation in stage 9 females with fully mature ovary. However, the ovary possessed vasa deferentia bilaterally connected to each gonad.

Further work on the Australian crayfish *Cherax quadricarinatus* showed that AG-implanted females demonstrated masculine behavior, as manifested by an increase in aggressive behavior in the presence of normal males, and the display of male courtship and false copulations in encounters with females (Karplus et al., 2003; Barki et al., 2006). In addition, the female secondary sexual characters were also reduced in size. More interestingly, AG implantation in the female also inhibited vitellogenin synthesis both at the transcriptional level and at the protein level, resulting in a sharp decrease in oocyte diameter and gonadosomatic index (Khalaila et al., 2001; Manor et al., 2004). All these experimental results indicate the antifeminizing effects of AG.

In the freshwater prawn *M. rosenbergii*, masculinizing effects of AG, when implanted in the female, were even more prominent. AG implantation in the *M. rosenbergii* females led to the appearance of appendix masculina and the masculinization of the chelipedes. Ovotestis was observed in females implanted with AG at relatively early life stages (Nagamine et al., 1980). On the contrary, if the AG removal was made in the adult males, redifferentiation of the amputated male appendix masculina and the chelipeds did not occur at the next molt cycle. Instead, they regenerated as immature forms, suggesting that the AG is not only responsible for the differentiation of both primary and secondary male sexual characters but also is required for the maintenance of secondary sexual characters as well as the testicular activities. Reimplantation of the

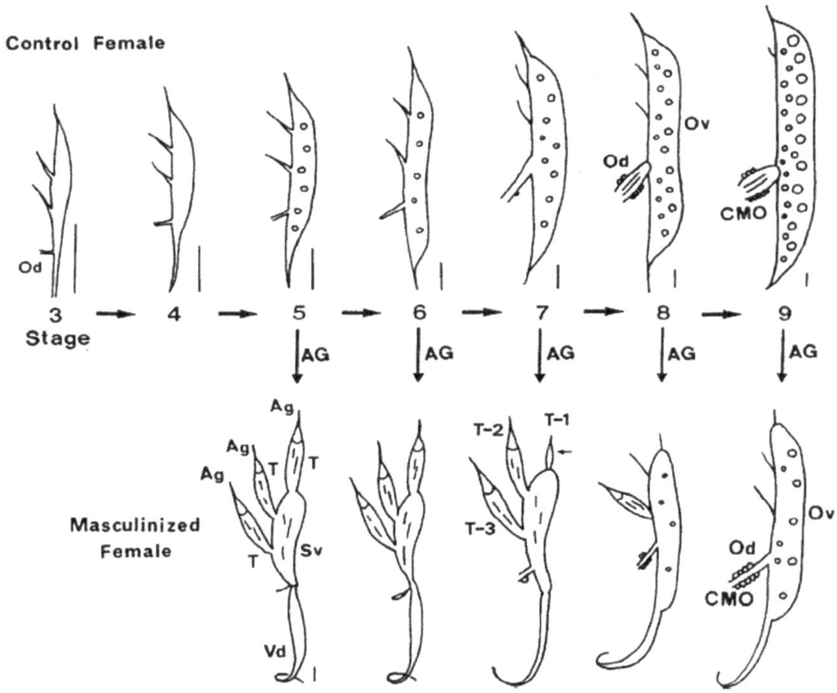

FIGURE 2.4 Gonadal masculinization induced by AG implantation. Schematic representation of gonadal masculinization two months after implantation of AG into stage 5–9 females. Control female shows normal female. *Ov*, ovary; *Od*, oviduct; *CMO*, cellular mass on oviduct; *T*, testis; *Ag (AG)*, androgenic gland; *Sv*, seminal vesicle; *Vd*, vas deferens. The numbers of T (T-1 etc.) show the order of differentiating time in normal male. Scale bars represent 200 μm. *Reproduced from Suzuki, S., Yamasaki, K., 1997. Sexual Bipotentiality of developing ovaries in the terrestrial isopod* Armadillidium vulgare *(Malacostraca, Crustacea). Gen. Comp. Endocrinol. 107, 136–146, Academic Press, Fig. 6, p. 143.*

ablated AG may significantly reverse the effect of androectomy, confirming the role of AG in male differentiation (Ventura et al., 2011). In *M. rosenbergii*, other studies showed that if the AG implantation was made before complete ovarian differentiation, a full and functional sex reversal to produce neomales could be produced (Malecha et al., 1992). These neomales, when crossed with normal females produced progenies, whose sex ratio provided the first evidence concerning genetic and endocrinological control of sex determination in the Malacostraca. Female-determining genes are apparently expressed in either sex unless repressed by male-determining factors.

M. rosenbergii, being a gonochoristic species, the sexual plasticity is restricted to a narrow window early in the life cycle. AG ablation at different life stages revealed its effect on sexual and morphotypic differentiation of the males. In *M. rosenbergii*, the mature males pass through a succession of three morphotypes: the small transforming into orange-clawed intermediate males, which in turn transformed into the dominant blue-clawed males. Ablation of AG in the small males did not elicit any transformation to next orange-clawed males suggesting that the AG regulates morphotypic differentiation. Conversely, AG ablation in the orange-clawed males did not affect their transformation into blue-clawed males. However, the reproductive systems of both the morphotypes atrophied after AG ablation (Sagi and Cohen, 1990). Interestingly, Aflalo et al. (2006) found that AG-ablated juveniles underwent complete sex reversal into neofemales that mated later with normal males to produce viable all-male progenies.

In the female crab *Eriocheir japonicus*, implantation of AG, however, produced partial masculinization, by developing male-like appendages. The crayfish, *C. quadricarinatus* intersex individuals are genetic females, but lack the ability to undergo sexual shift. In these intersexuals, the testis is active, but the ovary is permanently nonfunctional. On AG ablation, these intersexuals underwent a dramatic behavioral, morphological, and physiological sex shift to femaleness. The male reproductive organs regressed and there was vitellogenesis in the ovary.

The functional role of AG in male differentiation is further supported by studies on sequential hermaphroditic forms. In these forms, during the male phase the AG is highly active and starts regressing during the intersexual phase only to totally disappear during the female phase. Furthermore, using the sequential hermaphrodites, the sex can be changed in the inappropriate time either by removing the AG or by transplanting it into the female phase. Admittedly, AG takes control of male differentiation in both gonochoristic and hermaphroditic crustaceans. Further, the sex changing tendency of crustaceans is underlined by the easy way in which AG changes function during the ontogeny of a sequential hermaphrodite.

2.9 ANDROGENIC GLAND HORMONE

2.9.1 Chemical Nature

Experimental studies enumerated above established the fact that AG is crucial for the development and maintenance of male primary and secondary sex characters. Ironically, the chemical nature of the AG hormone (AGH) was elucidated much later than its functional role in sex differentiation was determined. From the ultrastructural studies of AG in the shore crab, *Pachygrapsus crassipes*, King (1964) indicated that the AGH might be a protein. However, biochemical studies of Férézou et al. (1978) suggested that the AGH in *Carcinus maenas* was lipoidal in nature. This substance was later purified and found to be farnesylacetone. On the contrary, the protein nature of AG was confirmed by chemical characterization and amino acid sequencing in the terrestrial isopod *A. vulgare* (Katakura et al., 1975; Okuno et al., 1997). In these studies, a loss of hormonal activity following proteolysis and amino acid reduction in the

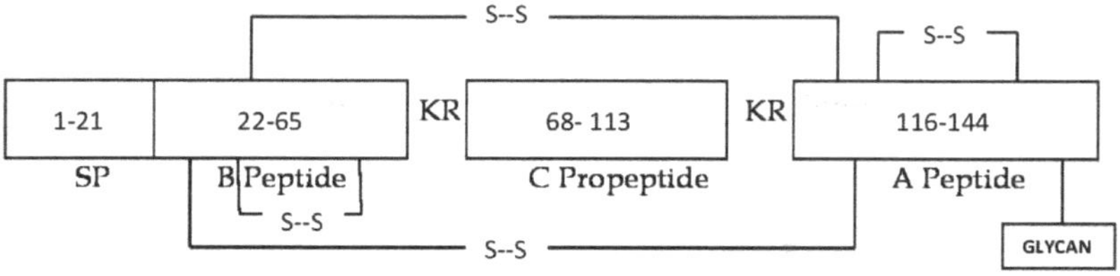

FIGURE 2.5 Schematic structure of the AGH of *A. vulgare* (Okuno et al., 1999; Martin et al., 1999). The Uniprot blast of *A. vulgare* revealed amino acid sequences of subunits A, B, and C, where C chain is propeptide flanked by KR cleavage sites. C-propeptide is cleaved while the protein gets matured. SP indicates signal peptide. KR = lys – arg. The numbers in the peptide chain represent amino acid residues of each domain. *Lines* indicate intra- and interchain disulfide bridges. In the A peptide, a glycan moiety attached to one amino acid as denoted.

hormone confirmed the protein nature of AGH. Hasegawa et al. (1987) analyzed the amino acid composition of two protein fractions that showed AG activity and determined its molecular weight to be 17 kDa. The complete sequence determination of *A. vulgare* AGH has been made by Martin et al. (1999) who also determined the glycosylation sites in this glycopeptide.

Okuno et al. (1999) cloned the cDNA of *A. vulgare*, encoding AGH, which comprised a linear preprohormone, organized according to the insulin-like superfamily of peptides. It starts with signal peptide which is cleaved off to give rise to the prohormone that starts with a B chain at the N′-terminus, then a connecting peptide (C-peptide), and finally the A chain (Fig. 2.5). The mature hormone has three disulfide bonds, two interchain bonds, and one intrachain bond (within the A chain) as in insulin-like peptides (Okuno et al., 1999; Martin et al., 1999). This peptide shows similarity with insulin/insulin-like growth factor/relaxin family peptides in its amino acid sequence, cys backbone positions, and the putative cleavage patterns (Ventura et al., 2011, Fig. 2.6).

Further characterization reveals that the A chain contains three cysteine residues which are present in the same position of the AGH characterized from other crustaceans such as *A. vulgare*, *Porcellio scaber*, *P. dilatatus*, *C. quadricarinatus*, *Cherax destructor*, *M. rosenbergii*, *Penaeus monodon*, *and Portunus pelagicus* (Table 2.2). Insulin-like AGH sequence comparisons have been made against a standard AGH sequence model of *A. vulgare* with accession number BAA86893.1 to obtain the percentage of similarity.

2.9.2 Isolation of Insulin-like Androgenic Gland Hormone

Ohira et al. (2003) first showed the similarity of two AG-specific protein sequences of isopod species, *P. scaber* and *P. dilatatus* with insulin-like peptides. The isopod sequences exhibited similarity in all features, including the overall linear sequence, a fourth intrachain disulfide bridge (in the B chain), and a predicted glycosylation site in the A chain. Among decapods, AG-specific insulin-like peptides were first isolated in the crayfish, *C.*

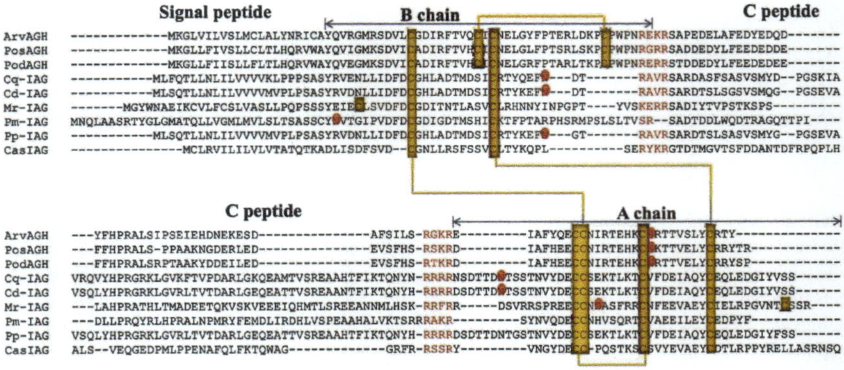

FIGURE 2.6 Multiple sequence alignment of IAGs. IAGs were aligned by using the CLUSTAL W algorithm, modified to fit predicted cleavage sites (RxxR, KxxR, or xR; shown in red). The most conserved feature is the backbone consisting of six cysteine residues (highlighted in orange), which gives rise to disulfide bridges *(orange lines)*. Also highlighted in orange are another conserved disulfide bridge that is found specifically in the isopod species and two other cysteine residues in Mr-IAG that might form a third interchain disulfide bridge. N-glycosylation sites (predicted in all but ArvAGH) with the sequence of NxS/T are highlighted with *red circles. Reproduced from Ventura, T., Rosen, O., Sagi, A., 2011. From the discovery of the crustacean androgenic gland to the insulin-like hormone in six decades. Gen. Comp. Endocrinol. 173 (3), 381–388, Elsevier, Fig. 2(A), p. 384.*

quadricarinatus. Manor et al. (2007) established a subtractive cDNA library from the AG of *C. quadricarinatus*. Screening of this library revealed an AG-specific gene, expressed exclusively in males. This gene was designated as *Cq-IAG* (insulin-like AG factor). In situ hybridization of *Cq-IAG* revealed the exclusive localization of its expression in AG. Cloning and complete sequencing of the gene showed that its cDNA was found to contain 1445 nucleotides encoding a deduced translation product of 176 amino acids. *Cq-IAG* was found to be expressed as early as eight days after male juveniles were released from the brood. The *Cq-IAG* of the crayfish had all characteristics of an insulin-like peptide of isopods, with the exception of the additional intra B-chain disulfide bridge. In addition, there were differences in the glycosylation sites of B and A chains, and the sequence similarity was also low. *Cq-IAG* extends sperm viability in the crayfish. Following this, Ventura et al. (2009) isolated insulin-like peptides (*Mr-IAG*) from *M. rosenbergii* by using an SSH cDNA library of the AG. IAG has also been isolated from several other decapod species and their sequence identity is given in Table 2.2. Ventura et al. (2011) produced a recombinant *Mr-IAG* and raised an antibody, which specifically recognized a 20 kDa protein in the AG extract. The *Mr-IAG* transcript was expressed in the juveniles as early as 20 days after metamorphosis, before the appearance of external sex characteristics. To be specific, the *Mr-IAG* expression starts with PL20 when the sex differentiation of both male and female

TABLE 2.2 IAG Accession Numbers of Crustaceans, Along With the % Identity to *A. vulgare*

Species	Accession No	Systematics	Identity (%)
Armadillidium vulgare (Arv)	Q9U8R2.1	Isopoda; Oniscidea; Armadillidiidae; *Armadillidium*; *vulgare*	100
Porcellio scaber (Pos)	AAO11675.1	Isopoda; Oniscidea; Porcellionidae; *Porcellio*; *scaber*	65
Porcellio dilatatus (Pod)	Q86SA8.1	Isopoda; Oniscidea; Porcellionidae; *Porcellio*; *dilatatus*	64
Cherax quadricarinatus (Cq)	ABH07705.1	Decapoda; Parastacoidea; Parastacidae; *Cherax*; *destructor*	36
Portunus pelagicus (Pp)	ADK46885.1	Decapoda; Brachyura; Portunidae; *Portunus*; *pelagicus*	36
Fenneropenaeus chinensis	AFU60546.1	Decapoda; Dendrobranchiata; Penaeoidea; Penaeidae; *Fenneropenaeus*; *chinensis*	31
Litopenaeus vannamei (Lv)	AIR09497.1	Decapoda; Dendrobranchiata; Penaeoidea; Penaeidae; *Litopenaeus*; *vannamei*	30
Penaeus monodon (Pm)	ADA67878.1	Decapoda; Dendrobranchiata; Penaeidae; *Penaeus*; *monodon*	29
Macrobrachium rosenbergii (Mr)	ACM181171	Decapoda; Caridea; Palaemonidae; *Macrobrachium*; *rosenbergii*	24
Macrobrachium nipponense	AJQ31851	Decapoda; Pleocyemata; Caridea; Palaemonoidea; Palaemonidae; *Macrobrachium*; *nipponense*	23
Cherax destructor (Cd)	ACD91988.1	Decapoda; Parastacidae; *Cherax*; *destructor*	22

starts. The expression level continues to rise through the juvenile stage, during which time, the external sexual differentiation is completed. In the adult males, *Mr-IAG* expression continues to be high in the adult males, to regulate the male morphotype differentiation. Taken together, these data provide convincing evidence that the AGH is the male sex hormone, influencing early sex differentiation and male phenotypic differentiation in *M. rosenbergii*.

2.9.3 Silencing of Mr-IAG

The cysteine backbone of the predicted mature *Mr-IAG* peptide (B and A chains) showed high similarity to that of other crustacean AG-specific insulin-like peptides. In vivo silencing of the gene, by injecting the prawns with *Mr-IAG* double-stranded RNA, temporarily prevented the regeneration of male secondary sexual characteristics. In terms of reproductive parameters, it resulted in the arresting of testicular spermatogenesis and spermatophore development in the terminal ampullae of the sperm duct, accompanied by hypertrophy and hyperplasia of the AGs when eyestalk ablated (Khalaila et al., 2001; Kim et al., 2002). Moreover, *Cq-IAG* silencing in intersex *C. quadricarinatus* resulted in the reduction of sperm production and testicular degeneration simultaneously with the onset of vitellogenesis (Rosen et al., 2010) suggesting that the AG-specific insulin-like factor not only regulates male sex differentiation but also plays the role of a gender switcher, balancing male/female components of the intersex crayfish.

While the presence of IAG in different decapods has been reported, and its expression in AG was blocked by RNAi, the pathway by which it exerts action on the control of sex differentiation has not been clearly understood. The detection of an insulin-like receptor gene (*Mr-IR*) in *M. rosenbergii* was possible, after the construction of a transcriptomic library (Sharabi et al., 2015). Silencing of *Mr-IR* expression in the young male prawns resulted in the hypertrophy of the AG and increased production of *Mr-IAG* in the AG. Associated with these changes, an unusual accumulation of immature sperm occurred in the distal vas deferens. However, silencing of *Mr-IR* gene did not bring about any sexual shift, as occurred when the *Mr-IAG* gene was silenced. It is suggestive that other IAG receptor genes may also be involved in the transduction pathway of IAG in controlling sexual differentiation and maintenance in crustaceans.

2.10 MOLECULAR BASIS OF SEX DIFFERENTIATION

It is clear from the foregoing discussion that AGH controls the sex differentiation process in the malacostracan crustaceans. However, molecular studies using genetic markers to trace the germline differentiation of the gonad in crustaceans are limited. Genetic information is transmitted from generation to generation by highly specified germline cells that are segregated as primordial germ cells, from somatic lineages during embryogenesis. Traditional cytological observations have failed to identify immature germ cells as they are morphologically indistinguishable from somatic cells.

In vertebrates, several genes have been identified during the course of gonadal differentiation. Recent gene expression studies relating to sex differentiation in crustaceans include mainly *vasa* gene. Vasa gene is a member of the DEAD-box gene family and has a restricted expression exclusively to the germ cell lineage in most animals (Mich et al., 2009). Zygotic expression of *vasa* mRNA is specific to germ cell lineage throughout the development of the embryo and the adult gonad, without any expression in the somatic tissue (Li et al., 2010).

In Crustacea, the *vasa* gene expression was investigated in *Daphnia magna* (Cladocera), *Parhyale hawaiensis* (Amphipoda), and *L. vannamei*, *Fenneropenaeus chinensis*, *M. japonicus* (Decapoda) (Sagawa et al.,, 2005; Extavour, 2005; Özhan-Kizil et al., 2009; Aflalo et al., 2007; Sellars et al., 2007). In *M. japonicus*, *Mjpl10 vasa* gene expression was observable from the 2-cell embryonic stage onward (30min postspawning), suggesting that it is maternally expressed. *Mjpl10* expressed continually at low levels throughout the embryogenesis. *Mjpl10* expression increases significantly in the first 25h after hatching (nauplii IV) and decreases gradually over the next 52-day period. The transcript levels of *Mjpl10* are also detected in the ovary and testes of mature adults. These results suggested that *vasa* gene control germ cell migration toward somatic gonad tissue sometime between larval stages nauplii I and IV (Sellars et al., 2007). This is in agreement with the previous histological studies on *M. japonicus* and *L. vannamei* that sex differentiation occurred in these shrimps early during postlarval development. (Nakamura et al., 1992; Campos-Romos et al., 2006; Garza-Torres et al., 2009; Zhao et al., 2009). In the Pacific white shrimp, *L. vannamei*, *Lv-vasa* gene is specifically expressed in male and female gonads. In the adult females, *Lv-vasa* gene is expressed during late oogenesis, suggesting a possible maternal contribution to the embryo (Aflalo et al., 2007). In the Chinese shrimp, *F. chinensis*, *vasa* gene expression occurs during the migration of primordial germ cells to the genital ridge (Feng et al., 2011). *Fc-vasa*-like mRNA was first localized in the cytoplasm of oocytes before and after spawning, and in the embryo, it was uniformly distributed in the early embryos from the 2-cell embryo to the blastula, suggesting that *vasa* mRNA was of maternal origin. Then, the expression of the *Fc-vasa* gene centralizes gradually and is restricted to the 5-cell clusters as precursor cells of primordial germ cells at the limb bud stage. These 5-cell clusters are located at the base of the antennules and antenna, as well as the cephalic lobe. Later on, these cells migrate to a dorsolateral position in the naupliar and zoel stages and gradually enter the genital ridge at the mysis 1 stage. The histological appearance of gonads in the PL1 is common to *F. chinensis*, *L. vannamei*, and *Penaeus japonicus*. Furthermore, *M. rosenbergii vasa* gene (*Mrvlg*) transcripts were detected in the cytoplasm of oogonia, previtellogenic, and vitellogenic oocytes and were also detected in the nucleoplasm of mature oocytes. In the testis, the *Mrvlg* transcript was detected in the cytoplasm of spermatogonia and primary spermatocytes but was detected in the nuclei of secondary spermatocytes and sperm (Nakkrasae and Damrongphol, 2007). In *D. magna*, vasa protein was first visualized in one of the blastomeres at the 8-cell stage; as cleavage progresses, the number of vasa-positive blastomeres increased. These blastomeres divided further and moved toward the site of future gonad (Sagawa et al., 2005). Similar expression of *vasa* gene was reported in *P. hawaiensis* (Özhan-Kizil et al., 2009) and its expression is schematically plotted in Fig. 2.7. Vasa-like gene expression in all the above penaeid shrimps, the cladoceran, *D. magna* and the amphipod, *P. hawaiensis*, as well as several insect species point to the possible phylogenetic linkage between insect and crustacean (Table 2.3)

Among the brachyuran crabs, *vasa* gene was characterized from the Chinese mitten crab *Eriocheir sinensis*, in which, the mRNA expression was specific to the gonad and its temporal expression in the ovary and testes were significantly different in various developmental periods. Especially, *Es-vasa* mRNA transcripts were at the highest levels during periods of rapid development in the gonads (stage III-2 in ovaries and spermatocyte stage in testes) and gradually decreased as the gonads matured (Wang et al., 2012). Similarly, in the green mud crab *Scylla paramamosain*, the *Sp-vasa* expression was higher in the growth stage of ovary than in the maturation stage, and in stages I and II of testis, the expression level of *SP-vasa* was higher than in stage III (Wang

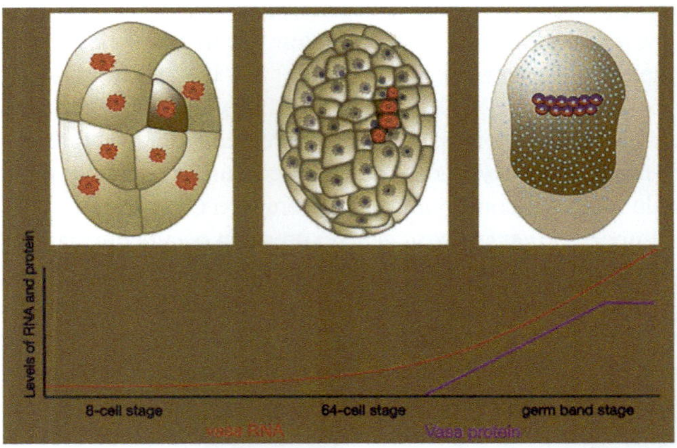

FIGURE 2.7 Model of RNA localization and protein expression in the germ cells of *P. hawaiensis*. Schematic drawings of *Parhyale* embryos at the 8-cell stage, the 64-cell stage, and the germ band stage. At the 8-cell stage, maternal *Ph-vasa* RNA are ubiquitous. At the 64-cell stage, maternal RNA becomes localized to the germ cells by its differential stabilization in that lineage. At this time, zygotic transcription starts to contribute to RNA levels. At the germ band stage, RNA of *Ph-vasa* is detected in the germ cell cluster together with zygotic Vasa protein. *Reproduced from Özhan-Kizil, G., Havemann, J., Gerberding, M., 2009. Germ cells in the crustacean* Parhyale hawaiensis *depend on Vasa protein for their maintenance but not for their formation.Dev. Biol. 327, 230–239, Elsevier, Fig. 7, 237.*

TABLE 2.3 Full-Length Vasa Ortholog cDNAs

Group	Common Name	Gene Name	Polypeptide Length	Gene Bank Accession
Parhyale hawaiensis	Amphipod crustacean	*Vasa*	676	ABX76969
Daphnia magna	Water flea	DmaVas	779	BAE00180
Litopenaeus vannamei	Pacific white shrimp	Vasa-like protein	703	AAY89069
Fenneropenaeus chinensis	Chinese white shrimp	Vasa	712	ABQ00071
Macrobrachium rosenbergii	Giant freshwater prawn	*Mrvlg*	710	ABC87271
Moina macrocop	Water flea	MmVas	843	BAD99524
Artemia franciscana	Brine shrimp	*AfVas*	726	BAD99523
Palaemon carinicauda	Ridgetail prawn	Vasa	600	AGF90963
Palaemon modestus	Siberian prawn	Vasa	602	AID54678.1

et al., 2012). To sum up, vasa gene expression location during the progression of embryogenesis in several crustaceans has shown the utility of this gene as a marker gene for germ cell specification and to the accurate tracking of PGCs during gonadogenesis.

2.10.1 *dmrt* Gene Expression

Several genes regulate sexual development in different animal taxa and one such is *dmrt1*, which is a distant homologue of mammalian testis specifying *sox9* and *sry* genes. The *dmrt* (doublesex- and mab-3-related transcription factor) family of genes is originally implicated in the sex differentiation of *D. melanogaster* and *Caenorhabditis elegans*, but known to be present in a variety of animal phyla. Farazmand et al. (2010) have recently described the expression pattern of *dmrt*-related genes in the differentiating gonads of two species of sexually reproducing and parthenogenetic *Artemia*, *A. wimiana*, and *A. parthenogenetica*. RT-PCR results combined with in situ hybridization studies revealed that *dmrt* genes are expressed only in female gonads in both the species with different reproductive modes. *dmrt* genes encode proteins with conserved DNA-binding domains that direct similar aspects of sex differentiation (Volff et al., 2003). In other animals, these genes also have sex-specific functions such as differentiation of copulatory structures, yolk protein expression, and control of male sexual behaviors. The expression of *dmrt* gene in the ovary of *Artemia* is intriguing, as this gene, except in *Drosophila*, is specifically expressed in testis in all other animals, including mammals, suggesting a male-differentiating role for it. However, earlier work of Kato et al. (2008) on another primitive cladoceran species, *D. magna* has shown that DM-domain genes belonging to *dmrt1* gene family are expressed only in the testis. *Daphnia* reproduces asexually when in an optimal environment for food, photoperiod, and population density. Three DM-domain genes that corresponded to *dmrt11E*, *dmrt93B*, and *dmrt99B* were identified in *D. magna*, but only one of them (*dmrt93B*) was found to be expressed in the testis, suggesting a male-specific role in sex determination. *dmrt1*, a DM-related transcriptional factor 1 and a DM-domain gene, is considered to be an important sex-determining gene in both vertebrates and invertebrates. In a protandrous marine fish, *Acanthopagrus schlegelii*, *dmrt1* is specifically transcribed in bisexual testis and not in ovary (Shin et al., 2009). However, *dmrt1* expression wanes in the bisexual and functional testis in the 3-year-old fish, when it sexually changed to females, suggesting a specific role in testis formation in the fish. Understandably, this gene might be targeted during the functional transformation of sex from male to female in this fish. In a similar way, multiple alternative splicing and differential expression of *dmrt1* occur during gonad differentiation from ovary via ovotestis to testis in the rice field eel, which undergoes natural protogynous sex reversal. 3′ UTRs of *dmrt1* isoforms have a role in the regulation of *dmrt1* expression. DM genes share a conserved cystine-rich, DNA-binding domain (DM) and each plays similar roles in male-specific sexual development. The expressions of DM genes from invertebrates to vertebrates are sex-specific and are associated with male-specific development. In crustaceans, male sex differentiation is under the direct influence of androgenic hormone, and in the absence of androgenic hormone, the ovary develops by default. The available literature, although fragmentary, would imply that *dmrt1* expression in the undifferentiated gonad as well as the testis of a few crustaceans may have a controlling role over the AG formation and function. However, this hypothesis needs further experimental proof by analysis of *dmrt1* expression during the differentiation of both gonochoric and hermaphroditic gonads in different species.

2.10.2 Doublesex (*Dsx*) Gene

The Doublesex (*Dsx*) genes play an important role in controlling sexual dimorphism in genetic sex-determining organisms such as nematodes, insects, and vertebrates. Doublesex gene was originally identified in *D. melanogaster* as a critical and terminal transcription factor in the fly sex-determining cascade, and it spatially and temporally transcribed into two sex-specific splice forms conferring sexually dimorphic traits during development (Pomiankowski et al., 2004; Burtis and Baker, 1989; Robinett et al., 2010). Later, *dsx* gene homologs were identified from primitive crustaceans such as *D. magna*, *Daphnia pulex*, *Daphnia galeata*, *Ceriodaphnia dubia*, and *Moina macrocopa* (Toyota et al., 2013).

Kato et al. (2011) reported the identification of two *Dsx* genes from *D. magna*. One of these genes, designated *DapmaDsx1*, is responsible for the male trait development when expressed during environmental sex determination (ESD). The domain organization of *DapmaDsx1* was similar to that of *Dsx* from insects, which are thought to be the sister group of branchiopod crustaceans. Intriguingly, the molecular basis for sexually dimorphic expression of *DapmaDsx1* is different from that of insects (Fig. 2.8). Rather than being regulated sex-specifically at the level of pre-mRNA splicing in the coding region, *DapmaDsx1* exhibits sexually dimorphic differences in the abundance of its transcripts. During embryogenesis, expression of *DapmaDsx1* increased only in males and its transcripts were primarily detected in male-specific structures. Knock-down of *DapmaDsx1* in male embryos resulted in the production of female traits including ovarian maturation, whereas ectopic expression of *DapmaDsx1* in female embryos resulted in the development of male-like phenotypes. Expression patterns of another *D. magna Dsx* gene, *DapmaDsx2*, were similar to those of *DapmaDsx1*, but silencing and over-expression of this gene did not induce any clear phenotypic changes. Based on these observations, Kato et al. (2010) concluded that *DapmaDsx1* is a key regulator of the male phenotype in *D. magna*. *Dsx* gene has not been reported in any malacostracan crustacean. Interestingly, AG has not been reported in lower crustaceans such as *D. magna*. It is tempting to suggest that, by virtue of the male-determining function of *DapmaDsx1*, this gene could act in the place of androgenic hormone in these cladocerans. Furthermore, the existence of these male-determining genes in the cladocerans, which is placed phylogenetically in between insects and higher crustaceans, could offer important insights on the commonality present in the genes responsible for gender determination in both these sister groups of arthropods (see Fig. 2.8).

2.10.3 Tra Homologue Genes

The transformer-2 gene (*tra-2*) plays a key role in the sexual differentiation regulatory hierarchy. Recently, tra-2 gene homologs designated as *Mntra-2* was cloned and characterized from *Macrobrachium nipponense* (Zhang et al., 2013). The expression levels of *Mntra-2* varied significantly during different developmental stages of embryo, larvae, and postlarvae and in various adult tissues. During embryogenesis, the expression level of *Mntra-2* was slightly higher at the cleavage stage than at the blastula stage and reached the highest level at the nauplius stage. In the larvae, the *Mntra-2* expression gradually increased from 1-day larvae posthatch (L1) to L10 and decreased to the lowest level at the end

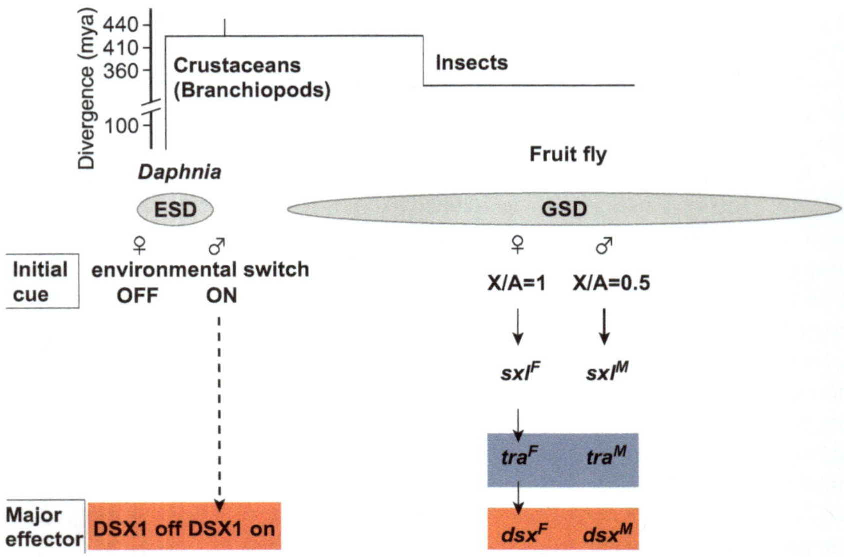

FIGURE 2.8 An environmental sex determination (ESD) pathway in *Daphnia* is compared with genetic sex determination (GSD) pathways in insect model species fruit fly. *mya*, million years ago. *Modified from Kato, Y., Kobayashi, K., Watanabe, H., & Iguchi, T., 2011. Environmental sex determination in the branchiopod crustacean* Daphnia magna*: deep conservation of a Doublesex gene in the sex-determining pathway. PLoS Genet 7, e1001345, Fig. 6, p. 8.*

of metamorphosis. During the postlarvae, the *Mntra-2* expression was higher at 5 days after the metamorphosis (P5). RT-qPCR showed the *Mntra-2* mRNA was expressed in ovary, testis, muscle, heart, abdominal ganglion, brain, and intestine with the highest level of expression in muscle and intestine (Zhang et al., 2013). In Chinese shrimp, *F. chinensis*, *tra-2* homologue (*FcTra-2*) are isolated as three alternatively spliced mRNA transcripts, designated as *FcTra-2a*, *FcTra-2b*, and *FcTra-2c*. In the early developmental stage of the shrimp, the expression levels of *FcTra-2c* increased rapidly, reaching a maximum level at the mysis stage. In the juvenile stage, *FcTra-2c* displayed a significantly higher expression level in female Chinese shrimp than in males. These data indicated that *FcTra-2* might be involved in female sex determination in Chinese shrimp (Li et al., 2012). Unlike Chinese shrimp, only one homologue Tra gene (*Pmtra 2*) was identified in *P. monodon* (Leelatanawit et al., 2009). Hence, *tra-2* homolog probably plays important roles in embryonic development and sex differentiation in penaeid shrimps.

In conclusion, it may be said that crustaceans conform to the arthropodan pattern of genetic sex determination. However, sex differentiation is controlled by a sex hormone secreted by AG with epigenetic influence on sex-determining factors. AG is present only in malacostracan crustaceans, influencing male sex differentiation and inhibiting female differentiation. In addition, AGH is structurally similar to insulin growth factors, having definitive control over spermatogenesis and the appearance and maintenance of secondary sex characters. Notwithstanding, several gene markers have been identified and used in germ cell specification and tracing the migration of primordial germ cell migration to genital ridge through *vasa* gene expression. Other genes involved in sex differentiation such as *dmrt1* have also been identified in crustaceans to uncover similarities in gene expression with insects and vertebrates. In particular, the occurrence of these genes in lower crustaceans has more relevance to their sex differentiation role in the obvious absence of AG. Furthermore, identification of *DSX* (doublesex) gene in *D. magna* has revealed its functional importance in the ESD of this parthenogenetically reproducing cladoceran branchipod. *Dapma Dsx1* is the key regulator of male phenotype both in *D. magna* and insects. Similarity in the domain organization of *Dapma Dsx1* of *D. magna* with *Dsx* of insects underscores the close phylogenetic kinship between insects and the basal and primitive crustacean species, like the branchiopods.

Chapter 3

Sexual Systems

3.1 INTRODUCTION

Sexual systems originate from the acquisition of sexuality by way of sex determination and sex differentiation. The evolution of sexual systems has relevance to mating success and fertilization accomplishment within a given population. Different sexual systems, as occurring in Crustacea, have a direct relationship with the reproductive adaptation of different species inhabiting diversified ecological niches, particularly of aquatic environments. As a result, they exhibit complex social structures in every conceivable aquatic habitat, in much the same way as insects have established them in terrestrial conditions. The occurrence of diversified sexual systems in crustacean taxa has a bearing to the mode of sex determination, which ranges from heterogametic to polyfactorial as well as environmental regulatory mechanisms (see chapter: Sex Determination). Such a range of sex determination patterns implies that the major regulatory genes of sex development are susceptible to epigenetic influences (Bull, 1983). Obviously, sex determination is seldom dictated by an individual's genetic constitution per se (Avise, 2011). As in several species of crustaceans, the individuals' gender is a complex genetic and epigenetic outcome of physiological and developmental processes that activate or repress multiple male-specific or female-specific genes during ontogeny (Charniaux-Cotton and Payen, 1985).

The majority of crustaceans are gonochoric with individuals in a population expressing only one sex during the lifetime. However, in view of the vast number of species having environmental or epigenetic sex determination patterns, a multiple of sexual systems occur in different taxa. In particular, primitive groups such as notostracans and conchostracans show an extensive variation in their reproductive strategy that may range from gonochorism to self-fertilizing hermaphroditism and parthenogenesis and to an extreme sexual type, androdioecy (Ombretta et al., 2005). Furthermore, the varied types of hermaphroditic condition warrant an inference that the origin of dioecy (separate sexes) may be through the pathway of hermaphroditism. Such a suggestion has been advanced for the evolution of dioecy in plants also (Charlesworth and Charlesworth, 1978). Fig. 3.1A and B illustrates the two possible pathways from which both the sexual conditions would have originated from each other.

Sexual Biology and Reproduction in Crustaceans. http://dx.doi.org/10.1016/B978-0-12-809337-5.00003-4

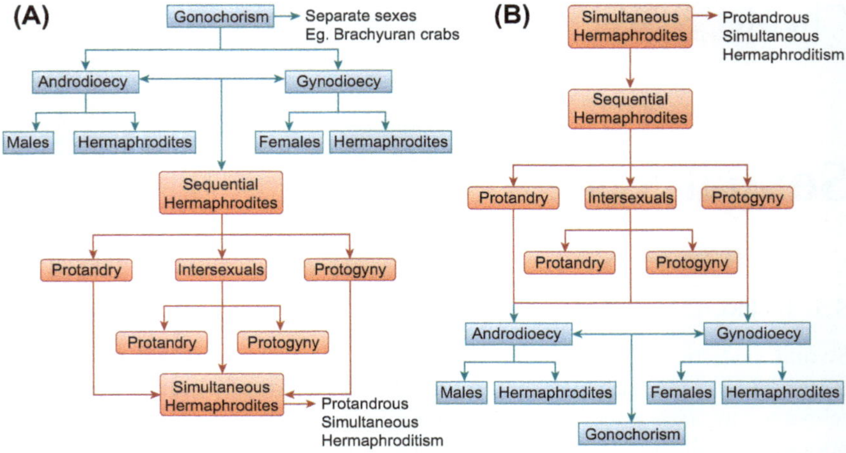

FIGURE 3.1 (A) and (B) Models showing evolutionary transition between gonochorism and hermaphroditism in Crustaceans. Gonochorism and hermaphroditism are on opposite evolutionary poles of a reproductive spectrum that includes various mixed sex modalities.

3.2 GONOCHORISM

Gonochorism describes sexually reproducing species in which individuals have one of at least two distinct sexes (see Subramoniam, 2013). This condition is also referred to as dioecy. In gonochorism, individual sex is genetically determined and does not change throughout the lifetime. Genetic sex determination systems are those in which the development of one sex or the other is triggered by the presence or absence of one or more critical genetic factors. Crustaceans are well known for their different modes of genetic sex determination. Thus, different kinds of male heterogamety are known in Crustacea, with Xo present in branchipods, isopods, and copepods, XY males in copepods such as *Tortanus gracilis*, several decapods, and in the isopod *Anisogammarus anandalei*. X_1,X_2O is known in ostracods and X_1X_2Y is found in the decapod *Cervimunida princeps* (Legrand et al., 1987). On the other hand, female heterogamety is represented by WZ chromosomal pattern in the anostracan *Artemia salina* (WZ) and the isopodan superspecies *Jaera albifrons* (W_1W_2Z).

3.3 HERMAPHRODITISM

The vast majority of malacostracan crustaceans are gonochoric, having separate sexes, although many groups exhibit hermaphroditism, a condition in which both the male and the female sexual characteristics are found in a functional way throughout its reproductive life. A true hermaphrodite or euhermaphrodite (also called simultaneous or synchronous hermaphrodite) is thus an animal that produces ripe sperm and ova at the same time and in which distinct male and female secondary sexual characters coexist or do not exist (Clark, 1977).

Examples for simultaneous hermaphroditism in crustaceans are the sessile thoracican barnacles. Other examples are found in cephalocarids (Charniaux-Cotton and Payen, 1985), notostracans (Ghiselin, 1969), and conchostracans (Sassaman and Weeks, 1993). The burrowing thalassinid *Calocaris macandreae* and the primitive cave-dwelling remipedes *Speleonectes benjamini* and *Godzilliognomus frondosus* are also found to be simultaneous hermaphrodites (Yager, 1991). All the simultaneous hermaphrodites except the notostracans practice cross-fertilization.

3.3.1 Nonfunctional Hermaphroditism

Apart from the functional hermaphroditism, the occurrence of nonfunctional hermaphroditism in males of gonochoristic species of several malacostracans has been reported. For example, in the amphipod *Orchestia mediterranea*, Charniaux-Cotton (1959) described the development of an anterior ovarian region before the onset of spermatogenesis in the males. By a series of experimental studies using androgenic gland ablation, she proposed the hypothesis of spontaneous ovarian differentiation or ovarian autodifferentiation of the gonadal rudiment in the absence of androgenic gland. In *Pandalus gracilis*, oogenesis has been observed in very young gonads before initiation of spermatogenesis and appearance of male external characters (Aoto, 1952). It may be noted here that oogenesis can only proceed in the testes up to primary vitellogenesis. However, the reverse situation of spermatogenesis occurring in the female gonads has never been observed.

3.3.2 Sequential Hermaphroditism

Many crustacean species, however, exhibit a range of true hermaphroditic conditions. The most common form of hermaphroditism is the sequential hermaphroditism, a sexual system in which the sexes are functionally separate and morphologically distinct, but where an individual produces both male and female gametes at different periods of its life. There are two types of sequential or successive hermaphroditism, viz., protandry and protogyny. Typically, protandric and protogynous forms do not exhibit the characteristics of both sexes simultaneously. They are usually either male or female at a time with a very brief transitional state called "intersexual" during the changeover phase.

3.3.3 Protandric Hermaphroditism

Functional protandric hermaphroditism, referring to a sex change from male to female is described in various species of certain decapod infraorders, such as Caridea (Hippolytidae and Pandalidae), Thalassinidae (Axiidae), and Anomura (Hippidae), as well as in three suborders of Isopoda, namely, Flabellifera, Epicaridea, and Oniscidea (see Table 3.1). Wenner (1972) proposed a size-related

TABLE 3.1 Hermaphroditism in Crustacea

Class	Order	Family	Genus	Species	Type of Hermaphroditism	References
Malacostraca	Decapoda	Atyidae	*Paratya*	*curvirostris*	PH	Allsop and West (2003a)
Malacostraca	Decapoda	Atyidae	*Atyoida*	*bisulcata*	PH	Allsop and West (2003a)
Malacostraca	Decapoda	Atyidae	*Atyoida*	*serrata*	PH	Allsop and West (2003a)
Malacostraca	Decapoda	Atyidae	*Caridina*	*richiersi*	PH	Allsop and West (2003a)
Malacostraca	Decapoda	Atyidae	*Austratya*	*striolata*	PH	Allsop and West (2003b)
Malacostraca	Decapoda	Campylonotidae	*Campylonotus*	*rathbunae*	PH	Allsop and West (2004)
Malacostraca	Decapoda	Campylonotidae	*Campylonotus*	*semistriatus*	PH	Allsop and West (2004)
Malacostraca	Decapoda	Campylonotidae	*Campylonotus*	*capensis*	PH	Almeida and Buckup (2000)
Malacostraca	Decapoda	Campylonotidae	*Campylonotus*	*vagans*	PH	Almeida and Buckup (2000)
Malacostraca	Decapoda	Processidae	*Processa*	*edulis*	PH	Nöel (1973)
Malacostraca	Decapoda	Hippolytidae	*Chorismus*	*antarcticus*	PH	Bauer and Holt (1998)

Malacostraca	Decapoda	Hippolytidae	*Hippolyte*	*inermis*	PH	Berglund (1986)
Malacostraca	Decapoda	Pandalidae	*Pandalopsis*	*dispar*	PH	Butler (1964)
Malacostraca	Decapoda	Pandalidae	*Pandalus*	*platyceros*	PH	Boddeke et al. (1991) and Brook et al. (1994)
Malacostraca	Decapoda	Pandalidae	*Pandalus*	*latirostris*	PH	Boddeke et al. (1991) and Brook (1994)
Malacostraca	Decapoda	Pandalidae	*Pandalus*	*eous*	PH	Boddeke (1991) and Brook et al. (1994)
Malacostraca	Decapoda	Pandalidae	*Pandalus*	*tridens*	PH	Boddeke (1991) and Brook (1994)
Malacostraca	Decapoda	Pandalidae	*Pandalus*	*curvatus*	PH	Boddeke et al. (1991)
Malacostraca	Decapoda	Pandalidae	*Pandalus*	*formosanus*	PH	Boddeke et al. (1991)
Malacostraca	Decapoda	Pandalidae	*Pandalus*	*prensor*	PH	Boddeke et al. (1991) and Brook et al. (1994)
Malacostraca	Decapoda	Pandalidae	*Pandalus*	*gurneyi*	PH	Boddeke (1991) and Brook (1994)

Continued

TABLE 3.1 Hermaphroditism in Crustacea—cont'd

Class	Order	Family	Genus	Species	Type of Hermaphroditism	References
Malacostraca	Decapoda	Pandalidae	*Pandalus*	*teraoi*	PH	Boddeke et al. (1991) and Brook et al. (1994)
Malacostraca	Decapoda	Pandalidae	*Pandalus*	*chani*	PH	Boddeke (1991)
Malacostraca	Decapoda	Pandalidae	*Pandalus*	*gracilis*	PH	Boddeke et al. (1991) and Brook et al. (1994)
Malacostraca	Decapoda	Pandalidae	*Pandalopsis*	*dispar*	PH	Butler (1964)
Malacostraca	Decapoda	Pandalidae	*Pandalopsis*	*longirostris*	PH	Carpenter (1978)
Malacostraca	Decapoda	Pandalidae	*Pandalopsis*	*lamelliger*	PH	Charniaux-Cotton (1956)
Malacostraca	Decapoda	Pandalidae	*Pandalopsis*	*japonica*	PH	Carpenter (1978) and Charniaux-Cotton (1960)
Malacostraca	Decapoda	Pandalidae	*Pandalopsis*	*rubra*	PH	Carpenter (1978)
Malacostraca	Decapoda	Pandalidae	*Pandalopsis*	*gibba*	PH	Charnov (1979a)

Malacostraca	Decapoda	Pandalidae	*Pandalus*	*borealis*	PH	Boddeke et al. (1991) and Brook et al. (1994)
Malacostraca	Decapoda	Pandalidae	*Pandalus*	*danae*	PH	Boddeke et al. (1991) and Brook et al. (1994)
Malacostraca	Decapoda	Pandalidae	*Pandalus*	*goniurus*	PH	Boddeke (1991) and Brook (1994)
Malacostraca	Decapoda	Pandalidae	*Pandalus*	*hypsinotus*	PH	Boddeke et al. (1991) and Brook (1994)
Malacostraca	Decapoda	Pandalidae	*Pandalus*	*jordani*	PH	Boddeke (1991) and Brook et al. (1994)
Malacostraca	Decapoda	Pandalidae	*Pandalus*	*montagui*	PH	Boddeke (1991) and Brook (1994)
Malacostraca	Decapoda	Pandalidae	*Pandalus*	*stenolepis*	PH	Boddeke et al. (1991) and Brook et al. (1994)
Malacostraca	Decapoda	Pandalidae	*Pandalus*	*kessleri*	PH	Buchanan (1963)

Continued

TABLE 3.1 Hermaphroditism in Crustacea—cont'd

Class	Order	Family	Genus	Species	Type of Hermaphroditism	References
Malacostraca	Decapoda	Pandalidae	*Pandalus*	*gracilis*	PH	Boddeke et al. (1991) and Brook et al. (1994)
Malacostraca	Decapoda	Pandalidae	*Pandalus*	*nippoensis*	PH	Boddeke et al. (1991) and Brook et al. (1994)
Malacostraca	Decapoda	Parastacidae	*Parastacus*	*brasiliensis*	PSH	Charnov and Hannah (2002)
Malacostraca	Decapoda	Parastacidae	*Samastacus*	*spinifrons*	PSH	Charnov and Skúladóttir (2000)
Malacostraca	Decapoda	Parastacidae	*Virilastacus*	*rucapihurlrnsis*	PSH	Chen et al. (1998)
Malacostraca	Decapoda	Penaeidae	*Melicertus*	*kerathurus*	PH	Heegaard (1971)
Malacostraca	Decapoda	Crangonidae	*Argis*	*dentata*	PH	Almeida and Buckup (2000)
Malacostraca	Decapoda	Crangonidae	*Crangon*	*franciscorum*	PH	Charnov and Bull (1989)
Malacostraca	Decapoda	Crangonidae	*Crangon*	*vulgaris*	PH	Allsop and West (2003a)

Malacostraca	Decapoda	Crangonidae	*Notocrangon*	*antarcticus*	PH	Charnov and Bull (1989)
Malacostraca	Decapoda	Hippidae	*Emerita*	*analoga*	PH	Barnes and Wenner (1968)
Malacostraca	Decapoda	Solenocendae	*Solenocera*	*membranacea*	PH	Heergaard (1967)
Malacostraca	Isopoda	Cymothoidae	*Emetha*	*audouinii*	PH	Montalenti (1941)
Malacostraca	Isopoda	Cymothoidae	*Anilocra*	*physodes*	PH	Montalenti (1941)
Malacostraca	Isopoda	Cymothoidae	*Anilocra*	*frontalis*	PH	Legrand and Juchault (1970)
Malacostraca	Isopoda	Cymothoidae	*Philoscia*	*elongata*	PH	Arcangeli (1925)
Malacostraca	Isopoda	Cymothoidae	*Nerocila*	*californica*	PH	Brusca (1978)
Malacostraca	Isopoda	Hemioniscidae	*Hemioniscus*	*balani*	PH	Kozloff (1987)
Malacostraca	Isopoda	Liriopsidea	*Liriopsis*	*pygmaea*	PH	Kozloff (1987)
Malacostraca	Isopoda	Bopyridae	*Munidion*	*pleuroncodis*	PH	Markham (1975)
Malacostraca	Isopoda	Oniscidea	*Rhyscotus*	*ortonedae*	PH	Jackson (1928)
Malacostraca	Amphipoda	Lysianassidae	*Acontiostoma*	*marionis*	PH	Lowry and Stoddart (1986)

Continued

TABLE 3.1 Hermaphroditism in Crustacea—cont'd

Class	Order	Family	Genus	Species	Type of Hermaphroditism	References
Malacostraca	Amphipoda	Lysianassidae	*Acontiostoma*	*tuberculata*	PH	Lowry and Stoddart (1986)
Malacostraca	Amphipoda	Lysianassidae	*Stomacontion*	*pungapunga*	PH	Lowry and Stoddart (1986)
Malacostraca	Amphipoda	Lysianassidae	*Scolopostoma*	*prionoplax*	PH	Lowry and Stoddart (1986)
Malacostraca	Amphipoda	Lysianassidae	*Ocosingo*	*borlus*	PH	Lowry and Stoddart (1986)
Malacostraca	Amphipoda	Lysianassidae	*Ocosingo*	*borlus*	PH	Lowry and Stoddart (1986)
Malacostraca	Amphipoda	Stegocephalidae	*Stegocephalus*	*infiatus*	PH	Steele (1967)
Copepoda	Calanoida	Calanidae	–	–	PH	Fleminger (1985)
Cirripedia	Ascothoracica	–	*Synagoga*	*sandersi*	PH	Newman (1974)
Cirripedia	Ascothoracica	–	*Gorgonolaureus*	*muzikae*	PH	Grygier (1981)
Malacostraca	Isopoda	Anthuridae	*Cyanthura*	*carinata*	PG	Legrand and Juchault (1963)

Malacostraca	Isopoda	Anthuridae	*Cyanthura*	*polita*	PG	Burbanck and Burbanck (1974)
Malacostraca	Isopoda	Anthuridae	*Cyanthura*	*profunda*	PG	Kensley (1982)
Malacostraca	Isopoda	Sphaeromatidae	*Paraleptosphaeroma*	*glynni*	PG	Buss and Iverson (1981)
Malacostraca	Isopoda	Sphaeromatidae	*Gnorimosphaeroma*	*luteum*	PG	Brook et al. (1994)
Malacostraca	Isopoda	Sphaeromatidae	*Gnorimosphaerone*	*oregonense*	PG	Brook et al. (1994)
Malacostraca	Amphipoda	Tanaidacea	*Heterotanais*	*oerstedi*	PG	Bückle-Ramirez (1965)
Malacostraca	Amphipoda	Tanaidacea	*Hageria*	*rapax*	PG	Modlin and Hams (1989)
Malacostraca	Amphipoda	Tanaidacea	*Leptochelia*	*neapolitan*	PG	Ishimaru (1984)
Malacostraca	Amphipoda	Tanaidacea	*Leptochelia*	*dubia*	PG	Highsmith (1983)
Malacostraca	Amphipoda	Tanaidacea	*Leptochelia*	*forresti*	PG	Stoner (1986)
Malacostraca	Decapoda	Diognidae	*Clibnarius*	*vittatus*	PG	Turra (2004)
Malacostraca	Decapoda	Diognidae	*Clibnarius*	*antillensis*	PG	Turra (2004)
Malacostraca	Decapoda	Diognidae	*Clibnarius*	*sclopetarius*	PG	Turra (2004)
Malacostraca	Decapoda	Hippolytidae	*Lysmata*	*seticaudata*	PSH	Fiedler et al. (2010)

Continued

TABLE 3.1 Hermaphroditism in Crustacea—cont'd

Class	Order	Family	Genus	Species	Type of Hermaphroditism	References
Malacostraca	Decapoda	Hippolytidae	*Lysmata*	*nitida*	PSH	Fiedler et al. (2010)
Malacostraca	Decapoda	Hippolytidae	*Lysmata*	*californica*	PSH	Bauer and Newman (2004)
Malacostraca	Decapoda	Hippolytidae	*Lysmata*	*grabhami*	PSH	Wirtz (1997)
Malacostraca	Decapoda	Hippolytidae	*Lysmata*	*nitida*	PSH	d'Údekem d'Acoz (2003)
Malacostraca	Decapoda	Hippolytidae	*Lysmata*	*bahia*	PSH	Baeza (2008)
Malacostraca	Decapoda	Hippolytidae	*Lysmata*	*intermedia*	PSH	Baeza (2008)
Malacostraca	Decapoda	Hippolytidae	*Lysmata*	*nayaritensis*	PSH	Baeza and Thiel (2007)
Malacostraca	Decapoda	Hippolytidae	*Lysmata*	*ankeri*	PSH	Fiedler et al. (2010)
Malacostraca	Decapoda	Hippolytidae	*Lysmata*	*amboiensis*	PSH	Fiedler et al. (2010)
Malacostraca	Decapoda	Hippolytidae	*Lysmata*	*wurdermanni*	PSH	Fiedler et al. (2010)
Malacostraca	Decapoda	Hippolytidae	*Lysmata*	*debelius*	PSH	Fiedler et al. (2010)
Malacostraca	Decapoda	Hippolytidae	*Lysmata*	*pederseni*	PSH	Fiedler et al. (2010)

Malacostraca	Decapoda	Hippolytidae	*Lysmata*	*bogessi*	PSH	Fiedler et al. (2010)
Malacostraca	Decapoda	Hippolytidae	*Lysmata*	*rafa*	PSH	Fiedler et al. (2010)
Malacostraca	Decapoda	Hippolytidae	*Lysmata*	*acicula*	PSH	Fiedler et al. (2010)
Malacostraca	Decapoda	Hippolytidae	*Lysmata*	*trisetacea*	PSH	Fiedler et al. (2010)
Malacostraca	Decapoda	Hippolytidae	*Lysmata*	*galapagensis*	PSH	Fiedler et al. (2010)
Malacostraca	Decapoda	Hippolytidae	*Lysmata*	*moorei*	PSH	Fiedler et al. (2010)
Malacostraca	Decapoda	Hippolytidae	*Lysmata*	*nilita*	PSH	Fiedler et al. (2010)
Malacostraca	Decapoda	Hippolytidae	*Lysmata*	*holthuisi*	PSH	Fiedler et al. (2010)
Malacostraca	Decapoda	Hippolytidae	*Lysmata*	*seticaudata*	PSH	Fiedler et al. (2010)
Malacostraca	Decapoda	Hippolytidae	*Lysmata*	*ternatensis*	PSH	Fiedler et al. (2010)
Malacostraca	Decapoda	Hippolytidae	*Exhippolysmata*	*ophloporoides*	PSH	Fiedler et al. (2010)
Malacostraca	Decapoda	Hippolytidae	*Lysmata*	*liplei*	PSH	Fiedler et al. (2010)
Malacostraca	Decapoda	Hippolytidae	*Lysmata*	*anchisteus*	PSH	Fiedler et al. (2010)

Continued

TABLE 3.1 Hermaphroditism in Crustacea—cont'd

Class	Order	Family	Genus	Species	Type of Hermaphroditism	References
Malacostraca	Decapoda	Hippolytidae	*Lysmata*	*hochi*	PSH	Fiedler et al. (2010)
Malacostraca	Decapoda	Barbouriidae	*Parhippolyte*	*mistica*	PSH	Fiedler et al. (2010)
Malacostraca	Decapoda	Hippolytidae	*Lysmata*	*wundermanni*	PSH	Fiedler et al. (2010)
Malacostraca	Decapoda	Hippolytidae	*Lysmata*	*debelius*	PSH	Fletcher et al. (1995)
Malacostraca	Decapoda	Hippolytidae	*Lysmata*	*debelius*	PSH	Fiedler et al. (2010)
Malacostraca	Decapoda	Hippolytidae	*Lysmata*	*amboiensis*	PSH	Fiedler et al. (2010)
Malacostraca	Decapoda	Hippolytidae	*Thor*	*manningi*	PPH	Bergstrom (1997)
Malacostraca	Decapoda	Parastacidae	*Parastacus*	*nicoletti*	PPH	Rudolph (1995)
Malacostraca	Decapoda	Alpheidae	*Arete*	*kominatoensis*	PPH	Baeza (2006)
Malacostraca	Decapoda	Alpheidae	*Arete*	*indicus*	PPH	Baeza and Thiel (2007)
Malacostraca	Decapoda	Crangonidae	*Crangon*	*crangon*	PPH	Charnov (1979a)

Malacostraca	Decapoda	Hippidae	*Emerita*	*asiatica*	PPSH	Gavio et al. (2006)
Malacostraca	Decapoda	Calocaridae	*Ambiaxius*	*aberrans*	SH	Chiba et al. (2004)
Malacostraca	Decapoda	Calocaridae	*Ambiaxius*	*faveolatus*	SH	Chiba et al. (2003)
Malacostraca	Decapoda	Calocaridae	*Ambiaxius*	*flanklinae*	SH	Chiba et al. (2000)
Malacostraca	Decapoda	Calocaridae	*Ambiaxius*	*japonicus*	SH	Chiba et al. (2003)
Malacostraca	Decapoda	Calocaridae	*Ambiaxius*	*surugensis*	SH	Cipriani and Collin (2005)
Malacostraca	Decapoda	Calocaridae	*Calastacus*	*laevis*	SH	Collin et al. (2005)
Malacostraca	Decapoda	Calocaridae	*Calastacus*	*mexicanus*	SH	Conides et al. (2006)
Malacostraca	Decapoda	Calocaridae	*Calastacus*	*colpos*	SH	Conides et al. (2006)
Malacostraca	Decapoda	Calocaridae	*Calastacus*	*crosnieri*	SH	Correa et al. (2003)
Malacostraca	Decapoda	Calocaridae	*Calastacus*	*stilirostris*	SH	Chiba et al. (2000)

Continued

TABLE 3.1 Hermaphroditism in Crustacea—cont'd

Class	Order	Family	Genus	Species	Type of Hermaphroditism	References
Malacostraca	Decapoda	Calocaridae	*Calaxiopsis*	*felix*	SH	Dohrn (1950)
Malacostraca	Decapoda	Calocaridae	*Calaxiopsis*	*manningi*	SH	d'Údekem d'Acoz (2003)
Malacostraca	Decapoda	Calocaridae	*Calaxiopsis*	*mclaughlinae*	SH	Fiedler (1998)
Malacostraca	Decapoda	Calocaridae	*Calaxiopsis*	*serrata*	SH	Dohrn (1950)
Malacostraca	Decapoda	Calocaridae	*Calocaris*	*barnardi*	SH	Chiba et al. (2004) and Dohrn (1950)
Malacostraca	Decapoda	Calocaridae	*Calocaris*	*caribbaeus*	SH	Conides et al. (2006)
Malacostraca	Decapoda	Calocaridae	*Calocaris*	*macandreae*	SH	Chiba et al. (2004), Collin et al. (2005) and Dohrn (1950)
Malacostraca	Decapoda	Calocaridae	*Calocaris*	*templemani*	SH	Chiba et al. (2004) and Dohrn (1950)
Malacostraca	Decapoda	Calocaridae	*Eucalastacus*	*torbeni*	SH	Fricke and Fricke (1977)

Malacostraca	Decapoda	Calocaridae	*Lophaxius*	*investigatoris*	SH	Chiba et al. (2004) and Dohrn (1950)
Malacostraca	Decapoda	Calocaridae	*Lophaxius*	*rathbunae*	SH	Chiba et al. (2004)
Malacostraca	Decapoda	Calocaridae	*Lophaxius*	*granulosa*	SH	Dohrn (1950) and Fukuhara (1999)
Malacostraca	Decapoda	Calocaridae	*Lophaxius*	*sagamiensis*	SH	Fukuhara (1999)
Malacostraca	Decapoda	Hippolytidae	*Lysmata*	*amboiensis*	SH	Fiedler (1998)
Malacostraca	Decapoda	Hippolytidae	*Lysmata*	*wundermanni*	SH	Bauer and Holt (1998)

PH, Protandric hermaphrodites; *PPH*, Partial protandric hermaphrodites; *PSH*, Protandric simultaneous hermaphrodites; *SH*, Simultaneous hermaphrodites.

sex ratio for crustaceans to explain anomalous growth pattern, in which a close overlap between males and females in the mid-size classes would predict a sequential hermaphroditism. While early workers used this method to suggest sex reversal in decapods, direct observations on both secondary and primary sexual characteristics during the ontogeny of the organisms indicated a vast number of sex changers in malacostracans. Estimation on the occurrence of sex changers in crustaceans revealed a total number of 105 hermaphrodites, of which the decapods alone account for 74. Among the decapods, the caridean prawns dominate the list, having as many numbers as 59. In an earlier review, Bauer (1986a) also reported that out of the 32 decapod species, 26 are caridean shrimps. Obviously, carideans are the most extensively described hermaphroditic crustaceans in the literature. Incidentally, they occur in various forms, giving valuable insights into the evolution of this interesting sexual system in the whole of invertebrate species.

Protandric hermaphroditism has been recorded most frequently in caridean shrimps (Bauer, 1986a). Earlier workers on caridean shrimps described them as protandrous, in which individuals first mature as males and then, with increase in size and age, change sex to females. Subsequent studies nevertheless indicated variations existing in this sexual system of protandry (Fig. 3.2). For example, in species such as *Pandalus platyceros* and *Pandalopsis dispar* (Butler, 1964, 1980), as well as *Lysmata seticaudata* (Dohrn, 1950), all individuals in the population are reported to transform from males into females. Conversely, in other protandric carideans, a variable proportion of the population mature as primary females without passing through a male phase (MP) (Butler, 1980; Charnov et al., 1978; Bergstrom, 1997; Fréchette et al., 1970;

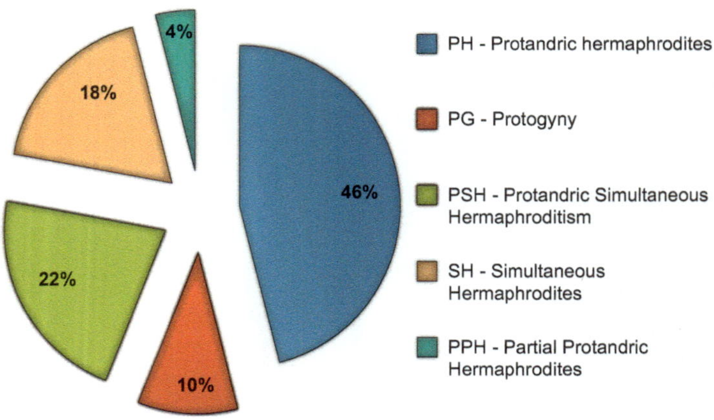

FIGURE 3.2 Showing percentage distribution of different types of hermaphroditism in caridean shrimps. PH (46%) is the dominant form of hermaphroditism in carideans. PSH (22%), followed by SH (18%), are the next dominant types. PG (10%) is comparatively rare in crustaceans. *Reproduced from Subramoniam, T., 2013. Origin and occurrence of sexual and mating systems in Crustacea: A progression towards communal living and eusociality. J. Biosci. 38, 951–969.*

Noël, 1976; Boddeke et al., 1991). In other species, such as *Thor manningi*, of the family Hippolytidae, the population is composed of 50% primary males and 50% protandric hermaphrodites (Bauer, 1986a). Yet again, in the alpheid species, *Athanas kominatoensis* (Nakashima, 1987), the population consists of both primary males and sex changers, whereas in other species like *Athanas indicus*, a mix of primary females, primary males, and sex changers are found (Gherardi and Calloni, 1993). Another variation in the sexual system of caridean shrimps was reported for the two genera, *Lysmata* and *Exhippolysmata* (Hippolytidae). In this system, the sequential (protandric) hermaphroditism has evolved one step further into protandric simultaneous hermaphroditism, in which the sex changed "females" are capable of mating both as male and female (Bauer and Holt, 1998; Fiedler, 1998; Bauer and Newman, 2004).

3.3.3.1 Caridean Shrimp

Early descriptions of protandry were made in the caridean shrimp *Processa edulis* (Nöel, 1973). The sequential appearance of sexual characters during its life cycle of the shrimp is given in Fig. 3.3. After passing through the sexually indeterminate juvenile phase, the shrimps begin to acquire the male secondary sexual characters in the first year which is followed by the appearance of functional testes lasting for several months. After serving as a functional male

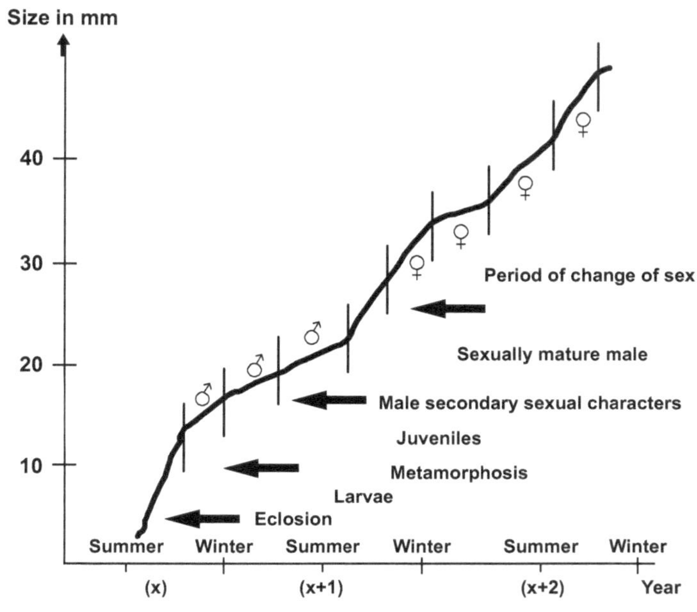

FIGURE 3.3 Sequential appearance sexual characters in Crustacea. *Adapted from Nöel, P., 1973. Cycle biologique et inversion sexuelle du crustacé décapode Natantia Processa edulis. Cah. Biol. Mar. 14 (2), 17–227.*

for a few months the shrimp begins to exhibit female secondary characteristics, followed by the conversion of testes into ovotestes, thus serving as an intersexual for a few days. The intersexual loses the male characters and begins to produce oocytes in the gonad and breed as female in the next year. Simultaneously, the testicular part either regresses or stops producing sperm cells, thus acquiring a functional female phase (FP) which is retained till the end of the reproductive life. Biometric study on the population of *P. edulis* indicated that individuals mate as males in the first year and then change sex at the end of summer and breed as female next year before disappearing at an appropriate age of two years. *P. edulis* is, however, not a strict protandric species in that protandric males coexist with primary females in the same population (Noël, 1976). This kind of protandry with primary females in the population has also been reported in genera such as Crangonidae (Boddeke et al., 1991) and Hippolytidae (Zupo, 1994). In *Hippolyte inermis*, a shrimp inhabiting sea grass meadows of the Mediterranean and Atlantic coasts of Spain is a protandric species with a small number of primary females in the population. However, the process of sex reversal is different from other protandric hermaphrodites in that the disappearance of appendix masculina is abrupt and could be lost in a single molt, resulting in a rapid sex change, without an intermediate ovotestis phase. The testis degenerates completely after the MP and a new ovary is formed from the preexisting gonial cells. Interestingly, feeding the shrimp *H. inermis* with a diatom *Cocconeis* resulted in the apoptosis of androgenic gland, followed by the complete disruption of the testis (Zupo and Messina, 2007).

Protandric hermaphroditism has also been described in the freshwater shrimp, *Paratya curvirostris*, belonging to Atyidae, Decapoda (Carpenter, 1978). In this prawn, size-specific sex ratio studies (Wenner, 1972) showed a sigmoid curve, typical of protandrous species. As in other true protandric forms, the male secondary sex characters, including the appendix masculina, become reduced and finally disappear during the changeover intersex stage. Concurrently, there is an increase in the volume of the brood cavity under the abdomen. The testis of the functional MP is a convoluted structure, showing no sign of oogenesis. However, in the large intersexual males, the distal and peripheral gonad is testis, but proximally and in the centre, it is ovarian. Histological examination of the androgenic gland revealed a high activity during the MP, but in the large intersex males, the gland has become completely atrophied. The frequent occurrence of such intersexuals along with a low frequency of primary females in the population confirmed protandry in *P. curvirostris*.

In addition to the caridean freshwater prawns, high incidence of protandric hermaphroditism has been reported in the deep-sea caridean prawns coming under the family Pandalidae (Berkeley, 1929, 1930). In this family, almost all the individuals pass through a functional male stage of 1–3 years before undergoing transformation into a functional female stage. However, during the male stage, the gonad is ovotestis, with only the testicular elements functional. In the female stage, the gonad is the true ovary and the testes have degenerated during the sexual transmission. Species like *P. dispar* and *P. platyceros* are dominantly protandric, whereas other species such as *Pandalus jordani* and *Pandalus hypsinotus* are partially protandrous (Allen, 1959; Carlisle, 1959). In the ovotestes of *P. platyceros*, the cortical layer of epithelial cells gives rise to the male gametes, whereas oocytes and follicle cells originate from the medullary region.

Many of the early studies on sequential hermaphroditism in the caridean prawns have been predicted from the size-specific sex ratio investigations, followed by histological studies on the gonadal transformation. Okumura et al. (2005) studied sex reversal mechanisms in a protandric deep-sea pandalid shrimp, using laboratory-maintained *P. hypsinotus*. In the early immature males, at the age of 16 months the gonad contains only ovarian tissues consisting of oogonia and oocytes and no testicular tissues. However, they possessed ejaculatory ducts. When these shrimps attain 19–31 months of age, they develop testicular tissues in the peripheral parts of the gonads, which begin to produce spermatozoa at the age of 34 months. After reaching sexual maturity, these male shrimps follow an annual reproductive cycle. They have a distinctive annual reproductive cycle, starting from April to November. During this MP, the androgenic gland activity remains high throughout the reproductive cycle. Interestingly, eyestalk ablation induced hypertrophy of the androgenic glands and accelerated sexual maturity. This study revealed that 16.7% of these laboratory-maintained male shrimps changed sex. The size of the sex-changed shrimps was observed to be 38.2–42.6 mm carapace length.

After the MP, there is a degeneration of spermatogonia and spermatocytes, with the gonadal functions completely taken over by oocyte development and maturation, followed by oviposition, during the rest of its reproductive life. Interestingly, the oocytes were present right from the beginning of the MP, but the oocyte development stood completely arrested at the previtellogenic stage. Understandably, the androgenic gland that becomes active during the MP could exert its inhibitory effect on vitellogenesis. Only at the cessation of the MP, with the concomitant disappearance of androgenic gland, the secondary vitellogenesis resumes to produce mature oocytes. The appearance of vitellogenin in the hemolymph could be considered as an internal female secondary sexual characteristic to predict the stage of sex change in these protandrous shrimps. Other species of *Pandalus* such as *P. platyceros* and *Pandalus goniurus* exhibit strict protandry (Butler, 1980; Bergström, 2000).

Crangon crangon, a caridean shrimp belonging to the family Crangonidae is another example for protandric hermaphroditism. Males and primary females of *C. crangon* develop from the larval stage, while secondary females develop after sex reversal of spent males (Boddeke et al., 1991). A sex change from male to female in *C. crangon* has been suggested, based on the structure of gonads. Externally, differences in the length and shape of the endopods of the first pair of pleopods and the appendix masculina, a small squamiform protuberance of the second pleopod in the males, are used to distinguish the male and FPs of the hermaphrodites. Both male and female gonads are paired organs, consisting of two longitudinal tubes with closed ends. While the ovarian part occupies the anterior cephalothoracic region, the testicular portion is present posteriorly extending up to the first abdominal segment. The extension of the gonad in longitudinal and lateral directions depends on sex and stage of development. The oviducts arise from the ovarian lobes at about one-third of the length from its anterior end, whereas the vas deferens originates from the middle of both testicular lobes. During the spawning season, the protandric males copulate several times and then change sex. Further field studies, together with laboratory rearing have indicated only a few males changed sex, suggesting that *C. crangon* is a facultative protandric hermaphrodite (Hufnagl et al., 2010).

In the genus *Thor*, several species displayed various sexual systems of sequential hermaphroditism, along with gonochoristic forms. Species such as *Thor dobkini* and *Thor floridanus* are exclusively gonochorists (Bauer and VanHoy, 1996). On the contrary, another species, *T. manningi*, exhibits partial protandry with populations composed of primary males, primary females, and protandric individuals (Bauer, 1986a). Yet another species *Thor amboinensis* is also a protandric hermaphrodite (Baeza and Piantoni, 2010). These shrimps show a symbiotic social relationship with sea anemone and, while changing the sex, they also change external phenotypic characteristics, including the coloring patterns distinctive of males and females. The existence of different types of protandry within the genus *Thor* may reflect their independent evolutionary origins from different ancestral gonochoric species, although socially mediated sex change in different environmental conditions, as suggested for *Lysmata* species cannot be ruled out (Bauer, 2002; Baeza and Piantoni, 2010).

3.3.3.2 The Anomuran Crab Emerita asiatica

Many decapods, other than the caridean shrimps are gonochoristic, although a few species have revealed sequential hermaphroditic tendencies. The anomuran mole crabs, belonging to the genus *Emerita* is a case in study. Based on size-specific sex ratio studies, Barnes and Wenner (1968) suggested protandric hermaphroditism in the American species *Emerita analoga*. However, Wenner and Haley (1981) withdrew this claim by finding no direct evidence for its occurrence in this species. Almost at this time, Subramoniam (1979b, 1981) reported unequivocal evidence for the existence of protandric hermaphroditism in a tropical species *E. asiatica* (=emeritus). He studied the ontogenic development of protandric hermaphroditism in this intertidal mole crab inhabiting the sandy beaches of Chennai in the east coast of India. Males in *Emerita* species are characteristically neotenic; with the juvenile acquiring precocious sexual maturity soon after the metamorphosis from the megalopa larva (Efford, 1967; Subramoniam, 1977b). The size-related sex ratio studies on the population did not suggest a possible sex reversal in this species, as the overlap in size range between males and mature females is too wide. Further, males with distinct genital papillae and juvenile females devoid of genital papillae are found in almost equal proportion among the postlarvae collected during megalopa settlement in the beach. In *E. asiatica*, the males achieve sexual maturity soon after metamorphosis from megalopa at 3.5 mm CL, whereas the females attained maturity only after considerable body growth at 19–22 mm CL. Subsequent observations, however, indicated that some neotenous males continue to grow further after serving as active males and then lose male secondary and primary sexual characters and mating functions, and undergo sex reversal by acquiring female characters around 19 mm CL. The disappearance of genital papillae around 15 mm CL is the first visible sign of sex reversal. Spermatogonial activity in testis ceases concurrently. A hyperactivity of the mesodermal cells in the testis is correlated to its inactivity. In the size range of 19–22 mm CL, the male's gonad comprises inactive testicular and active ovarian portions. Fig. 3.4 shows the comparison of gonads of male, female, and hermaphroditic stages of *E. asiatica*. During sex reversal, a separate ovarian anlagen is formed on the dorsal region of the inactive testis, which spread along the mid-dorsal line of the paired testis. In the hermaphroditic gonad, the median ovarian limb beyond the fused posterior extremity of the testis lacks testicular elements. The ovarian structure contains typical follicle cells that surround the vitellogenic oocytes mediating yolk protein uptake. The vas deferens in these intersexuals is intact, but its opening is occluded. They begin to possess a pair of functional oviducts and have three pairs of pleopods to which are attached a bunch of eggs, thus constituting secondary females in the population. Androgenic glands, active in the neotenous males, show signs of degeneration in the larger males. At 15 mm CL and above, the androgenic gland was not detected in the males.

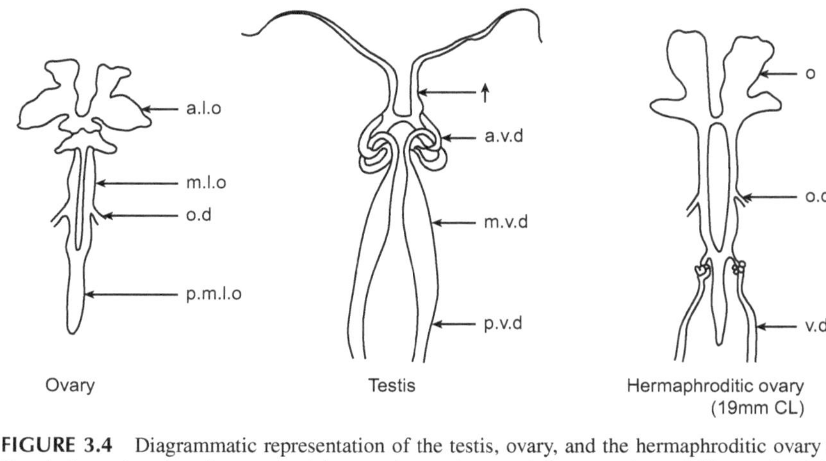

FIGURE 3.4 Diagrammatic representation of the testis, ovary, and the hermaphroditic ovary of *E. asiatica*. *a.l.o.*, Anterior ovarian lobe; *a.v.d.*, anterior vas deferens; *m.l.o.*, middle ovarian lobe; *m.v.d.*, middle vas deferens; *o*, ovary; *o.d.*, oviduct; *p.m.l.o.*, posterior median ovarian limb; *p.v.d.*, posterior vas deferens; *t*, testis; *v.d.*, vas deferens. *Redrawn from Subramoniam, T., Gunamalai, V. 2003. Breeding biology of the intertidal sand crab, Emerita (Decapoda: Anomura). Adv. Mar. Biol. 46, 91–182.*

In addition to these intersexual females, instances of incomplete transformation of sex, as found in large females of *E. asiatica* have also been observed in the population. In these crabs, histological examination indicated that a separate ovarian portion is never found above the nonfunctional testis. However, the anterior half of the paired tubular gonad is dominated by oocytic differentiation, whereas the posterior half possessed inactive testicular tissues. The mature oocytes in the anterior portion undergo oosorption, accomplished by the infiltrating follicle cells, which on entry into the ooplasm, turn phagocytic. These intersexuals, however, retain the vas deferens, but the genital papillae are absent. Paired oviducts are also present but are incompletely differentiated, compared to the functional oviducts of secondary females, arising from successful sex reversal. The occurrence of such incomplete sex-transformed *E. asiatica* may imply that the androgenic gland has not completely regressed during the sex reversal to bring about total transformation. On the other hand, the low titer of androgenic gland hormone may cause such incomplete inversion in the reproductive tissues seen in *E. asiatica* (Subramoniam, 1981).

3.4 PROTANDRIC SIMULTANEOUS HERMAPHRODITISM

An interesting aspect of protandric hermaphroditism in *P. platyceros* (described above) is that, at the completion of MP, degeneration of the testicular elements has occurred in the cortex of the ovotestes, which is followed by the onset of oogenesis in the medullar region, along with the formation of the oviduct (Hoffman, 1972). At the same time, the vas deferens containing fully matured spermatozoa as well as the male secondary sexual characters are all present in a functional way, indicating their behavioral ability to mate as males. Since the oocytes are fully mature and the oviducts are functional, these hermaphroditic individuals can also act as females. This kind of simultaneous hermaphroditism originating from a protandric MP was subsequently discovered in another caridean genus *Lysmata* by Bauer (2002), who rechristened it as protandric simultaneous hermaphroditism (PSH). PSH is defined as the sexual pattern, in which individuals first mature as a male and, with increasing size, it molts to a simultaneous hermaphrodite or "euhermaphrodite," which can reproduce both as a male and female without self-fertilization. Unlike strict protandric carideans, *Lysmata* PSH retains reduced male gonadal tissues and ducts and are able to mate nonreciprocally as males as well as to reproduce as females. Although all the *Lysmata* species are functional protandric simultaneous hermaphrodites, most of their reproductive efforts are devoted to embryo production and incubation (Bauer, 2007).

As early as 1912, Spitschakoff described this kind of simultaneous hermaphroditism in *L. seticaudata*. It possesses an ovotestis, in which the testicular part is well developed in smaller individuals, while the ovarian part becomes prominent in older, larger (female-phase) individuals. Extensive histological studies on this species have confirmed that the anterior region of male-phase individuals is the ovarian portion of the ovotestis, while the posterior part of the gonad is testicular. Furthermore, not only the male ducts are retained in the FP but also spermiogenesis and sperm production continue after the sex change (Charniaux-Cotton and Payen, 1985). That these FP individuals can outcross with other FP shrimps has been demonstrated by time-lapse video observations in another species *Lysmata wurdemanni* (Bauer and Holt, 1998). In this shrimp, all smaller individuals possessed characters typical of caridean males, including male gonopores, appendices masculinae, ejaculatory ducts, and cincinnuli (coupling hooks) on the first pleopods. The gonad of the male-phase individuals has an ovotestis with an undeveloped ovarian part. However, the female-phase individual, which spawns eggs and incubates embryos, also has male gonopores and an ovotestis terminating in ejaculatory ducts containing sperm. In the FP, male secondary characters are reduced or absent, but have expanded pleopod flanges, a characteristic egg-brooding character of the females. In this species, not all the small males transform into FP hermaphrodite.

That the PSH may be a common feature in the genus *Lysmata* is revealed from further work on other species such as *Lysmata bahia* and *Lysmata intermedia* (Baeza, 2008). In the tropical Caribbean coast of Panama, the population of each species consisted of functional males and functional simultaneous hermaphrodites. Males have cincinnuli and appendices masculinae on the first and second pair of pleopods, respectively. Gonopores are located at the coxae of the third pair of walking legs, and the ovotestis has a well-developed male portion full of sperm, but the female portion is undeveloped. Hermaphrodites lacked appendices masculinae and cincinnuli, although they have male gonopores. However, their ovotestes have well-developed ovaries, full of mature oocytes and testes with sperm. When these hermaphrodites were maintained in pairs, both of them molted and spawned eggs that initiated development, showing that they can reproduce as males, by inseminating the other hermaphrodites, acting as females. PSH is the only sexual system present in all species examined so far under the genus *Lysmata* (Table 3.2). Bauer (2007) predicted that a change in the environment would have caused the transition from a protandric *Lysmata* ancestor to the extant PSH species.

Protandric simultaneous hermaphroditism appears to occur in a paraphyletic genus *Exhippolysmata*. Braga et al. (2009) recently provided morphological evidence in *Exhippolysmata oplophoroides* for the occurrence of PSH. As in *Lysmata* species, the gonad of the MP consisted of paired ovotestis with an undeveloped ovarian portion. In the hermaphroditic stage, most individuals have small appendices masculinae, both male and female gonopores, and ovotestis with vitellogenic oocytes and spermatozoa. The anterior ovarian region is formed by two lobes, each with an oviduct that opens at the base of the third pair of pereiopods. In the testicular region, two lobules with a lateral sperm duct in each of them opens at the base of the fifth pereiopods. Spermatozoa were recorded in the testes and in the ejaculatory duct of all analyzed shrimps. Interestingly, in the early stage male-phase ovotestis a few oogonia and a few

TABLE 3.2 Different Types of Sexual Systems in Crustaceans

Categories	Ontogeny	Examples
Gonochorism	Juvenile → Primary male Juvenile → Primary female	*Heptacarpus sitchensis* (Bauer, 1976)
Androdioecy	Juvenile → Primary male Juvenile → Hermaphrodites	*Eulimnadia texana* (Hollenbeck et al., 2002)
Gynodioecy	Juvenile → Primary male Juvenile → Hermaphrodites	
Sequential hermaphroditism		
Simple protandry	Juvenile → Male phase → Female phase	*Pandalus goniurus* (Bergström, 2000)
Protandry (with primary females)	Juvenile → Male phase → Female phase Juvenile → primary female	*Processa edulis* (Nöel, 1973)
Protandry (with early maturing females)	Juvenile ⇄ Male phase → Female phase "early maturing" female phase	*Pandalus borealis* (Charnov, 1981)
Protogyny	Juvenile → Female phase → Male phase	*Cyanthura carinata* (Legrand and Juchault, 1963)
Simultaneous hermaphroditism	Juvenile → Both functional male and female in the same individual	*Calocaris macandreae* (Buchanan, 1963)
Protandric simultaneous hermaphroditism	Juvenile → Male phase → Simultaneous hermaphrodites	*Lysmata seticaudata* (Fiedler et al., 2010)

Intersexuality		
Normal intersexuals (existing as a part of life cycle)	Juvenile → Male phase → Intersexuals (Transitory) ↓ Female phase (as in protandric hermaphroditism) Juvenile → Female phase → Intersexuals (Transitory) ↓ Male phase (as in the case of protogyny) Juvenile → Male phase → Intersexuals (Due to incomplete feminization by parasites)	*Parastacus varicosus* (Rudolph et al., 2001) *Cyanthura carinata* (Legrand and Juchault, 1963) *Gammarus deubeni* (Kelly et al., 2004)
Epigenetic intersexuals (arising from epigenetic factors)	Juvenile → Female phase → Intersexuals (Due to androgenic gland implants) Juvenile → Sex intermediates (Hormonal disruption by methyl farnesoate)	*Callinectes sapidus* (Johnson and Otto, 1981) *Daphnia pulex* (Frey, 1965)
Sex intermediates	Embryo → Bilateral gynandromorphism (Due to the chromosomal mutation in the early embryonic stage)	*Artemia franciscana* (Campos-Ramos et al., 2005)
Gynandromorphism	Embryo → Bilateral gynandromorphism (Due to chromosomal loss during early embryonic cleavage)	*Homarus gammarus* (Farmer, 1972)

From Subramoniam, T., 2013. Origin and occurrence of sexual and mating systems in Crustacea: A progression towards communal living and eusociality. J. Biosci. 38, 951–969.

previtellogenic oocytes were observed. However, the ovarian portion containing vitellogenic oocytes is present only in individuals of the simultaneous hermaphroditic phase, whereas the spermatozoa in the ejaculatory duct are detected in all individuals of different sizes. Collectively, the smaller shrimps of *E. oplophoroides* would reproduce as males while larger shrimps would be capable of reproducing both as males and females, as is the rule in all PSH (Bauer and Holt, 1998; Fiedler, 1998; Baeza and Anker, 2008). On the contrary, in all protandric hippolytids, such as *T. manningi*, MP individuals gradually lose their male characters and consequently, with the increase in size, the sex characters change completely to the female (Bauer, 1986a). Obviously, these sex-changed females could not mate as males.

While social interaction could be the cause for the protandric simultaneous hermaphroditism in *L. wurdemanni*, this type of sexual system would have possibly evolved from a basic protandric ancestor (Bauer and Holt, 1998).

In another instance, the intersexuals among the hermit crabs exhibiting protogynous hermaphroditism also shows a similar kind of simultaneous hermaphroditism. Turra (2004) had postulated that the intersexuals can act as males as well as reproductive females and develop both functional gonads, suggesting that an individual can reproduce both as male and female. This indicates a simultaneous hermaphroditism before female gonopore closure and ovary resorption, if a protogynic process is occurring.

3.5 PROTOGYNY

Protogyny occurs in organisms that are born female and at some point of their life span change sex to males. In general, the occurrence of protogyny is associated with the following three features: low mobility of females, low abundance of males, and intense competition among males for access to females. In the protogynous sexual system, the relative fitness of males increases faster with age than it does for females (Charnov, 1993). Interestingly, Isopoda is the only order in which both forms of sequential hermaphroditism occur. All protogynous isopods are free-living, whereas the protandrous isopods are mostly parasitic. Among the 63 sex-changing species in Crustacea, Brook et al. (1994) found 11 species (17%) were protogynous. Indeed, protogyny is the reproductive norm in several isopod taxa. Incidentally, in all the protogynous isopods, males guard their mates, and hence older and larger males probably have a reproductive advantage over smaller males by virtue of an enhanced ability to thwart competitors (Abe and Fukuhara, 1996). In consequence, protogyny arises when small males are prevented from mating with females by larger males, making it advantageous to become male when a competitive larger size is reached (Warner, 1988).

Brook et al. (1994) described protogynous sex change in the free-living marine isopod *Gnorimosphaerone oregonense*. These are sexually dimorphic intertidal isopods, with 31% of the females having rudimentary penes, suggesting sex change from female to males. Laboratory growth experiments also confirmed that females produced one brood of juveniles, and then passed through a variable number of molts as immature males, before becoming sexually mature males. However, not all juveniles are protogynous: only a small percentage of them developed to protogyny in the laboratory, others directly develop into mature males. These individuals first develop to breeding females and later change to reproducing males (termed secondary males), whereas others (termed primary males) remain as males, lacking any female characters, throughout their lives (Highsmith, 1983; Brook et al., 1994). Protogynous sex change is socially mediated in fishes (Avise, 2011), but in the isopods, it appears to be facultative protogyny, because a large percentage of laboratory-reared juveniles developed directly into males. However, Brook et al. (1994) noted an evident association between protogyny and precopulatory guarding in isopods.

This follows that the size advantage theory, proposed by Ghiselin (1969), should be favored in species in which male reproductive success is dependent on large body size. In *G. oregonense*, precopulatory mate guarding is considered to be the primary selective force favoring protogyny. Larger males may guard females or a resource more effectively than a smaller competitor, thus giving a higher reproductive success to larger males. An interesting dichotomy in the existence of protogyny in crustaceans is that size advantage theory of bigger body size is correlated to female's higher fecundity in those species that lay heavily yolk-laden eggs. Hence, the advantage gained by larger males in the protogynous crustacean mating system overrides the selective force of increased female body size in isopods.

3.6 ECOLOGICAL AND EVOLUTIONARY IMPORTANCE OF SEQUENTIAL HERMAPHRODITISM

Sequential hermaphrodites that change sex as a normal part of their life history occur in a wide range of taxa among Crustacea. By far, protandrous hermaphroditism is the most prevalent form among other sexual patterns inasmuch as the adaptive significance of this system could be easily explainable. In most species, the size and age of the individuals are positively correlated, as the crustaceans tend to grow continuously during their lifetime. Warner (1988) advanced the size advantage theory to explain the adaptive significance of sequential hermaphroditism in crustaceans. This theory predicts that sex change is favored by natural selection when an individual reproduces most efficiently as one sex when young or small, but as the other sex, when large and older. Depending on the biology and ecology of a particular species, males might have the reproductive advantage when small and females when large, in which case protandry could be favored by natural selection. Conversely, in many other species, individuals might maximize their fitness by functioning as females when small or young but as males when larger and older, in which case protogyny will be favored. In explaining the size advantage hypothesis further, sex allocation theory relating to the energy cost of reproduction should be considered. Charnov (1979a, 1982) refined the size-advantage hypothesis and predicted qualitatively the optimal timing of sex change within the overall theory of sex allocation. According to him, the optimal age of sex change t is that which maximizes the product $F(t) \times M(t)$, where $F(t)$ is the fitness gained through female function and $M(t)$ is the fitness gained through male function.

Sex allocation is defined as the distribution of an organism's resources to male versus female function (Warner, 1988). Therefore, sex allocation has much relevance in the production of egg versus sperm in hermaphrodites, thereby influencing the direction and timing of gender switching in the sex changers. Male gametes are usually small and numerous relative to those of females, and female gamete production often entails energy-demanding activities such as production and storage of yolk in eggs for embryonic nutrition and larval development. In the *Lysmata* species, *L. wurdemanni*, the size at which sex changes from the smaller MP to the larger simultaneous hermaphrodite FP is variable. Below a certain minimum size, individuals are MPs and cannot function as females because they do not have the energy resources to produce the large yolky eggs typical of caridean shrimps. Therefore, most MPs do not change sex until they are larger in size. Bauer (2002) further observed that a delay in sex change in *L. wurdemanni* could also be due to the high abundance of FPs in the population to increase the opportunity to mate as a male. Furthermore, differential growth or mortality between the sexes could result in the natural selection for a sex changing life history. If one sex grows faster than the other, and fecundity increases with size, it may be advantageous to be the faster-growing sex first and to change sex later.

Timing of sex change and growth rate of hermaphrodites at latitudinal populations of pandalid shrimps such as *Pandalus borealis* has been well worked out (Charnov, 1979a; Bergström, 2000). From the American coastal region, age or size at maturity and sex change of southern populations is roughly earlier (smaller) than that of northern populations. Such a latitudinal pattern of the life history variation could be caused by latitudinal differences in water temperature and growth rate. Furthermore, Charnov and Hannah (2002) indicated that the timing of sex change in the *pandalid* shrimp population is a flexible response by individuals to local conditions. Individual plasticity in the timing of sex change can generate significant structure in the mean size or age of sex change at large spatial scales (Munday et al., 2006). Notwithstanding, some protandrous shrimps do not alter the timing of sex change in response to local environmental conditions (Bergstrom, 1997). Equally intriguing is the fact that in some shrimp species a proportion of the population matures directly into the other sex, whereas other individuals never change sex because favorable conditions do not arise.

In the anomuran intertidal sand crab *E. asiatica*, the time for sex reversal to female take considerable body growth from the functional MP (Subramoniam, 1977b, 1981). In this crab, the males are neotenous, acquiring the male function soon after their metamorphosis from megalopa larva at around 3.5–5 mm carapace length. However, many of the males grow to a bigger size, but do not participate in mating, although they possess enlarged vas deferens with spermatophores. They grow still further to bigger size without any male function and only at 15 mm CL, they lose external genital pappillae, being the first visible sign of sex reversal. Thereafter, the female characters, both primary and secondary, appear concurrently, with the testicular inactivity. The completion of sex reversal occurs with the formation of separate ovarian anlagen outside the testicular tissue, and hence the ovigerous hermaphroditic females are found only from 19 mm CL onward. This clearly suggests that the sex reversal has to wait till the males grow to a size at which the normal primary females attain sexual maturity. *E. asiatica* is a tropical species having a year-round reproduction and hence any environmental force might not be a causative factor to fix the sex reversal time (Subramoniam and Gunamali, 2003). Obviously, an epigamous factor exerted during male somatic growth at a specific time corresponding to the onset of female sexual maturity may bring about the disappearance of androgenic gland, followed by autodifferentiation of a new ovary. The evolutionary implication of the attempted protandric hermaphroditism in *Emerita* species is the overall reproductive adaptation of these intertidal burrowing crabs to maximize the egg-laying female population from the secondary females introduced by way of protandry (Subramoniam, 1981).

3.7 ANDRODIOECY

Androdioecy is a condition in which hermaphrodites and males coexist within a species (Avise, 2011). This rare but specialized sexual system is well documented in Notostraca and Conchostraca among branchiopods. In the conchostracan genus *Eulimnadia*, the hermaphrodites lack clasping appendages for mating and are physically unable to mate with one another. Therefore, hermaphrodites have to self-fertilize or mate with males (Sassaman and Weeks, 1993; Weeks et al., 2005). Androdioecy has evolved as a response to reproductive assurance that experience episodic low densities (Weeks et al., 2006). In the tadpole shrimp, *Triops* (Notostraca) androdioecious reproduction has been described by Sassaman (1991) and Sassaman et al. (1997). As in the clam shrimp *Eulimnadia*, androdioecy in *Triops* involves the participation of few males, amphigenic hermaphrodites, and monogenic hermaphrodites. The monogenic as well as amphigenic hermaphrodites produce offspring either by crossing with a male or by selfing (García-Velazco et al., 2009). *Eulimnadia texana*, a clam shrimp under the family Limnadiidae is an androdioecious species coexisting with hermaphrodites of two phenotypically similar but genetically different types: "amphigenic" and "monogenic" hermaphrodites. Males of *E. texana* have a higher mortality rate than hermaphrodites, but increased male mating opportunities extended the male lifespan. Hermaphrodites produce primarily female gametes, with only a limited portion of the ovotestis devoted to sperm production (Zucker et al., 2001). However, the amount of sperm produced is sufficient to fertilize all of the hermaphrodite's eggs, if a male is unavailable to supply sperm (Zuker et al., 1997). Nevertheless, hermaphrodites behave in ways that increase the likelihood of outcrossing. Receptive hermaphrodites are attracted to males when available. Hermaphrodites will even delay moving their eggs to their brood chambers when males are unavailable, suggesting that they can extend their "window of opportunity" for outcrossing when not in the presence of males (Zucker et al., 2002).

Cirripedes are another lower crustacean order that exhibits a mixture of sexual systems. For example, the families within the Thoracica exhibit dioecy, hermaphroditism, and androdioecy. Crisp (1983) has traced the sequence of sexual steps in the androdioecious barnacle, *Chelonibia patula*. *C. patula* is protandrous, where male organs develop first and female organs develop later to produce a simultaneous hermaphrodite. The development of sex organs is size specific. Male organs arise when the organism reaches a basal diameter of 2.5 mm and become functional at 5 mm in basal diameter. Female organs begin development at about 4 mm in diameter and become mature at 7 mm in basal diameter, which is the point at which *C. patula* becomes a fully functional, simultaneous hermaphrodite. In this species, the males settling on the opercular region of the hermaphrodites reaches only 2–4 mm in size and never develops female functions. According to Crisp (1983), these males are different from complemental males of other barnacles. In another species, *Chelonibia testudinaria*, complemental males settle in depressions in hermaphroditic shell plates (Rawson et al., 2003). Zardus and Hadfield (2004) suggest that the small complemental males facilitate a higher rate of reproduction by producing sperm at a young stage.

3.8 INTERSEXUALITY IN CRUSTACEA

In humans, there is considerable stability in sex determination involving the sex chromosomes X and Y. However, in crustaceans, there is relative plasticity in the sex determination mechanisms. The labile nature of sex determination in crustaceans has resulted in the sporadic occurrence of mixed sex or intersex individuals in many species. Imprecise sex determination leads to the production of faulty sex organs and consequently to reproductive impairment. Intersexuality refers to an individual having sexual characteristics intermediate between those of a typical male and a typical female. It also occurs naturally as a transitional phase during sex change in sequential hermaphroditic malacostracan crustaceans. Thus, intersexuality is common in certain groups of caridean shrimp as a result of protandrous sex reversal (Yaldwyn, 1966; Bauer, 1986a).

In several gonochorists, intersexuality has been reported throughout the adult stage. In them, two phenotypes of intersexuals are reported: intersex male and intersex female. They, however, possess sex characteristics of the opposite sex. Within Crustacea, intersexes have been reported from many groups such as isopods, amphipods, cladocerans, copepods, and decapods. In crustaceans, production of intersexes has been ascribed to both genetic as well as epigenetic causes including temperature, photoperiod, parasitic infection, and environmental pollutants that induce endocrine disruption. Because intersexuals are produced due to alterations in environmental, parasitic influences, and/or hormonal imbalances, different phenotypic variations occur. Hence, for a given intersexual, as exemplified in the sergestid shrimp *Acetes sibogae*, different types or subdivisions of the intersexes could be discerned (Hanamura and Ohtsuka, 2003). Upregulation or downregulation of sex-determining genes, induced by unusual temperatures, may also be possible causes of intersexual development. In addition, in the intersexual gammarid amphipod, *Gammarus duebeni*, photoperiods after mating determine the sex of the offspring (Legrand et al., 1987). Crustacean intersexes behave and reproduce either as males or females, but their reproductive output is always lower than the normal male and female (Charnov, 1982). In a population of the amphipod *G. duebeni*, Bulnheim (1965) noticed that intersexes were frequently sterile and were typically rare.

In the primitive branchiopod clam shrimp *E. texana*, two types of intersexes, viz., (1) a "morphological" intersex, possessing primarily male characteristics such as claspers and an egg-producing gonad and (2) a gonadal intersex possessing primarily male traits (eg, male secondary sexual characters and male gamete production) but also producing low levels of abortive female gametes. These intersexes are likely products of limited crossing over between the sexes determining chromosomes that results in the array of mixed sexual phenotypes (Weeks et al., 2006).

3.8.1 Copepods

Sex determination in copepods is under strong environmental control. Inadequacies in such sex controlling mechanisms bring about a variety of mixed sexes in the population. In *Acrocalanus gracilis*, sex determination occurs late in the life cycle, ie, at the fifth copepodite (C5) stage (Gusmão and Mckinnon, 2009). An important consequence of this kind of sex determination is the increased risk of intersex production. Copepods are another crustacean family, in which intersexuality is widespread. However, the presence of intersexes in several copepod species occupying several environmental niches suggests that environmental sex determination (ESD) may be a primary cause of intersexuality. Furthermore, in planktonic copepods, intersexuality is a manifestation of late sex change during development, in response to environmental cues. More precisely, intersex conditions could arise during the sex switching period in copepods. In the calanoid species, the intersex individuals originate from genotypic copepodite (C5) males which have changed sex. Late sex determination in the life cycle of copepods might be an adaptation to the determinate model of development in this group.

3.8.2 Amphipods

It is well understood now that ESD and the consequent plasticity in sex determination leads to sexual abnormalities such as intersexes in crustaceans. Amphipods are another crustacean group, which exhibits wide-ranging types of intersexuals, caused primarily by environmental factors, including photoperiod, parasitism, pollution, and endocrine–disruptive factors present in the aquatic media. Some of the different types of inter sexuality are listed in Table 3.3.

ESD in the amphipod *G. duebeni* is adaptively cued to photoperiod. Therefore, intersexuality occurs more significantly and frequently in populations under the influence of photoperiodic conditions which cue male development (Dunn et al., 1993). Intersexuality is heritable: intersex mothers produce more intersex offspring than do true females. Subsequent work on *G. duebeni*, however, showed that intersexuality results from incomplete feminization by the protist parasitic sex ratio distorter, *Nosema granulosis* (Kelly et al., 2004). In this amphipod, feminization of males is the result of parasite-induced inhibition of androgenic gland activity, thus preventing male development (Bandi et al., 2001). Hence, it is possible that intersexuals are those that did not inherit sufficient parasite numbers to allow complete feminization. The combinatorial effects of environmental temperature could also influence the production of intersexuals in the population. Kelly et al. (2002) reported lower efficiency of feminization by *N. granulosis* in correspondence with the appearance of more intersexuals in the population. Nevertheless, photoperiod-induced intersexuality is found only in populations free from any parasitic infestations. Conceivably, intersexuality is induced by both environmental and parasitic factors in populations of *G. duebeni*. Intersexuality in another amphipod species *Orchestia gammarellus* has been shown to be related to the presence of another parasitic protozoan *Paramarteilia orchestiae*. Transfection of infected tissue from this species was found to induce intersexuality in other amphipod species like *O. mediterranea* and *Orchestia aestuarensis* (Ginsburger-Vogel, 1991). Taken together, it may be said that in populations where sex is environmentally determined, intersexes result from both inappropriate development under ESD and from parasitism.

Intersexuality in the wild populations of amphipods is linked to anthropochemicals, which cause reproductive disorders via disruption to endocrine systems. In the marine/estuarine amphipod *Echinogammarus marinus*, high incidence of intersexuality is reported (Ford et al., 2004). Subsequent studies on this amphipod have, however, indicated feminizing microsporidian parasites as the most probable cause of the observed intersex (Ford et al., 2006). An alternative explanation might be that the contaminant in the environment can affect the ability of feminizing parasites to change the gender of the host, by way of endocrine disruption. Any incomplete feminization in this process could result in intersex production.

TABLE 3.3 Occurrence of Different Types of Intersexuality in Crustaceans

1. Normal Intersexuals		
Species	**Mechanism**	**References**
Decapods		
Uca pugilator	Androgenic gland mediation	Zou and Fingerman (2000)
Potamon fluvitaile		Micheli (1991)
Chionoecetetes opilio		Taylor (1986)
Cancer irroratus		Moriyasu et al. (1988)
Quadrella coronate		Galil and Tom (1990)
Clibnarius sp.		Turra (2004)
Homarus americanus		Chace and Moore (1959)
Parastacus varicosus		Rudolph et al. (2001)
Samastacus spinifrons		Rudolph (2002)
Upogebia stellata		Pinn et al. (2001)
Penaeopsis sp.		Farfante (1978)
Pandalus borealis		Ivanov and Sokolov (1997)
Graspus albolineatus		
Coenobita rugosus	Seasonal variation of reproductive period	Gusev and Zabotin (2007)
Acetes sibogae	Mechanism not given	Hanamura and Ohtsuka (2003)
Acetes erythraeus, Acetes sibogalis	Intersex may be due to some environmental factor	Achuthankutty (1973)
Copepods		
Paramphiascella hyperborean, Stenhelia gibba	Occurrence of intersex due to pollution caused by sewage outfall	Moore and Stevenson (1991, 1994)
Mysidacea		
Acanthomysis mitsukurii	Pollutants like nonylphenol and bisphenol-A are the possible cause for the occurrence of intersex	Yamashita et al. (2001)

Continued

TABLE 3.3 Occurrence of Different Types of Intersexuality in Crustaceans—cont'd

1. Normal Intersexuals		
Species	**Mechanism**	**References**
Harpacticoid copepods	Polluting chemicals can be the cause for the presence of intersex	Moore and Stevenson (1991)
Neomycin integer	Mechanism not given	Hough et al. (1992)
Amphipod		
Gammarus deubeni	Prevalence of a parasite called *Nosema granulosis*, a feminizing microsporidian parasite resulted in intersexuals among amphipods	Kelly et al. (2004), Dunn et al. (1993)
	Influence of both hormonal and parasitic factors	Bulnheim (1977)
	Parasite-induced sex reversal mechanisms	Rodgers-Gray et al. (2004)
Echinogammarus marinus	Endocrine-disrupting pollutants cause intersexuality by interfering with the function of the androgenic gland	Ford et al. (2004)
Gammarus chevreuxi	Mechanism not given	Sexton (1924)
Orchestia gammarellus	Action of a feminizing epigenetic factor which is temperature sensitive	Ginsburger-Vogel (1975)
Orchestia aestuarensis	Presence of intersex due to the parasite *Paramarteilia orchestiae*	Gingsburger-Vogel (1991)
Cherax destructor	Occurrence of intersex due to the presence of internal parasites or symbionts such as the microsporidian *Thelhonia*	Mills and McCloud (1983)

TABLE 3.3 Occurrence of Different Types of Intersexuality in Crustaceans—cont'd

1. Normal Intersexuals		
Species	**Mechanism**	**References**
Gammarus minus	Specific environmental factors, such as salinity and temperature may indirectly affect the incidence of intersexuality in amphipods	Kelly et al. (2002)
Echinogammarus marinus, Geothelphusa dehaani, Orchestia aestuarensis	Industrial contamination of water caused the infection by microsporidic parasites which might be the reason for the occurrence of intersexuality	Ford et al. (2006), Ford et al. (2004), Takahashi et al. (2000)
Cherax quadricarinatus	Insulin-like gene Cq-IAG silencing induced dramatic sexual alterations which disrupted the androgenic hormones thus resulting in intersexuals	Rosen et al. (2010)
Isopoda		
Armadillidium vulgare	Intersexuality due to the presence of *Wolbachia*	Rigaud and Moreau (2004)
	Androgenic hormone (ACH) mediation	Azzauna et al. (2004)
Mesidosia sibirica	Mechanism not given	Korczynski (1988)
Armadillidium vulgare	Temperature induces the intersexuality	Juchault et al. (1980)
Asellus communis	Mechanism not given	Smith (1967)
Idotea balthica	Influence of temperature difference	Mocquard et al. (1978)
Mysida		
Siriella japonica	Infestation of parasites results in intersexuals	Ohtsuka et al. (2003)

Continued

TABLE 3.3 Occurrence of Different Types of Intersexuality in Crustaceans—cont'd

1. Normal Intersexuals		
Species	**Mechanism**	**References**
Branchiopoda		
Daphnia magna	Intersex individuals can arise when maternal daphnids are exposed to marginal concentrations of methyl farnesoate	LeBlanc (2007)
Eulimnadia texana	Intersexes are likely the products of low frequencies of crossing over between the sex-determining chromosomes	Stephen et al. (2006)
Daphnia magna	Longer exposure to high temperature	Mitchell (2001)

3.8.3 Decapods

Among the dioecious decapods, intersexuality has been reported in several crayfish species (Table 3.3). The red claw crayfish *Cherax quadricarinatus* is genetically female, but morphologically and functionally male (Parnes et al., 2003). The intersexual individuals have both male and female genital openings, a testis and sperm duct with attached androgenic gland, and an ovary containing oocytes, arrested at previtellogenic stage. Sagi et al. (2002) found no vitellogenin gene expression in the hepatopancreas of the intersexuals. However, these intersexuals possess externally visible oviductal opening. When both male and female openings are present on one side, an ovary is absent on that side. Conversely, an ovary with an oviduct is found in the individuals, in which a female opening is present in the absence of a male opening on the same side. When placed with receptive females, intersex individuals are able to mate and fertilize the eggs, producing viable progeny. Unlike true hermaphroditism, intersex crayfish has a stable intersexual status, rather than a transient state, as no sex changes were reported in the laboratory-maintained intersexuals (Sagi et al., 1996). The behavior of intersex crayfish is also different from males in that the duration of fights of intersex individuals were intermediate between those of males and females in contests against males (Barki et al., 2006). However, on androgenic gland ablation, these intersex individuals did not exhibit male mating behavior. Furthermore, experimental removal of androgenic gland from an early stage juvenile intersex resulted in de-masculinization of the intersex and the appearance of female secondary and primary sex characters.

3.9 GYNANDROMORPHISM IN CRUSTACEA

Gynandromorphism is an abnormal reproductive condition in which both female and male characteristics are displayed in one and the same individual. Gynandromorphs, otherwise called sexual mosaics, are chimeric individuals consisting of genetically male and genetically female tissues (Narita et al., 2010). In each cell of a gynandromorph, the genetic sex (eg, sex chromosome constitution) is consistent with the sexual phenotype. On the contrary, intersexes are genetically uniform (ie, complete male, complete female, or intermediate in every tissue). Gynandromorphs are of rare occurrence among crustaceans but have been reported sporadically among several genera such as Decapoda, Anostraca, and Cladocera. The rarity of gynandromorphism in Crustacea could be understood from the record of Hartnoll (1960), who found only one gynandromorphic form of the spider crab, *Hyas coarctatus*, out of 2500 specimens examined. In addition, Farmer (1972) identified only one gynandromorphic lobster, *Nephrops norvegicus* among 40,000 specimens examined. In all these forms, division of gonads into testis and ovary was found to be bilateral, with corresponding external secondary sexual characteristics. Both the ovary and testis appear normal and contain mature gametes.

The underlying causes of gynandromorphism are many. In *Drosophila melanogaster*, this sexual mosaic condition is generated by loss or damage of a sex chromosome during zygotic mitosis (Hinton, 1955; Narita et al., 2010). Such genetic origin in the form of loss of a chromosome during early embryonic cleavage has also been suggested for crustaceans. Table 3.4 enlists the occurrence of many bilateral gynandromorphs in different crustacean species. Furthermore, if the initial embryonic cleavage were determinate, a single abnormal division involving sex chromosomes would be sufficient to cause gynandromorphism (Charniaux-Cotton, 1975). In crustaceans, in spite of the genetically based sex determination, the sex differentiation is deeply affected by sex hormone, like androgenic hormone (see chapter: Sex Differentiation). In addition, some crustaceans are subject to ESD. Imperfect influence of sex-determining hormones such as androgenic gland hormone could also result in gynandromorphism. Similarly, infection of feminizing endosymbionts like *Wolbachia* as well as the additional effects of temperature also causes gynandromorphism, especially in several lower crustacean orders.

3.9.1 Gynandromorphism in Decapods

Among decapod crustaceans, an early report of Veillet (1945) described bilateral mosaic in the swimming crab *Carcinus maenas*, with a testis on one side and a mixed gonad on the other. However, Johnson and Otto (1981) gave a full description of gynandromorphism in the blue crab *Callinectes sapidus*, in which a perfect bilateral division into male and the female sides (mixed gonad) exists. Externally, sexually dimorphic male characters were present on the right side, whereas the female characters occurred on the left side. The male half of the

TABLE 3.4 Occurrence of Gynandromorphism in Crustaceans

Subphylum (Class) Order Family	Species	References
Malacostraca		
Decapoda		
Parastacidae	*Samastacus spinifrons*	Rudolph (1995)
Potamidae	*Geothelphusa dehaani*	Araki and Matsuura (1995)
	Potamon fluviatile	Micheli (1991)
Brachyura	*Callinectes sapidus*	Johnson and Otto (1981)
Brachyura	*Hyas coarctatus*	Hartnoll (1960)
Penaeiodea	*Metapenaeus monoceros*	George (1963)
Nephropidae	*Nephrops norvegicus*	Farmer (1972)
Nephropidae	*Homarus gammarus, Homarus americanus*	Farmer (1972)
Nephropidae	*Palinurus frontalis*	Farmer (1972)
Nephropidae	*Palaemonetes varians*	Charniaux-Cotton and Cazes (1979)
Palaemonidae	*Macrobrachium rosenbergii*	Nagamine et al. (1980)
Isopoda		
	Armadillidium vulgare	Juchault et al. (1991)
Branchiopoda		
Anostraca		
Artemiidae	*Artemia franciscana*	Campos-Ramos et al. (2006)
	Artemia parthenogenetica	Campos-Ramos et al. (2006)
Chirocephalidae	*Artemia salina*	Bowen and Hanson (1962)
	Branchinecta lindahli	Sassaman and Fugate (1997)

TABLE 3.4 Occurrence of Gynandromorphism in Crustaceans—cont'd

Subphylum (Class) Order Family	Species	References
Cladocerans		
Daphniidae	*Daphnia pulex*	Kuttner (1909)
	Daphnia schaefferi	De La vaulx (1919)
	Daphnia atkinsoni	Banta et al. (1939)
	Daphnia longispina	Banta et al. (1939)
	Sinocephalus expinosus	Banta et al. (1939)
	Daphnia longispina	Bykova and Markevich (1979)
	Daphnia magna	Olmstead and LeBlanc (2007) and Tatarazako et al. (2003)

abdomen was narrow, but the female half was broad, with a convex lateral margin. The gonad of *C. sapidus*, as in other brachyuran crabs, is "H"-shaped with an anterior and a posterior lobe on each side and a central crossbridge. The male side was morphologically and histologically normal. The testicular lobules on the male side contained primary and secondary spermatocytes, spermatids, and developing sperm. The anterior vas deferens contained fully encapsulated spermatophores and the mid and distal vas deferens contained seminal secretions. However, on the female side, the gonad is of mixed nature; the anterior portion of the gonadal arm contained seminiferous tubules consisting of primary spermatocytes, spermatids, and developing sperm, just like the left arm. Anterior to the crossbridge, a vas deferens is also present. Only the distal part of the anterior lobe of the gonad contained vitellogenic oocytes and a few gonia. Many oocytes were found to be in degenerating condition. The apex of the posterior lobe of the gonadal arm contained only previtellogenic oocytes. In general, the vitellogenic oocytes in the female side of the gonad were fewer than those in the normal crab ovary. The seminal receptacle appeared to be like that of a mature female and contained remnants of a sperm plug and a few sperm, suggesting recent mating with a male. The reproductive morphology of the blue crab gynandromorph clearly suggests that it has performed mating both as a male and female, albeit with less success in sperm transmission and storage. Apparently, gynandromorph is less fecund, due to the presence of limited number of vitellogenic oocytes in the ovarian part of the gonad.

An interesting feature of gynandromorphism in the American lobster, *Homarus americanus* is that the division into the male and female sides is matched by a bilateral color difference (Farmer, 1972). The male left side was mottled and spotted with dark greenish brown on a ground color of orange, while the female side was similarly mottled and spotted but in shades of blue over a lighter blue ground color. Despite a perfect bilateral gynandromorphism along with bilateral color variations in the lobsters, no clear cut explanation has been offered toward the underlying factors controlling this sexual mosaic. Farmer (1972) suggested that initial embryonic cleavage in these species may be determinate, and a single abnormal division involving the sex chromosomes in the first embryonic cleavage could result in this perfect bilateral asymmetry.

3.9.2 Branchiopoda: Anostracans

Gynandromorphism in Anostraca is generally thought to arise from mitotic aberrations occurring during early embryogenesis that create somatic-cell mosaics (Sassaman and Fugate, 1997). Bowen and Hanson (1962) explained the first perfect gynandromorphism in *A. salina*, caused by chromosomal loss in one of the two daughter cells during the first cleavage of a heterogametic (female) zygote. Consequently, each cell inherently expressed either as a male or female phenotype, irrespective of positional or hormonal influences (Bowen et al., 1996). Additionally, in a fairy shrimp, *Branchinecta*, influence of epigenetic factors on segmental gender differentiation during larval development has been proposed (Sassaman and Fugate, 1997). The existence of sexual alterations in appendages of some of the *Artemia* mosaics is correlated to chromosomal instability and/or high rates of mitotic recombination. High mutational rates in these mosaics are explained by their exposure to high doses of X-rays or from lines that were inbred on a rare eye-color mutation (Bowen et al., 1966). Obviously, the change of phenotypic expression in these *Artemia* species is due to the altered cellular genotype. The genetic or developmental influence on the sexually dimorphic characters in gynandromorphs of anostracans is in contrast to the peracarids and decapods, in which the androgenic gland hormone plays pivotal role in the sexual inversion of both primary and secondary sex characters (Charniaux-Cotton, 1965).

In a gonochoristic diploid population of *A. salina*, Bowen et al. (1996) identified 11 sex mosaics, among which three were perfect bilateral gynandromorphs. Sex mosaics were internally female or male; externally they had structures differentiated into female and male as spatial sex mosaics. Such mixed conditions could be attributable to mixtures of cells with male or female genotypes in the epidermis. Interestingly, in each of the 11 sex mosaics, the internal organs were male or female, rather than intersexual in character. However, externally they had structures (such as the antennae and genitalia) differentiated into spatial sex mosaics. Correspondingly, each cell in the mosaics is either male or female in phenotype. In anostracans, sex is determined autonomously and hence, any sex hormone, if present, may not have influence over the differentiation of cells with a chromosome constitution of the opposite sex.

Most of the descriptions on gynandromorphism are made on anostracans that reproduce sexually as zygogenetic forms (Campos-Ramos et al., 2006). In addition to wild populations, laboratory strains kept under UV irradiation also produced gynandromorphs. However, Bowen and Sterling (1978) described gynandromorphism in a parthenogenetic diploid brine shrimp, *Artemia parthenogenetica*, which reproduce asexually, with expected progeny composed of parthenogenetic females. The gynandromorphic specimens were bright red and had long antennules, abdomen, and cercopods, characters that distinguish *A. parthenogenetica* from females of sexually reproducing *Artemia franciscana*. The occurrence of parthenogenetic gynandromorphs is best explained in the light of sex determination mechanisms, the high recombination rate of sex chromosomes, the origin of rare nonrecombinant ZZ males, and the restoration of diploidy (Stefani, 1967). Further, a chromosomal loss during embryonic development gives rise to a WZ-OZ spatial mosaic gynandromorphy. The unique appearance of a long, mobile penis-like structure in a parthenogenetic gynandromorph may be due to chromosomal mutation, combined with aberrant embryonic development.

The freshwater fairy shrimp belonging to the genus *Branchinecta* also showed gynandromorphism (Belk, 1978). Sassaman and Fugate (1997) obtained gynandromorphs from individuals of *Branchinecta lindahli*, reared from natural sediment samples as well as from a series of laboratory crosses of adults. Interestingly, *Branchinecta* gynandromorphs differed strikingly from the pattern of bilateral dimorphism reported for other genera (Artemiidae and Chirocephalidae) in that the characteristics of different gender are expressed along the length of the body, with the transition from one gender to the other occurring at the cervical groove. Virtually, all of them have a male anterior region, but with female genital morphology. This morphological pattern is referred to as axial, in contrast to bilateral gynandromorphous *Artemia* species. The more common forms, with male heads and female trunks, could reproduce as functional females, as inferred from the observation that shelled eggs were found inside the ovisac, implying successful mating and fertilization (Belk, 1978). However, the gynandromorphs of another species, *Branchinecta packardi* that have female heads but male genitalia are nonreproductive, irrespective of their gonadal maturity. They are physically precluded from mating by the absence of the clasping organs necessary to accomplish pair formation.

Furthermore, Sassaman and Fugate (1997) found sequential expression of gender on successive limbs due to epigenetic factors influencing directly on segmental differentiation, resulting in adults carrying an axial array of gender differences. Epigenetic factors influencing morphogenesis and tissue differentiation could be environmental factors as well as nutritional status of the larvae. Since all *Branchinecta* gynanders have male heads and female ovisacs, transformation would be largely limited to feminizing influences acting on the male phenotype.

Sassaman and Fugate (1997) performed investigation on 35 gynandromorphic individuals of *Branchinecta lindahli* with reference to morphology, the frequency of occurrence, and their aggregation with the sibships. Among the three species of *Branchinecta* (*B. packardi*, *B. mackini*, and *B. lindahli*), the first one exhibited bilaterally female second antennae, but exhibited normal male morphology throughout the thoracic and the genital regions. The second individual exhibited bilaterally symmetric male second antennae, but all the thoracic and the genital segments expressed female morphology. In the third species, four samples were collected out of which the first two of them exhibited male second antennae and female thoracic and genital regions and the other one was entirely male except for the development of a partial ovisac on one side. The last sample had a normal male head but with female gonads and genital segments.

Gynandromorphism in anostracans is generally thought to arise from mitotic aberrations occurring during early embryogenesis that create somatic cell mosaics. It can also arise from segregational anomalies of chromosomes during embryonic development. The qualitative and quantitative support to this hypothesis reveals that gynandromorphism in *Branchinecta* results from incomplete transformation, by a cytoplasmically transmitted factor of one sex to the other during larval development.

3.9.3 Methyl Farnesoate Control of Gynandromorphism

In the cladoceran, *Daphnia magna*, Olmstead and LeBlanc (2007) found that the sex-determining hormone methyl farnesoate could have impact on gynandromorphic changes. Several studies have indicated that in branchiopod crustaceans, methyl farnesoate is a sex determinant (Olmstead and LeBlanc, 2007; Tatarazako et al., 2003). In *D. magna*, high methyl farnesoate levels program mitotic eggs resulting from parthenogenesis to develop into male offspring while in the absence of methyl farnesoate these eggs develop into females. In this cladoceran, gynandromorphism is produced when methyl farnesoate levels are intermediate between low levels that stimulate the production of broods containing all female offspring and high levels that stimulate the production of broods of all male offspring. However, incidence of methyl farnesoate-induced gynandromorphism was as low as 0.14%. The percentage of incidence increased significantly, when the animals were hormone-treated at 30°C, suggesting that higher temperature has a combined effect with methyl farnesoate in producing gynandromorphs in *Daphnia*. In addition, environmental chemicals such as the insecticide pyriproxyfen can stimulate the gynandromorphic phenotype via perturbations in methyl farnesoate signaling (Olmstead and LeBlanc, 2007).

3.10 PARTHENOGENESIS

Parthenogenesis is a form of asexual reproduction in which the development of an individual from an egg takes place without fertilization. This type of asexual reproduction is not prevalent in Crustacea; their occurrence is limited to the lower crustacean taxa such as branchiopods. Parthenogenesis, as a part of the life cycle has been well recognized in two forms, the water flea *Daphnia* and the brine shrimp *Artemia*. Interestingly, parthenogenesis, as an obligatory reproductive process, has been reported in a freshwater crayfish. Parthenogenetic development takes place in *Daphnia* from the eggs that do not undergo meiosis and fertilization. This type of parthenogenesis is termed as apomixes with all offsprings genetically identical to the parent. During meiosis I, the two half bivalents that have segregated to opposite poles in the egg, move back to the equatorial plate, thereby preventing the cytokinesis and skipping the first meiotic division. The second meiotic division restores the normal diploid condition necessary for the development of embryo without fertilization. Similarly, in the brine shrimp *A. salina*, meiosis is interrupted at some stage between the end of the first anaphase and the second metaphase. Then the two haploid plates reunite forming a new metaphase spindle and the second diploid meiosis is accomplished (Stefani R., 1960). Nevertheless, parthenogenetic offspring shows genetic variability because chromosomal recombination can take place between homologous chromosomes during the abortive meiosis. Parthenogenesis in *Daphnia* has been thought to have evolved from sexual reproduction by changing the meiotic program by way of skipping a division. This is supported by the finding that some of the meiosis-related and meiosis-specific genes are expressed differentially in sexual reproduction and parthenogenesis of *Daphnia pulex* (Schurko et al., 2009).

3.10.1 Parthenogenesis in *Artemia salina*

A. salina is a primitive group of branchiopod crustacean known to reproduce parthenogenetically, whereas other species are obligatory parthenogens, with alternation of sexual reproduction with parthenogenesis, like the water flea *Daphnia*. The obligatory type of parthenogenesis takes place when the environmental conditions such as crowding and high food availability are favorable for sexual reproduction. But, on the other hand, the environmental conditions which favor parthenogenetic type of reproduction include poor population density and impoverished food condition in the medium along with other favorable abiotic factors.

Barata et al. (1996) conducted studies to identify the demographic parameters for sexual and parthenogenetic species of *Artemia*. Samples of *Artemia tunisiana*, which reproduces sexually and samples of *A. parthenogenetica*, which reproduces parthenogenetically were used for laboratory rearing at three different temperatures (15°C, 24°C, and 30°C). It was found that there is a significant reduction in life expectancy as the temperature increases for all strains, with the sexual species' lifespan significantly smaller at 24°C and 30°C compared to parthenogenetic strains of *A. parthenogenetica*.

3.10.2 Parthenogenesis in *Daphnia* Species

The life cycle of *Daphnia* during the growth season is characterized by its asexual mode of reproduction (apomixis). A female produces a clutch of parthenogenetic (amictic) eggs, which are placed in the brood chamber where the development is completed. In addition to producing the diploid eggs which undergoes development without a resting phase, in *Daphnia* a different type of egg is produced for resting. These resting eggs are encapsulated in a protective, saddle-like structure called ephippium. Diploid resting eggs are also produced by obligate parthenogenetic *Daphnia* such as *D. pulex* which are excellent examples for geographic parthenogenesis, as these forms are found in the northern parts of America and Eurasia. As shown in the figure, the common sexual *Daphnia* produces resting eggs followed by the asexual production of diploid males which are needed to fertilize the haploid eggs. The male production is under environmental control.

During the parthenogenetic cycle, females produce diploid eggs that develop directly into females. The same female may produce diploid asexual eggs that develop into males. Male production is under environmental control. Furthermore, the same female may produce haploid eggs that require fertilization by males. These eggs are then enclosed in a protective shell (ephippia) and need to undergo diapauses before female offspring will hatch from them.

3.10.3 Facultative Parthenogenesis in the Crayfish *Orconectes limosus*

Asexual reproduction by parthenogenesis has rarely been reported in decapod crustaceans. Recently, facultative parthenogenesis has been reported in the spiny-cheek crayfish *Orconectes limosus* (Buřič et al., 2011). All European cambarid crayfishes, including *O. limosus* reproduces sexually, mate twice a year, in autumn and in spring, although they lay eggs only in spring. However, females, when isolated physically from males for 10 months, a vast majority of them successfully spawned and carried clutches of apparently viable offspring. Unlike other cambarid crayfish, in *O. limosus* long-term storage of sperm has not been reported. Therefore, the only possible mechanism of self-fertilization is by apomictic parthenogenesis, in which the first meiotic division is suppressed. Again, as expected, all the progeny resulting from this parthenogenetic development carried maternal alleles, and hence are all identical to their mothers. Genotyping of seven microsatellite loci clearly pointed to apomictic parthenogenesis in these isolated females. These results strongly suggest that *O. limosus* females are capable of facultative parthenogenesis under certain conditions. This type of asexual reproduction was also found in two other species of crayfishes, *Procambarus fallax* and *Procambarus clarkii*. Self-cloning in the freshwater crayfish *P. clarkii*, via parthenogenesis could be induced by specific environmental factors such as *Wolbachia* infection (Yue et al., 2008).

Understandably, parthenogenetic reproduction in *O. limosus* may be triggered under social circumstances, when males are unavailable for mating.

However, the factors inducing switching between these two alternative reproductive strategies are not known. It has been suggested that intrasexual chemical communication between numerous females in a confined space, indicating "crowding" of this sex could contribute to the switch to parthenogenesis in this freshwater crayfish. Adequate abiotic conditions such as light and temperature regimes corresponding to the mating season may be necessary for triggering parthenogenesis in both natural and artificial conditions.

3.10.4 Parthenogenesis in Marbled Crayfish

Another cambarid freshwater crayfish, called marbled crayfish is reported to reproduce parthenogenetically (Martin et al., 2007). The conclusion that the marbled crayfish propagates apomictically is based on the finding that 19 selected individuals of various generations of this crayfish showed identical allelic composition. In addition, histological studies of the ovaries also indicated the absence of meiosis. The absence of any genetic differences over a number of generations suggests that no recombination had occurred during oogenesis.

In summarizing the foregoing discussion on sexual systems, it becomes evident that crustaceans provide a good model system to study the relationship between sexual patterns, reproductive strategy, and the environmental adaptation. Although crustaceans are generally gonochoristic, several alternative sexual systems are found among different crustacean taxa. Assumption of different sexual systems reflects on the control mechanisms of sex differentiation. Unlike insects, the genetic factors involved in sex determination are greatly influenced by epigenetic factors such as the androgenic gland hormone. This hormone is responsible for the differentiation and maintenance of male sex characters. Any developmental disturbance in the activity of this sex hormone could bring about an array of sexual patterns. As a consequence, a variety of hermaphroditic conditions, ranging from sequential to simultaneous hermaphroditism has been realized among decapods. The increased frequency of hermaphroditism among certain groups such as the caridean shrimps point to their adaptive ability to the precarious environmental conditions to achieve greater success in their reproduction. Even more reproductive patterns and strategies are found among lower crustacean orders such as the branchiopods. ESD and the absence of androgenic gland to control sex differentiation in the nonmalacostracan crustaceans have resulted in peculiar sexual patterns such as androdioecy and parthenogenesis, in addition to the production of sexual anomalies like intersexuals and gynandromorphs. In the absence of androgenic gland, other hormonal factors such as methyl farnesoate play a role in the control of sex differentiation in forms such as the water flea, *Daphnia*. In Crustacea, there exists a strong interlink between sexual system and the formation of mating systems, leading to the evolution of several social systems in these aquatic arthropods.

Chapter 4

Mating Systems

4.1 INTRODUCTION

Mating system is defined as the general pattern by which males and females mate. In many malacostracans, mating systems involve dramatic premating behavior on the part of the male, such as courtship signaling, fighting among males, or defense of females before mating. In Crustacea, mating systems as well as factors that determine them are diverse. Life history patterns, ecological influence, functional morphology of copulation and insemination, and social determinants of both male and female behaviors are some key criteria employed in categorizing the mating systems found in Crustacea. Additional factors such as sperm competition and pre- and postcopulatory mate choice are also considered in shaping mating strategies. Ecological factors apparently contribute to the evolution of diversified mating systems as well as the attributed behavioral and morphological differences between the sexes. For example, crustaceans such as isopods, amphipods, shrimps, and crabs that have developed complex symbiotic relationships with other marine invertebrates like sponges, sea anemones, and corals have featured a wide array of social structures along with wide-ranging mating systems. Mating systems therefore refer to the procedure used in finding and securing a mate (usually a female), the number of mates an individual acquires, the type of pair bonds, and the nature of parental care (Subramoniam, 2013).

Sex attraction and sex recognition before mating also form an important component in the evolution of different mating systems in Crustacea. In addition, gamete morphology has a bearing on the establishment of different mating systems. For instance, the spermatozoa of malacostracan crustaceans are atypical and invariably nonmotile, and hence have to be delivered to the site of fertilization in the female body. This is in stark contrast to the condition found in many other marine invertebrates such as echinoderms, in which male and female gametes are simply released into the sea water to effect broadcast fertilization, without requiring any male–female contact. In addition, except the free-spawning dendrobranchiate penaeiodean shrimps, all other crustaceans incubate their yolky eggs in a brood chamber or attached to pleopodal hairs up until advanced larval development. Different crustaceans, therefore, resort to either internal or epizoic fertilization by delivering the sperm cells, packaged in special structures called spermatophores. The origin and occurrence of several mating systems correlate with environmental conditions as well as the

Sexual Biology and Reproduction in Crustaceans. http://dx.doi.org/10.1016/B978-0-12-809337-5.00004-6

lifestyle, adopted to live and reproduce in the diversified environmental niches. Similarly, crustaceans exhibit a wide array of mating behaviors, which have evolved in response to special prerequisites for protecting the females, which in many decapods and peracarides, undergo an obligatory molting before mating and spawning. Thus, the precopulatory mate guarding by the males is prevalent in aquatic brachyuran decapods and marine isopods, and in many instances, the mate guarding has extended beyond copulation; with the result, the egg-brooding females continue to receive guarding from the males.

In decapod crustaceans, mating strategy of males also depends on the timing and duration of female receptivity. The latter will influence the amount of time a male invests in staying with the receptive female by defending or guarding her, or in searching for other receptive females so as to achieve maximum reproductive success (Parker, 1970). Furthermore, restricted female receptivity caused a male-biased operational sex ratio, increasing male–male competition and the potential for sexual selection. According to Emlen and Oring (1977), ecological and behavioral potential to monopolize mates is a major element of the mating strategy of a species. Thus, habitat difference as well as behaviors needed to survive in them has profound influence on shaping the different mating systems. For example, mate guarding is present in the intertidal grapsid crab species, *Hemigrapsus sexdentatus*, whereas in another species, *Cyclograpsus lavauxi*, it is absent. Although these two crabs have similar duration of female receptivity, habitat differences have caused differences in their mating strategies. Interestingly grapsid crabs mate during intermolt stage (Anilkumar et al., 1999).

The important factors or conditions responsible for shaping various mating systems are enumerated below.

4.2 FACTORS INFLUENCING MATING SYSTEM

4.2.1 Anisogamy

Anisogamy, the bimodal distribution of gametic sizes (smaller in males, larger in females), is a characteristic feature of all sexually reproducing animals. It implies that males have an inherent capacity to produce vast numbers of small and energetically cheap gametes, whereas females can produce far fewer but energetically more expensive eggs. As a consequence, males have more reproductive potentials than the females in terms of producing more offspring. However, the female reproductive success is maximized by the choice of mates that confers material or genetic benefits, whereas male reproductive success is maximized by mating with as many females as possible (Clutton-Brock and Parker, 1992).

The evolutionary effects of anisogamy on mating systems include higher fecundity potential in males than in females, behavioral tendencies in males to seek multiple mates with greater inclination toward polygyny, greater investment by females in postzygotic care of progeny, greater competition for females among males than among females, and the elaboration of secondary sexual traits in males than in females. In Crustacea, the above characteristics influence markedly the formation of different mating systems as well as mating associations.

4.2.2 Availability of Breeding Females

A significant condition that influences the form of mating system is the availability of breeding females in a population. In general, female crustaceans produce fewer eggs and hence become the limiting resource in reproduction. On the other hand, males produce numerous sperm to be available for reproduction throughout their reproductive life. Hence, the ratio of sexually receptive females to sexually active males (operational sex ratio) is crucial in the structuring of the mating system in a given crustacean population.

4.2.3 Female Receptivity

Female receptivity is an important condition to determine the mating systems, contributing to the availability of active females in the mating process. In Brachyura, representing true crabs, female receptivity is restricted to a brief period of fresh molt, during which the vulval operculum become softened for intromission (eg, Cancridae, Portunidae) or during an intermolt period (hard-shelled condition), when a temporary decalcification occurs in the vulval operculum (eg, Grapsidae and Ocypodidae). In other species with mobile opercula (some Grapsidae and Ocypodidae) as well as those species with permanent soft vulval openings (eg, Majidae), there is no restriction for female receptivity. The latter species, in which the females can anatomically mate is referred to as continuously receptive, although their attractiveness to males depends on their breeding status (Diesel, 1991). Such variation in female receptivity could affect pre- and postcopulatory mate behavior of males, common among brachyuran crabs.

The American lobster, *Homarus americanus*, generally mate when they are newly molted, but Waddy and Aiken (1991) reported that mating could occur at any time during intermolt. Therefore, female receptivity in *H. americanus* is related to the ovarian condition, rather than the molt stage. In addition, a female loses receptivity, once inseminated, and males rarely attempt to mate with a female that is carrying stored spermatophores. However, receptivity is restored once the sperm supply is depleted. Therefore, mature uninseminated females often become more active as the time of spawning approaches and look for males for mating.

4.3 INFLUENCE OF MOLTING ON MATING SYSTEM

In many malacostracan crustaceans, molting continues into the reproductive adulthood, playing a pivotal role in timing the transfer of sperm/spermatophore via mating to the females. Hence, in females of most crustacean species, mating and fertilization are closely associated with female molting. The females molt first and mating occurs thereafter, followed by the release of egg, which will be fertilized by the sperm contained in the spermatophore. Importantly, females are receptive for several minutes to a few days after molting and hence mating must take place during this time period, and the rest of the eggs will be resorbed in the ovary. However, other species maintain the ability to mate throughout most of their molt cycle.

Hartnoll (1969, 2000) distinguished two types of mating in decapod crustaceans: "soft-female mating" and "hard-female mating." Soft-female mating was defined as copulation occurring immediately after molting of the females, usually preceded by a lengthy premolt courtship, including precopulatory guarding by the male. In the portunid crab, *Callinectes sapidus*, after a prolonged mate guarding of the premolt female, copulation occurs immediately following her pubertal molt (Gleeson, 1991). With this last molt, the female enters terminal anecdysis after reaching maturity for the first time. The female blue crab incidentally mates only once during her lifetime, as the one-time mating provisions adequate sperm, stored in its seminal receptacle, for fertilizing eggs from subsequent spawning.

Hard-female mating was defined as mating, in which the female copulates during intermolt stage after a relatively brief courtship. In species inhabiting terrestrial and semiterrestrial habitats, mature females copulate in the hard-shelled condition. Examples are the land hermit crabs of the genus *Coenobita* and the brachyuran crabs of the Grapsoidea and Gecarcinidae. Briefly, the observations of Hartnoll on brachyuran crabs led to the general preposition that a lengthy premolt courtship behavior is associated with soft-female mating, whereas a relatively brief courtship behavior happens with hard-female mating.

The situation in penaeid shrimps is somewhat different, the molting condition of mating female is determined according to the type of thelycum it possesses. In the open thelycum type, the females copulate in the hard-shell condition, whereas in the closed thelycum type, the female molts just before mating. In spite of this difference in the molting condition of these two types of penaeid shrimps, there is no difference in the mating behavior between the open- and closed-thelycum species, the copulation occurring after a brief interaction between a male and a female (Asakura, 2009). Nevertheless, several other malacostracans, including the peracarid species, such as isopods and amphipods follow molt-mate-spawning sequence during the reproductive cycle. It is tempting to suggest that the evolutionary origin of crustacean mating systems is interrelated to the occurrence of molting in the reproductive adults, almost alternating with the female reproductive cycle. In the continuously reproducing intertidal mole crab, *Emerita asiatica*, molting cycle precisely alternates with the reproductive cycle, yielding high fecundity and faster body growth (Gunamalai and Subramoniam, 2002).

Furthermore, continued molting in the adult female is also an intervening factor for reproductive activities, resulting in the relegation of both mating and egg-laying to postmolt and intermolt stages. Understandably, mating in the postmolt condition is a physiological necessity, as the female gonopore will be accessible to intromittent organs only in the soft-shelled condition. Since molting females are vulnerable to predation, they need physical protection, which is offered by the males that seek fresh-molt females for mating. As a result, a new behavioral trait, namely precopulatory mate guarding by the males has evolved in many crustacean taxa, in which molting precedes mating. Since the spawned eggs are attached to the pleopodal hairs in the ventral sternum to allow brood development, in many cases, precopulatory mate guarding extended to postcopulatory guarding of the brooding females. Thus mate guarding has emerged as the central behavioral component in crustacean reproduction.

4.4 LIFE HISTORY VARIATION AND MATING SYSTEM

Life history variation involving burrow-living and symbiosis with other marine invertebrates has given rise to pair-bonding monogamous mating system in several caridean shrimps. This system could further evolve into social mating systems, in which both males and females protect the habitat to overcome predation in a crowded environment. Several caridean shrimps have adapted to live in specialized bioshelters such as living sponges, providing the symbiotic shrimps both protection and food. Specialization of mating associations, as necessitated from this physiological and ecological adaptiveness has also resulted in advanced social formation of community living as well as eusocialization, primarily in aquatic environment (see below).

4.5 TYPES OF MATING SYSTEMS

From an evolutionary perspective, the type of mating behavior in a species is a product of sexual selection for increasing the reproductive success. As for the crustaceans, sexual selection, which arises from intraspecific competition for mates, takes the form of sexual phenotypes such as the large claws in gonochoristic forms (eg, fiddler crab *Uca* sp.), whereas in the hermaphroditic taxa, it is evidenced as elaborate courtship and copulatory behaviors. Thus, the assumption of a particular mating system in a population depends mainly on the reproductive potential of a particular male, facilitated by its access to females as well as the female's choice of the opposite sex. With reference to reproductive or fertilization success, mating systems are classified into social and genetic mating systems (Avise, 2007). Social mating system includes pair-bonding monogamous mating, whereas genetic mating system is more prevalent among polygamous males, which achieves more reproductive success than the social mating system. Nevertheless, different types of mating systems have evolved independently in different taxa in response to environmental conditions to which the animals have adapted to live and reproduce.

The basic mating systems in Crustacea are monogamy and polygamy; Table 4.1 summarizes various subdivisions under the two main categories—monogamy and polygamy.

TABLE 4.1 Mating Systems in Crustaceans

Type of Mating System	Definition	Species	References
1. Monogamy	Each male and female has only one mate	*Stenopus hispidus*	Johnson (1969)
1.a. Social monogamy	Social monogamy refers to the exclusive mating of one male with one female. This mating bonding is usually sustained through one mating season or could extend throughout the adult life of the individuals	*Pontonia margarita* (Caridea)	Baeza and Thiel (2007)
1.b. Persistent pairs	Males seek out, remain with and provide parental care for isolated, synchronously receptive females	*Spongicola levigata* (Stenopodidae)	Hayashi and Ogawa (1987)
1.c. Sequential pairs	After mating with one female, the male leaves the cavity after the female spawns and seeks out for subsequent females in a serial manner	*Gonadactylus bredini* (Stomatopoda)	Shuster and Caldwell (1989)
1.d. Eumonogamy	This is true monogamy. The male and the female form a cooperative pair to inhabit and guard a burrow together	*Hemilepistus reaumuri* (Isopoda)	Wickler and Seibt (1981)
2. Polygamy	A mating system in which at least some individuals have multiple mates. Four subtypes are recognized under polygamy	*Pandalus* sp. (Caridea)	Charnov (1982)

TABLE 4.1 Mating Systems in Crustaceans—Cont'd

Type of Mating System	Definition	Species	References
2.a. Polygyny	Particular males may have multiple mates, but each female typically has only one mate	*Uca paradussimieri* (Brachyura)	Murai et al. (2002)
2.b. Polyandry	Particular females have multiple mates but each male has only one mate	*Rhyncocinetis typus* (Caridea)	Thiel and Hinojosa (2003)
2.c.Polygynandry	Members of both genders typically have two to several mates each. Promiscuity is an extreme form of polygynandry in which each male and female has many mating partners	*Uca lactea* (Brachyura)	Kim et al. (2004)
3. Eusociality	Only a single reproductive female with eusocial colony organization having strong reproductive skew	*Synalpheus neptunus neptunus* (Alphidae); *Synalpheus regalis* (Alphidae)	Didderen et al. (2006) and Duffy and Thiel (2007)
4. Mass mating Semelparous mass mating	Both males and females are characterized by single reproductive cycles over the course of its lifetime	*Mancocuma stellifera* (Cumacea)	Guewuch and Croker (1973)
Iteroparous mass mating	Both males and females are characterized by multiple reproductive cycles over the course of its lifetime	*Geocarcinoides* (Brachyura)	Seeger (1996)
5. Aggregational mating by neotenous males	Several neotenous males clinging on to the ventral region of the female, depositing spermatophores at the same time	*Emerita asiatica* (Anomura)	Subramoniam (1977a)

4.5.1 Monogamy

Monogamy is defined as the mating system in which an individual reproduces sexually with only one partner of the opposite sex. Much against the common belief that monogamous mating system evolved from the male's involvement in brood care, in the monogamous crustaceans males do not normally participate in brood care. In Crustacea, heterosexual pair formation normally occurs after the males engage in mate guarding, when the female is close to molting or spawning a new batch of unfertilized eggs. However, variations do occur in which mating is not coupled to molting, as in the burrow-living stomatopods.

Wickler and Seibt (1981) distinguished two types of monogamy in Crustacea: (1) sexual monogamy in which an individual reproduces sexually with only one partner of the opposite sex; (2) social monogamy, meaning togetherness of two heterosexual adults as a social system. Thus, in many crustaceans, the male provides little assistance to the female partner except providing sperm to fertilize eggs, but remains with the female beyond insemination to ensure that he will have a mating partner when she again becomes receptive. This type of permanent pairing of monogamous crustaceans is called "mate guarding monogamy." Many shrimps living in symbiotic association with other marine invertebrates live in pairs and both members of the pair may cooperate in defense of their host, a resource necessary for their survival and production of offspring. In other conditions, called "extended monogamy," the male stays with one female and waits for her to molt. The male guards her fiercely from other males, as the female is a limited source for males. However, the male is not faithful; if another nearby female molts, the male immediately abandons its partner and will mate with the other female.

Monogamous mating system in crustaceans is well known in stomatopods, isopods, and amphipods, and a variety of coral reef-associated caridean shrimps and crabs that live in symbiosis with other marine invertebrates such as echinoderms and sponges.

4.5.1.1 Stomatopods

Stomatopods live in some sort of refuge, such as the preexisting cavities in rock or coral rubble, or in self-excavated burrows in sandy flats or mud in the subtidal region. They are strongly adapted for living in burrows, and their burrowing habit has resulted in the evolution of diverse mating systems. Prominent among them is the monogamous pair-living, a characteristic feature of many burrowing crustacean species. In these burrowing decapods, risk of predation or loss of a burrow while searching for a mate may have favored monogamy. However, mating in stomatopods is not coupled to molting. In other words, reproduction and molting in stomatopods are mutually exclusive events (Caldwell, 1991).

Different species belonging to the genus *Lysiosquilla* are burrowing stomatopods found primarily on sandy substrates. For example, *Lysiosquilla maculata* occurs as male–female pairs, occupying U-shaped burrows on sandy substratum (Caldwell, 1988). Once the pair formation is established, the male guards the burrow by defending the cavity against male and female intruders. Males are larger than females, doing all hunting of fishes from the burrow entrance. In this stomatopod, heterosexual pair formation appears to form before or at the time of sexual maturity. The pairs stay together for months and remain in the same burrow for life. During molting and probably brooding, the burrows are sealed with thick plugs, both animals remaining inside. When the females die or are removed from the burrow, the males typically leave the burrow. However, if the male is lost, the female might look for a new mate, often smaller than its previous mate. One of the reasons attributed for the monogamous mating tactics of stomatopods is the fact that there is a high cost involved in the construction of the burrow and also there is a risk involved in moving from one burrow to another. Moreover, *Lysiosquilla* are strongly adapted for living in burrows and their legs are best suited for holding position in a burrow rather than for walking. Thus, risks of predation and/or loss of burrow while periodically searching for a new mate may be the reason for favoring monogamy in this group.

Monogamy is also observed in another stomatopod species *Gonadactylus bredini* (Caldwell, 1988). They occupy cavities found in rocks, coral rubbles, and calcareous algae. The males occupy the cavity and it is only the receptive female that goes in search of the males. The females develop their ovaries and cement glands one to two weeks before the onset of full moon and then the pair formation occurs, followed by mating. The oviposition takes place thereafter at the full moon. In these stomatopods, the males leave their burrows after the female spawns and goes in search of other receptive females. Even if the male locates a female which is ready to mate, she will not allow the male into its cavity. The male, which is normally larger than the female, has to fight its way in and, if it is sufficiently persistent to gain entry, the female usually mates and remains with the male in the cavity.

4.5.1.2 Pair-Forming Parasitic Pea Crabs

Pair-forming and monogamy have been reported in unusual circumstances, where pea crabs are found to be internal parasites/symbionts with echinoderms and marine mollusks. Hamel et al. (1999) described a peculiar parasitic association between a pea crab species, *Pinnotheres halingi* and a holothurian host, *Holothuria scabra*. Infestation of the holothurians occurred at the megalopa stage of the pea crab into the right respiratory tree, where the larvae grow and get encysted within a thick wall of cellular materials derived from the host tissue. Young males of the pea crab were particularly attracted to hosts that sheltered a single large female. Copulation was observed within the female cysts, preceding oviposition. After mating, the males leave the female cyst and form its own cyst in the deeper part of the respiratory tree of the host. After this heterosexual pair formation, no further male enters the host, implying that the pair formation is permanent.

4.5.1.3 Desert Isopod Hemilepistus reaumuri

Isopods are well adapted to live in a wide variety of aquatic and terrestrial habitats. They also exhibit highly variable mating systems. In general, aquatic isopods practice mate guarding in almost all free-living taxa, whereas in terrestrial forms, this mating behavior is abandoned. Nevertheless, the burrow-living desert isopod *Hemilepistus reaumuri* follows a strict monogamy. The sexual as well as social monogamy in this isopod is associated with the guarding of the burrows which are costly to produce in the harsh conditions prevailing in desert. Therefore, the male and the female form a cooperative pair to inhabit and guard a burrow together, against intra- and interspecific competitors (Linsenmair, 2007). The females are semelparous (having a single breeding season in a lifetime), and the offspring cohabit in the burrow with the males participating in the brood care, making the family a strictly closed unit inside the burrow. An extremely variable chemical recognition system exists among them to discern the family members from aliens. In the harsh desert environment, there is always competition for the burrows which are a prerequisite for survival and reproduction for both males and females. Hence, these environmental constraints also make monogamy the best option for the male as well.

4.5.1.4 Caridean Shrimps

Although monogamy is practiced in several crustacean taxa such as stomatopods and isopods, this mating system has reached an advanced level in the caridean shrimps, living in symbiosis with other marine invertebrates such as sponges, sea anemones, and echinoderms. Certain crab species that live in peculiar symbiotic/parasitic relationship with echinoderms have also adopted monogamy (Duffy and Thiel, 2007). Caridean shrimps inhabiting the coral reef niches have evolved different grades of monogamous mating systems. Further evolution of mating systems is found in the snapping shrimps which live in burrows in the shallow intertidal region. In the *Alpheus* species, there is a temporal extension of male mate guarding to encompass the entire female reproductive cycle, thus transforming sexual monogamy into social monogamy (see below).

An additional factor that has contributed to the evolution of this mating system in the snapping shrimp is the territorial cooperation between partners. While social monogamy has become a characteristic lifestyle in *Alpheus*, its occurrence in the closely related genus *Synalpheus* has been considered to be an ancestral condition leading to more complex assemblages including eusociality (Duffy, 1996).

The origin of monogamy in the caridean shrimps is based on the fidelity of both sexes to stay, defend, and share a specific refuge (Bauer, 2004). The invertebrate hosts of these symbiotic shrimps are the sea anemones, crinoids, as well as other microhabitats such as a cavity in corals or burrows in gravel substrata. The pair formation in the host occurs between a male and female in the gonochoristic species and protandric simultaneous hermaphrodites of species belonging to Hippolytidae. Heterosexual pair formation normally occurs in crustaceans after the males engage in mate guarding, when the female is close to molting or spawning a new batch of unfertilized eggs. The mate guarding males may abandon the females soon after the eggs are fertilized. However, in pair-bonding species, males cohabit with females, independent of their reproductive status or of the stage of development of the brooded embryos. Some of the well-documented examples for the stable pairing and the individuals recognizing each other include the banded shrimp *Stenopus hispidus* (Johnson, 1969), the scarlet cleaner shrimp *Lysmata amboinensis* (Fiedler, 1998), and the harlequin shrimp *Hymenocera picta* (Seibt and Wickler, 1979). Monogamous pair formation is also common in several species of snapping shrimps such as *Alpheus angulatus*, *Alpheus heterochelis*, *Alpheus armatus*, and *Alpheus roquensis* (Asakura, 2009). An important contributing factor to the monogamous lifestyle in these caridean shrimps is that both males and females are likely to benefit by sharing the labor of refuse defense, maintenance, and food acquisition. Hence, the mating system of these caridean species is also termed as resource-defense monogamy. Low population densities, dispersed distribution of refuges, and high predation pressures are the factors that limit encounter probabilities between males and extra-pair females, promoting continuous fidelity toward mates (Thiel and Baeza, 2001).

In addition to these symbiotic crustacean species living in special habitats, free-living species such as stenopodid shrimps inhabiting rocky subtidal zones and many alpheid shrimps living in rock crevices or in burrows of their own in mudflats and other soft bottoms are known to practice this mating system (Correa and Thiel, 2003). The monogamous mating systems described above nevertheless exhibit a number of generalized features. The monogamous shrimps are territorial and cooperatively defend their habitats against other conspecifics. In addition, the monogamous pair always chooses larger partners, as the female size is positively correlated with fecundity in shrimps. For females, sharing a host with a large male would result in indirect benefit (good genes) or direct benefits (increased protection against predators or competitors).

4.6 SOCIAL MONOGAMY

Wickler and Seibt (1981) defined social monogamy as the togetherness of two heterosexual adults as a social system, where the mating bonding is usually sustained throughout one mating season or could extend throughout the adult life of the individuals. Social monogamy or the heterosexual pairing behavior has evolved many times in a broad range of animal taxa, including mammals, birds, reptiles, amphibians, fish, and insects, in which shared parental care has a contributory role in the formation of such sexual behavior (Lack, 1968). In crustaceans, social monogamy has evolved primarily in line with their symbiotic living of caridean shrimps with other marine invertebrates such as corals, sponges, and echinoderms. For example, alpheid shrimps, especially belonging to the genus *Synalpheus* have symbiotic lifestyle, with the majority of them practicing social monogamy. Hermaphroditic shrimps, like *Lysmata pedersoni*, living in high-density aggregations, lead a symbiotic life, and practice social monogamy (Bauer, 2007). Furthermore, Baeza and Thiel (2007) advanced a hypothesis that pair-bonding monogamy evolved, when the hosts are small and relatively rare, and also when predation risk is high. Eventually, movements between hosts are highly restricted; resulting in the monopolization of hosts by the heterosexual pair. As spatial restriction could allow only a few adult symbiotic individuals to cohabit in the same host, selection of extended mate guarding behavior also became a selective force toward the advancement of social monogamy in the caridean shrimps. Again, cooperation in territorial defense by the heterosexual pair can be the driving force to the origin of social monogamy (Kleiman, 1977).

4.6.1 Symbiotic Shrimp *Pontonia margarita*

Pontonia margarita, a symbiotic shrimp, dwells as heterosexual pairs in the mantle cavity of the pearl oyster *Pinctada mazatlanica* (Baeza, 2008). Since the males inhabit host individuals with a single female for a long period of time covering one or several reproductive cycles, the condition is considered to be socially monogamous. However, it has not been determined whether the males and females have a single sexual partner during their entire lifetime. *P. margarita* shows low sexual dimorphism in body size and weaponary, which are the characteristics of polygamous crustaceans. The smaller size of the male partner helps in lower consumption of food, thereby making it available for egg production in the female.

4.6.2 Snapping Shrimp

Most alpheids, the commonly known snapping shrimps, exhibit social monogamy, in which adults live in heterosexual pairs. They monopolize the host by guarding them against intruders (Correa and Thiel, 2003). In view of the fact that the female mating receptivity is limited to a short period after molt and that the sperm storage in the female is absent, the female must mate each time she ovulates. Social monogamy in pair-forming alpheids is maintained both by benefits to males of mate guarding and by cooperation between mates in defending the common burrow (Duffy and Thiel, 2007). *Alpheus heterochelis* is a benthic species living on reefs and oyster beds, presumably burrowing into the mud underneath. Like most other alpheid shrimps, *A. heterochelis* individuals are most often encountered as mate pairs and hence social monogamy appears to be a widespread phenomenon among this shrimp species. Female *A. heterochelis* copulates only after molting and hence the female breeding cycle could favor male mate guarding, with the male remaining with her until the molt, ensuring a breeding opportunity for both. Furthermore, females copulate with only one male per molt cycle, whereas males mate readily with several females in succession. This behavior could account for the sequential monogamous mating by the males.

Heterosexual pairing has been reported in the snapping shrimp *Alpheus armatus*, which lives symbiotically outside of its sea anemone host, *Bartholomea annulata* (Knowlton, 1980). These shrimps are highly territorial; the juveniles defend their anemones against all conspecifics, while adults generally tolerate the presence of another adult of the opposite sex. Unlike the other monogamous shrimps, *A. armatus* shows sexual dimorphism; the males have proportionately larger major chelae and more conspicuous uropod spines. These characters in the males are helpful to increase their fighting ability during agonistic and courtship situations. On the contrary, the females also act territorially toward other females, making harem formation impossible. The females normally stick on to its anemone host and rarely move to other hosts in search of males. Males improve their mating success by not only making their courtship of females more effective but also defending their mates against other males. However, the snapping shrimp increase their reproductive potentials by visiting other anemones in search of other mates, thus exhibiting polygamous tendencies.

In another species, *Alpheus heterochelis*, mate guarding by the male, rather than shared territoriality or biparental care of the brood, appears to be the key factor leading to social monogamy (Rahman et al., 2003). The female molt cycle is highly cryptic and the females are receptive only for a few hours immediately after molting. However, males did not discriminate among females at different intermolt stages for pairing, resulting in extended mate guarding. Therefore, mate guarding through male–female pairing maximizes mating opportunities. This is also advantageous to the female because it minimizes the need to search for a mate during the vulnerable soft-bodied receptive phase. According to Baeza and Thiel (2007), social monogamy in the symbiotic shrimp evolved when hosts are small enough to support few individuals and are relatively rare, and when predation risk away from the host is high. Hence, special constraints allow only a few adult symbiotic individuals to cohabit in the same host and allow adult male and female to maximize their reproductive success by sharing their dwelling with a member of the opposite sex.

In another *Alpheus* species (*Alpheus angulatus*), Mathews (2003) posited that the mate guarding behavior would play a pivotal role in the evolution of social monogamy. A Y-maze experiment testing for distance chemical communication revealed that males of *A. angulatus* were attracted to water treated by exposure to premolt females and engaged in precopulatory mate guarding. Males also had a tendency to pair with females closer to sexual receptivity, when offered a chance, indicating that chemical cues emanating from the receptive premolt females alone elicited mate searching behavior in the males. However, the males responded differently to water treated with both intermolt females as well as premolt males, suggesting that males use different chemical cues to gather information about a conspecific's gender and molt status.

While mate guarding behavior would have given way to pairing behavior, ecological factors such as population density and sex ratio might have played an important role in the temporal extension of mate guarding to social monogamy in these burrow-dwelling snapping shrimps. Taken together, it may be surmised that selection of mate guarding behavior became a selective force toward the advancement of social monogamy in the caridean shrimps, unlike the other socially monogamous animals like birds and mammals, in which behavioral factors such as biparental care and territoriality are the driving forces to the origin of social monogamy (Lack, 1968; Kleiman, 1977).

4.6.3 Crab *Planes major*

Symbiotic living of decapod crustaceans is not limited to the coral reef-associated caridean shrimps alone. Some brachyuran crabs such as the Columbus crab, *Planes minutes*, are reported to live in commensalism with the loggerhead sea turtles in the North Atlantic Ocean (Dellinger et al., 1997). Another species, *Planes major*, is a facultative symbiont of loggerhead sea turtle, *Caretta caretta*, living in the safe refuges of supracaudal/inguinal spaces of the turtle. The host turtle does not support more than two crabs of heterosexual nature. Such cohabitation of a heterosexual pair in a turtle host facilitates social monogamy pointing to a relationship between obligate symbiosis and the evolution of social monogamy. Males pair with females regardless of their reproductive activity, and the male does not abandon the female after copulation in search of another receptive female (Pfaller et al., 2014). Hence, host switching and male–male competition have rarely been observed, suggesting that the crabs remain on the same host, presumably with the same partner, for a long period of time. Nevertheless, a considerably high percentage of solitary females brooding eggs has also been noticed in the sea turtles (Pfaller et al., 2014). Evidently, such a solitary ovigerous female would have already mated with a male that would have left the host probably in search of another potential female. This observation would further indicate that pairing in *P. major* is not always long term and certain degree of host switching is possible at least among males. Short-term monogamy also exists in the pea crab, *Austinixa aidae*, in which heterosexual pairing is truncated and males are found only with receptive females (Peiro, 2012).

4.7 EUSOCIALITY IN SPONGE-DWELLING SNAPPING SHRIMP *SYNALPHEUS*

As a sequel to social monogamy in synalpheid shrimps, complex social organizations have been evolved through cooperative living as a colony inside their sponge hosts. Unlike most *Alpheus* and *Athanas* species, which live either singly or in heterosexual pairs, the social *Synalpheus* species are in frequent contact with nestmates, often huddling closely with several other individuals.

Interaction with other shrimp involves the mutual recognition by antennulation, typical of carideans (Duffy et al., 2002; see chapter: Sex Pheromones). Clearly, social monogamy, developing into well-coordinated aggregations in the *Synalpheus* species, seems to be on the cusp of eusociality.

Eusociality, the apex of animal social organization, has been well recognized and reported only among social insects and the African mole rats (Alexander, 1991; Jarvis et al., 1994). This social system is defined as the condition of multiple generations organizing into groups by means of an altruistic division of labor, cooperative care of offspring, and reproductive skew, by which reproduction is relegated to a single mother queen (Wilson, 1971). Eusociality in crustaceans inhabiting coral reef environment is first recognized by Duffy (1996) from his lifestyle description of sponge-dwelling alpheid shrimp species, belonging to the genus *Synalpheus* (Duffy, 1996). The defining features of eusociality in *Synalpheus* species are cohabitation of different adult generations, substantial reproductive skew (reproductive division of labor), and cooperative care of the young (Duffy et al., 2002). *Synalpheus* (Decapoda: Alpheidae) is the most species-rich and abundant genera of decapod crustaceans, dominating the coral reef fauna throughout the world. Many species are symbiotic with sessile invertebrates such as sponges and crinoids, often showing tremendous host specificity. The symbiotic shrimps apparently live their entire life within their host sponges.

Social organization in the single genus *Synalpheus* ranges from heterosexual pairs, to groups with multiple breeders, to eusocial colonies containing a single breeding female, along with other sterile workers (Tóth and Bauer, 2007). Social *Synalpheus* species are very similar to termites, in that they are diploid, generally monogamous, and have direct development. They live in cloistered environments, which they must defend from competitors (Thorne, 1997). On the other hand, the host sponges provide a stable environment for the shrimps in terms of protection from the predators as well as providing food for the colony, thus satisfying the main condition for the evolution of eusociality. The shrimps in turn guard their invertebrate hosts fiercely with powerful weaponry such as their snapping claws. *Synalpheus* shrimps invariably exhibit social monogamy, living as heterosexual pairs inside their burrows or sponges. As the female receptivity is restricted to a brief period after molting, males practice precopulatory mate guarding and do not allow other males into their territory. Since *Alpheus* do not have the facility of sperm storage, the female must maintain a constant association with males by way of pair-bonding that helps in the mating of females each time she ovulates.

4.8 SOCIAL MONOGAMY TO EUSOCIALITY

The social monogamy in these sponge-dwelling shrimps apparently have originated from a primitive form of mating system, called "pure searching," as observed in many fast swimming caridean shrimps, living in dense populations, in which mate finding is not a difficult proposition (Erdman and Blake, 1987).

More complex level of communal living is also observed in the sponge-living species such as *Synalpheus longicarpus,* in which hundreds of heterosexual pairs cohabit in a host with multiple breeding females (Duffy et al., 2002). The breeding females, however, retain the large snapping claws, which are useful in defending the colony. The females from these multiqueen colonies were, however, smaller and less fecund. Multiple breeding queens also exist in other species such as *Synalpheus chacei*.

However, it is only in *Synalpheus regalis* that full complements of eusocial characters are reported (Duffy, 1996). In *S. regalis*, the colonies (up to 350 individuals) consist of full-sib offspring of a single breeding pair, as determined by genetic data and reproductive allometry. The presence of a single breeding queen in the colony is called monogyny. Nevertheless, Duffy et al. (2002) have observed a large male in association with the queen, behaving aggressively when other large males approach her. Interestingly, *S. regalis* queen shrimp loses the large chela and acquires a small chela, as found in juveniles, thereby becoming less aggressive. In the primitively eusocial wasps, the breeder queen aggressively suppresses reproduction in subordinates in the colony (Michener and Brothers, 1974). As there is no evidence of aggression or behavioral dominance in the breeding queen of *S. regalis*, reproductive monopoly over the subordinate females should be accomplished by other means such as pheromones (see Subramoniam, 2013).

The occurrence of minor chelae in the queens is also reported in two other eusocial species, *Synalpheus rathbunae* and *Synalpheus crosnieri* (Chace, 1972; Banner and Banner, 1983). However, *S. regalis* colonies are large, stable, nuclear families with consistently high relatedness, whereas *S. chacei* colonies are smaller, more dynamic, and genetically heterogeneous (Duffy et al., 2002).

The eusocial shrimps have direct development, and hence the offspring never leave the sponge nest and remain with the parent shrimps. Offspring remaining at the nest of their birth and assisting their parents in rearing additional broods is another criterion for the development of eusociality in the social insects, such as the termites. The offspring of the eusocial shrimps also delay reproduction on their own, while increasing fecundity of their parents. The colony members of *S. regalis* also exhibited behavioral differentiation among classes of individuals. The large males defended the colony against intruders, allowing the juveniles and the queen to feed and grow unmolested. This altruistic behavior of the offspring in maintaining and protecting the colony of *S. regalis* is explained by the Inclusive Fitness Theory of Hamilton (1964). Kinship plays a central role in the origin of social behavior, and hence called kin selection theory: the more closely related individuals in a group are, the more likely they are to be altruistic and cooperative. Inclusive fitness is a product of kin selection, the means by which an individual influences the reproduction of its collateral relations, such as siblings and cousins (Hamilton, 1964). Clearly, such groups of animals are meant to evolve into advanced sociality such as eusociality. Table 4.2 summarizes the occurrence of eusocial condition in different species of the genus *Synalpheus* of Alpheidae.

TABLE 4.2 Occurrence of Eusocial Developmental Traits in the Genus *Synalpheus* of Alpheidae

S. No.	Name of the Species	Criteria Conferring Eusociality			Location	Host Sponge Species	References
		Single/ Multiple Queen Colonies	Larval Development Direct/ Dispersive	Claw Development in Breeding Females			
1.	*S. regalis* Duffy 1996	Single	Direct	Smaller claw	Carrie Bow Cay, Belize	*Xestospongia* and *Hyattella* sp.	Duffy (1996) and Duffy et al. (2002)
2.	*S. filidigitus* Armstrong 1949	Single	Direct	Smaller claw	Carrie Bow Cay, Belize	*Oceanapia* sp. and *Xestospongia* sp.	Duffy and Macdonald (1999)
3.	*S. chacei* Duffy 1998	Multiple	Direct	Larger snapping claw	Carrie Bow Cay, Belize	*Niphates amorpha*	Duffy (1998)
4.	*S. agelas* L. H Peguegnat and Heard 1979	Multiple	Direct	Larger snapping claw	Caribbean coast of Panama	*Agelas clathrodes*	Duffy (1992)
5.	*S. rathbunae* Courtiere 1909	Multiple	Direct	Larger snapping claw	Caribbean coast of Panama	*Xestospongia rosariensis*	Duffy (1992)
6.	*S. brooksi* Courtiere 1909	Multiple	Direct	Larger snapping claw	Caribbean coast of Panama	*Spheciospongia vesparium*	Duffy (1992, 1996)
7.	*S. paraneptunus* Courtiere 1909	Single	Direct	Unknown	Florida, Belize, and Caribbean Panama	*Xestospongia rosariensis*	Duffy (1992) and Duffy et al. (2000)
8.	*S. neptunus* Dana 1852	Multiple	Unknown	Unknown	Australia and Philippines	Coral	Banner and Banner (1975)

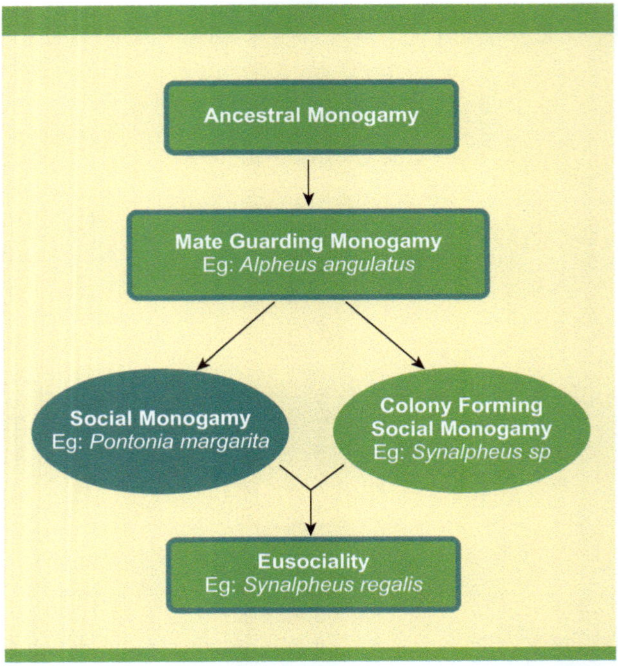

FIGURE 4.1 Evolution of eusociality in coral reef caridean shrimps.

Apparently, direct development that fosters the formation of close kin groups, as well as the consequential long-term parent–offspring associations are the two important preconditions in the evolution of eusociality among the synalpheid shrimps (Fig. 4.1). Evidently, in the synalpheid shrimps, the defining eusocial characteristics include (1) symbiotic social living in sponge which provides shelter and food, (2) mate guarding and shared resource defense, (3) nondispersive larvae, and (4) social monogamy. The genetic relatedness among the colony members of *S. regalis*, as revealed from allozyme data, together with demographic evidence of natal philopatry, would further indicate that colonies in this species represent close kin groups, as a necessary prerequisite for evolution into eusocial condition (Duffy et al., 2002).

4.9 BROMELIAD CRABS

Life history traits such as direct development with nondispersive larvae as well as extended parental care have also been observed in other crustacean species, indicating a subsocial route for the evolution of eusociality in crustaceans. In the Jamaican bromeliad crab, *Metopaulias depressus*, with the potential to become a eusocial species, only the colony mother is reproductive, whereas other subordinate females with physical attributes to reproduce, will not carry eggs in the pleopod. However, the subordinate females that leave the colony, start reproducing by becoming solitary breeding foundresses of new colonies (Diesel et al., 2007). Furthermore, subordinate females could refrain from early dispersal and reproduction for the benefit of staying at home and taking over the colony, once the colony mother dies, or acquiring a new adjacent bromeliad clone. Apparently, the subordinates are under reproductive suppression by social cues emanating from the breeding colony queen. Whether the suppression of reproduction in these subadults is caused by any pheromone released by the mother crab of the colony or impacted by any social cues from the breeding dominant, and in this case, the aggressive female, is not clear.

Although the genetic relatedness of the siblings in the colony of this crab has not been determined, the retention of the subordinate females in the colony may provide some inclusive fitness benefits by their altruistic behavior and perhaps by improving survival and growth of younger siblings. Strangely, unlike the synalpheid queens, which have lost their fighting claws, the bromeliad queen crabs can attack and expel intruding adult females as well as smaller individuals. However, adult female colony members are allowed to stay, probably because they may be the daughters of the breeding queen. In the bromeliad crabs, parental care for offspring is also a special case of kin selection, since a breeding female vigorously protects the young with its powerful chelae. Thus, bromeliad crabs meet the criteria described in Crespi (1994) for the origin of eusociality: a clustral habitat that combines food and shelter, strong selection for defense, and the ability to defend via weaponry.

4.10 POLYGAMY

According to the conceptual model of Emlen and Oring (1977), the abundance and distribution of resources and receptive females in space and time determines the environmental potential for polygamy. The widespread occurrence of polygamy among crustacean taxa may also depend on additional factors such as sperm competition (Parker, 1970) and pre- and postcopulatory mate guarding by males and female choice (Eberhard, 1996). However, observations on polygamy or multiple mating in the natural conditions of the primarily aquatic crustaceans is rather difficult. Nevertheless, mating frequencies of both males and females (multiple mating) are determined by microsatellite loci analysis in the egg mass as well as the larvae to infer polygamous mating systems existing in the free-living, parasitic, and colony-living crustaceans (Jossart et al., 2014). In the ectoparasitic crab, *Dissodactylus primitivus*, both sperm analysis of spermathecae as well as microsatellite loci analysis revealed a polygamous mating system, where males and females move between hosts for mate search. Use of molecular markers like microsatellite polymorphism has yielded information on multiple paternities in the thalassinidean ghost shrimp, *Callichirus islagrande*, although mating in this deep-burrowing shrimp under natural conditions has never been observed (Bilodeau et al., 2005). The occurrence of polyandry has also been determined by genotyping the male alleles from the brooded offspring of the mangrove land crab, *Ucides cordatus* (Baggio et al., 2011). In this land crab, multiple mating occurs by forced mating; the males use their pereiopods to subdue females and drag them inside the burrow for mating (Costa, 1979).

Polygamy, as a result of repeated mating behavior, has also been reported in the mantis shrimp, *Squilla holoschista*, under laboratory conditions (Deecaraman and Subramoniam, 1983c). Under laboratory observations, both male and female *S. holoschista* involve themselves in repeated mating (Fig. 4.2A–C and Fig. 4.3A–C). Once mated, the male can mate with other nonmated and mated females. Similarly, mated female can also mate with a mated or unmated male. Mating entails the transfer of a sperm cord through an intromittent organ. Along with the transfer of a sperm cord, secretions from a pair of male accessory gland are also delivered into the female oviduct. These proteinaceous accessory gland secretions ultimately reach the ovary for final incorporation into the yolk (Deecaraman and Subramoniam, 1980; see also chapter: Accessory Reproductive Glands).

Polygamous mating thus seems to be a common mating system among malacostracan crustaceans. Even in mate guarding marine isopods, polygamy, coupled with multiple mating, has been reported. In the supralittoral

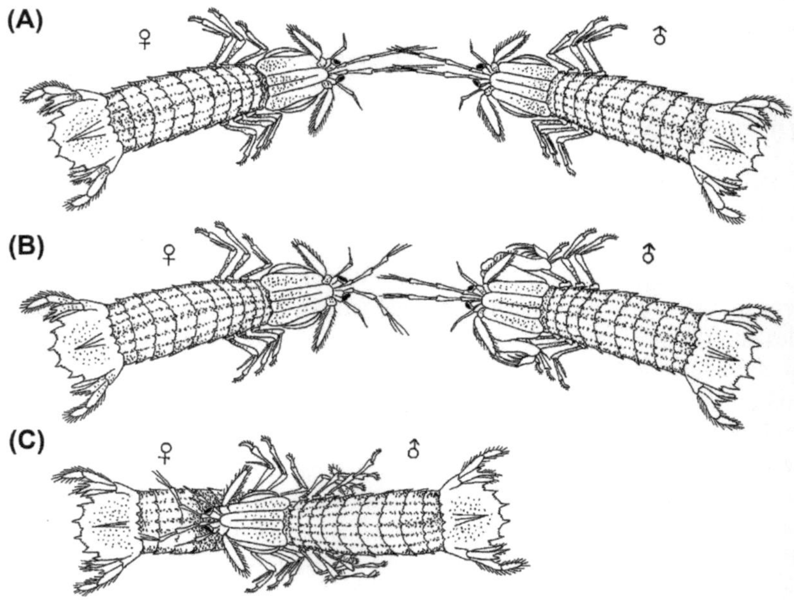

FIGURE 4.2 *Squilla holoschista*. Mating behavior. (A) Antennule contact between the male and female; (B) spreading of the raptorial meri by the male; (C) movement of the male over the female.

isopod, *Ligia dentipus*, the molting is biphasic, with the posterior body portion molting first, exposing the ovipores at the coxa of the fifth pereon segment (Santhanakumar et al., 2014). The mate guarding male then copulates with the female but leaves immediately in search of other receptive females. The female receptivity lasts even after the male leaves the female, and if allowed, the female mates with other males, without any premate guarding.

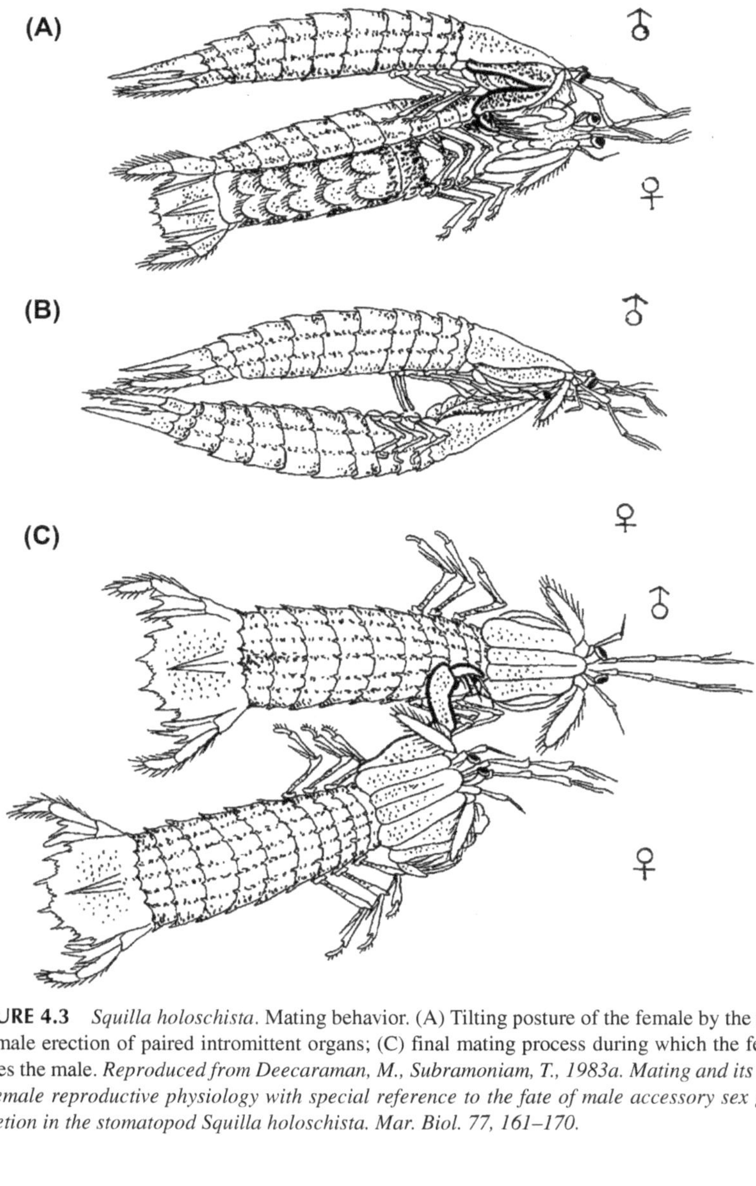

FIGURE 4.3 *Squilla holoschista*. Mating behavior. (A) Tilting posture of the female by the male; (B) male erection of paired intromittent organs; (C) final mating process during which the female strikes the male. *Reproduced from Deecaraman, M., Subramoniam, T., 1983a. Mating and its effect on female reproductive physiology with special reference to the fate of male accessory sex gland secretion in the stomatopod Squilla holoschista. Mar. Biol. 77, 161–170.*

In general, multiple mating (polygamy) occurs in those species that lack sperm storage facility; when the sperm is stored in spermatheca and used for multiple spawning, repeated mating is discouraged by the female. The polygamous mating system is further classified into polygyny, polyandry, and polygynandry.

4.11 POLYGYNY

It is a form of polygamy where one male has an exclusive relationship with two or more females. Polygyny is expected to occur under conditions in which the females are widely dispersed and males compete by trying to outrace rivals in acquiring receptive females. In this type of polygamy, males leave the females immediately after mating and resume searching for further females. Polygyny among crustaceans exists in three forms, viz., scramble-competition polygyny, female-centered polygyny, and host-defense polygyny.

4.11.1 Scramble-Competition Polygyny

Anostracans such as *Artemia* adopt a mating pattern in which males avoid aggressive encounters with their mates but concentrate their reproductive efforts on finding, courting, and inseminating receptive females. In this mating pattern, designated as scramble-competitive polygamy (Alcock, 1980), females obtain sperm from a different male for each clutch, thereby increasing the genetic variability of the offspring (Belk, 1991). Anostracan females do not store sperm and hence a separate copulation is required to fertilize each clutch. However, anostracan females typically accept intromission, only when they have ripe eggs in their lateral pouches. In *Artemia*, the female accepts the precopulatory clasping, called amplexus, only when they are in receptive condition.

Scramble-competition polygyny has also been described by Murai et al. (2002) in the fiddler crab, *Uca paradussimieri.* The fiddler crab lives in individual burrows in mixed sex colonies on intertidal mud flats. Females are generally sedentary and wander through the population to search for mates (Christy, 1983; Murai et al., 1983). The fiddler crabs mate underground during full moon and new moon days. As soon as the tide recedes, males enter the burrows of the females. Males copulate and guard these females until they ovulate. When interrupted by an intruding male, the first male to reach the female is usually able to defend it successfully and mate with it. In this species, aggression was common between males that courted the same female.

4.11.2 Host Defense Polygyny

In the symbiotic crustaceans living with diverse marine invertebrates, the type of polygynous mating system varies in accordance with host characteristics. In host-defense polygyny, the hosts are morphologically simple, and there is moderate to high risk of predation of the symbionts. Hence, males defend the host as well as the relatively stable aggregation of females within the host. The males aggressively exclude male intruders, creating an opportunity for them to mate with more than a single female in the guarded host. In the ascidian-dwelling amphipod *Anamixis hanseni* and the sponge-dwelling isopod *Paracerceis sculpta,* a single male guards the entrance of the relatively large host, permitting cohabitation of several females (Thomas and Bernard, 1983; Shuster and Wade, 1991). In *P. sculpta*, the dominant male (alpha male) necessarily practices polygyny as the females have asynchronous receptivity.

4.11.3 Female-Centered Polygyny

Another type of polygyny, the female-centered polygyny, exists among the symbionts when the host body size supports a moderate number of symbiotic individuals. The host abundance and the predation pressure are also moderate. The females are randomly distributed among hosts with moderate density. In this mating system, males roam within hosts and among hosts in search of receptive females. When found, the males guard and protect the females against other males until they become receptive for mating. However, the males leave the females soon after mating and go in search of additional females in the same or different hosts. The females may also resist copulation, thereby selecting high-quality males to increase their reproductive success. Examples for female-centered polygyny are reported in the anemone-dwelling crab, *Inachus phalangium* (Diesel, 1986b) and the sea urchin-dwelling shrimp, *Athanus kominatoensis* (Nakashima, 1987). These two species exhibit marked sexual dimorphism and in the males, the chelae are well developed as weapons. Interestingly, small males of *A. kominatoensis* change sex to females, if they are unsuccessful in copulating with females. On the other hand, small males of *I. phalangium* attempt sneaking copulations with females even if they are guarded by large dominant males.

4.12 POLYANDRY

This is another form of polygamy where one female has an exclusive relationship with two or more males. In species with male ontogenetic stages, multiple mating through convenience polyandry is common. The convenience polyandry has a follow-up of cryptic female choice. Thiel and Hinojosa (2003) had observed the case of convenience polyandry in the female caridean shrimp *Rhyncocinetis typus*. The mature rock shrimp of this group exists as several ontogenic stages that differ in their morphological and physiological capabilities. In natural populations, the majority of the males are in early ontogenic stages (termed typus), many are in intermediate stages (intermedius), and a few are in terminal molt stage (robustus). Robustus males have a higher potential than the subordinate typus males to defend receptive females against other males and fertilize the entire clutch of a female. Females show preference only to robustus males even though they receive sperm from typus males, through multiple mating. The females avoid being seized by the typus males in the absence of the robustus males. The subordinate typus males transfer their spermatophores to females before they start spawning but the robustus males wait until the females begin to spawn before they transfer their spermatophores. In a competitive environment, typus males transferred spermatophore within a minute of seizing a female, which was never found in robustus males. Such behavioral difference between typus and robustus males is responsible for the female preference of dominant males for mating. Their biological fitness gives assurance of fertilizing the entire egg clutch. Under experimental conditions, the female rock shrimp evaded seizure by typus males in the absence of dominant males. Nevertheless, by accepting sperm from typus males, the females may avoid further harassment (convenience polyandry), but subsequently may discriminate these typus or subordinate males by delaying spawning or removal of their attached spermatophore (Bauer, 2004).

Similarly, in the freshwater prawn, *Macrobrachium rosenbergii*, the female attempted to evade subordinate males, but readily mated with dominant males (Ra'Anan and Sagi, 1985). In the female crayfish, *Orconectes rusticus*, Snedden (1990) reported a high tendency to evade mating attempts from small-clawed males than from large-clawed males. The blue crab *C. sapidus* female may also exert subtle preferences by exhibiting resistance to small males during premating interactions, which effectively leads to mating with large males (Jivoff and Hines, 1998). In the isopod *Lirceus fontinalis*, females may perceive the condition of males and prefer males in good condition (Sparkes et al., 2000). In short, females engaging in convenience polyandry choose sperm from preferred males to fertilize their eggs.

4.13 POLYGYNANDRY

Polygynandry has evolved as an offshoot of "pure search mating system." In pure search polygynandry, the shrimps inhabit morphologically complex hosts, which can support many symbionts at a time. The hosts are also abundant and free from predation risk. Both males and females move freely within the host. There is neither host guarding nor monopolization by the symbiotic males. The males are smaller than the females, resulting in the increased agility and encounter rate with potential mating partners. Under such circumstances, males maximize their reproductive success by roaming within and among hosts in search of females. After finding receptive females, the males mate and abandon the female immediately to continue searching for other receptive females. Simultaneously, multiple mating of females also takes place due to female choice or male coercion. The porcellanid crab, *Liopetrolisthes mitra*, which inhabits an abundant, large, morphologically complex sea urchin, has been shown to be another example for pure search polygynandry (Baeza and Thiel, 2000).

Baeza and Thiel (2007) have described yet another type of polygynandry found among the sedentary females, living symbiotically with marine invertebrate hosts. For example, in the crab *Zebrida adamsii*, the female lives solitarily with the sea urchin host, whereas the males roam around the hosts in search of sedentary females. On finding a receptive female, the male mates with it, but does not appear to guard the female and leaves immediately in search of another female (Yanagisawa and Hamaishi, 1986).

4.14 AGGREGATIONAL MATING IN SAND CRABS

Different species of the anomuran sand crab belonging to the genus *Emerita* have neotenous males, with reproductive organs attaining precocious maturity soon after the megalopa settlement on the beach (reviewed in Subramoniam and Gunamalai, 2003). In the Indian species *E. asiatica* (= *E. emeritus*), the neotenous males show a general simplicity of the appendages associated with their small size. However, the testis and vas deferens are well developed, with the latter swollen with spermatophoric components. The neotenous males also possess a pair of conspicuous genital papilla or sperm sac, situated at the inner side of the base of the coxa on the fifth thoracic leg of *E. asiatica*. In the American species, *Emerita analoga*, the neotenous males apparently gather around the egg-laying females, 2–5 days before molting and remain in contact with the female (MacGinitie, 1938). The males attach to the females by the dactyls of their fourth legs, which are equipped with a sucker pad surrounded by stiff hairs. This kind of mating association has been termed as "incipient parasitism" (Wharton, 1942). In the continuously breeding tropical species *E. asiatica*, small functional males in the size range of 3.75 and 5 mm carapace length (CL), occur throughout the year and some of them were observed to grow to a large size of 11 mm CL. However, these larger females were not found to take part in mating with larger egg-laying females, although they possessed well-developed spermatophores in the swollen vas deferens. Subramoniam (1977b) found that the sexually active smaller males, once metamorphosed from the megalopa, not only attained precocious maturity without body growth but also undergo a certain degree of growth regression during subsequent molts. Since these males are inside the burrow, clinging on to the females, they have no chance of active feeding by antennal filtering, thus resulting in growth retardation. The smaller size is also advantageous for their hideout on the ventral region of the burrowing females. Therefore, the females accept these smaller males, rather than the bigger males, for mating since they would not disturb normal activities such as burrowing. Subramoniam (1979b) reported a peculiar mating behavior of the larger males of *E. asiatica*. A male of 8.5 mm CL forcefully deposited a spermatophoric ribbon on the ventral side of a freshly molted immature female indicating indiscriminate copulation, amounting to heterosexual raping in these intertidal anomuran crabs. The aggregation behavior of the neotenous males coupled with their "parasitic" mode of group mating have all contributed to a highly adaptive mating system in these intertidal filter-feeding sand crabs, evolving as a means to increase the chances of fertilization in the unstable habitat. Aggregational mating of *Emerita* by neotenic males is a reflection of the animal's lifestyle adaptation to the precarious intertidal environment.

4.15 MATING SYSTEMS IN HERMAPHRODITES

A distinguishing feature in the sexuality of crustaceans from that of insects is the prevalence of hermaphroditism in different taxa. Fertilization occurs in hermaphroditic crustaceans both by selfing (self-fertilization) and by outcrossing (mating with separate individuals). When selfing extends across successive generations, genetic variation (heterozygosity) within each genetic lineage quickly declines. Nevertheless, many crustacean species resort to self-fertilization, enforced by environmental constraints. In others, individuals may outcross exclusively, as in the simultaneously hermaphroditic barnacles, in which sperm are deposited in the mantle cavity of the female-acting hermaphrodite (pseudo-copulation), often by more than one adjacent hermaphrodite (Charnov, 1979b). The spermatozoa of barnacles such as *Balanus balanoides* are usually immobile in the seminal fluid but become extremely active after being discharged through the penis into the mantle cavity of another individual. Oviposition normally takes place in these barnacles as a result of the stimulus of copulation (Crisp and Davies, 1955). When these balanoids were separated by a distance of 5 cm or more from its neighbors, no fertilization of eggs could happen in these isolated barnacles. The distance of 5 cm represents the maximum extension of the penis in these cirripedes. Although such obligatory cross-fertilization might be the rule for all cirripedes, isolated individuals of species like *Chthamalus stellatus*, *Verruca stroemia*, and *Balanus perforates* have been found to have self-fertilization (Barnes and Crisp, 1956). Nevertheless, self-fertilization is less efficient, as the eggs are less viable most of the time.

In the androdioecious species *Eulimnadia texana*, the hermaphrodites either undergo self-fertilization or mate with free-living males. The hermaphrodites cannot pair with another hermaphrodite for reciprocal fertilization because they do not possess the claspers. Only the males possess them to facilitate mate guarding of the hermaphrodite until mating (Sigvardt and Olesen, 2014).

Chapter 5

Mating Behavior

5.1 INTRODUCTION

Behavioral aspects of crustacean reproduction are mainly concerned with many modes of mating, which accomplish union of both male and female gametes during fertilization. Mating behavior found in a particular animal species is a product of sexual selection to increase reproductive success for both the sexes. In Crustacea, sexual selection acts through exaggerated morphological traits, such as large size and possession of weaponry in males, combined with peculiar behavioral traits. As a result, male mating competition (intrasexual selection) as well as mate choice (intersexual selection) forms the important components of sexual selection in decapod as well as other malacostracans.

Mating behavior in crustaceans comprises initial interactions between males and females by way of chemical communications followed by mate guarding of the receptive females and finally mating. Premating behavior of males also involves courtship signaling and fighting among males to gain access to receptive females. In addition to the well-known sex pheromones, hormones such as the molting hormone and the male-specific androgenic gland hormones play a significant role in sex attraction and the mating behavior of several crustacean species. Copulation in many crustaceans is the direct transfer of spermatophores onto the sternum of the females or insemination of fluid semen containing tiny spermatophores into the seminal receptacle or spermatheca by well-developed copulatory organs. Transfer of free sperm and seminal plasma occurs only in a few species such as cirripedes. Notwithstanding, a variety of sperm transporting mechanisms have evolved in both marine and terrestrial crustacean species (Subramoniam, 1993). This is reflected in the equally diversified mating habits or behaviors, which are categorized into mating associations. In recent years, much attention has been paid to the elaborate mating sequence of premate guarding in decapods and isopods, but simple transfer of spermatophores/sperm with little courtship or male–female interactions has also been found in many species. Mating behavior in crustaceans could be best studied from several types of mating associations found in different taxa.

The following is a broad classification of mating associations (Table 5.1) in decapod crustaceans, according to Christy (1987).

1. Pure searching or scramble competition
2. Female-centered competition

Sexual Biology and Reproduction in Crustaceans. http://dx.doi.org/10.1016/B978-0-12-809337-5.00005-8

TABLE 5.1 Mating Association in Crustacea

Species	Mating Association	References
Malacostraca		
1. Decapods (brachyuran crabs)		
I. Female-Centered Competition		
Corystes cassivelanus	**A. Search and defend** Males search for and aggressively defend mobile females that release waterborne pheromones. Prolonged (days) of mate guarding	Hartnoll (1968)
Carcinus maenas		Berrill and Arsenault (1982)
Portunus pelagicus		Fielder and Eales (1972)
Callinectes sapidus		Gleeson et al. (1984)
Cancer oregonensis		Knudsen (1964)
Cancer productus		Knudsen (1964)
Cancer magister		Snow and Nielson (1966)
Cancer irroratus		Elner and Elner (1980)
Cancer borealis		Elner et al. (1985)
Cancer pagurus		Edwards (1966)
Inachus phalangium	**B. Patrol and defend** Dominant males temporarily defend receptive females for one or more days	Diesel (1986a,b)
Cataleptodius floridans	**C. Capture and defend** Males may guard the females for several days, but females may avoid or resist capture and thereby choose large aggressive males	Engstrom and Lucenti (1984)
Neopanope sayi		Swartz (1976a,b, 1978)
Dotilla mictyrodies		Tweedie (1950)
Scopimera globosa		Wada (1981, 1983)
Uca deichmanni		Zucker (1983)
Uca stenodactylus		Crane (1975)
Macrophthalmus japonicus	**D. Attract and defend** Guarding may last several days and females are free to choose males on the basis of location and courtship signals	Wada (1984)
Ilyoplax pusillus		Wada (1981, 1983)
Ocypode ceratophthalamus		Brook (1981)
Ocypode saratan		Linsenmair (1967)
Ocypode ryderi		Vannini (1980a,b)
Ocypode kuhii		Jones (1972)
Ocypode guadichaudii		Crane (1941)
Uca musica		Zucker (1984)

TABLE 5.1 Mating Association in Crustacea—Cont'd

Species	Mating Association	References
II. Resource-Centered Competition		
Gecarcodia natalis	**A. Breeding site defense** Males compete directly for sites (burrows) where females mate and incubate eggs	Hicks (1985)
Gecarcinus lateralis		Bliss et al. (1978)
Cardisoma guanhumi		Henning (1975)
Uca pugilator		Christy (1980, 1982, 1983)
Uca rapax		Greenspan (1980)
Uca tangeri		Muller (1983)
Uca lactea		Murai et al. (1987)
Trapezia sp.	**B. Refuge defense** Male defends areas containing refuges (hosts) which are occupied by females. Guarding by males directly during brief mating associations	Huber (1985)
Menippe mercenaria		Sinclair (1977)
Pachygrapsus transversus		Abele et al. (1986)
Sesarma reticulatum		Seiple and Salmon (1982)
III. Encounter Rate Competition		
Helice crassa	**A. Neighborhoods of dominance** Mobile males interact aggressively with others in shifting areas where they search for, court, and mate with females	Nye (1977)
Aratus pisoni		Warner (1967, 1970)
Goniopsis cruenta		Salmon (1984)
Hermiplax latifrons		Griffin (1968)
Heloecius cordiformis		Griffin (1968)
Uca vocans		Salmon (1984)
Uca thayeri		Salmon (1987)
Rhynchocinetes typus		Bauer and Thiel (2011)
Macrobrachium rosenbergii		Bauer and Thiel (2011)
Hemigrapsus oregonensis	**B. Search/interception** Males seek mates freely to establish temporary sites from which they intercept and court mobile females	Lindberg (1980)
Hemigrapsus rudus		Knudsen (1964)
Graspus graspus		Kramer (1967)
Sesarma cinereum		Seiple and Salmon (1982)
Pachygrapsus crassipes		Bovbjerg (1960)
Macrophthalmus hirtipes		Beer (1959)
Scopimera proxima		Silas and Sankarankutty (1967)

Continued

TABLE 5.1 Mating Association in Crustacea—Cont'd

Species	Mating Association	References
2. Amphipods		
Gammarus lawrencianus	**Precopulatory mate guarding** Females carry fertilized eggs and occasionally brood juveniles in a ventral brood pouch. The male's decision to adopt and guard a female for mating purposes is influenced by the contents of the brood pouch. Males are positioned dorsal to the females and grasp the female's integument with prehensile thoracic appendages called gnathopods	Dunham et al. (1986)
Gammarus pulex		Plastow et al. (2003)
Hyalella azteca		Cothran (2003)
3. Isopods		
Thermosphaeroma thermophilum	**Precopulatory mate guarding** Before the female mating, males engage in guarding behavior usually by positioning them ventrally and carrying them for an extended period of time	Jormalainen and Shuster (1999)
Idotea baltica		Jormalainen et al. (1999)
Asellus aquaticus		Benesh et al. (2007)
Venezilla evergladensis	Males orient to receptive females using olfactory lobes	Johnson (1985)
Cymbodetta gambosa	**Extended mate guarding** The male grasps the female with its pereopods. The female responds by rolling together, either firmly or loosely held depending on her readiness to copulate	Bowman (1973)
Branchiopods		
4. Spinicaudata		
Limnadia badia	**Precopulatory mate guarding**	Benvenuto et al. (2009)

3. Resource-centered competition
4. Encounter rate competition
5. Mate guarding
6. Alternative mating strategy

5.2 PURE SEARCHING

"Pure searching," a term coined by Wickler and Seibt (1981), refers to a male mating strategy, commonly found in all penaeoidean species, living in mobile aggregation, called "schooling" as well as the caridean shrimps, living in congregation in the coral reef niches. In the caridean shrimps, the males are small in size and do not develop sexually dimorphic features such as large cheliped. There is no precopulatory mate guarding or aggressive defense of females by the males that practice pure search mating. Smaller male size has an obvious advantage for searching of females, when the latter are scarce and dispersed. Interestingly, many of the pure searching (promiscuous polygyny) caridean shrimps with small-sized males are protandric hermaphrodites (ie, *Rhynchocinetes uritai*) (Bauer and Thiel, 2011). The frequency of males contacting the females for mating is rather high and hence, the mobile males quickly identify postmolt receptive females by visual cues or contact chemoreception. The males, being nonterritorial and highly mobile, are continually on the prowl for a receptive female and when one is encountered, she is copulated with minimum premating interactions. The male then sets off to search for another receptive female.

Unlike other mating systems, the males do not fight for or defend females, although there is an intramale competition in the form of rapid and efficient searching for females. Pure searching is a primitive mating strategy in the evolution of mating systems of Crustacea in as much as the males make no investment in guarding or protecting the female beyond spawning (Bauer, 1986a). Understandably, conditions like high-population density and high frequency of spawning in these caridean shrimps favor the evolution of a pure searching, rather than a female-guarding strategy by males. Christy (1987) compared this kind of promiscuous mating system with that of "encounter rate polygyny," described for the fiddler crab.

Deviations from the pure search mating are also found in two species of *Athanas* (*Athanas kominatoensis* and *Athanas indicus*), leading a solitary symbiotic life on sea urchins. Adult males change their host in search of receptive females, living solitarily on the sea urchins. There, they form heterosexual pairs. Such a behavior of changing hosts for pair forming is also seen in the spider crab *Inachus phalangium*. After mating, they depart to search for new mates. However, when males have limited chances of getting extra-pair copulations, due to low-population densities, they attend females for long periods, resulting in monogamy. This habit of switching over for mate finding is not found in the solitary females (Correa and Thiel, 2003). Furthermore, small males which could not compete with large males in the accession to the receptive females change sex to partial protandric hermaphroditism, a process mediated through social interaction (Nakashima, 1987).

In the penaeoidean shrimp, *Sicyonia dorsalis*, Bauer (1996) studied the pure search mating behavior. In this species, males are smaller than the females and possess small chelipeds. Males of *S. dorsalis* are not aggressive among themselves, even in the presence of copulatory females (ie, receptive females with empty seminal vesicles). Although they do not have any elaborate precopulatory mating behavior, the males exhibit a "following" behavior, in which a male upon contact with a female moves behind it, placing one antennal flagellum on each side of the female abdomen, and the pair walk in tandem for a few seconds or minutes. During copulation, which always occurs after the female molts, the male moves forward of the female, tilts the female abdomen upward, in such a manner that the male genitalia can be opposed to that of the female. Although sperm storage occurs within the paired seminal receptacle, mating usually takes place after the female molts, because the stored sperm from previous mating is lost along the old cuticle during ecdysis. In this shrimp species, multiple mating also takes place. Receptive females are the ones that have emptied the spermathecae by molting and hence, the males increased the following behavior with newly molted females and copulated almost exclusively with them. As the male invariably leaves the female after mating, this mating strategy is considered to be "pure searching."

5.2.1 Female-Centered Competition

In general, all female crustaceans produce less numerous, but energetically expensive eggs, whereas males produce an overabundance of sperm. Consequently, energy investment for gamete production differs in between males and females, leading to higher reproductive potentials in males than in females. In addition, mating in many crustacean species is linked to molting, restricting the female receptivity period. As a result, operational sex ratio (OSR) is always male biased in comparison with sexually receptive females in a population. Breeding females thus become the limiting sex and critical resource for which males compete.

Female-centered competition is a type of male mating competition that takes many forms in crustaceans. It could vary from intense aggressive fights over females to brief male–male aggressive interaction, and from prolonged mate guarding to mating without attending the female longer than necessary for copulation. Among decapods, male–male aggressiveness is well known in crayfish and lobsters, whereas mate guarding behavior is exhibited extensively in brachyuran crabs as well as the amphipods and isopods. In the crayfish, males are always dominant over the females by possessing larger chelae, which are sexually selected for increasing their fighting ability. Furthermore, male–male agonistic interactions are more intense than are male–female or female–female interactions.

In the Cambaridae species, males have alternating reproductive morphotypes, with a reproductive form (form I) and a nonreproductive form (form II). Form I males have larger chelae relative to body length than do form II males. Hence, form I males are always dominant over form II males, even when males have similar carapace lengths (Guiasu and Dunham, 1998). In addition, form I individuals exhibited a higher level of aggression toward form II individuals. Similarly, reproductive status also alters the dominance order in the intraspecific competition with males. Hence, female maternal crayfish (with brood carrying eggs or first instars larvae) showed more aggression toward males and nonmaternal females than did nonmaternal female crayfish (without developing brood). Such differences in the reproductive status of both males and females also formed a determining factor in the social hierarchy of the crayfishes.

In many crustaceans, copulation takes place shortly after molting, when the female is soft-shelled. The restricted female receptivity is the root cause for development of mate guarding practices in brachyuran crabs, in which competition among males to gain access to females is heavy. For example, in the portunid crab, *Callinectes sapidus*, males start guarding the females even before the parturial molt, but copulate only after molting. The male also extends guarding of the females up to spawning (Jivoff, 1997).

However, restricted female receptivity could also be seen in those species in which females copulate hard-shelled, during the intermolt stage. Hartnoll (1969) found local decalcification of the hinges of immobile opercula sealing the gonopores, which makes them temporarily flexible and mobile in those brachyuran crabs that mate in hard-shelled condition. The temporal distribution of such receptive females also results in male–male competition for mate finding. In three New Zealand grapsid crabs, *Cyclograpsus lavauxi*, *Helice crassa*, and *Hemigrapsus sexdentatus*, the females mate during the intermolt stage, but the female receptivity is restricted to a short period when the gonopore opercula becomes movable due to local decalcification (Brockerhoff and McLay, 2005a,b,c). Interestingly, *C. lavauxi* males do not mate guard females, but in *H. sexdentatus*, postmate guarding lasts for several days. This difference is mate guarding could be explained in terms of different habitats that they occupy. Whereas *C. lavauxi* occupy the rocky high intertidal zone, where the harsh conditions do not permit prolonged mate guarding, *H. sexdentatus* inhabits the mid-intertidal zone, which is free from wave action and predation, thus encouraging prolonged mate guarding and defending of receptive females.

Breeding ecology of the female is also a determining factor for female-centered competition among the active males. In two species of cancrid crabs, *Cancer magister* and *Cancer gracilis*, Orensanz et al. (1995) found differences in their mating behavior, despite their similar reproductive ecology. Both the species have soft-shell mating and the males defend the mobile females with pre- and postcopulatory mate guarding. *C. gracilis* produces multiple broods during the extended breeding season. Since the female receptivity is asynchronous, *C. gracilis* males fight fiercely for the scarce receptive females, and takeovers of the guarded females by dominant males is rather common. On the other hand, *C. magister* females produce only one annual brood, and therefore, the female receptivity is synchronous. Hence, mate finding in this species is not difficult, and the males invest less in aggressive contests, tending toward scramble competition.

5.2.2 Resource-Centered Competition

Crustaceans are known for their varied life history patterns to live and reproduce successfully both in water and on land. Although most crustaceans are free-living, several others have adopted burrow-living, occupying crevices or cavities in rocky or coral reef environments, in addition to making their own burrows in the intertidal sand or mud flats. Several species of caridean shrimps have evolved to live in association with a variety of invertebrate hosts, making either symbiotic or parasitic relationships. This type of lifestyle has a role in the evolution of new strategies in mating competition. In the resource-centered competition, males are involved in building or protecting such refuges and making them breeding sites for the females. Resource-centered competition is prevalent among the burrow-living decapods such as the brachyuran crabs, lobsters, and crayfishes.

The defense of breeding burrows is commonly found among the fiddler crabs, belonging to *Uca* species (Christy, 1983; Morrell et al., 2005). Fiddler crabs are semiterrestrial, found in dense mixed-sex colonies on intertidal sand flats and mud flats, where they dig individual burrows. There are two general modes of mating in the genus. In the first mating type, as found in several American species, the male occupies a burrow and defends it. The male, standing at the entrance of the burrow, makes waving displays with its enlarged major claw to attract females for mating and breeding. On the contrary, the receptive females search for mates and breeding burrows, by leaving their own burrows. Mate choice is indicated when the female stays in a male's burrow, into which the female is guided by the male. After the female enters the burrow, the male plugs the entrance. The copulation takes place underground, and the female ovulates and attaches the fertilized eggs to her abdominal appendages. The female will stay in an enlarged terminal chamber of the burrow for about two weeks until she releases her planktonic larvae. The male then leaves the burrow in search of another female. Sexually receptive females respond sequentially to visual and acoustic courtship signals of several males before choosing their mates. However, the final female choice appears to be based primarily on the quality of the burrows, defended by males as breeding sites.

Thus, male defense of breeding sites arises when females require access to specific microenvironments to breed successfully and when they are unable to create or defend such sites themselves. Females have to do foraging effectively to produce energy-rich eggs during breeding, whereas males could afford to expend energy in finding and guarding females, in addition to building burrows for females to breed. Anisogamy and the economical way of producing enormous numbers of male gametes could underscore this conflict of interest in the breeding behavior of male and female fiddler crabs. Most of the species showing this mode of mating is small to medium size, living at high densities in the supratidal zone. *Uca pugilator* males compete for burrows located higher on the beach and large males exclude small ones from these burrows (Christy, 1983). In other species, the males may even help to defend the burrows of their neighbor against an intruder. Male competitive ability is largely determined by size. High-population densities coupled with female mobility and breeding site requirements might have led to the evolution of resource-defense behavior in *Uca* species. They are also best examples for polygynous mating strategy. Sexual selection favors males that increase their mating rate by decreasing the time they spend with each female. For example, the male *U. pugilator* can mate with up to three females in a single breeding cycle, at the same burrow in separate incubation chambers (Christy, 2007).

Contrary to this pattern of female searching, in many of the Indo-Pacific *Uca* species, it is the males that search for mates among nearby female burrow residents (Christy, 2007). Courtship may not include male claw waving, but includes tactile stimulation and possibly seismic signals transmitted through substratum. The pair usually mates on the surface at the entrance of the female's burrow, and the female subsequently breeds alone in its burrow. Many of the species with this type of mating are relatively large and live at low densities in muddier sediments in the upper intertidal zone.

5.3 MATE GUARDING

From an evolutionary perspective, a particular mating behavior has a bearing on the reproductive success of each sex. In Crustacea, mating behavior has evolved around the percept that molting in the adult female is an intervening factor for reproductive activities (Subramoniam, 2013). Female reproduction alternates with molting activities in such a way that mating and egg laying are relegated to postmolt condition. Mating in the postmolt condition is therefore a physiological necessity, as the female gonopores will be accessible to intromittent organs only in a soft-shelled condition. A new behavioral trait, namely precopulatory mate guarding by the males has evolved in many crustacean taxa, in which molting precedes mating. Precopulatory mate guarding is considered to be male's competitive strategy to monopolize a female before copulation, to ensure his priority of access to the female at the moment she is ready to copulate, and thus enhancing his mating success (Grafen and Ridley, 1983). In particular, this strategy of mate guarding has evolved in species in which female receptivity for males is time-limited and predictable.

Among crustaceans, precopulatory mate guarding behavior is prevalent among decapods, amphipods, and marine isopods. It typically takes the form of male carrying the female, sometimes for the period of days, as in brachyuran crabs, or the male performing noncontact guarding by closely attending the female and in some cases caging her between the chelipeds, as in the rock shrimp *Rhynchocinetes typus* (Correa et al., 2003). The obvious benefit of guarding for a male is mate monopolization when the probability of encountering receptive mates is low (Parker, 1974). The probability of finding mates depends on the availability of mates that are ready for mating, a condition known as OSR (Emlen and Oring, 1977). When OSR is less than one, females are rare and competition for mates is presumably intense. This condition could cause sexual selection in the form of increased male size, as in isopods to favor mate guarding behavior.

In peracarids, OSR is highly male biased, because of time-limited female receptivity to copulation, which starts at the parturial molt when the exoskeleton is still soft. In addition, receptivity ends with immediate oviposition following copulation, when the oostegites forming the external egg pouch and developing eggs close access to oviducts (Johnson et al., 2001). Thus, precopulatory mate guarding has functional significance in ensuring the male's presence at the moment of receptivity for copulation. This also applies to other crustaceans in which mate guarding is linked to restricted female receptivity at the time of its reproductive molting, paving the way for the evolution of precopulatory mate guarding in Crustacea (Jormalainen, 2007). In fact, Ridley (1983) found a taxonomic association between precopulatory guarding and time-limited opportunity for fertilization in crustaceans.

Interestingly, mate guarding is commonly practiced as a mating behavior among dominant males of species, possessing different morphotypes. For example, in the freshwater prawn *Macrobrachium rosenbergii*, the large blue-clawed male morphotype invariably guards the female before molting and mating, whereas the other subordinates perform sneak mating by depositing their spermatophores quickly, without any mate guarding (Ra'anan and Sagi, 1985). The same is also true with other *Macrobrachium* species like *Macrobrachium amazonicum* (Moraes-Riodades and Valenti, 2004). Many species of the rock shrimp *Rhynchocinetes* also exhibit mate guarding by dominant robustus male morphotype, while the female-like typus and intermedius types mate stealthily and deposit spermatophore without any mate guarding. Table 5.2 enlists the presence of male morphotypes in different malacostracan crustaceans.

Precopulatory mate guarding is considered as a time investment strategy that permits males to maximize their reproductive success, when mating is restricted to a relatively brief period during the female's reproductive cycle. Species with male-biased sex ratios and/or very restricted opportunities to copulate are assumed to guard females for longer periods of time than species in which the female resource is more plentiful (Ridley, 1983). Thus, in snow crabs and isopods, males guarded females significantly longer, if the OSR became more male biased. In an experimental study, Jormaleinen and Shuster (1999) showed that in the isopod *Thermosphaeroma thermophilum*, the mate guarding duration was longer under male-biased than under female-biased sex ratios.

TABLE 5.2 Male Morphotypes and Mating Behavior in Caridean Shrimps

Family	Species	Male Morphotypes	Mating Behavior	Mate Guarding	Fertilization Success (Ovigerous)	Remarks	References
Palaemonidae	*Macrobrachium rosenbergii*	Small males (SM), orange claw (OC) males, and blue claw (BC) males	OC males showed reduced reproductive activities in the presence of BC males. Sneak mating was observed in SM. Repeated sperm transfer events were not observed	Dominant males (BC) actively court and protect the females	93.3% in BC males, 37.5% in OC males, and 33.3% in SM	Experiments conducted in both competitive and noncompetitive environments	Ra'anan and Sagi (1985)
	Macrobrachium amazonicum	Translucent claw (TC), cinnamon claw (CC), green claw 1 (GC1), and green claw 2 (GC2)	Unknown	Unknown	Unknown	–	Moraes-Riodades and Valenti (2004)
	Cryphiops caementarius	Large males and small males	Unknown	Unknown	Unknown	–	Rojas et al. (2012)

Continued

TABLE 5.2 Male Morphotypes and Mating Behavior in Caridean Shrimps—Cont'd

Family	Species	Male Morphotypes	Mating Behavior	Mate Guarding	Fertilization Success (Ovigerous)	Remarks	References
Rhynchocinetidae	*Rhynchocinetes typus*	Typus males (TM), intermedius males (IM), and robustus males (RM)	All males behaved in similar manner. Touching, overlapping, and caging behaviors were observed. Multiple sperm transfer events were also observed. Stimulating and checking behaviors also followed the sperm transfer events	Prolonged mate guarding by RM were observed in cage position	All males have same reproductive potential to mate with the receptive females	Experiments were conducted in noncompetitive environment	Correa et al. (2003)
	Rhynchocinetes brucei	Typus males (TM), intermedius males (IM), and robustus males (RM)	All males behaved in similar manner. Touching, overlapping, and caging behaviors were observed. Multiple sperm transfer events were also observed	Prolonged mate guarding by RM were observed in cage position	All males have same reproductive potential to mate with the receptive females	Experiments were conducted in noncompetitive environment	Thiel et al. (2010)

	Rhynchocinetes durbanensis	Typus males (TM), intermedius males (IM), and robustus males (RM)	All males behaved in similar manner. Touching, overlapping, and holding behaviors were observed. Sperm transfer event was observed only once	Prolonged mate guarding was not observed	TM has showed reduced reproductive potential compared to IM and RM	Experiments were conducted in noncompetitive environment	Prakash et al. (2015)
Hippolytidae	*Saron marmoratus*	Large males and small males	Unknown	Unknown	Unknown	–	Tirmizi and Kazmi (1971)

Several theoretical models have been advanced to explain mate guarding behavior in crustaceans. A central question addressed in these models concerns the timing and duration of guarding and the factors that influence it. From a male's perspective, guarding duration reflects a balance between a selective force favoring prolonged guarding to ensure his monopolization of the female and a contradictory selective force favoring minimization of guarding duration, by way of approaching a female, nearer to its sexual molt or receptive period (Jormalainen, 1998). This will not only decrease the guarding costs but also increases the male's mating chances. In other words, the decision to start guarding is based on the ability of males to assess the female's reproductive status.

Obviously, the main factor that influences the practice of precopulatory mate guarding in Crustacea is the restricted female receptivity during the reproductive or parturial molt of the female, when the soft cuticle around the ovipores is permissible to intromission. Therefore, mating should be accomplished during this short receptive period, lest the breeding female could not oviposit fertilized eggs. Precopulatory mate guarding behavior thus assures fertilization insurance in a wide variety of crustacean species, in addition to offering paternity assurance by effectively preventing sperm competition. Nevertheless, in the absence of mate guarding, alternative forms of paternity assurance by way of sperm plug formation in certain brachyuran crabs could evolve (Diesel, 1991).

Mate guarding in many species commences well before the onset of molting to gain priority to the prospective female's sexual molt and receptivity period. Therefore, the decision to start guarding is based on the ability of the males to assess whether the female is in an appropriate reproductive state (Jormalainen, 1998). Guarding duration has been shown to be influenced by the availability of receptive females, intensity of competition among males, and female mate choice.

Mate guarding by males also imposes a fitness cost on females, resulting in intersexual conflict over guarding duration (Jormalainen, 1998). The sexual conflict over mate guarding is well demonstrated in amphipods and isopods. Cothran (2004) demonstrated that costs of precopulatory guarding due to predation by fish are higher for females than for males in the amphipod species he studied. In a way, female resistance to the guarding attempts of males in amphipods helps to shorten the guarding duration, thereby reducing the guarding costs on females. Under conflicting interests, a compromised guarding duration would be set somewhere between the optimal durations for the male and the female.

While the precopulatory mate guarding has been witnessed in different crustacean species, in which mating is associated with female molting, it is not uncommon to find the guarding to be extended beyond copulation and female egg laying in many species. In Brachyura, the portunid and cancrid crabs, exhibiting postcopulatory mate guarding, invariably mate after female molt (Jivoff, 1997), whereas the grapsid crab, which copulates in the intermolt stage, does not practice postcopulatory guarding of females (Brockerhoff and

McLay, 2005b). In decapods, postcopulatory mate guarding is considered to be a behavioral adaptation in males for the avoidance of sperm competition. While several brachyuran crabs have contrived mechanisms such as sealing of female spermatheca by sperm plugs to prevent sperm competition (Diesel, 1991), the continued mate guarding by males is an important method to prevent supernumerary mating by other males. Wilber (1989) concluded that the extended postcopulatory mate guarding, found in the male stone crabs even under conditions of predation risk, could be driven by sperm competition.

5.3.1 Mate Guarding in Decapoda

Brachyuran and anomuran crabs exhibit a variety of mate guarding behaviors. In the blue crab *C. sapidus*, courtship and mating involve a sequence of behaviors that extends over a period of several days (Gleeson, 1991). Male crabs attend or guard pubertal females that are within approximately 10 days of molting. The male exhibits a distinctive courtship-display behavior, which includes the extension of the chelae toward a lateral position, extension of the walking legs such that the male is elevated high off the bottom, and rhythmic lateral waving of the swimming appendages (fifth pereopods), which are rotated to an anterodorsal position above the carapace. The male orients the display toward the female and slowly approaches her. In response, the female moves toward the male, maneuvering to a position under him. The male then embraces the female and, using his first pair of walking legs, holds her beneath him in what is termed the cradle-carry position. In this condition, the crabs may remain coupled for several days until the female undergoes her pubertal molt. During molting, the female is released; however, the male continues to stand over her. At the completion of the molt, the pair once again assumes the cradle-carry position. Mating is initiated by the female within a few minutes to an hour after she molts. The female positions herself upside-down beneath him and extends her abdomen, allowing the male to insert his paired copulatory appendages into her genital pores. Spermatophores are deposited within the spermathecae over a 5–12 h period after which the cradle-carry position is re-established and maintained for two or more days.

In crab species with precopulatory mate guarding, males mate more frequently by obtaining females that require short mate guarding times. Large males have advantages in aggressive interactions with other males for access to females and in struggles with females to capture or physically control them. In general, large males spend less time guarding each female than do small males (Grafen and Ridley, 1983). In the blue crab *C. sapidus*, female behavior also changes as they progressed through the final pubertal molt cycle, such that early molt-stage females avoided males, but late molt-stage females initiated pair formation. For blue crab males, longer periods of mate guarding offer a higher probability of mating than increased searching for females that require short periods of guarding. The resistance of mate guarding attempts by the premolt females indicates that prolonged guarding increases the probability of aggressive encounters with unpaired males, which can result in female injury and reduces the feeding opportunities needed for the final molt (Christy, 1987). Females resist males during their early premolt mating, thereby preventing mate guarding altogether; but females approaching the time of molting resist less and court males more often (Strong, 1973).

Although in many brachyuran crabs mating occurs soon after the female molts, several species inhabiting terrestrial or semiterrestrial environments mate in the hard-shelled intermolt condition. In these crab species, both visual and/or acoustic signals are used for mating communication. Nevertheless, in marine crabs belonging to Majidae, such as the spider crab *Inachus dorsettensis*, mating occurs in the postpubertal intermolt females carrying eggs nearing hatching or which have recently hatched their eggs (Jones and Hartnoll, 1997). It thus seems possible that not only the female's molting, but also the egg-brooding condition is attractive to the males in the spider crab.

Premolt mate guarding is also seen in the anomuran crabs. In the hermit crabs, belonging to the genus *Diogenes*, the male with its unusually large chelipeds, grasps the rim of the shell inhabited by a mature female, and guards her for one to several days before copulation, and fights with other males approaching him using the major cheliped (Wada et al., 1997).

Mate guarding assumes a new dimension in the spiny lobsters. Contrary to the assumption that the females are passive partners in mating, the premolt females of the American lobster *Homarus americanus* actively "forage" for dominant males and share the male's shelter during molt and copulation (Dunham and Hurshman, 1991). Molting is such an important prerequisite for mating in *H. americanus* that the receptive females staggered the timing of their molts in such a way that each molted and subsequently mated inside the shelter of a dominant male. Such a behavioral control over molting by the female means that it can even regulate the timing of its egg-laying by postponing the molting (Cowan and Atema, 1990). In the lobsters, the release of pheromones by the females to attract the males could thus be regulated (Christofferson, 1978).

While mate guarding in brachyuran crabs belonging to Portunidae and Cancridae is associated with molting of the female partners, in the spider crabs of Majidae family, mate guarding by the male is associated with the prespawning conditions. Spider crabs are known for their multiparous spawning behavior and multiple mating and practicing sperm competition (Diesel, 1991). Like other brachyuran crabs, mating behavior in spider crab includes "grasping," "mating embrace," or "guarding," collectively described under precopulatory mate guarding in Crustacea. In spider crabs, precopulatory mate guarding behavior is observed both during the preparturial molt in the primiparous females or during the prespawning period of the multiparous females. In species such as *Chionoecetes opilio* males are attracted to and mate with prespawning females, which they guard until spawning. Males may also mate with females that have recently spawned without mate guarding. Evidently, egg hatching may provide attraction for the males to court and mate with the females, as inferred from the fact that females with eggs in the early stages of development could not attract the males (Diesel, 1991). The female snow crab *C. opilio* has a preference for large males for mating as it could prevent takeover attempts by smaller rivals, although the fertilization capacity of small males' ejaculate is no different from that of large males. In captivity, small males copulated successfully only with soft newly molted primiparous females, which are less fecund than multiparous females (Conan et al., 1988).

Prespawning mate guarding has also been reported in the ghost spider crab, *I. phalangium*. The spider crabs probably recognize the prehatching condition of eggs carried on the pleopods and as many as four males may visit, mate, and guard a prespawning female which has taken refuge under the stinging tentacles of sea anemones. Frequently a guarding male is replaced by a larger one on a subsequent day until finally the largest dominant male in an area guards her after copulation until spawning (Diesel, 1986b). The beneficial aspects of both pre- and postcopulatory mate guarding for the brachyuran crabs are related to ensuring a prospective mate during the restricted female receptivity period, together with the assurance of paternity by preventing supernumerary mating. In another sea anemone spider crab, *Anemone sulcata*, the males guard the adult females with the eggs in the brood at least two days before the hatching of the egg and aggressively guard the female from the smaller males. Interestingly the males leave the females after mating.

5.3.2 Mate Guarding in Amphipods

In the euryhaline amphipod *Gammarus lawrencianus*, mate guarding behavior is highly stereotyped (Dunham and Hurshman, 1991). When a male comes into contact with an acceptable free-swimming female a few days before her molt, he pulls her close to his ventral surface in such a way that the female is placed into a more efficient carrying position with the female underneath the male and their body axes parallel. The males possess prehensile thoracic appendages called gnathopods to grasp the female's integument. The pair swims together in this "precopula" position until the female molts, at which time ova are passed into the female's external brood chamber, where they are fertilized by the guarding male. In these amphipods, the males discontinue pairing soon after fertilization and the females do not store sperm (Sutcliffe, 1992). Amphipod mating is characterized by intrasexual competition and by complex intersexual interactions reflecting a tension between cooperation and conflict (Trivers, 1972). In freshwater amphipods, time-limited and asynchronous female receptivity give rise to male-biased sex ratios. With the result, there is intense competition among males for mating opportunities. This condition obviates an energetically costly male searching behavior, resulting in longer mate guarding duration and intersexual conflict. Evidently, pairing females use resistance to counter the costs of prolonged precopulatory mate guarding by males. This resistance behavior may also impose sexual selection on male traits such as large size, which is advantageous to the competing males in scramble competition or takeover attempts from small males (Bollache and Cezilly, 2004). In another freshwater amphipod, *Hyalella azteca*, females through their resistance behavior appear to control pair formation (Strong, 1973).

5.3.3 Mate Guarding in Isopods

Isopods are known for their varied mating strategies, ranging from strict monogamy to promiscuity, and from serial mate guarding of free-ranging iteroparous females to semelparous harem polygynandry (Johnson et al., 2001). Precopulatory guarding is a common behavior in all aquatic isopod groups with the possible exception of the terrestrial suborder Oniscidea. As in other crustaceans, mate guarding is linked to molting of the females. However, the pattern of molting in isopods is biphasic, the posterior portion of the cuticle is shed first, and the anterior portion molts with a time gap. The males guarding the females sometimes help removing the cuticle of the posterior portion. The molting of the posterior part exposes the soft genital pores at the base of each fifth walking leg. Female receptivity to copulation usually starts at the parturial molt, when the exoskeleton is still soft and the receptivity ends with immediate oviposition following copulation when the oostegites, close access to oviducts. Therefore, the presence of ripe ovary during molting is a combined requirement for the mate guarding to occur. The functional significance of precopulatory mate guarding in isopod lies in ensuring presence of the males at the moment of receptivity for copulation.

In two isopod species, *Thermosphaeroma milleri* and *T. thermophilium*, mate guarding duration has a relationship with sex ratio. Guarding lasted longer with a male-biased sex ratio in both the species, and the difference in guarding duration was due to males adjusting their guarding criteria rather than to female responses to the level of male harassment (Jormalainen and Shuster, 1999). In the marine isopod *Idotea baltica*, the mate guarding time adjustment occurs in a size-specific manner: large males in the mating population postpone the start of guarding when put together with a female, compared to smaller males, implying that males are able to collect information on the composition of the mating population and that they respond to it in a size-specific way. Large males in this species are capable of taking over females from small ones, thereby postponing the start of female guarding (Grafen and Ridley, 1983).

In the isopods, an observed association between protogyny and precopulatory mate guarding may also suggest that the benefits for large males may be the primary selective agent favoring protogyny. Interestingly, precopulatory mate guarding has not been reported in protandric hermaphroditic crustaceans (Brook et al., 1994). In isopods, the molting is of two types: one common molt and the other reproductive molt. In most isopods, though mate guarding may commence long before the posterior molt, copulation occurs soon after the molting of the posterior part, followed immediately by oviposition. Thus, guarding duration is primarily adapted to the length of female reproductive cycle, as the females lack sperm storage facility in the oviduct. The sequence of reproductive activities during the molting cycle along with the mate guarding duration for a supralittoral isopod *Ligia dentipes* is represented in Figs. 5.1 and 5.2 along with the timing of each reproductive molting event (Santhanakumar et al., 2014). It may be noted that mate guarding takes place only before the reproductive molt, with a ripe ovary. No mate guarding occurs before or after the common molt, when the females are nonreceptive with an undeveloped ovary.

A possible exception is found in the thermal-spring isopod species, *T. thermophilium* and *T. milleri*, where there is a temporal separation of reproductive molt and oviposition (Jormalainene and Shuster, 1999). In both the species, mate guarding commences after the parturition of the previous molt and the

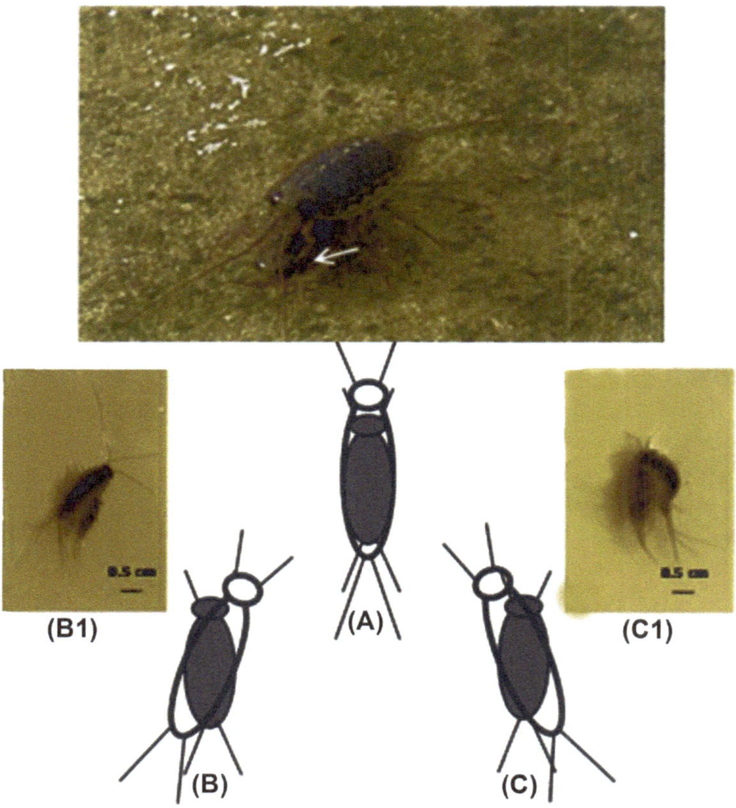

FIGURE 5.1 (A–C) Mate guarding positions of male and female in *L. dentipes*. B1 and C1 photographic representation. *Reproduced from Santhanakumar, J., Mohanraju, R., Kirubagaran, R., Subramoniam, T., 2014. Mate guarding behaviour in the supralittoral isopod,* Ligia dentipes *(Oniscidea) from the Andaman and Nicobar Islands. Invertebr. Repr. Dev. 58, 128–137.*

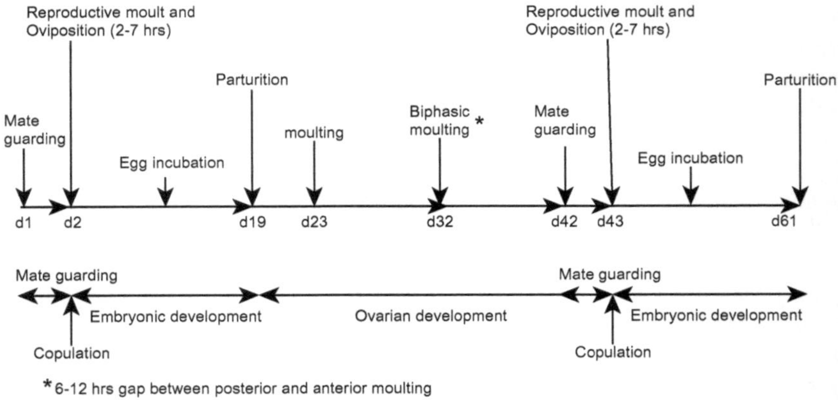

FIGURE 5.2 Sequence of reproductive activities during molting in *L. dentipes. Reproduced from Santhanakumar, J., Mohanraju, R., Kirubagaran, R., Subramoniam, T., 2014. Mate guarding behaviour in the supralittoral isopod,* Ligia dentipes *(Oniscidea) from the Andaman and Nicobar Islands. Invertebr. Repr. Dev. 58, 128–137.*

copulation or insemination occurs soon after the posterior molt. The males invariably terminate female guarding after insemination. However, oviposition takes place long after the insemination, because ovarian development itself starts after the molting in *T. milleri* and *T. thermophilium*. In both the species, oocyte development extends up to intermolt stage, when the oviposition is possible. Interestingly, many females of *T. milleri*, underwent secondary molt instead of oviposition, thus separating oviposition from insemination. Evidently, this isopod is capable of transmolt sperm storage in the oviduct. Since guarding during ovarian development may interfere with their ability to provision the developing ova, this strategy of temporally separating copulation and oviposition is considered to be an evolutionary female trait arising in the context of sexual conflict.

Precopulatory guarding and copulation may sometimes be followed by short postcopulatory guarding. *I. baltica* and *T. thermophilium* males guard and mate with the female after it has molted the posterior part of the exoskeleton, but the male continues to guard the female even after the shedding of the anterior cuticle, when the female has already oviposited (Jormalainen and Shuster, 1999). Such postcopulatory guarding probably functions as paternity assurance when there is a risk of sperm competition, especially in the species with continuous receptivity. In addition to males physically attaching to females during guarding, guarding may take the form of cohabitation of the pair. For example, in the wood-boring isopod *Limnoria tripunctata*, the male and female may spend extended periods in a burrow, during which they mate and produce multiple broods (Wilson, 1991). In another wood-boring isopod, *Sphaeroma terebrans*, the male cohabits with a female having ripe ovaries and leaves the burrow after egg laying, implying that copulation has taken place during the temporally restricted female receptivity.

Isopods also show subtle variation in their guarding strategies. Although the commonest practice is the precopulatory mate guarding, in certain Jamiridae species, guarding of the newborn mancae in the female brood has been reported (Thiel, 2002). In *Iais pubescens*, males manipulate the marsupium of females to obtain virgin mancae and carry them for a week until the first postmarsupial molt; then they copulate and release the mancae. The fourth pereopod of the adult males is reduced and is used to grasp a manca tightly underneath. Franke (1993) has described a similar kind of guarding behavior in the commensal species *Jaera hopeana*, in which males guard the mancae up to 12 days until the first postmarsupial molt, at which point copulation takes place. The immature mancae store the sperm until they reach maturity but remain continuously sexually receptive and may engage in further copulations. However, as the females have the tendency to refuse additional copulations, sperm precedence of the mating with the first manca stage is assured. Thus, both the sperm storage capability and early receptivity to mating may be adaptations of the female mating strategy to enhance colonization success.

While prolonged mate guarding is the rule in aquatic and semiterrestrial isopods, land isopods belonging to the suborder Oniscidea are an exception (Zimmer, 2001). However, they may have a short, approximately 1-min "nuptial ride" before copulation. Carrying females for prolonged periods may be costly for males in a terrestrial environment. There can be increased energetic cost of carrying or the maneuverability of precopulatory pairs may be difficult in habitats in crevices or under logs and stones. Evidently, terrestrialization would have contributed to the loss of guarding because of the costs of mate guarding being higher than the reproductive gains. Furthermore, mates guarding might have been the ancestral condition in isopods, by its prevalence in aquatic species and in the semiterrestrial forms such as *Ligia*. A transitional condition in the female receptivity and mate guarding time has also been noticed in the supraterrestrial isopod *L. dentipes* (Santhanakumar et al., 2014). In *L. dentipes* females show restricted receptivity for up to 12 h as compared to nearly six days in *Ligia oceanica* (Nicholls, 1931). In this species, mate guarding starts much earlier than that of *L. dentipes*.

The abandonment of prolonged precopulatory mate guarding in the terrestrial oniscoid isopod is associated with the loss of restricted female receptivity for males during parturial molt. Instead, the females have extended receptivity, permitting as many as 30 complete copulations in different *Porcellio* species (Linsenmair, 1989). The difference in mating tactics is also reflected on other characteristics such as prolonged sperm storage in the land isopods. Therefore, in terrestrial isopods, each oviposition is not preceded by mating, whereas in the marine isopods, each reproductive molt is followed by mating and then oviposition, as long-term sperm storage in females is lacking. It appears that the loss of temporal restrictions of female receptivity may have forced males to give up costly precopulatory mate guarding (Jormalainen, 2007).

Laboratory investigations on the premate guarding behavior of the semiterrestrial isopod *L. dentipes* suggested that the male and female always had a tendency for immediate heterosexual pairing, in spite of intersexual conflicts, before pair formation (Santhanakumar et al., 2014). The successful male guards the female by clinging on to the dorsal side of the female by holding it firmly with its first pereopod (Fig. 5.1). The guarding male helps the female partner, aiding in the removal of the molted exoskeleton. The mating occurs immediately after the posterior body exuviates. The male leaves the female after copulation and goes in search of another receptive female, demonstrating a polygamous mating behavior, existing in these isopods. In a similar way, the mated females also remate with several other males without any premate guarding. Intrasexual encounters among the males were witnessed before the dominant male captures the receptive female. In general, the larger males are successful in the competition.

5.4 ALTERNATIVE MATING STRATEGY

Alternative mating tactics is a mating strategy that differs from the prevailing strategy used by male or female. Many crustaceans employ a predominantly common mating behavior in a mating system, yet several of them exhibit discontinuous variations in mating behavior and morphology, known as alternative mating tactics. Intrasexual selection in male crustaceans favors development of weapons such as large chelae or other traits such large body size to improve monopolization of females. Sexual selection of this type gives rise to occurrence of different morphotypes in males, favoring alternative mating strategies, especially by subordinate males (Baeza and Thiel, 2007).

5.5 MALE MORPHOTYPES AND ALTERNATIVE MATING TACTICS

The occurrence of different male morphotypes in isopods and different species of caridean shrimps with differences in size and associated morphological features illustrates the practice of alternative mating strategy in crustaceans. The marine isopod *Paracerceis sculpa*, which breeds inside the spongocoel, has three genetically discrete male morphotypes, α male, β male, and γ male (Shuster, 1987). The largest α male possesses elongated uropods and defends harems within spongocoels. The smaller β male resembles females in morphology and size and invades the spongocoels by mimicking the behavior of females. The γ male is tiny and secretive and enters the harems by stealth. Although the mating potentialities of these three male morphs seem to be the same, the relative fertilization success among them varied significantly. If the spongocoel contains only one female, the α males could effectively defend and mate with it, in the presence of more β male or γ male. However, when more females are present in the spongocoel, the success rate of mating of both β and γ males increased, even in the presence of a α male. Thus, the aggregation of females in the spongocoel allowed the coexistence of both β and γ males, creating a mating niche for these subordinate male morphs.

In the spider crab, *Libinia emarginata*, two male morphotypes have been described—one small that possesses unabraded small claws and the other type includes males with fully or partly abraded exoskeleton with large claws (Sagi et al., 1993). The males with large claws have very well-developed testis and show aggressive mating behavior. Interestingly, such aggressive mating behavior in the large-clawed male morphotypes is correlated to a high level of methyl farnesoate in the hemolymph.

In natural populations of caridean shrimps, presence of several male ontogenetic stages with differences in morphology as well as physiological capacities is documented in the freshwater prawn, *M. rosenbergii* and the rock shrimp *Rhynchocinetes*. In *M. rosenbergii*, there are three morphotypes, viz., small male, orange-clawed, and blue-clawed coexist within the population. These morphotypes develop sequentially from the small male to blue-clawed males which represent the dominant mating type. The large or dominant blue-clawed males actively court and protect the females before mating. The subordinate or intermediate type stages display reduced activities in the presence of larger males. The orange-clawed small males, which are similar to females, always show sneak mating behavior with high mobility (Ra'anan and Sagi, 1985). The developmental expression of phenotypic plasticity in *Macrobrachium* is sensitive to changing social environmental cues in such a way that males adjust their mating phenotypes in response to the changing social environment (Shuster and Wade, 2003).

Three species belonging to the genus *Rhynchocinetes*: *R. typus* (Correa et al., 2000), *Rhynchocinetes brucei* (Thiel et al., 2010), and *Rhynchocinetes durbanensis* (Prakash et al., 2015) also possess male morphotypes. Table 5.2 gives different types of male morphotypes along with their behavioral characteristics. Males first become mature as the female-like typus morphotype, after which they molt through various intermediate stages ("intermedius morphotypes") to the final robustus morphotype. In *R. durbanensis*, there is strong sexual dimorphism in that the robustus morphotypes are greater in size and with larger appendages (extended third maxillipeds and major chelipeds) than females. As typus males grow bigger, their maxillipeds and major chelipeds grow proportionately larger than those of females. The number of corneous spines on the terminal segment of the maxillipeds is also reduced. As regards the body size, there are no marked differences between typus and intermedius males and females, although robustus males are substantially larger than females (Fig. 5.3). However, considerable size difference exists among different male morphotypes as well as female in another species, *R. typus* (Correa et al., 2000).

In these caridean shrimps, the presence of more than one male morphotype reflects phenotypic plasticity in reproductive growth resulting in alternative mating behaviors. Rhynchocinetid males establish a linear morphotype-related dominance hierarchy, in which the largest male type dominates. Males of these species have evolved complex and stereotyped (or ritualized) agonistic behaviors accompanied by morphological adaptations for fighting (Karplus and Harpaz,

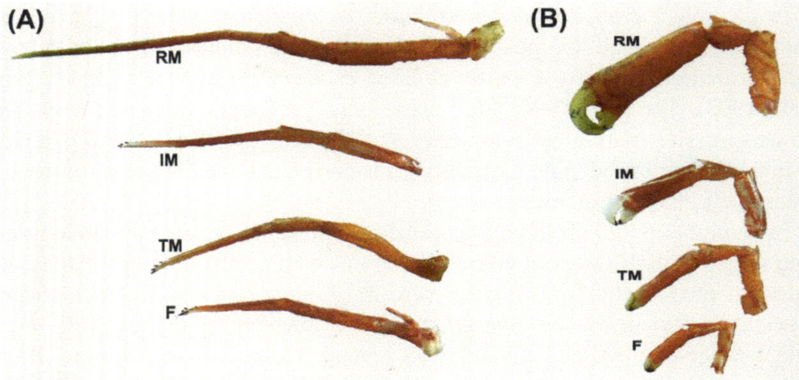

FIGURE 5.3 Variation in body parts of male and female morphotypes: (A) third maxillipeds; (B) major chelipeds. *F*, female; *TM*, typus males; *IM*, intermedius males; *RM*, robustus males. *Reproduced from Prakash, S., Ajithkumar, T. T., Bauer, R., Thiel, M., Subramoniam, T., 2015. Reproductive morphology and mating behaviour in the hingebeak shrimp* Rhynchocinetes durbanensis *Gordon, 1936 (Decapoda: Caridea: Rhynchocinetidae) in India. J. Marine Biol. Assoc. U.K. 1–10.* http://dx.doi.org/10.1017/S0025315415001083.

1990; Correa et al., 2003). When the dominant morphotype courts and defends receptive females, the subordinate males exhibit a sneak mating behavior and rapidly attach spermatophores on the female sternum (Correa et al., 2003). Such alternative mating tactics obviously result in multiple matings (Thiel et al., 2010). Recent investigations on the mating behavior of *R. durbanensis* in a noncompetitive environment showed that the robustus males are faster in approaching the receptive females for mating compared to other subordinate males (intermediate and typus morphotypes). These results suggest a dominance hierarchy in these shrimps in which females appear to prefer mating with the larger males. Thus, male size and weaponry is positively correlated with their mating success as reported for other caridean shrimps, eg *R. typus* (Correa et al., 2000) (Fig. 5.4) and *R. brucei* (Thiel et al., 2010).

In another instance, alternative mating strategies have been witnessed in the partially protandric alphid shrimp, *A. kominatoensis* (Nakashima, 1987). In this species, individuals mature as males and then change sex as in other protandric hermaphrodites. However, only the smaller and subordinate males change sex to females, whereas the larger ones remain males throughout their lives. Since the small males have low mating success in the presence of big males, they adopt the alternative strategy of changing sex to females.

Although it is expected that females favor dominant males under these circumstances, because of genetic advantages or for female protection during spawning, female preference for certain males has not been clearly demonstrated (but see Hinojosa, 2001). Smaller males, unable to succeed in direct competition with dominants, have shown a high degree of behavioral

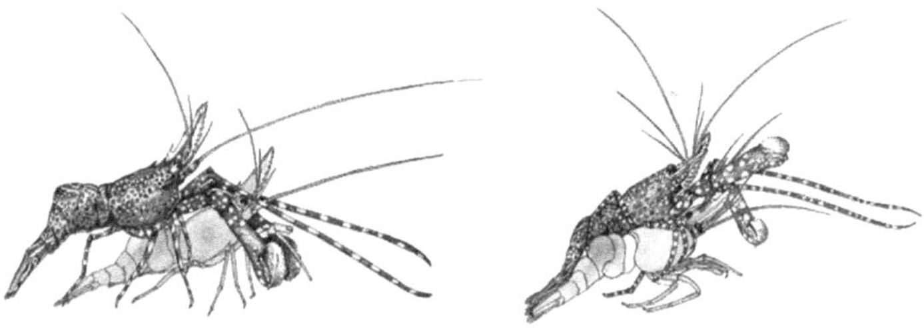

FIGURE 5.4 Caging position of dominant males on females during mating in *R. typus. Reproduced from Correa, C., Baeza, J. A., Dupré, E., Hinojosa, I. A., Thiel, M., 2000. Mating behavior and fertilization success of three ontogenetic stages of male rock shrimp* Rhynchocinetes typus *(Decapoda: Caridea). J. Crust. Biol. 20, 628–640.*

plasticity. Making the best of a bad situation they are able to shift their mating behavior toward a simplified mating tactic (similar to "pure searching" see below) when exposed to a limiting social environment (Ra'anan and Sagi, 1985; Correa et al., 2003).

5.6 MATING CONTESTS

5.6.1 Intersexual Conflict of Reproductive Interests

Prolonged mate guarding in several crustacean species described above provokes intersexual conflicts. Parker (1970) proposed this conflict of interest theory for mating, when the reproductive strategy or behavior of one sex imposes a cost on the fitness of the other sex. The differences in sexual features such as investment in gametes and differential reproductive rates are responsible for different male and female mating strategies. In the isopod *I. baltica*, a typical conflict situation occurs when males attempt to mate and females resist the mate attempts. In this species, there is a period of several days preceding the start of the guarding phase, during which the males continuously initiate guarding and the female resists the male's attempt. The female may assume a scrolled or hooked position, making firm attachment by the male difficult or the female may vigorously kick the male that attempts to guard (Jormalainen, 1998, 2007).

Intersexual conflict also arises during the mate guarding of the freshwater isopod, *Lirceus fontinalis*, found in stream habitats. In this species, whenever there is an encounter between the male and the female, there is struggle and the male attempts to overpower the female. If the male is successful, precopulatory guarding will follow for one to three days, and copulation and egg laying takes place immediately following the parturial molt. Mating contests in this isopod involves male preference and female resistance, contributing to different components of the mating sequence, although female resistance ultimately determined whether or not pair formation occurred.

Males expressed a preference for females that appeared to be close to molt, using 20-hydroxyecdysone, as the waterborne distance pheromone. Female resistance appears to function in both mate avoidance and mate choice. In the aquatic isopods, males exhibit precopulatory mate guarding as a strategy to ensure their presence during the short fertilizable period that follows female molt. Males and females generally adopt strategies that minimize the mate guarding costs: males prefer females that are close to molt and females are less resistant as molt approaches. The sexual conflict that arises from female mate choice leads to the development of male traits via sexual selection (Sparkes et al., 1996).

In Crustacea, mate guarding by males imposes some energetic costs on females. In the marine isopod, *Idotea pelagica*, mate guarding restricts female feeding thereby affecting the fecundity (Jormalainen, 1998). Further, guarding costs on female fitness imply intersexual conflict. Prolonged mate guarding may also entail lost mating opportunities, energetic of carrying cost and restricted male feeding during guarding, thereby decreasing growth and mating success. On the contrary, prolonged mate guarding may offer certain other benefits such as assured paternity for the male. In the amphipod *G. lawrencianus*, Robinson and Doyle (1985) showed that mate guarding lowered growth rates for males in comparison to other males that were not allowed to guard.

When mate guarding is costly for females, it leads to intersexual conflict and female sexual selection in terms of female behaviors and traits that reduce guarding costs. As exemplified in amphipods and isopods, the prominent sexual behavior involved in the sexual conflict is female resistance and mate choice. In *I. baltica*, female resistance to guarding attempts effectively decreases actual guarding duration (Jormalainen and Merilaita, 1995). Furthermore, female resistance leads to differences in size-specific guarding duration: large males are expected to guard both large and small females for longer periods than the small males. Several studies in crustaceans have indicated that female behavior plays a more significant role in determining mating outcomes.

In another freshwater isopod, *L. fontinalis*, female resistance also acts as a mechanism of mate choice. The females could first resist all males equally and later selectively mate with males that can overcome resistance. The females can also directly assess some qualities of males during contests and choose to either resist or cooperate with particular males. Thus, female *L. fontinalis* that are close to growing molts avoid males and the costly contests and prefer to live in the microhabitats of the stream where males are scarcely found (Sparkes et al., 2000).

Intrasexual selection has also improved male behaviors and traits that are used in locating, achieving. and guarding mates. For instance, during their breeding period, *I. baltica* males swim more actively than females, braving the risk of predation (Jormalainen et al., 1995). In mate guarding amphipods, sizes of the second antennae and the second pair of gnathopods are sexually dimorphic, indicating their enhanced sensory ability and aggressiveness in male–male interactions or their enhanced use in carrying capacity of the females during mate guarding. Since molting of the males disrupts guarding, males can even postpone their own molting during guarding in the amphipods. Alternatively, males of *I. baltica* may even delay the start of guarding after their own molt, resulting in longer molt cycles in males (Wickler and Seibt, 1981). Furthermore, intrasexual selection for large males is a common feature and a general rule in all mate guarding crustaceans.

5.6.2 Female Mate Choice

Although female crustaceans have a tendency to be passive partners in copulation, they do exhibit a mate choice in selecting a male that could give her fitness benefits and genetic benefits. It is commonly observed in Crustacea that large dominant males are more successful in contests over females and that females themselves prefer such dominant males for mating. Differential mating success, arising from male–male competition among males, as well as female mate choice, is considered to be the driving forces for sexual selection in crustaceans. Thus male characteristics and female preferences could co-evolve and, in fact, sexual selection may be stronger through mate choice than through male–male competition (Shuster and Wade, 2003). However, male morphological traits that influence a male's competitive ability such as large claws could be used by the females as mate choice criteria. By choosing certain mates over the others, females derive genetic benefits in terms of offspring viability and attractiveness, as well as direct fitness benefits. By choosing a large male, for example, the American lobster *H. americanus* obtained direct benefits in terms of the fertilization rate of ova (from more ejaculate) and better protection (longer cohabitation) during the vulnerable postmolt period (Gosselin et al., 2003). In a different way, the female isopod, *L. fontinalis* discriminated against energy-depleted (ie, recently mated) males for mating, thereby increasing fertilization success (Sparkes et al., 2002). In the rock shrimp, the female choice is always on the large robustus male morph over the small typus morph, sensed through chemical signals (Diaz and Thiel, 2004).

The females often use various criteria in their selection of mates. Aquatic crustaceans employ chemical cues to locate potential mates. In the blue crab *C. sapidus*, both male and female emit pheromones to attract mates, but when at close proximity, both mates engage in mutual displays, suggesting use of visual cues for mate choice (Jivott and Hines, 1998; see chapter: Sex Pheromones).

Female mate choice may also take other forms such as the selection of a resource that a male defends. Thus, in the fiddler crab *U. pugilator*, females base their choice on burrow stability and not on male size (Christy, 1983). *Uca crenulata* also selected mates on the basis of burrow characteristics that are important for successful incubation and release of larvae. Understandably, in all *Uca* species, in which the burrow is critical for reproductive success, female mate choice depends primarily on the physical features of the burrow (De Rivera, 2005). In a way, female resistance to mate guarding attempts by males is suggested to act as a means for mate choice because the females could directly obtain information regarding the male's vigor and condition. Isopods and amphipods are the best examples for their female resistance in selecting a high-energy large male for mating (Jormalainen and Merilaita, 1995).

In another form of mate choice, known as cryptic female choice, the female exercises the mate choice at sperm level after the copulation has occurred. In the rock shrimp *R. typus*, the subordinate males often seize a receptive female and attach spermatophores before the female is taken over by a dominant male of the robustus morph. Under these circumstances, the female delays spawning or actively removes a large part of the spermatophores of the subordinate male, thus giving priority for dominant male's sperm in fertilizing the eggs (Thiel and Correa, 2004).

In some crustaceans, the male also increases its reproductive success by exercising its mate choice toward the females that are temporally closer to becoming fertilizable, thereby reducing the cost of prolonged mate guarding. Hence, in the marine isopod, *I. baltica*, the mate preference is not toward a larger female but toward the one that matured earlier for parturial molt, indicating that maturity was more important than female size (Jormalainen et al., 1994). The circulating level of molting hormone could be the basis for such male mate choice, as found in the isopod *L. fontinalis* (Sparks et al., 2000).

Chapter 6

Sex Pheromones

6.1 INTRODUCTION

Pheromones are evolved chemical signals that are used to communicate within a species. By the definition of Karlson and Luscher (1959), pheromones are chemicals which are secreted to the outside of an individual and which act on another animal of the same species. In general, pheromones released in the terrestrial environment (eg, insects and vertebrates) are far more studied than are marine pheromones. Nevertheless, aquatic environment is ideal for the exchange of chemical signals and hence chemical communication is more developed among crustaceans which are primarily aquatic in habitation. In decapod crustaceans, the existence of pheromones has been recognized mostly from behavioral evidence. Further, they have an important role in attracting sexual partners, signaling aggressive motivations, larval hatching or settling, and for individual recognition (see Dunham, 1978).

In crustaceans, sex pheromones have been shown to be stimulators or modulators in sexual communication and copulation (Gleeson et al., 1987). Crustaceans use a variety of stimuli and senses to attract and recognize the opposite sex: visual, acoustic, tactile, and chemical. Among them, chemical stimuli are the most extensively studied, and the chemical compounds that serve to attract and stimulate sexual activity are called sex or reproductive pheromones. Sex pheromones are defined as odors, produced by either males or females that stimulate one or more behavioral reactions in the opposite sex, bringing the males and females together for the purpose of mating. Sex pheromones not only facilitate encounters between the sexes but also modulate mate choice once potential mates come into close proximity. In the majority of the crustacean species, it is the female who releases the long-distance sex pheromone to attract potential mates, although in some species, it is the receptive male who emits pheromones to exhibit their readiness for mating. However, in highly mobile gregarious species (penaeoidean and caridean shrimps), encounters between males and females occur by chance. In such cases, direct contact is necessary to reveal the reproductive status of the female. Obviously, sex pheromones of different types (distance and contact) may play significant roles in accomplishing such mating activities (Atema and Steinbach, 2007).

Sexual Biology and Reproduction in Crustaceans. http://dx.doi.org/10.1016/B978-0-12-809337-5.00006-X

6.2 TYPES OF PHEROMONE

Based on the characteristics such as solubility, concentration, and mode of dispersion, pheromones are categorized into olfactory (distance) pheromones and contact pheromones in crustaceans. Olfactory pheromones are soluble in water and could be dispersed some distance from the emitting individual. Contact pheromones, on the other hand, are insoluble in water and could be perceived by the chemoreceptors by touch. In crustaceans, chemical information in the form of pheromones is broadcast in urine expelled from a small anterior nephropore opening near the base of the antennae. In *Callinectes sapidus*, reproductive pheromones are released from fully mature females as a mating attraction to conspecific males (Gleeson et al., 1987). In the Chinese mitten crab *Eriocheir sinensis*, males are known to receive distance and contact pheromones from the molted females, which stimulate their mating behavior (Herborg et al., 2006). Lobsters and other decapods can control the timing and direction of the urine stream (Aggio and Derby, 2011). In crustaceans, the olfactory pheromone receptors and setae are reported to be present on antennules, especially the lateral aesthetasc hairs, but not on antennae (Schmidt and Derby, 2005; Hallberg and Skog, 2011). Contact pheromones are present predominantly in the caridean shrimps which live in high congregations, whereas distance pheromones are reported in crabs, lobsters, and crayfishes. The occurrence and mode of action of sex pheromones in various crustacean groups are given in Table 6.1.

TABLE 6.1 Occurrence and Mechanism of Action of Sex Pheromones in Crustaceans

Species	Mechanism of Action	References
Caridean Shrimps		
Palaemon paucidans *Hymenocera picta* *Macrobrachium rosenbergii* *Rhynchocinetes typus*	Female olfactory sex pheromones are emitted upon or after the mating molt. Such pheromones stimulate obvious mate search or courtship behavior in males	Kamiguchi (1972) and Díaz and Thiel (2004)
Alpheus angulatus *Athanus kominatoensis*	Olfactory pheromones. Males which live paired with females in burrows get attracted to water exposed to premolt females	Mathews (2003) and Nakashima (1987)
Palaemonetes pugio *Lysmata wurdemanni*	Olfactory pheromone. The male-phase individuals approach and remain in contact with a prespawning female-phase individual about an hour before the latter's molt, when mating occurs. The males detect the females by a water-soluble substance	Bauer and Abdalla (2001) and Zhang and Lin (2004)

TABLE 6.1 Occurrence and Mechanism of Action of Sex Pheromones in Crustaceans—Cont'd

Species	Mechanism of Action	References
Lysmata boggessi	Lipophilic cuticular hydrocarbons as contact sex pheromones	Zhang et al. (2011)
Lysmata wurdemanni	Olfactory pheromone—UDP	Zhang et al. (2010)
Palaemonetes pugio	Cuticular hydrocarbons, glycoproteins, or other compounds secreted on the surface of the cuticle of the postmolt parturial females might serve as sex attractants	Caskey and Bauer (2005)
Litopenaeus vannamei	Olfactory pheromone. Males are stimulated to precopulatory chasing of other individuals (not necessarily females) when "female water" from a reproductively mature female is introduced into their environment	Yano et al. (1988)
Rhynchocinetes typus *Macrobrachium australiense* *Macrobrachium rosenbergii*	The premolt parturial females seek out dominant males. The male apparently attracts the female by regularly or continuously releasing his water-soluble pheromone. The dominant male may then recognize the female (and vice versa) by an interchange of olfactory and/or contact pheromone cues before allowing her into the mating nest	Díaz and Thiel (2004), Karplus et al. (2000) and Lee and Fielder (1982)
Hymenocera sitchensis *Hymenocera paludicola*	Contact pheromone. Males are not noticeably stimulated at a distance by parturial (prespawning) females. It is only when males touch a newly molted parturial female with the long antennal flagellum that they react dramatically, seizing the female and copulating with it. A "nondiffusible" substance on the female exoskeleton was perceived by the male, which stimulated it to copulate	Bauer (1976, 1979)
Heptacarpus paludicola	Distance and contact pheromone. Newly molted female with full ovaries evokes a copulatory response from males upon contact. Males recognize a female for copulation by apparent contact chemoreception with the antennal flagellae. The receptive females of this species emit a pheromone that increases activity of males	Bauer (1978)

Continued

TABLE 6.1 Occurrence and Mechanism of Action of Sex Pheromones in Crustaceans—Cont'd

Species	Mechanism of Action	References
Decapod Crabs		
Callinectus sapidus	Distance pheromone. The male crabs of this species produce distinctive courtship behavior when they smell a pubertal female. A pheromone is present in the pubertal female's urine which is detected by sensors in the aesthetasc sensilla on the antennules of males	Gleeson (1980, 1982, 1991)
Carcinus maenas	Distance pheromone. Precopula females of the crab *C. maenas* released a pheromone that stimulated mating behavior of the males	Hardege et al. (2002)
Carcinus maenas	UDP, a nucleotide accumulated from UDP-*N*-acetyl glucosamine during chitin biosynthesis, is the major component of the female crab pheromone and that stimulated mating behavior of the males	Hardege and Terschak, 2011
Telmessus cheiragonus	The female crab produces a distance pheromone in the urine that stimulated only tracking and courtship behavior but not copulation. There was an additional pheromone released from postmolt females that evoked copulation in male crabs	Kamio et al. (2002)
Eriocheir sinensis	The mate recognition in this species occurs after the physical contact between the male and the female via a contact pheromone	Herborg et al. (2006)
Portunus sanguinolentus	Males of the species display a behavioral response to the presence of premolt females which is the same as their behavior when they are exposed to water in which premolt females have been kept. Release of a sex-attractant pheromone is indicated	Ryan (1966)
Heterozius rotundifrons	Distance pheromone. Females mate immediately after molting. Females remain attractive to males for almost 3 days. A female-derived waterborne pheromone seems to be the basis of this attraction	Thompson and McLay (2005)

TABLE 6.1 Occurrence and Mechanism of Action of Sex Pheromones in Crustaceans—Cont'd

Species	Mechanism of Action	References
Decapod Crayfish		
Pacifastacus leniusculus	Distance pheromone. The sexually mature female crayfish released a pheromone that stimulated mating behavior in males	Stebbing et al. (2003)
Copepod		
Centropages typicus	When males of this copepod cross a pheromone trail left by a female, they increase their swimming speed, zigzag rapidly along the exact path taken by the female, rapidly overtake the female and mate with it	Bagøien and Kiørboe (2005)
Trigriopus japonicus	Surface glycoprotein cleaved by trypsin function as a contact pheromone. These lectin-binding glycoproteins also differ in their locations on the body surface	Ting et al. (2000)
Amphipod		
Gammarus palustris	The most important stimulus for male reproductive behavior is the female exoskeleton and suggests the existence of a contact pheromone	Borowsky (1986)

6.2.1 Sex Pheromones in Brachyuran Crabs

The occurrence of elaborate courtship behaviors, coupled with a variety of mating systems in decapods, would indicate an involvement of a chemical communicative system, which is ideal for the exchange of chemical signals between genders, especially in aquatic habitats. Following the demonstration of a female sex pheromone in the swimming crab *Portunus sanguinolentus* (Ryan, 1966), several other species of brachyuran crabs have been shown to use the sex pheromones in their courtship behavior and mating. However, in many instances the presence of sex pheromones has been inferred from the behavioral traits of male and female during mating, rather than from their identification and chemical characterization. Nevertheless, sex pheromones have been identified mainly in the urine of many decapod crustaceans. In the blue crab, *C. sapidus*, these chemical signals are present in the pre- and postpubertal molt females, inciting courtship and mate-carrying behavior in males (Gleeson, 1991). Since females are the limiting sex in the blue crabs, female pheromones also stimulated intrasexual competition among males for mates. In *C. sapidus*, the pheromone-sensitive chemoreceptors reside in the aesthetasc sensilla found on the outer flagellum of the biramous antennules (Gleeson, 1980). Similarly, the pubertal female blue crabs also showed a preferential attraction to male blue crab odors, as indicated by a choice test (Gleeson, 1991). Earlier studies of Kittredge et al. (1971) attributed this sex pheromone activity to the molting hormone, crustecdysone, as it formed a high proportion in the urine of premolt crabs. However, the pheromonal bioactive fraction of *C. sapidus* urine did not contain the molting hormone.

The female blue crabs release a species-specific sex pheromone in their urine that is composed of two functional classes of molecules, both of which are small and polar. One class distinguishes females from males and thus is a sex-specific signal and a second class distinguishes blue crab from other species and thus constitutes a species-specific signal (Kamio and Derby, 2011). In the green crab, *Carcinus maenas*, the premolt female urine elicited a display of searching behavior in the male conspecifics, in addition to provoking male crabs to fight intensely among them. The pheromone-responsive males even attempted to mate with inanimate objects, conditioned with this chemical signal (Hardege et al., 2002). In *C. maenas*, the release of pheromone is under the control of receptive female and as such, a postmolt female may release more pheromone so that the male will continue to protect her during this vulnerable period (Sneddon et al., 2003). In this crab, detecting the female pheromone may excite the male and increase energy turnover in preparation for mate-guarding behavior and sperm production. On the other hand, a female in a pair may release large amounts of pheromone in the presence of a solitary, larger, more attractive male. As a result, aggression between the males ensues, with the large male obtaining the female.

The involvement of endocrine factors in the modulation of the chemically evoked sexual behavior and the actual control of mating process has not been fully understood. In *C. sapidus* receptive males, stimulation of antennular pheromone receptor normally activates the CNS "centers" controlling courtship display. Neural pathways in the eyestalk ganglia are important links between pheromone receptors on the antennules and CNS centers controlling courtship behavior of the males. An eyestalk neuropeptide is obviously implicated in transmitting or processing of the pheromonal information received via the antennules. Disrupting the neural connections between the eyestalks and the brain blocks the chemosensory activation of the courtship display center in the central nervous system (Gleeson, 1991).

6.2.1.1 Chemical Nature

Hardege et al. (2002) identified the female sex pheromone, as a nucleotide compound, uridine diphosphate (UDP) in the HPLC-separated urine sample of premolt female shore crab, *C. maenas*. UDP not only attracted the males but also induced the mating behavior. Interestingly, males responded to these female cues only during the summer mating season (Hayden et al., 2007). UDP accumulates as a by-product of chitin biosynthesis and is released predominantly into the urine during new cuticle formation at molting (Mansecal, 1999). Testing of synthetic UDP and the nucleotide analogs using both olfactometry and guarding stance assays revealed behavioral responses from males at biologically appropriate concentrations of 10^{-5} M (Hardege and Terschak, 2011).

The fact that UDP is released as a metabolic product from UDP-*N*-acetyl glucosamine to chitin production adduces further evidence that pheromone production is related to the molting process. In addition, urine produced in the postmolt crab contains ecdysteroids, as an excretory product as well as a "bouchet" of bioactive compounds, including UDP and other small molecules with a purported function in male attraction. This system of pheromone production in the brachyuran crabs is in accordance with the percept that molting, female receptivity, and copulation happen in precise succession.

As this behavioral sequence is common among many other brachyuran crabs, Bublitz et al. (2008) examined the interspecific activity of the sex pheromone UDP with other crabs. Bioassays using synthetic pheromone UDP at concentrations of 10^{-3}–10^{-4} M elicited significant sexual behavior in the snow crab *Chionoecetes opilio* and the yellow arrow crab *Stenorhynchus seticornis* toward UDP-treated objects. *C. opilio* is a hard-shell mating crab but still the mating is preceded by precopulatory mate guarding, mediated by a female sex pheromone (Bouchard et al., 1996). The positive response of this crab to UDP is intriguing and may suggest an alternate source for UDP release probably from the ovary (Fittschen, 2001). These results suggest that the female sex pheromone UDP is not species-specific but is present and active in other decapod crustaceans too. Interspecific behavioral activity induced by female sex pheromone has been demonstrated in *C. maenas* and *Macropipus depurator*, whereby males display sexual behavior toward a freshly molted female of the other species, when already stimulated by conspecific females (Seifert, 1982). However, the American lobster *Homarus americanus*, zanthoid crab *Rhithropanopeus herissi*, and *C. sapidus* did not show any response to UDP, suggesting that other pheromonal substances may also play a role in the attraction of males in decapods. For example, in the hair crab *Erimacrus isenbechii*, the active substance in the urine was found to be a mixture of eight unique ceramides, which elicited mate guarding behavior of male (Asai et al., 2000). Ceramides are metabolites of lipid secondary messengers used in intracellular signal systems. The chemically synthesized ceramides were found to be similar to the natural pheromone with respect to eliciting male mating behavior, similar to the natural female pheromone (Masuda et al., 2002). However, these ceramides were not detectable at biologically relevant concentrations in urine raising doubts about its pheromonal activity (Asai et al., 2000).

In the helmet crab *Telmessus cheiragonus*, Kamio et al. (2002) showed two female sex pheromones: one found in the urine of premolt females that induced male mate guarding and the other secreted by the postmolt female (not found in the urine) provided the signal that induced copulation. Understandably, different pheromonal factors are produced by the receptive females to control mate guarding and copulation independently. The chemical nature of these pheromones are not, however, clear.

6.2.2 Lobsters and Crayfishes

The courtship and mating of the American lobster *H. americanus* are based on male dominance and female choice. Chemical signals emanating from the urine of male and female provide the necessary information for establishing and maintaining male dominance and pair bonding with the females (Atema and Steinbach, 2007). Through this chemical signal, the males broadcast their sex, status, identity, and location both to rival males and to potential mates. The females also convey their sex, identity, and state of receptivity to the males, through such chemical signals. Lobsters have different chemoreceptor organs that could be involved in detecting pheromones. They are present in the lateral and medial flagella of the antennules, the antennae, the dactyl and propodus segment of the chelated walking legs, and the endopodite of the third maxillipeds. Of them, the lateral and medial flagella of the antennules are important in the perception of the odorous signals involved in sex attraction. The antennules and the antennae are grouped as the cephalic receptor organs with neural connections into anterior brain centers, transmitting chemical information from the receptors to the brain (Sandeman et al., 1992). Cephalic chemoreceptor organs function primarily to perceive the odors in the free stream around the animals head. The lateral filaments of each antennule flicks to sense the odor directly. Obviously, this organ is involved in the detection of sex pheromones (Johnson and Atema, 2005).

For lobsters, shelters and mating opportunities are limiting resources. Dominance is established by behavioral displays followed by physical fights among males. The male dominance is maintained by the chemical signals, which also provide the individual recognition. Furthermore, males release their sex signals more or less continuously, causing females to be attracted from a distance. In choice flume assays, females are more attracted to any male-occupied shelter than to an empty shelter, suggesting that males release their sex pheromones almost continuously (Bushmann and Atema, 2000). On the other hand, females control the release of pheromones in such a way that they use it in male proximity and to enter male shelters. Males also attract other males, leading to fights between them. The dominant male occupies a shelter sufficiently large for two animals, making the shelter a focus of social interactions. There seems to be an exchange of chemical signals between the males and females to understand the male's readiness as well as the female's receptivity. In odor choice tests, females prefer to associate with dominant males over the subordinates (Bushmann and Atema, 2000).

Once the female is allowed to enter into the male's shelter, the two animals cohabit until the female molts, after which they mate. The males also exhibit pereopods fanning causing water current that helps in carrying the pheromone, released by the cohabiting pair. This activity is seen as male "advertising," since it attracts more visitors, including premolt females (Cowan and Atema, 1990). Female chemical signals may also play a role in courtship and mating during the cohabitation with the male. Just before molting, the female faces the male and places her claws on the male's anterior carapace and claws. This curious behavior is accompanied by urine release, which is an indication for the imminent molting followed by copulation. These chemical signals may be transmitted to the male during copulation when the male turns over the recently molted female to insert his gonopodia into her seminal receptacle. The male appendages used to turn over the female (first two pairs of pereopods and third maxillipeds) are supplied richly with chemo- and mechanoreceptive sensilla that may possibly be involved in responding to the contact pheromone used in the mating of *H. americanus*. The chemical nature of the pheromone released through the urine is still unknown in lobsters, but its high-protein content may suggest a pheromonal role (Atema and Steinbach, 2007).

Just like lobsters, crayfishes, like *Astacus astacus* also create a forward current of pheromone to attract the males and control the timing and direction of the urine stream (Atema and Steinbach, 2007). Mature *Pacifastacus leniusculus* releases a chemical signal that induces sexual behavior in males (Stebbing et al., 2003). However, in *Austropotamobius pallipes*, males respond only to visual and chemical cues together (Acquistapace et al., 2002). However, crayfish possesses many chemoreceptors on all areas of the body but are concentrated on the antennules, antennae, mouthparts, and at the tip of the pereopods (Breithaupt and Thiel, 2011). Although the mechanism of pheromone perception is similar to lobsters, information on their chemical nature is nonexistent in crayfish.

6.2.3 Caridean Shrimps

In caridean shrimps, the use of distance pheromone is rather uncommon, by virtue of their gregarious living and the associated mating habits like pure searching, without conspicuous mate guarding behaviors. Burkenroad (1947) was the first to propose a "nondiffusible coating of the integument of the female" involved in sex attraction of a caridean shrimp, *Palaemonetes vulgaris*. Yet, among caridean shrimps, the presence of dispersing (olfactory) sex pheromones was demonstrated in the freshwater shrimp, *Palaemon paucidens* (Kamiguchi, 1972). The males of this prawn became agitated into frenzied searching, when a parturial female molted in the aquarium. Laboratory experiments of Kamiguchi with females of varying reproductive and molting stages revealed that only females with mature ovaries emit a sex pheromone, which is released after its molt. During this time, the females become receptive to mating, lasting up to spawning. This sex attractant is water soluble, as it elicits male searching without the presence of a receptive female. However, other shrimp species such as *Heptacarpus sitchensis* and *Heptacarpus paludicola* do not respond to females without touching them with the long antennal flagella (Bauer, 1976). Evidently, the stimulus for this overt male reaction is provided by contact pheromone and/or tactile stimulus. Nevertheless, experimental studies on the same shrimp revealed a role for olfactory sex pheromones in the attraction of the males (Bauer, 1979). In the rock shrimp, *Rhynchocinetes typus*, Díaz and Thiel (2004) suggested that waterborne pheromones attracted the dominant (robustus) males toward the receptive females. In addition, the rock shrimp males are capable of recognizing the presence of a reproductive female by visual cues from aggregations of males attempting to mate with the female. In the pair-bonding Harlequin shrimp, *Hymenocera picta*, chemical cues are conveyed through pheromones by females to the males for the formation of pair bonding (Seibt, 1973). Clearly, both olfactory and contact pheromones are important in sex identification, sex discrimination, as well as sexual attraction between males and females in caridean shrimp.

In the male *Macrobrachium rosenbergii*, the olfactory system is thought to be the main pathway for modulating sexual behavior through pheromone perception. Kruangkum et al. (2013) described the structures of the olfactory receptors (sensilla setae), their neural pathways, and possible role in modulating mating behavior. The lateral antennules contain special setae, called aesthetascs. They are connected to the olfactory receptor neuron, which in turn are connected to the olfactory neuropil of the brain. These olfactory receptor neurons convey sensory inputs from the aesthetascs to the olfactory neuropil of the deutocerebrum. The neurotransmitter GABA may be one of the sensory signals that modulate olfactory perception in *M. rosenbergii*. Bilateral ablation of tufted aesthetasc setae resulted in the significant reduction of mating behavior, indicating that this organ is crucial for sex pheromone perception.

We studied the morphology of sensillary setae of major claw and antennules of the male prawn *Macrobrachium idella* using field emission scanning electron microscope. Fig. 6.1 shows the presence of sensilla on the finger of the major claw. Fig. 6.2 is the magnified view of aesthetascs tuft, visible from the anterolateral region of the claw. The surface of the setae is smooth under higher magnification. Presence of hooklike structure on the tip of the aesthetascs are also seen (Fig. 6.2). The nonaesthetasc portion on the mid-lateral region of major claw indicates the distribution of spinelike structures on the lateral surfaces (Fig. 6.3). Fig. 6.4 shows the tip of the antennules, which is a cylindrical tubelike structure. Sensillary setae are present on the rim of the tip, folded inward. They probably play a role in the pheromonal perception received from the receptive females (Rabel et al., unpublished data).

Many observations on caridean shrimps, however, suggest that during courtship, males usually recognize a receptive female upon contact implicating chemotactile signals (Ra'Anan and Sagi, 1985; Correa et al., 2000). The occurrence of a contact sex pheromone in the caridean species has been well documented in highly gregarious shrimp species belonging to the genus *Palaemonetes*. In *Palaemonetes pugio*, for example, males recognize females by contact but only during a brief period preceding the female's parturial molt (Bauer and Abdalla,

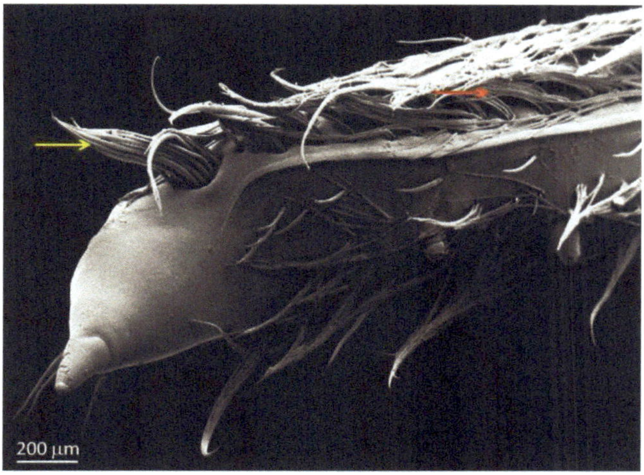

FIGURE 6.1 Image of field emission scanning electron microscope of *M. idella* showing the sensilla in the finger region of the major claw (1,12,000×). *Yellow arrow* indicates the tuft of aesthetascs generated from the anterolateral region. *Red arrow* indicates nonaesthetasc portion on the mid-lateral region of major claw.

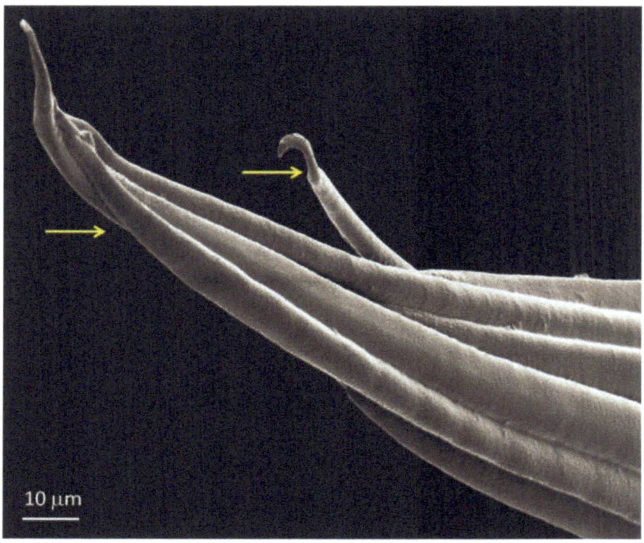

FIGURE 6.2 Magnified region (1,50,000×) of aesthetasc tip from the anterolateral region of major claw showing smooth surface on the lateral sides. *Yellow arrow* indicates hooklike structure.

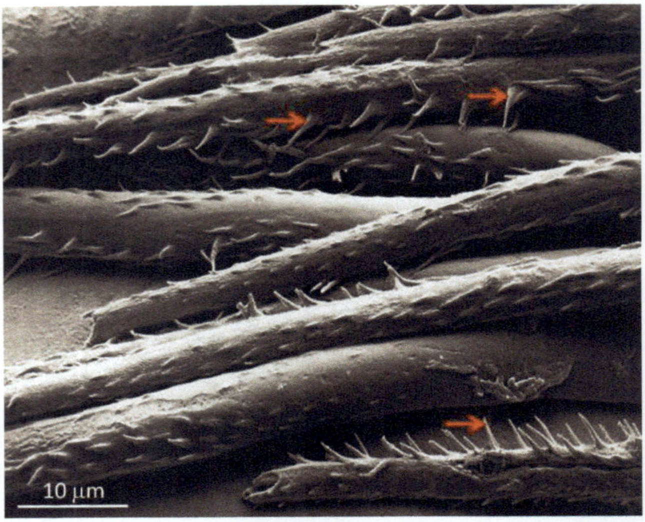

FIGURE 6.3 Higher magnification of nonaesthetasc region (3,50,000×) on the mid-lateral region of major claw. *Red arrow* indicates spinelike structures on the surface.

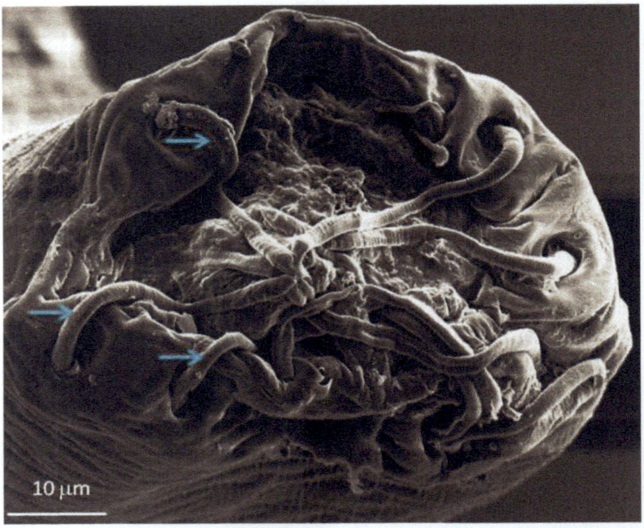

FIGURE 6.4 Field emission scanning electron microscope image of antennular region (3,00,000×). It is a cylindrical tubelike structure with the tip surrounded by sensory setae *(blue arrow)*. The setae are serrated on their lateral surface.

2001). Similarly, in hippolytid shrimp *H. paludicola*, Bauer (1979) observed that males needed to contact females with their antennae to recognize their receptive state. Although a nondiffusible coating over the cuticle has been considered to be the contact pheromone in the shrimps (Burkenroad, 1947), the chemical composition of the contact pheromones is still inconclusive.

Alphid shrimps are known for their symbiotic living with other invertebrate hosts such as sponges. Communication among them involves visual, mechanical, and chemical modalities (Herberholz and Schmitz, 2001). In the eusocial species, *Synalpheus regalis*, living in dense colonies inside the sponges, the individuals are in continuous contact with each other. It has been suggested that nest mate discrimination in this social shrimp species involves both waterborne and contact chemical signals (Duffy et al., 2002). Resident *S. regalis* are able to discriminate between foreign conspecifics and heterospecifics from their own inmates. The mutual antennulation that initiates nearly all contacts in alpheids indicates that contact pheromones have an important role to play in mediating recognition in these shrimps. Interestingly, intruders of the sponge nest contact the egg-laying queen much more than the other individuals, suggesting that the fertile female may release a sex pheromone which might be nonspecies-specific. In social insects, normal function is facilitated by exchange of pheromones that are produced by the queen and distributed among other colony members by frequent touching of her body (Holldobler and Wilson, 1990). However, no sex pheromones released by the synalpheid queens have been reported so far.

6.2.3.1 Chemical Nature of Contact Pheromones

In the copepods, surface glycoproteins that can be targeted by lectins were found to be important in mate recognition (Lonsdale et al., 1998), whereas Sparkes et al. (2000) suggested that 20-hydroxy ecdysone could coact with the cuticular glycoproteins in attracting the males for mating. In the harpacticoid copepod, *Tigriopus japonicus*, cuticular glycoproteins are proposed as contact cues (Ting et al., 2000). Similarly, in the caridean shrimp *P. pugio*, glucosamine-containing glycoproteins are potentially involved in mate recognition, as demonstrated by competitive inhibition studies (Caskey et al., 2009). GC-MS analysis of extracts from the cuticle of this shrimp also showed that cuticular composition of post-molt parturial females is different from that of males and intermolt females. Cuticular hydrocarbons function as inter- and intraspecific chemical signals in a variety of insects (Ayasse et al., 2001). Many insect species use a blend of cuticular hydrocarbons as their sex pheromones (Howard and Blomquist, 2005).

In the peppermint shrimp *Lysmata wurdemanni*, both distance and contact pheromones operate together for successful copulation. Whereas distance pheromone led to precopulatory approach before the female molted, contact pheromones induced copulatory behavior (Zhang and Lin, 2004). Removal of both antennae and antennules inhibited detection of distance and contact pheromones, resulting in the prevention of mating between males and receptive females. Ablation of outer flagella of the antennules inhibited only the detection of distance pheromones. Interestingly, a compound similar to UDP was identified in the distance pheromone (Zhang et al., 2010). Aesthetascs on the outer flagella could be the site of detection for distance pheromones, as they are innervated by olfactory receptor neurons (Zhang et al., 2012). However, the site of synthesis of the olfactory pheromones is still unknown, although they are released through the nephropores (Breithaupt and Thiel, 2011).

In the lysmatid species, *Lysmata boggessi*, the contact pheromone, found on the euhermaphroditic (EP) female phase shrimp, is lipophilic in nature and resemble insect cuticular hydrocarbons (Zhang et al., 2011). In *L. boggessi*, EP shrimp contains (*Z*)-9-octadecenamide as the major component of the cuticular hydrocarbon contact pheromone in the "bouchet," in addition to methyl linoleate and hexadecanamide, contributing to the sex-specificity bioactivity. Behavioral evidence suggests the existence of both distance and water-insoluble contact pheromones in these lysmatid species. The male phase shrimp uses distance pheromone to trace and locate receptive females. The male phase shrimp also use contact pheromones to recognize newly molted euhermaphrodites.

6.2.4 Amphipods

In the amphipod *Gammarus palustris*, two different types of waterborne pheromones are produced: one is a species-specific attractant, produced by both sexes and the other is produced only by females during the receptivity period to attract the potential males for pair formation and copulation (Borowsky, 1991). When offered waterborne cues in choice experiments with conspecific females, males show a strong preference for receptive females (Krång and Baden, 2004). The female-specific pheromone, to which only the males are attracted, could be considered as a contact pheromone, although it is diffusible, and acts in the physical presence of the female. The occurrence of both distance and contact pheromones are advantageous in mate location and pair formation and the subsequent mating. Perception of chemical stimuli happens via specific sensillae that are concentrated on the antennae but the sensillae could occur on other parts of the body (Hallberg and Skog, 2011). Although several experimental studies have implicated a role in sex attraction, no serious attempts have been made to understand the chemical nature of the pheromones in amphipods.

6.3 CONCLUDING REMARKS

Crustaceans have evolved an effective communication system in the aquatic medium, involving sex pheromones to attract the mates to ensure mating success and enhance reproductive fitness. However, as yet, most of our understanding of crustacean sex pheromones comes from experimental studies on mating behavior. Since mating in crustaceans involves complex and elaborate behaviors, the use of sex pheromones in attracting the mates in murky environments such as the benthic regions assumes great importance. Crustaceans employ two types of pheromones: the distance or olfactory pheromone and the contact pheromone. Distance pheromones are released through the urine to attract the males toward the receptive females. Conversely, contact pheromone refers to the water-insoluble compounds found on the surface of the cuticle of postmolt females. Special chemoreceptor organs are employed in perceiving the sex pheromones. Recent findings identified UDP as the chief olfactory pheromone, whereas a blend of cuticular hydrocarbons acts as the contact pheromone. There is a possibility that more than one molecule can act in conjunction with others present in the "pheromone bouchet." Several sensory setae present on the cephalic region of males are implicated with the chemoreception of the distance as well as contact pheromones. These sensory setae have neuronal connection with the olfactory neuropil in the brain. However, the mechanism underlying the neural transmission of pheromonal signals to the central nervous system is poorly understood in crustaceans.

While pheromones play a pivotal role in the sexual attraction of aquatic crustaceans by offering chemical cues for species recognition, the involvement of endocrine factors in the modulation of the chemically evoked sexual behavior and the actual control of mating process has not been fully understood. In *C. sapidus* receptive males, stimulation of antennular pheromone receptor normally activates the brain "centers" controlling courtship display. Neural pathways in the eyestalk ganglia are important links between pheromone receptors on the antennules and brain centers controlling courtship behavior of the males. Either eyestalk ablation or severance of the optic tracts connecting eyestalk ganglia and the brain resulted in the failure of males to respond to a pheromone stimulus (Gleeson, 1991). An eyestalk neuropeptide is obviously implicated in transmitting or processing of the pheromonal information received via the antennules. Disrupting the neural connections between the eyestalks and the brain blocks the chemosensory activation of the courtship display center in the central nervous system. While so much information has been generated over the years on the eyestalk neuropeptides in controlling the reproductive and molting activities of crustaceans, no studies have been made on the identification of the peptide controlling perception of sex pheromones. So much so, identification of pheromones involved in the species recognition in the eusocial colony of synalpheid shrimps has not been attempted.

Crustaceans are very good model arthropods to investigate the olfactory pathways in modulating sexual behavior through pheromonal perception. Chemical communication in the aquatic medium could be as challenging as that in aerial medium. More work is, however, needed to find out the small molecular composition of the urine, which forms the main medium to release the pheromone in decapod crustaceans.

Chapter 7

Reproductive Cycle and Environmental Control

7.1 INTRODUCTION

In general, sexual reproductive activities are cyclic within individuals in a population. They normally proceed through a repetitive series of reproductive events, of which gametogenesis is the central event. Giese and Pearse (1974) delineated different stages in the gametogenic cycle for a typical marine invertebrate: (1) accumulation of nutrients to be used during gametogenesis, (2) proliferation of gonial cells and their differentiation into gametes, (3) accumulation of ripe gametes, (4) release of gametes, and (5) a reproductively quiescent or spent stage. Gametogenic cycles within individuals are controlled by endogenous factors, whereas extrinsic or external factors, such as temperature, salinity, day length, food availability, and chemical factors, may initiate and synchronize the timing of reproductive events with environmental changes. Temporal patterns of the reproductive cycle may result through a complex coordination of a number of endogenous factors with respect to the interacting exogenous factors at a given time or over a period of time by individuals in a population (Sastry, 1983).

Crustaceans are in general fast growing but slow-breeding arthropods. Hence, gametogenesis, in particular oogenesis, takes several months for completion in many species of decapod crustaceans. Therefore, quite different environmental signals may be employed to regulate the progression of the cellular events that culminate in the successful reproduction. These environmental signals, referred to as proximate factors, are perceived by the neural, neuroendocrine, or endocrine systems to coordinate the successive events of the reproductive cycle (see Chapter 9 for details).

The physiology of reproduction in crustaceans is greatly influenced by somatic growth, permitted by periodic molting in the adults. As a result, there is an alternating cyclicity of female reproduction and molting in several species. Obviously, the timing of reproduction is affected by the programming of somatic growth and molting. In the American lobster, *Homarus americanus*, molting occurs usually in the summer season, with oviposition occurring in the next. Clearly, molting and reproduction are two temporally separated physiological processes in the lobster, regulated by definitive environmental factors (Waddy and Aiken, 1995). However, in the brachyuran field crab *Paratelphusa hydrodromous*, both

Sexual Biology and Reproduction in Crustaceans. http://dx.doi.org/10.1016/B978-0-12-809337-5.00007-1

reproductive and molting seasons are temporally separated within the same year (Pillai and Subramoniam, 1984). On the other hand, in the mole crab, *Emerita asiatica* inhabiting the sandy beaches of the Indian coast, the reproductive cycle is invariably followed by the molt cycle, without any pause. Here again, brooding of the eggs in the pleopods inhibits the onset of the next molt cycle, which commences only after hatching of the embryos (Subramoniam and Gunamalai, 2003). This sequence of molting and reproduction occurs almost all through the year in the actively reproducing female size groups. Synergism between the ovarian and molt cycles are well exemplified in a freshwater prawn *Macrobrachium nipponense* (Okumura, 1992) as well as penaeid shrimps during reproductive season (see Subramoniam, 2000 for a discussion).

7.2 PATTERNS OF REPRODUCTIVE CYCLE

In crustaceans, reproductive cycle includes a series of events, starting with (1) proliferation of gonial cells, (2) differentiation and growth of gametes to maturation, (3) reproductive behaviors associated with mating, (4) release of gametes by transfer of sperm during copulation by males and ovulation or oviposition by females, and (5) incubation of embryos until hatching to release the larvae or juveniles (Sastry, 1983). While this cycle is basic to all crustaceans, the time relationship between certain events and duration of events vary widely among different species. The majority of crustaceans brood their embryos by attaching them to abdominal appendages and retaining them till hatching. Only a few crustaceans such as penaeid shrimps (Omori, 1974) and some euphausiids (Mauchline and Fisher, 1969) shed their fertilized eggs into the environment without brooding. Occurrence of ovigerous females in the population helps in the easy estimation of reproductive cycles of many crustaceans. The time when females of a population are ovigerous is referred to as the breeding period of the species.

In general, the crustacean reproductive cycle takes different patterns in populations of the same or different species. For example, in populations living in areas of little seasonal change such as deep sea and parts of the tropics, continuous reproduction has been commonly encountered. Even here, gametogenesis may be staggered among different individuals during the year, so that reproduction appears to occur continuously throughout the year for the entire population (Giese and Pearse, 1974). Other patterns of reproductive cycle among both aquatic and terrestrial crustaceans occur in accordance with the changing environmental parameters like water temperature, photoperiod, salinity, and food supply.

7.3 CONTINUOUS REPRODUCTION

Most of the crustaceans living in tropical waters have been reported to reproduce continuously. Tropical species in general breed for longer periods due to narrow variations in temperature, rainfall, and nutrient inputs, which are important for reproduction and larval growth. Extended breeding also indicates that individuals produce several successive broods during the year or breed asynchronously. Variations in the intensity of reproduction, however, do occur during certain periods of time, reflecting minor environmental fluctuations. Closer examination of the gonads is, however, necessary to know the periods of intense reproduction in a year. The intertidal sand crab, belonging to the genus *Emerita* has a worldwide distribution in the tropical and temperate seas. Two tropical species, *Emerita portoricensis* from Jamaica and *E. asiatica* inhabiting the east coast of peninsular India, breed continuously throughout the year (Goodbody, 1965; Subramoniam, 1977a). In *E. asiatica*, the continuous breeding activity has been determined by the incidence of ovigerous forms in the Madras (India) population. The percentage of ovigerous females varied from 73 to 100 in the size class between 22 and 33 mm carapace length; such a high breeding rate, though unusual for the continuously breeding animal, occurs all through the year in *E. asiatica* (Subramoniam, 1977a). A year-round occurrence of zoea larva of *E. asiatica* in plankton collected from the nearshore waters of Madras coast corroborates the continuous breeding activity of this anomuran crab (Menon, 1933).

Complementary methods to estimate the reproductive activity, such as gonad and hepatic indexes (Giese and Pearse, 1974), and egg mass index (Eickstaedt, 1969) add further evidence to the existence of continuous reproduction in *E. asiatica*. Egg mass index, representing the "realized reproductive potential" of the brooding crab, has also been determined for this crab population in Madras coast (Fig. 7.1). As seen from the figure, there is a steady rise in the gonad and egg mass indices from January to May, followed

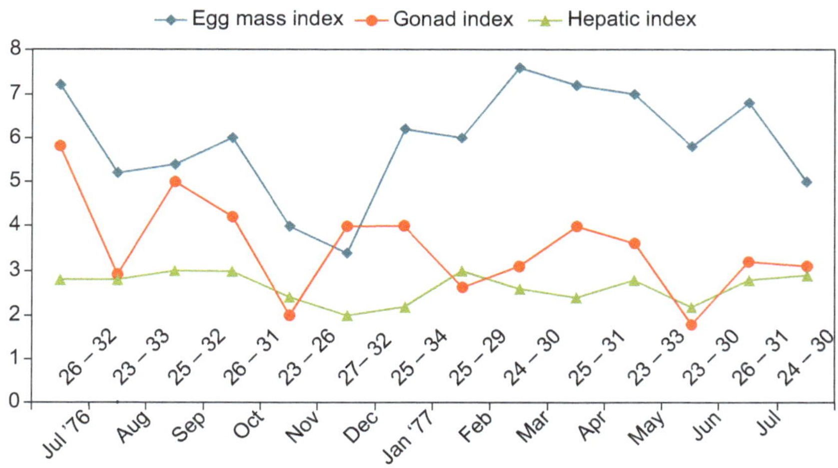

FIGURE 7.1 The annual fluctuations in the gonad, egg mass, and hepatic indices of *E. asiatica* with reference to carapace length.

by a fall in June and August, and then a steep fall in November and December (year 1977). The dip in reproductive activity occurs during the monsoon rainy months, whereas the high intensity of reproduction is maintained during premonsoon summer months in the southeast coast of India. Unlike the gonad index, the hepatic index of *E. asiatica* does not exhibit any significant fluctuation throughout the year. Hepatopancreas, being the central organ of intermediary metabolism in Crustacea, is expected to supply precursor materials both for egg production and the new cuticle formation during molting. In view of the year-round reproduction and molting in *E. asiatica* (Gunamalai and Subramoniam, 2002), there will be a steady synthesis and release of protein and other substrates for these two energy-requiring processes, thereby maintaining a low hepatic index all through the year.

Interestingly, another species of *Emerita*, *E. holthuisi* breeds more or less continuously in the northwest coast of India, with two peaks of breeding females, one in March and another in September (Nagabhushanam and Kulkarni, 1977). These two peaks coincided with the pre- and postmonsoon season on the northwest coast. However, the same species inhabiting the southwest coast breeds in the postmonsoonal months from July to December, indicating the strong influence of heavy rainfall on reproductive activity (Murugan, 1985).

In *E. asiatica*, when the eggs are undergoing development in the pleopods, a concurrent ovarian development progresses for the next egg laying. Determination of mean egg developmental rate of the brooding eggs, following the method of Eickstaedt (1969) also indicated the continuous release of zoea larvae throughout the year into the seawater (Subramoniam, 1979a). Taken together, these observations indicate that *E. asiatica* is not only a continuous breeder but also a repetitive breeder in the Madras coast. Continuous reproduction in the strictest sense is justified for *E. asiatica*, in which there seems to be no pause between two gametogenic cycles in the individuals. Goodbody (1965) also reported continuous breeding in *E. portoricensis* inhabiting the sandy beaches of Jamaica. He suggested that seawater temperature and the primitive way of filter feeding could be attributed to the continuous reproduction, all through the year. In the offshore waters of Madras coast, there is little or no seasonal variation in seawater temperature, the mean monthly range during the study (1976–1977) was only 4.7°C (25.5–30.2°C).

A comparative study on the breeding patterns of crustaceans in the east and west coasts of India further illustrates the influence of local factors such as rainfall on the breeding activity. In the west coast of India, the penaeid shrimp, *Penaeus indicus*, and the crabs, *Portunus pelagicus* and *Uca annulipes* all breed continuously with breeding peaks between October and January at Cochin (Pillay and Nair, 1971). However, on the east coast of India, the same species have much more extended peak activities during warmer months (Panikkar and Aiyar, 1939; Subrahamanyam, 1963; Rahaman, 1967). In the east coast, there is stable environmental temperature and high nutrients in the seawater due to upwelling during the rainy season. Therefore, there is no pause in reproduction. On the contrary, in the mid-west coast, monsoon rain has clear influence on breeding activity.

Continuous reproductive activity was also witnessed in the anomuran hermit crab, *Clibanarius clibanarius* from the southeast coast of India (Varadarajan and Subramoniam, 1982). Combining gonad and hepatic indices with percentage of ovarian stages of females plotted against time, a continuous breeding pattern with scattered peaks, particularly between August and May from the year 1975 to 1977 was delineated. Although *E. asiatica* and *C. clibanarius* occur on the same coast, reproductive intensities differ between them. Ovigerous crabs never exceeded 66% in *C. clibanarius*. However, in another continuously breeding tropical intertidal hermit crab, *Calcinus laevimanus*, the breeding frequency was as high as 80% during their restricted peak breeding season (Reese, 1968). In all other months of the year, the individuals breed less synchronously so that the intensity of breeding is far less. Continuous reproduction has been reported in hermit crabs, occurring in other geographical locations also. *Diogenes brevirostris*, inhabiting the southern coast of Mozambique breed all through the year with three peak spawning peaks in March, August, and December (Litulo and Tudge, 2005). In another hermit crab, *Paguristes tortugae*, found on the island of Anchieta, Brazil, Mantelatto and Sousa (2000) reported continuous breeding with peak reproduction in March–December. Furthermore, differences in the reproductive peaks of these hermit crabs have been shown to be influenced by variations in environmental parameters, including competition in shell selection (Mantelatoo and Dominciano, 2002). Continuous reproduction in the swimming crab *Arenaeus cribrarius* has been reported from the Brazilian coast near 23°30′ South (Pinheiro and Fransozo, 2002).

7.4 SEMIANNUAL REPRODUCTIVE CYCLE

When an extended breeding cycle is interrupted by a quiescent period in most or all the individuals in a population, a semiannual reproductive pattern is realized. In the tropics, such breeding seasons are influenced by semiannual monsoons. In the west and east coast of southern India, a number of studies have reported on semiannual breeding for the penaeid shrimps (George, 1962) in the west coast and for the sessile barnacle in the east coast (Paul, 1942; Antony Raja, 1963). In these cases, the fall in salinity in the estuaries and coastal waters are not suitable for spawning and larval release and therefore the animals skip this season for reproduction. In a burrowing anomuran crab, *Albunia symnista*, coexisting with *E. asiatica* in the Madras coast of south India, Subramoniam and Panneerselvam (1985) reported a semiannual breeding pattern. This crab breeds continuously with two distinctive peaks, one in January and another in July. During the rainy months of October to

December, the ovarian activity is minimal and in many females the ovary is in a degenerative condition. Frequency of ovigerous females is as low as 25% even during the peak reproductive season. Although *A. symnista* is an example for semiannual breeding, the environmental conditions responsible for this discontinuous pattern are not known. Perhaps, its life history pattern and the poor adaptability to live in this precarious intertidal environment may have an impact over their poor reproductive performance.

7.5 ANNUAL REPRODUCTION

The most common breeding pattern is the annual breeding season, characteristic of temperate seas. The timing of the breeding season varies from species to species and between localities for the same species, but generally breeding activity is concentrated in the summer months. An excellent example for annual breeding is found in different species of the sand crab *Emerita* which enjoy a worldwide distribution in the tropical and temperate seas. Breeding records available for different species inhabiting both tropical and temperate regions indicate clearly that those species found in the tropical locations breed continuously, whereas all the other species in the temperate regions follow an annual breeding pattern (Fig. 7.2). Evidently, species inhabiting tropical seas are continuous breeders, and those reproducing in temperate waters are all annual breeders. One such species is *Emerita analoga*, with the distribution in the Pacific coast of Americas from Alaska to Chile. Eickstaedt (1969) studied the reproductive cycle of *E. analoga* from Monterey Bay, California, employing several methods such as incidence of ovigerous females, gonad and hepatic indices, and mean developmental stages of the eggs attached to pleopods. He stressed the

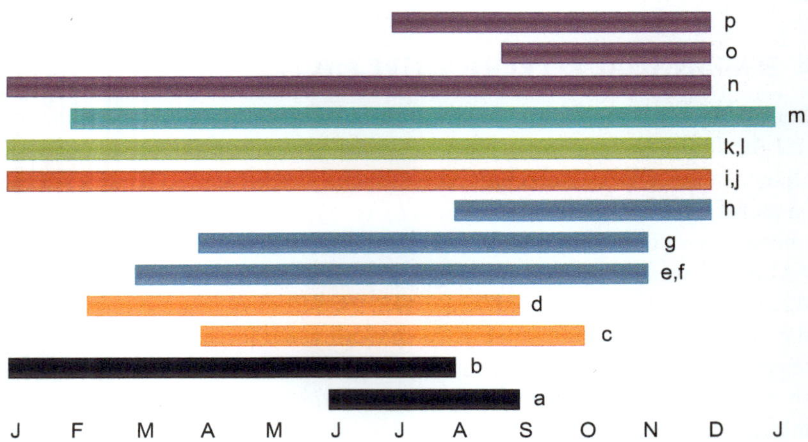

FIGURE 7.2 Bar diagram representing the breeding seasons of different *Emerita* species, listed in Table 7.1. Color bars indicate different *Emerita* species.

importance of determining egg developmental stages in accurately ascertaining the annual period of active breeding. In the Monterey Bay, the annual active breeding period extended from March to September, although ovigerous females were present throughout the year. However, analysis of the developmental stages of the egg revealed that no reproductive activity took place during the winter months. During the winter months, the water temperature is too low for the egg development in the brood. Furthermore, gonad indices indicate that most of the breeding occurs between the months of March and November. Eickstaedt (1969) from his field studies and experimental results concluded that water temperature and food availability are the major determining factors in timing the reproductive cycle of these sand crabs. The significant rise in the egg production in the month of April corresponded with the maximum phytoplankton levels in the seawater, suggesting the importance of food in the realization of the reproductive potentials. The rising temperature in the beginning of summer only triggers the reproductive activity. Similar patterns of breeding cycle of *E. analoga* have been reported by other workers from Monterey Bay (Boolootian et al., 1959), San Diego, California (Cox and Dudley, 1968), and El Tabo, Chile (Osario et al., 1967). The latitudinal differences in the annual breeding peaks are represented for different *Emerita* species in Fig. 7.2 and Table 7.1. It may be surmised from the above studies that the requirements of a minimum temperature for gametogenesis and an abundance of food are the major environmental parameters to determine the reproductive periodicities of these filter-feeding intertidal sand crabs.

Another type of annual breeding, in which temperature is not the only limiting factor, occurs in Indian peninsular regions, where the breeding of decapod crustaceans in the estuarine and coastal waters is affected by lowering of salinity, making it unsuitable for larval release and dispersion (Pillay and Nair, 1971). In the tropics where marine invertebrates tend to breed continuously, estuarine forms avoid the rainy months as the salinity changes affect the survival of decapod larvae and hence spawning should be so timed that the larvae are released from the brood at optimum salinity conditions (Costlow and Bookhout, 1970).

Monsoon rains have an even more marked effect on decapod crustaceans that depend on freshwater for breeding, egg development, and subsequent dispersal of juveniles. In the freshwater field crab, *P. hydrodromous*, Pillai and Subramoniam (1984) determined the annual reproductive cycle by calculating the percentage of ovigerous females in a Madras (India) population for a 3-year period. In this crab, gametogenic activity starts in December and vitellogenesis is completed during January and May. Spawning commences in the last week of June or first week of July with the release of brood in the month of September (1979), October (1978), or November (1980) during northeast monsoon. Therefore, the two exogenous factors influencing the breeding of *P. hydrodromous* are temperature (egg production) and rainfall (spawning and larval release) on an annual basis.

TABLE 7.1 Breeding Season of Different *Emerita* Species

Serial no.	Species	Location	References
a	*Emerita talpoida*	34.7182° N, 76.6638° W (Beaufort, NC (USA))	Wharton (1942)
b	*E. talpoida*	34.6946° N, 76.8147° W (Bogue Banks, NC (USA)	Diaz (1980)
c	*E. analoga*	36.6177° N, 121.9166° W (Pacific Grove, CA (USA))	Boolootian et al. (1959)
d	*E. analoga*	32.7157° N, 117.1611° W (San Diego California, CA (USA))	Cox and Dudley (1968)
e	*E. analoga*	33.4557° S, 71.6669° W (El Tabo, Chile)	Osorio et al. (1967)
f	*E. analoga*	36.6002° N, 121.8947° W (Monterey, CA (USA))	Eickstaedt (1969)
g	*E. analoga*	34.0259° N, 118.7798° W (Malibu, CA (USA))	Perry (1980)
h	*E. analoga*	33° 1′ 22″ S, 71° 34′ 9″ W (Caleta Abarea (Chile))	Conan (1978)
i	*E. portoricensis*	18° 15′ N, 77° 30′ W (Green Bay (Jamaica))	Goodbody (1965)
j	*E. portoricensis*	10°30′N 61°15′W (Trinidad (West Indies))	Quesnel (1975)
k	*E. asiatica*	13.0827° N, 80.2707° E (Madras (India))	Menon (1933)
l	*E. asiatica*	13.0827° N, 80.2707° E (Madras (India))	Subramoniam (1977a)
m	*E. emeritus*	8.5241° N, 76.9366° E (Trivandrum (India))	Murugan (1985)
n	*E. holthuisi*	16.9902° N, 73.3120° E (Rathnagiri (India))	Nagabushanam and Kulkarni (1977)
o	*E. holthuisi*	8.5241° N, 76.9366° E (Trivandrum (India))	Murugan (1985)
p	*E. holthuisi*	8.5241° N, 76.9366° E (Trivandrum (India))	Murugan (1985)

7.6 IMPACT OF CLIMATE CHANGE ON REPRODUCTIVE CYCLE

Recent studies have revealed global warming has affected all biological processes including growth and timing of reproduction in marine invertebrates (Lawrence and Soame, 2004). Of the several environmental cues, photoperiod and temperature have been shown to moderate reproduction either individually or in combination. In particular, they are shown to affect reproductive parameters, including sex determination, gametogenesis, and spawning. Gonadotrophic and spawning hormones are known to be the transducers between the environment and gamete production. Since the majority of crustaceans are marine, inhabiting different ecosystems, climate change would have a serious impact on fecundity, spawning success, and larval dispersal and recruitments. In a limited study on the crustaceans, it is predicted that increasing sea surface temperature may alter the timing of gonadal development, spawning, and food availability and cause changes in the larval dispersal and recruitment success (Przeslawski et al., 2008). Global warming is likely to uncouple and alter the phase relationship between temperature and photoperiod. This is likely to have significant consequences for animals that develop gametes during winter and spawn in the spring in temperate northern latitudes (Hines et al., 2010). In the blue crab, *Callinectes sapidus* inhabiting Chesapeake Bay, Florida, brooding in the population at lower altitudes, begins 3–4 months earlier than at high altitudes, allowing more broods per season (Jivoff et al., 2007). In addition, climate change reducing the severity of the winter is also predicted to promote rapid growth and reproduction. At the same time, warmer temperatures may reduce the size at maturity of the crabs (Hines et al., 2010).

Another climate changing factor affecting the reproduction of marine crustaceans is ocean acidification, which occurs as a consequence of sequestration of increasing atmospheric CO_2 into the ocean. In the benthic broadcast spawners, such as echinoderms and mollusks, ocean acidification is shown to have a negative impact on fertilization, larval development, and successful settlement. In the echinoderms, experimental studies have revealed that exposure to elevated CO_2 has resulted in decrease in fertilization rate, delay in the embryonic cleavage, as well as malformation in skeletogenesis in larval development, resulting in size reduction and impairment in larval settlement (Dupont et al., 2010). However, CO_2 exposure has no effect on the success rate of hatching in the copepods, *Acartia steueri* and *Acartia erythraea*. In these species, there is little sensitivity to elevated CO_2 during any stage in their early development (Kurihara et al., 2004). On the contrary, long-term exposure to lower pH (7.89 and 7.64) has an adverse effect on egg production and early larval stages of the shrimp, *Palaemon pacificus* (Kurihare et al., 2008). As calcification of larval cuticle is affected by ocean acidification, there could be a prolonging of the larval stages, reducing larval dispersal and settlement in the sessile barnacles (Ross, 2001).

7.7 ENVIRONMENTAL CONTAMINATION AFFECTING REPRODUCTION

A wide variety of chemicals that enter the aquatic environment are capable of disrupting endocrine functions by interfering with the actions of the sex hormones in several crustaceans, leading to reproductive dysfunction (LeBlanc, 2007). There is increasing evidence from ecotoxicological research that endocrine disruptive chemicals chiefly affect the sexuality of crustaceans such as isopods and amphipods (Mazurova et al., 2008). Ford et al. (2005) reported the increased rates of intersex and female-biased sex ratio in the marine amphipod, *Echinogammarus marinus* from the Scottish coast. Sexual anomalies including intersex conditions were found in the amphipod *Hyalella azteca*, exposed to ethinylestradiol (Vandenberg et al., 2003). Changes in sex ratios as well as feminization effects have also been observed in the freshwater amphipod, *Gammarus pulex*, after exposure to the environmental estrogen, 17 α-ethinylestradiol, nonylphenol, and diethylstilbestrol (Watts et al., 2002). This may not only affect the reproductive success but also leads to unbalanced population growth. Variations in the size and development of secondary sexual characters in different populations of the shore crab, *Carcinus maenas* in northern England are also indicative of endocrine disruption (Brian, 2005).

Endocrine disruptive compounds, contaminating surface waters, could cause adverse effects on development and reproduction in aquatic organisms by unintended estrogenic stimulation. Synthetic xenoestrogens, used as insecticides, plasticizers, and polychlorinated biphenyls, are known to mimic or enhance the effects of estrogens (Majer-Baranyi et al., 2015). In vertebrates, yolk precursor protein, vitellogenin, is widely used as a biomarker to study the effects of xenoestrogens. In the freshwater shrimp, *Palaemon elegans*, Sanders et al. (2005) detected vitellin in the larvae after exposure to nonylphenol and 17 β-estradiol. Stimulation of vitellin, otherwise called cypris major protein, was also noticed in naupliar and cypris larval stages of the barnacle *Balanus amphitrite* (Billinghurst et al., 2000). Similarly, in the grass shrimp, *Palaemonetes pugio*, pyrene exposure produced large eggs with enhanced vitellin levels. However, increased mortality was observed during hatching of the embryos (Oberdorster et al., 2000). This lipophilic pyrene is easily transported through the hemolymph into the eggs, where it becomes toxic.

Another endocrine-disruptive chemical, 4-nonylphenol, is shown to have embryo toxicity in *Daphnia magna*, through its interference with the metabolic elimination of testosterone (LeBlanc et al., 2000). Direct exposure of maternal daphnids to testosterone caused developmental abnormalities, including partial arrest of early embryonic development and abnormalities in the larval structures in the neonates. Exposure of maternal daphnids to concentration of

4-nonylphenol also produced such developmental abnormalities, suggesting that it is a developmental toxicant in *Daphnia*. Similarly, exposure of neonate *D. magna* to synthetic estrogen DES as well as estrogenic pesticide endosulfan has specific effects on the development of secondary sexual characters (Zou and Fingermann, 1999). Heavy metals like cadmium in the ambient water also inhibited ovarian growth in the fiddler crab *Uca pugilator* (Rodriguez et al., 2007). In another study, Medesani et al. (2004) showed that cadmium and copper inhibited ovarian growth by their interference with the transduction pathway of the eyestalk hormones in the crab *Carcinus granulates*.

Chapter 8

Oogenesis

8.1 INTRODUCTION

Crustaceans produce large numbers of yolk-laden eggs and brood them externally for extended periods. Therefore, vitellogenesis, the process of yolk formation is central to oogenesis. In accordance with the diversity found in the morphology and organization of ovary, vitellogenic system also shows remarkable variation among crustaceans. In the lower crustacean orders, the morphology of the ovary is highly variable often showing resemblance to those found in lower insects. The occurrence of nurse cells in the branchipod crustaceans clearly demonstrates their closeness with hexapods. In general, vitellogenesis is a biphasic event consisting of autosynthesis and heterosynthesis (Adiyodi and Subramoniam, 1983). Early electron microscopic observations on oogenesis revealed the existence of yolk protein synthesizing capacity in many malacostracans. Furthermore, recent molecular studies demonstrate the yolk protein gene expression both in ovary and hepatopancreas in the penaeiodean shrimps. In general, the mechanisms involved in the vitellogenesis is similar to those of insects in that there is commonality in the synthesis, transport, and sequestration of vitellogenin (Vg) into the ovary of both the arthropod classes. However, the complexity of yolk molecules makes the crustacean an ideal organism to study the cellular dynamics in vitellogenesis among invertebrates.

8.1.1 Morphology of Female Reproductive System

Crustacean female reproductive system consists of ovary, oviduct, and the associated accessory sex glands. The functional morphology of the female reproductive system is primarily concerned with the production, maturation, and expulsion of the eggs for fertilization, which could happen externally or internally. In Crustacea, the ovaries are located either dorsal or dorsolateral to the gut. The ovary is usually a paired organ, with its limbs fused either anteriorly, as in peracarids or in the midline as in macrurans and stomatopods (Kaestner, 1970). In the crayfish, belonging to Cambaridae, the ovary is Y shaped, whereas in Parastacidae, the ovary is H shaped or found in the form of parallel strands (Vogt and Holdich, 2002; Vazquez et al., 2008). In the brachyuran crabs, the ovary consists of paired lateral limbs, connected by a central bridge of ovarian tissues, giving it an H shape (Fig. 8.1). In the fully mature state, the ovary completely fills the dorsal region

Sexual Biology and Reproduction in Crustaceans. http://dx.doi.org/10.1016/B978-0-12-809337-5.00008-3

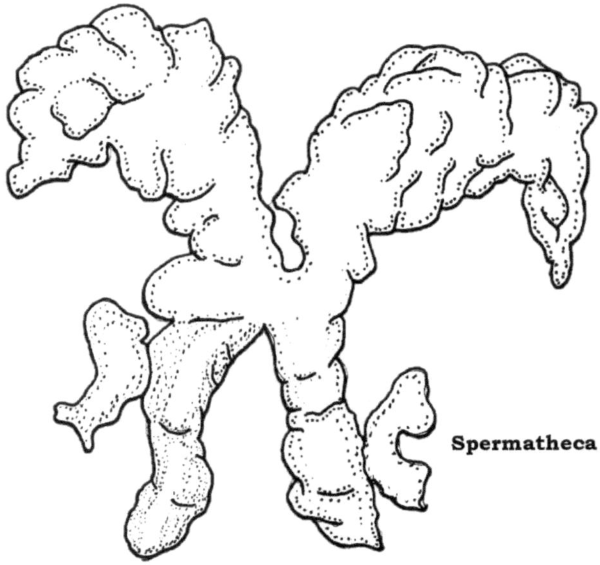

FIGURE 8.1 H-shaped ovary of a brachyuran crab.

of the body cavity within the cephalothoracic region (Ryan, 1967). As an exception, in the red-clawed mangrove tree crab, *Goniopsis cruentata*, the maturing gonads extend as far as the third abdominal segment (De Souza and Silva, 2009). Whereas the ovaries are restricted to the much broader and convex cephalothorax in the Brachyura, in the anomuran families such as Paguridae and Thalassinidae, they are located mainly in the rather elongated and cylindrical abdomen (see Kaestner, 1970). However, in hermit crabs, the ovary is asymmetrical due to the peculiar lodging of its soft abdomen inside the molluscan shell, with its consequent twisting of ovarian limbs and oviducts (Carayon, 1941). Thus, in the hermit crab *Clibanarius clibanarius*, the ovaries are paired elongated organs occupying the posterior two-thirds of the abdomen and lie on the hepatopancreas; the right ovary is shorter than the left, due to abdominal torsion caused by their lodging in the molluscan shell (Varadarajan and Subramoniam, 1980). In the anomuran sand crab, *Emerita asiatica*, the two ovarian limbs behind the interconnecting bridge run backward and join to form a ring in the mid region; from this ring extends a median cord that enters the abdominal segment (Subramoniam, 1981). In another hippid sand crab, *Albunia symnista*, belonging to Albunidae, the posterior limb of the ovary extends right into the abdomen, thus sharing the true anomuran as well as the some of the brachyuran features (unpublished observation). In the shrimps and lobsters, with cylindrical and elongated body, the posterior lobe of the ovary extends to the abdomen. In the caridean prawn, *Macrobrachium acanthurus* and the spiny lobster *Jasus frontalis*, the ovary reaches the second abdominal segment (Carvalho and Pereira, 1981; Silva and Cruz-Landim, 2006).

In the bivalve freshwater ostracods, the paired ovary shows bilaterality (Ferguson, 1967). In the parasitic copepods, the ovary is a paired structure, but in the free-living copepods, the ovary comprises only a single median lobe. In the freshwater Cladocera, the ovaries are simple tubes, with no diverticula to give rise to ovarioles, as in insects. In the sessile barnacles such as *Balanus amphitrite* and *Balanus eburneus*, the ovary consists of branched tubules, called ovarioles (similar in structure to insect ovarioles), which are located in the connective tissue between the mantle cavity and the basal membrane (Fyhn and Costlow, 1977). The elongated mature ovarioles are arranged along the ovarian ducts, which open into the mantle cavity. New ovariole buds continue to originate along the ovarian ducts in all intermolt stages. Evidently, there is a diversity of ovarian morphology among different crustacean taxa.

However, in the primitive state, the ovary of crustaceans is in the form of two longitudinal columns, extending the entire length of the cephalothorax and the abdomen. In the penaeid shrimp, *Penaeus setiferus*, the ovaries are paired, partly fused, bilaterally symmetrical bodies, extending in the mature condition almost its entire length, from the cardiac region of the stomach to the telson (King, 1948). In the anterior cephalic region, each ovary bears a slender anterior lobe and seven finger-like lateral projections. A pair of lobes, one from each ovary, extends the length of the abdomen. But in stomatopods, the ovary gives off lateral pouch-like projections throughout the length of the ovary. The lateral lobes are located dorsally to the large mass of hepatopancreas and ventrally to the pericardial chamber. The abdominal extensions lay dorsolateral to the intestine and ventrolateral to the dorsal abdominal artery. In the caridean shrimps, the gonads are located in the cephalothorax, situated dorsal to the gut and hepatopancreas. However, in the mature condition, the paired ovaries extend into the space above the cardiac stomach in the anterior cephalothorax and may extend back into the first few abdominal segments (Bauer, 2007).

8.1.2 Structure of the Ovary

The wall of the ovary in the penaeid shrimp, *P. setiferus*, is composed of three layers: a thin outer layer of pavement epithelium, a relatively thick layer of connective tissue, and an inner layer of germinal epithelium (King, 1948). However, the germinal epithelium is not distributed uniformly over the inner wall of the ovary, but confined to certain regions, which are termed as the zone of proliferation. The germarium represented by the zone of proliferation is more evident in the abdominal extension of the ovary as a streak running along the medial-ventral wall of each lobe. As oogonial proliferation progresses, both primary and secondary oocytes move as a column of cells from the germarium toward the center of the ovarian lobe. The proliferative zone in the lateral lobes is strictly ventral in position. In the brachyuran crabs, the germarium is centrally placed in both limbs of the ovary.

In the anomurans, the mature ovary is cylindrical with its germinal zone taking a ventral position. The immature ovary of the hermit crab *C. clibanarius* is characterized by the uniform disposition of the germinal zone, although a slightly ventral disposition is already apparent (Varadarajan and Subramoniam, 1980). As ovarian development proceeds, there is further shift in their placement to the ventrolateral positions, owing to torsion consequent to their lodging in the molluscan shell. In the advanced stage of vitellogenesis, the germinal zone becomes completely surrounded by overhanging vitellogenic oocytes giving it an apparent internal position. The crustacean ovary does not possess any strictly compartmentalized germarium and vitellarium, as in insects. Generally, a germinal zone or germinal epithelium is distinguishable during ovarian development in most Crustacea. In the primitive cladocerans, the germarium appears as a syncytium at one end of the ovary, whereas in the copepod *Calanus*, dividing oogonia occur at the posterior end of the ovary and the oocytes produced from them fill the rest of the ovary in ascending stages of development as they approach the oviduct (Matthews, 1962). In the halocyprid ostracod *Conchoecia imbricate*, the germarium is placed at the terminal of the saclike paired ovary, which runs along the alimentary canal (Ikuta and Makioka, 2004). In the remipede *Godzilliognomus frondosus*, the paired ovaries are located in the anterior part of the body and are composed

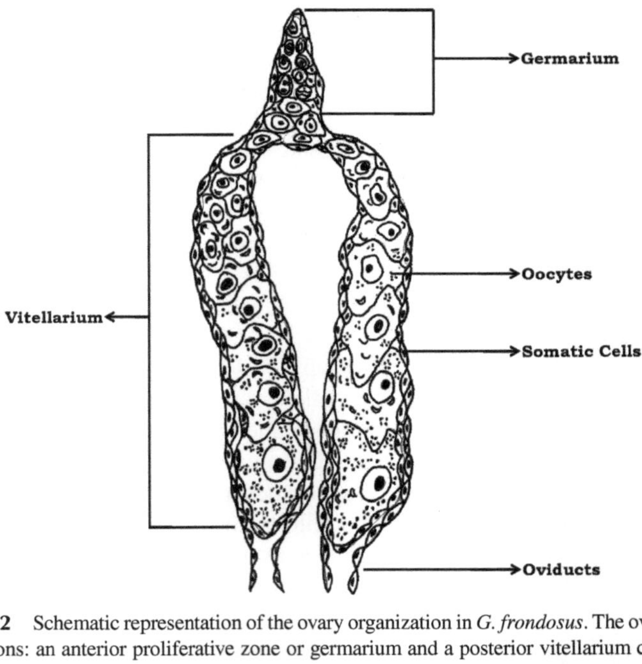

FIGURE 8.2 Schematic representation of the ovary organization in *G. frondosus*. The ovary consists of two regions: an anterior proliferative zone or germarium and a posterior vitellarium composed of two ovarian tubes. Each ovarian tube houses oocytes at subsequent stages of development (previtellogenic and vitellogenic), progressively more advanced toward the posterior end. The ovarian tubes pass into separate oviducts. *Adapted from Kubrakiewicz, J., Jaglarz, M.K., Iliffe, T.M., Bilinski, S.M., Koenemann, S., 2012. Ovary structure and early oogenesis in the remipede,* Godzilliognomus frondosus *(Crustacea, Remipedia): phylogenetic implications. Zoology (Jena) 115 (4), 261–269.*

of single anterior germarium and paired ovarian tubes (vitellarium) (Fig. 8.2) (Kubrakiewicz et al., 2012). Similar type of paired ovary with the anterior germarium has also been reported in cephalocarid *Hutchinsoniella macracantha* (Hessler et al., 1995). A well-defined anterior germarium located in the head region and the paired elongated tubelike vitellaria in both Remipedia and Cephalocarida (Xenocarida) share common features in the ovarian morphology with the basal hexapod lineage, grouped within the Entognatha, confirming the phylogenetic relationship between these primitive crustaceans and hexapods (Kubrakiewicz et al., 2012). In the thoracic barnacle, *B. amphitrite*, the germarium has no strictly defined localization but may be found in the periphery as well as in the center of the ovarioles. The oogonia are closely packed cells making up a continuous string of cells in the ovarioles. The oogonia are rounded with pycnotic nuclei. In the amphipod *Orchestia gammarella*, the germarium is a longitudinal zone attached to the basal lamella (Meusy, 1968).

There seems to be wide variation in the placement of the germinal zone in the ovary among crustaceans, as revealed from early histological studies. Adiyodi and Subramoniam (1983) have distinguished five categories of germarium placement among Crustacea. They are (1) peripheral, (2) peripheral, but confined to lateral or ventral regions as a thin band, (3) central as a germinal cord, (4) as germ nests, and (5) peripheral with germinal nests (Table 8.1). Peripheral placement of the germinal zone might be considered as a primary feature, inasmuch as many other invertebrate as well as vertebrate species possess a peripheral germinal zone. The majority of decapod crustaceans develop a large number of yolky eggs in the ovary and hence it is reasonable to assume that variations in the placement of the germinal zones found in different species are brought about by tension exerted by the developing ova. As seen in many crustacean species, the ventral or ventrolateral position might be acquired by the biased growth of the oocytes in the opposite direction due to the tubular nature of the body cavity and the availability of more space in the dorsolateral regions. The ventrally restricted germinal zone that makes inroads into the ovary owing to the pressure of the radiating cordons of developing oocytes has been illustrated in the hermit crab, *C. clibanarius* (Varadarajan and Subramoniam, 1980). The internalization of the germarium is more pronounced in brachyuran crabs with a relatively short and laterally spread-out ovary. Interestingly, Laulier and Demeusy (1974) described the germinal zone in the prepubertal females as peripheral, and the central germarium found in the adult is only secondarily acquired from an originally peripheral germinal zone. In the ovary of the swimming crab, *Portunus sanguinolentus*, Ryan (1967) presented evidence for the existence of a central tubular shaft of tissues that give off lateral lobes. The oogonial cells are located on the wall of the central tubule shaft and the dividing oogonial cells lack nuclear membrane. From these germinative zones, oocytes are proliferated and give off cordons of cells into the lobes. These radially placed lobes may be homologous to the ovarioles of insects.

TABLE 8.1 Location of Germinal Zone in the Crustacean Ovary

Species	Name of Location	References
Paguridae		
Pagurus bernhardus	Longitudinal ventral band	Jackson (1913)
Hippidae		
Emerita asiatica	Syncytial peripheral germinal region	Menon (1934)
	Peripheral with oogonial clusters in the interior	Subramoniam (1981)
Diogenidae		
Diogenes pugilator	Ventrolateral position	Bloch (1935)
Clibanarius isanthropus	Germinal zone medioventral	Carayon (1941)
Clibanarius olivaceous	Germarium as a narrow white band running along the ventral side	Kamalaveni (1947)
Clibanarius clibanarius	Ventrally located, but it may become central surrounded by overhanging vitellogenic oocytes	Varadarajan and Subramoniam (1980)
Penaeidae		
Litopenaeus setiferus	Inner layer of germinal epithelium in the ovarian wall. Zone of proliferation restricted to the medioventral region of abdominal wall; in the cephalic region the lateral lobes have strictly ventral position	King (1948)
Penaeus indicus	Innermost layer of ovarian wall, but not uniformly distributed. Distinct zones of proliferation are present ventrally and laterally	Subrahmanyam (1963)
Triopsidae		
Triops granarius	Germinal regions are at the tip of follicular ducts	Longhurst (1955)
Triops longicaudatus	Germinal regions are at the tip of follicular ducts	Longhurst (1955)
Triops australiensis	Germinal regions are at the tip of follicular ducts	Longhurst (1955)
Lepidurus arcticus	Germinal regions are at the tip of follicular ducts	Longhurst (1955)

TABLE 8.1 Location of Germinal Zone in the Crustacean Ovary—cont'd

Species	Name of Location	References
Chirocephalidae		
Chirocephalopsis bundyi	Germinal cells along the entire length of the ovary, on the side of the alimentary canal as thin, dense strands of tissues	Linder (1959)
Atyidae		
Caridina laevis	Ventrolateral aspect of each ovarian lobe	Pillai (1960)
Pandalidae		
Pandalus borealis	Single germinal zone runs along the gonad attached to the basal lamella	Charniaux-Cotton (1965)
Limnadiidae		
Limnadia lenticularis	Germinal region is seen at the tip of saclike dilations of ovarian tubules	Zaffagnini (1968)
Bathynellidae		
Bathynella natans	Germinal cells at the posterior extremity of the tubelike ovary	Kaestner (1970)
Palinuridae		
Jasus lalandii	Germinal epithelium at the innermost layer of the ovarian wall. This is pushed away from the wall to form a strand. From this condition in the immature ovary a lumen develops, which is clearly detectable in the spent ovary	Fielder (1965)
	A prominent central germinal strand (germinal epithelium) with a distinct lumen in immature stage, which disappears in the ripe ovary	Heydorn (1969)
Nephropidae		
Homarus sp.	Smaller oocytes are present in the center of the ovary and the larger oocytes toward the periphery	Kessel (1968)
Nephrops norvegicus	Germinal strand passing through each lobe of the ovary along its axis (central)	Farmer (1974)
Cambridae		
Procambarus clarkii	Germarium located in the ovarian epithelium near the bases of the oogenetic pouches	Ando and Makioka (1998)

Continued

TABLE 8.1 Location of Germinal Zone in the Crustacean Ovary—cont'd

Species	Name of Location	References
Gecarcinidae		
Gecarcinus lateralis	"Germ nests" distributed throughout the ovary	Weitzman (1966)
Portunidae		
Portunus sanguinolentus	A central hollow shaft, occasionally with an empty lumen, extending the length of the ovary with lobes projecting outward from this shaft. Germinative zones are located along the sides of the central shaft	Ryan (1967)
Portunus pelagicus	Germinal zone is in the form of a central shaft running the entire length of the paired ovary	Dhas et al. (1980)
Epialtidae		
Libinia emarginata	Syncytial mass of centrally located gonial cells	Hinsch (1970)
Cancridae		
Cancer pagurus	Oogonia and early vitellogenic oocytes in the central region of the ovarian tubules	Eurenius (1973)
Panopeidae		
Rhithropanopeus harrisii	Germinal zone as central, longitudinal cord	Payen (1974)
Menippidae		
Menippe mercenaria	Germinal zone as central, longitudinal cord	Payen (1974)
Cymothoidae		
Mothocya epimerica	Germarium as a longitudinal ridge on the ventrolateral side	Thampy and John (1974)
Talitridae		
Orchestia gammarella	Longitudinal germinal zone attached to the basal lamella	Zerbib (1976)
Chthamalidae		
Chthamaltus challengeri	Clusters of oogonia occur in the center of the ovarian tube	Iwaki (1975)

TABLE 8.1 Location of Germinal Zone in the Crustacean Ovary—cont'd

Species	Name of Location	References
Balanidae		
Balanus amphitrite	The germarium has no strictly defined location but may be found toward the periphery as well as in the center of the ovariole	Fyhn and Costlow (1977)
Balanus eburneus	The germarium has no strictly defined location but may be found toward the periphery as well as in the center of the ovariole	Fyhn and Costlow (1977)
Halocyprididae		
Conchoecia imbricata	Nipple-shaped germarium in each terminal of the ovary	Ikuta and Makioka (2004)
Cyprididae		
Cyprinotus uenoi	Located on the terminal region of the paired ovary	Ikuta et al. (2007)
Argulidae		
Argulus japonicas	A long germarium is embedded in the dorsal ovarian wall along the median longitudinal line of the ovary	Ikuta et al. (1997)
Remipedia		
Godzilliognomus frondosus	Germarium located in the anterior part of the paired ovarian tubules	Kubrakiewicz et al. (2012)

Within the lobes, the oocytes undergo differentiation and vitellogenesis. The ovarian wall is composed of two layers of connective tissue and all lobes are further covered by an additional connective tissue layer. In the brachyuran crabs aggregation of the primordial cells in the center of the ovarian tubules may be advantageous for the easy displacement of the primary oocytes to the growth zones in later stages.

8.2 CHELICERATE AND MANDIBULATE TYPE OF OVARY

Lopez-Greco (2013) classified the crustacean ovary into two morphological types. In type I ovaries, the developing oocytes that leave the germarium grow on the inner surface of the ovarian wall facing the ovarian lumen. This is called mandibulate type and found in most of the malacostracan species and all mandibulate arthropods. In the chelicerate type (type II), growing oocytes protrude to the hemocoel and the developing oocytes of various sizes are attached to the outer surface of the ovarian epithelium (Fig. 8.3). Chelicerate type of ovary is uncommon in malacostracans, but present in many lower crustaceans. As an example, in the halocyprid ostracod, *C. imbricate*, the very early previtellogenic oocytes leave the germarium placed in the terminal part of the ovaries and do not enter the ovarian lumen but bulge out into the outer surface of the ovarian wall, sandwiched between the ovarian epithelium and its basement membrane (Ikuta and Makioka, 2004). The oocytes grow to maturity, supported by a few tall ovarian mesodermal epithelial cells and raising the basement membrane of the ovarian epithelium. The larger oocytes lie away from the germarium on the ovarian wall. Further growth of oocytes on the outer surface of the ovarian wall is similar to the pattern found in the myodocopid ostracods, the brachyurans, pentastomids, and all the chelicerate arthropods (Itaka and Makioka, 1999, 1997; Nørrevang, 1983; Makioka, 1988).

However, in another family of ostracods, Podocopida, the ovary is of mandibulate type. In *Cyprinotus uenoi*, the germarium is located in the terminal region of the ovary, like the halocyprid ostracod. Previtellogenic oocytes are arranged according to their size, with the largest vitellogenic oocyte found nearer the ovarian lumen. Most of the growing oocytes as well as fully mature

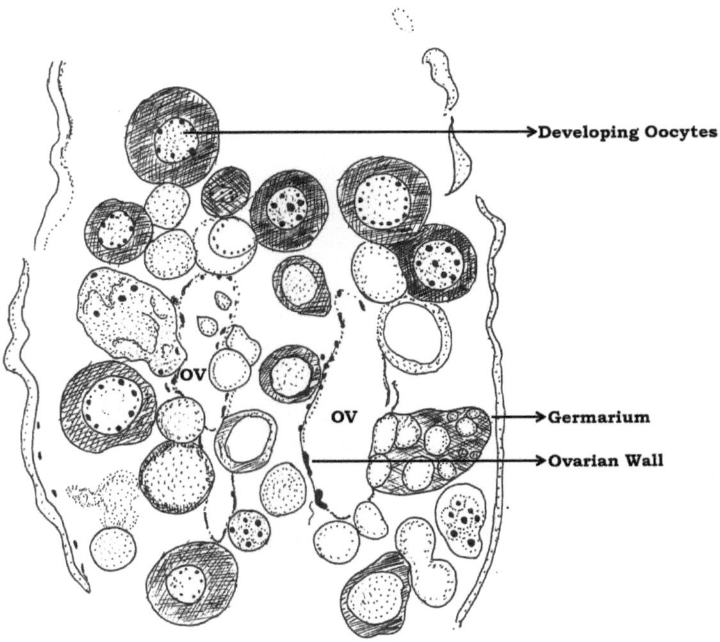

FIGURE 8.3 Transverse section of a pair of *Conchoecia imricata* ovaries, *OV*, Ovary. *Modified from Ikuta, K., Makioka. T., 2004. Structure of ovary and oogenesis in the halocyprid ostracod* Conchoecia imbricata. *J. Crust. Biol. 24, 72–80. BioOne.*

oocytes are found in the ovarian lumen (Ikuta et al., 2007). Evidently, there is diversity in the structural morphology of ovary among the lower crustaceans, particularly branchiopods and ostracods, and in many cases, they resemble the ovary of the apterygote insects, suggesting their phylogenetic relationship with Hexapoda.

8.3 OOGENESIS

In crustacean oocyte development, oogenesis is a period of rapid growth during which large quantities of yolk materials are accumulated within the ooplasm. The initial phase of oocyte growth is in general quite slow, while the period of vitellogenesis is characterized by rapid growth. Further, the early slow growth encompasses the previtellogenic stage, including the endogenous yolk formation. With reference to vitellogenesis, oocytes of various organisms vary markedly in the mechanism of yolk formation and acquisition. In many reproductively primitive invertebrates, the ooplasm possesses the necessary synthetic machinery and organelles required for intraooplasmic synthesis and packaging of yolk materials. On the other hand, in oviparous vertebrates and higher invertebrates such as insects, yolk materials are largely synthesized in extraovarian sources and then incorporated into the ooplasm by micropinocytosis where subsequent processing of yolk occurs. In many decapod crustaceans, producing large numbers of yolk-laden eggs, a combination of intra- and extraoocyte synthesis of yolk constituents occurs (Adiyodi and Subramoniam, 1983).

Based mainly on different phases of yolk synthesis and accumulation, the period of oocyte development and differentiation is divided into premeiotic, previtellogenic, and vitellogenic phases. Yet, the primordial germ cell has to pass through a proliferative phase that occurs in the germinal zone to increase the number of oogonial cells by mitotic multiplication. The resulting secondary oogonial cells are transformed into the primary oocytes, which finally differentiate into typical egg cells (see Adiyodi and Subramoniam, 1983).

8.3.1 Oocyte Proliferation

The germinal zone in crustacean ovaries is active continuously throughout the reproductive life of the female. Mitotic divisions of oogonial cells occur in rapid succession and hence it is difficult to observe them in the ovary. However, the clustered secondary oogonial cells in the germinal zone could be observed easily in the mature ovary. The oogonial cells have large round vesicular nuclei surrounded by a thin rim of cytoplasm and occur as "germ nests," deep into the ovarian lobe in the land crab, *Gecarcinus lateralis* (Weitzman, 1966). Oogonial cells could also be identified by their deeply staining chromatin granules which are closely applied against the nuclear membrane. Once the gametogenesis is reinitiated after spawning or after a sexual repose, these resting oogonial cells are transformed into primary oocytes and are displaced to the growth zone, where they immediately enter the prophase of meiosis. In the continuously and repetitively breeding tropical sand crab, *E. asiatica*, each ovulation is closely followed by oogonial proliferation (Subramoniam, 1977a). After passing through different phases of spirene or bouquet stages, chromosomal changes come to a temporary arrest at pachytene stage, the metabolic activity of the primary oocyte seem to override the meiotic changes. The secondary oogonia remain in the germinal zone until spawning.

8.3.2 Follicle Cells

Follicle cells are the somatic mesodermal cells present in crustacean ovary. As in insects and other animals, they are associated with the maturing oocytes, mediating yolk protein uptake from the hemolymph and also purportedly involve in the secretion of egg layers. These mesodermal cells are present even in the undifferentiated ovary at the time of hatching in *O. gammarella*, distributed at random among the gonocytes (Hort-Legrand et al., 1974). In the hermit crab, *C. clibanarius*, they occur in disorderly strands in undeveloped and spent stage ovary, mainly confining to germinal zone (Varadarajan and Subramoniam, 1980). They are seen as a "mesodermic web" in the non-gametogenic phase of the adult ovary. However, the follicular cells show significant streaming movements into the growth zone to surround the growing oocytes entering into vitellogenesis. The follicular cells are spherical or oval when in aggregation, but become flattened on attachment to the oocyte periphery. Charniaux-Cotton (1985) introduced the term "folliculogenesis" referring to the process by which follicle cells come to surround early vitellogenic oocytes. She also divided folliculogenesis into primary and secondary. In *O. gammarella*, when the primary vitellogenic oocyte leaves the germarium, mesodermal tissue surrounds each oocyte forming a follicular multilayered epithelium. This is termed as primary folliculogenesis. When the oocyte enters secondary vitellogenesis, the follicle cells change shape to become single layered columnar cells to surround the grown-up oocytes. This stage is termed as secondary folliculogenesis. She considered that folliculogenesis is a prerequisite for the uptake of yolk protein from outside of the oocyte and is probably stimulated by low ecdysterone titer in the amphipod *O. gammarella*. In this amphipod, follicular cells become endocrine after establishing contact with the oocyte and serve to stimulate synthesis of sex-specific protein in the fat body. However, crustacean follicle cells, unlike those of insects, remain as a permanent tissue after ovulation, and there is a process of renewal of these tissues before the next gametogenesis commences (Zerbib, 1976; Charniaux-Cotton, 1980).

That the follicular epithelial cells can also synthesize Vg was shown in the kuruma prawn *Marsupenaeus japonicus* by Yano and Chinzei (1987). Tsutsui et al. (2000) localized Vg mRNA in the follicle cells of the ovary of *M. japonicus* by in situ hybridization. In the amphipod, *O. gammarella*, ultrastructural studies by Rateau and Zerbib (1978) indicated protein synthetic activity in the endoplasmic reticulum (ER) and Golgi complex regions of follicle cell during vitellogenesis. But, the authors related this activity to the secretion of ovarian hormones. More work on the protein synthetic activity of the follicle cells has been reported in the prawn *Palaemon serratus* and the lobster *Homarus americanus*, but they are implicated with the formation of chorionic egg membrane, rather than contributing to yolk protein (Papathanassiou and King, 1984; Talbot, 1981). In addition, follicle cells participate in the oosorptive process in several crustacean species, acting as phagocytes (Baffoni, 1962; Subramoniam, 1981).

8.3.3 Nurse Cells

Nurse cells are totally absent in higher crustacean orders such as decapods, but in the lower crustacean orders, they are commonly present. Unlike insects, in which the rRNA synthesized by the nurse cells directly flows into the previtellogenic oocytes (Engelmann, 1970), in the lower crustacean orders, such as anostracans, the nurse cells actually break down to be ingested by the definitive oocytes in the germinal region. In the brine shrimp, *Artemia*, the nurse cells originate in the germinal zone, along with oogonial cells (Criel, 1991). In fact, most of the dividing oogonia and the primary oocytes differentiate into nurse cells and in a cluster, only one gonial cell differentiates into an oocyte. In *Artemia salina*, nurse cells and the oocyte are arranged in the form of a string. But there is no apparent morphological distinction between nurse cells and oocyte at the beginning of differentiation. Only at the end of previtellogenesis, wherein cytoplasmic volume increase occurs, oocyte–nurse cell differentiation is discernible. Nurse cells nuclei become polyploidy, increasing rDNA content, a requirement for massive accumulation of rRNA in the oocytes (Cave, 1982). However, nurse cells do not produce any yolk protein. Nurse cells in *Artemia* are arranged in rows, each cell closely pressed against the other, with the end cell of the row intimately associated with the surface of the oocyte. After differentiation, the nurse cell nucleus is larger and more intensely chromatic than the oocyte nucleus (Iwasaki, 1970). Electron microscopic studies have indicated that nurse cells are connected to the oocyte and to each other by intracellular bridges from incomplete cytokinesis, facilitating transport of materials to the oocyte (Criel and Macrae, 2002). *Artemia* oocytes grow at the expense of nurse cell activities, and the cytoplasm transferred from the adjacent nurse cells contains mainly free ribosomes, mitochondria, lipid droplets, ER, and annulate lamellae. Concurrently, signs of apoptosis are apparent in the nurse cells located at ribbon ends. The nurse cell ribbon detaches from the oocyte upon termination of vitellogenesis and is phagocytosized by the somatic (follicle) cells.

In the primitive ostracods, nurse cells, providing nourishment to the previtellogenic oocytes, have been described as early as 1894 in *Erythrocypris*

(Müller, 1894). In a freshwater ostracod, *C. uenoi*, small, spherical nurse cells are found among the early previtellogenic oocytes, arranged in a single file in order of size in the ovarian lumen (Ikuta et al., 2007). Nurse cells originate from the incomplete cytokinesis of oogonial cells in the germarium during mitotic multiplication, as evidenced by the presence of cytoplasmic bridge in between the nurse cells and oocytes. Cytoplasmic filaments connect the nurse cell with the oocyte. Cytoplasmic bridges in between the nurse cells and oocytes have also been reported in some branchiopod crustaceans (Martin, 1992) and insects with meroistic ovarioles (Büning, 1994). As in insects, nurse cells have large nuclei with distinct nucleoli, indicating a high rate of RNA and protein synthesis. For example, in the notostracan tadpole shrimp, *Triops cancriformis*, the nurse cells are bigger than the oocytes and have several nucleoli (Trentini and Scanabissi, 1978). In contrast, the nurse cells of ostracods are smaller than the oocytes and have small nucleoli in their nucleus.

Clearly, branchiopod crustaceans possess many similarities in the morphological structure and organization of the ovary with that of insects. Particularly, the presence of nurse cells to assist the oocyte growth by supplying nucleic acids and probably proteins could be considered as a common feature in the evolution of vitellogenic system in these two arthropod groups. However, the lack of nurse cells in the advanced crustacean orders could be perceived as a reproductive strategy to increase the number of primary oocytes produced during the mitotic multiplication of oogonial cells in the germinal zone. The production of nurse cells from the germ cell division will affect the total number of eggs produced in the ovary. In compensation, the oocyte nucleus becomes hyperactive in producing RNA materials, which are released into the ooplasm to enable protein synthesis during previtellogenesis.

8.4 OOCYTE DIFFERENTIATION

Oocyte growth and differentiation commence once the primary oocyte is pushed into growth zone. It has to be mentioned here that in crustacean ovary, there is no distinct growth zone or vitellarium, as in insects. Oocyte differentiation in the crustacean ovary is characterized by an extended period of vitellogenesis, as the majority of malacostracan crustaceans develop a large number of yolky eggs.

From a detailed cytomorphological study on the ovary of the amphipod *O. gammarella* and the brachyuran crab, *Callinectes sapidus*, Charniaux-Cotton (1980) classified vitellogenesis into three major divisions: (1) previtellogenesis, in which the meiotic prophase chromosomal changes occur, (2) primary vitellogenesis, characterized by the formation of numerous vesicles of the granular ER containing glycoproteins (endogenous vitellus), and (3) secondary vitellogenesis, in which yolk globules consisting of a lipoglycocarotenoprotein are formed from pinocytocized yolk precursor protein from the hemolymph. Fig. 8.4 illustrates all the cytomorphological changes that occur in different stages of oogenesis. In another brachyuran crab, *Eriocheir sinensis*, during primary vitellogenesis, the

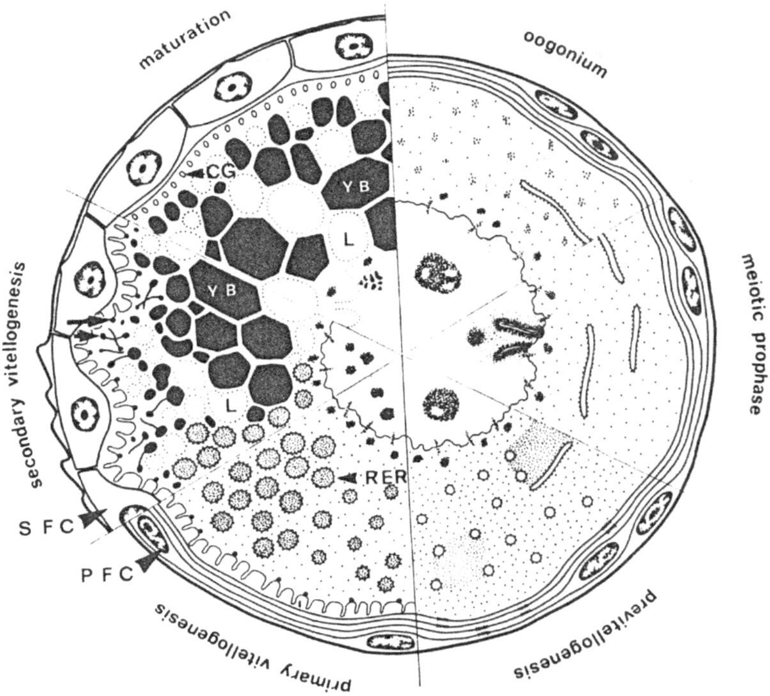

FIGURE 8.4 Schematic representation of the different steps of oogenesis in amphipods and decapods. During genital rest, oogenesis is blocked at the end of primary vitellogenesis. *CG*, Corticle granule; *L*, Lipid globule; *PFC*, Primary follicular cell; *RER*, Endoplasmic reticulum containing endogenous granule; *SFC*, Secondary follicular cell; *YB*, vitellin globule. *Adapted from Payen, G.G., 1981. Aspects fondamentaux de l'endocrinologie de la reproduction chez les crustaces marins. Oceanis 6, 309–339.*

oocytes synthesize a glycoprotein product which is stored in cytoplasmic vesicles. This is followed by the secondary vitellogenic stage, in which extraovarian, lipid-containing Vg is taken up at the oocyte surface and sequestered into a distinct type of yolk granule. Cytochemistry of the oocyte vitelline inclusions in *O. gammarella* has also revealed their sequential origin from inside and outside of the ovary. Zerbib (1976) distinguished four types of yolk in their sequential order of their appearance in the oocyte: (1) small ergastoplasmic cisternae containing glycoprotein, (2) yolk spheres containing lipoglycocarotenoprotein complex derived from Vg circulating in the hemolymph, (3) lipid spheres containing triglycerides, and (4) cortical granules of glycoprotein nature, visible at the end of vitellogenesis.

In the freshwater prawn *Macrobrachium rosenbergii*, Okumura and Aida (2000) followed the histological changes of the ovary during its development within a single intermolt. They classified ovarian development into previtellogenic and vitellogenic stages, based on cytological characteristics, gonadosomatic index (GSI), and hemolymph Vg level in different ovarian developmental stages. During the postmolt stage, the GSI was low (0.49%) and the ovary contained oogonia in the center, surrounded by previtellogenic and endogenous vitellogenic oocytes. This stage can be considered as the resting stage, as the same condition exists in the ovary of nonreproductive molt cycle. Oocytes in the previtellogenic stage exhibited hematoxylin-stained cytoplasm, which possessed oil globules and PAS-positive vesicles, characterizing endogenous vitellogenesis. The previtellogenic ovary is observable from the C_0 to C_1 molt stage. The vitellogenic stage starts from C_1 molt stage and is characterized by the appearance of yolk globules, containing yolk protein derived from sequestration of Vg from hemolymph, and hence it is termed as exogenous vitellogenesis. As the oocytes progress through D_0–D_3 molt stage, the oocytes are filled with yolk granules and globules. The ovary showed an increase in size and the GSI is very high. Immunohistochemistry with antivitellin serum has revealed that only exogenous vitellogenic stage showed positive reaction. The hemolymph Vg was undetected in the premolt stage but sharply increased during the exogenous vitellogenic stage. This stage is followed by the spent stage that occurs at A_1 stage of the next molt cycle, after oviposition. A comparison of ovarian stages between reproductive and nonreproductive molts of *M. rosenbergii* revealed that the previtellogenic and endogenous vitellogenic stages of oocytes persist all through the molt cycle stages in the nonreproductive molt cycle, indicating that oocyte development up to this stage is not under any environmental or endocrine stimuli. Invariably, exogenous vitellogenesis of the oocytes occurs synchronously in the ovary. After oviposition, the spent stage ovary contains only oogonia and primary oocytes at the previtellogenic and endogenous vitellogenic stage.

In the coonstriped shrimp, *Pandalus hypsinotus*, Okumura et al. (2004) classified the developing oocytes into four stages: previtellogenic oocytes (with homogenous hematoxylin-stained cytoplasm), endogenous vitellogenic oocytes (oocytes with PAS-positive vesicles in the cytoplasm), exogenous vitellogenic oocyte (oocytes with eosin-positive yolk globules), and maturing oocytes (germinal vesicle breakdown in the center of the oocyte). These four stages in the oocyte development have been integrated into three ovarian developmental stages, namely, inactive, early developing, and late developing stages.

The inactive ovary contains oogonia, previtellogenic oocytes, and endogenous vitellogenic oocytes. Surrounding the centrally placed germinal cells are the previtellogenic cells and the endogenous vitellogenic oocytes are found at the periphery. At electron microscopic level, the ooplasm of the previtellogenic oocytes contains ribosomes and mitochondria. The endogenous vitellogenic oocytes contain unstained vesicles, containing lipid droplets. At the electron microscopic level, the ooplasm is filled with mitochondria and rough ER, which contains dense homogeneous materials as well as small electron-dense granules at the periphery. This cytoplasmic architecture is indicative of yolk synthesis within the oocytes.

In the early developing stage, the exogenous vitellogenic oocytes are present at the periphery, and the endogenous vitellogenic oocytes are present in the center. In the beginning of this stage, small eosinophilic globules appear, followed by yolk globules that fill the entire ooplasm. When treated with antivitellin antibody, only the exogenous vitellogenic oocytes were stained, whereas the endogenous oocytes and the oogonial cells do not take this stain. This shows that only exogenous oocytes represent the actual vitellogenic stage, and the protein synthesis during endogenous oocytes is not related to the yolk protein. Electron microscopic observations revealed the presence of electron-dense yolk globules together with mitochondria, lipid droplets, and rough ER in the ooplasm. There is further evidence for the active endocytosis to occur at the oolemma, suggesting exogenous yolk protein uptake. Follicle cells completely surround these oocytes.

In the late developing stage of the ovary, the exogenous vitellogenic oocytes continue to accumulate yolk globules and reach a size of 875–950 μm. In the ripe stage, the oocytes are still enveloped by follicle cells, and a vitellin envelope with microvilli projecting toward the follicle cells is formed at the oocyte periphery. In the spent stage ovary, oogonia, previtellogenic oocytes, and the endogenous vitellogenic oocytes are present in the center, with ovulation traces present in the periphery of the ovary. The ovary also contains a few regressing mature oocytes.

At the end of vitellogenesis, the microvilli surrounding the oocytes disappear and the plasma membrane becomes smooth. At this time, the cortical granules, rich in glycoprotein, appear at the oocyte periphery. In the penaeid shrimp, *Penaeus monodon*, cortical rods appear first as spherical bodies near the periphery of the oocyte cytoplasm and elongates toward the nucleus as maturation proceeds. In *P. monodon*, cortical rods contain only glycoproteins and no lipids; in *Penaeus vannamei*, the cortical rods are in the form of feathery matrix, made up of polypeptides, originating from dilated ER and transported through vesicular bodies (Rankin and Davis, 1990). In other penaeiodean shrimp like *Sicyonia ingentis*, cortical rods lie in extraoocytic crypts, formed in the cytoplasmic membrane of the oocytes. The crypts are formed by the fusion of materials found in the pinocytotic vesicles secreted by the oocytes. The oocytes of the lobster *H. americanus* lack cortical rods but possess an equivalent structure called bottle brushes. These bodies constitute the innermost portion of the external oocyte envelope, namely the endochorion (Talbot, 1981). In the crab oocyte, cortical granules are present as equivalent structures of cortical rods, found in penaeid shrimp. Since cortical rods are so characteristic of penaeid shrimp oocyte development, many authors used this as the cortical rod stage in their classification of ovary (Mohamed and Diwan, 1994). Table 8.2 summarizes staging of oocyte development in a penaeid shrimp, as compared with those of other species.

TABLE 8.2 Distinctive Changes in the Color and Cytology of the Ovary in Relation to Vitellogenic Stages in *P. indicus*

Stage, Color, and Appearance of Ovary	Oocyte Phase	Cytoplasm	Nucleus	Nucleoli	Follicle Cells
Stage I Immature ovary: translucent smooth	Stage I Previtellogenic	Homogeneous, basophilic due to the presence of RNA; mitochondria and rough ER present	Vesicular with numerous nuclear pores	5–10; electron dense; arranged along peripheral margin of the nucleus	Tall with conspicuous nuclei
Stage II Early maturing ovary: pale cream smooth-granular	Stage II Early vitellogenic (primary vitellogenic/ vitellogenesis 1)	Granular; perinuclear halo of nucleolar RNA; mitochondria and rough ER present; vesicular primary yolk	Vesicular with nuclear pores	5–10; arranged in a circular ring along the nuclear wall	Decrease in cell height; hypertrophied nucleus and nucleoli
Stage III Late maturing ovary: light green granular	Stage III Late vitellogenic (secondary vitellogenic/ vitellogenesis 2)	Granular and acidophilic; lipid globules and yolk platelets present. ER and mitochondria not apparent; oolemma with micropinocytotic vesicles	Faintly stained	Nucleoli number reduced to 2–3	Flattened around oocyte; rough ER and mitochondria present
Stage IV Mature/ripe ovary: dark green granular	Stage IV Mature	Full of yolk platelets; disulphide-rich cortical bodies with feathery matrix	Faintly stained	Not apparent	Very flattened and stretched around oocyte
Stage V Spent/spent-recovering ovary: pale cream granular-flaccid	Stage V Spent/recrudescent	Deeply basophilic with signs of pycnosis	Vesicular	5–10; arranged along the perinuclear margin	Intensely PAS positive and hypertrophied

Modified from Sunil, K.K.M., Diwan, A.D., 1994. Vitellogenesis in the Indian white prawn *Penaeus indicus* (Crustacea: Decapoda: Penaeidae). J.Aqua. Trop. 9, 157–172.

8.5 BIOGENESIS OF YOLK

Under primitive conditions, oocytes were probably able to accumulate sufficient yolk through their own metabolic efforts. This kind of autosynthesis is more prevalent in marine invertebrates such as cnidarians, annelids, echinoderms, and the tunicate, *Ciona*, as observed by ultrastructural investigations (see Wourms, 1987). The other mode of yolk formation is heterosynthetic, in which yolk is synthesized by cells other than the oocyte and is then transported to the oocyte. For instance, insects depend totally on the metabolic reserves of fat body for the production of yolk substances (Engelmann, 1970). A third method, autoheterosynthesis also exists in many animals, in which a combination of the autosynthetic and heterosynthetic modes is employed in vitellogenesis. Early electron microscopic as well as recent biochemical investigations in a variety of crustacean species have indicated that autosynthetic yolk occurs first in the oocytes followed by uptake of yolk precursor molecules from the hemolymph (Adiyodi and Subramoniam, 1983). Nevertheless, crustaceans display various degrees of heterosynthetic yolk production, inasmuch as various extraovarian organs are employed in the synthesis of yolk precursor proteins. In those species, in which yolk proteins have a dual origin, the relative contributions of each source also varies, depending upon the requirement of nutrients for embryonic development (Wilder et al., 2002; Subramoniam, 2011).

8.6 ENDOGENOUS YOLK SYNTHESIS

Early electron microscopic studies on crustacean vitellogenesis focused on the intraoocytic yolk-synthesizing ability of decapod ovary (Beams and Kessel, 1962). Nuclear emissions from the early vitellogenic oocytes are common in autosynthetic oocytes. Transfer of nuclear material to the ooplasm, achieved by the nuclear emission, is a prelude to endogenous yolk synthesis in several oviparous animals. This process has been investigated electron microscopically in the spider crab, *Libinia emarginata* by Hinsch (1970), who showed that nucleolar vacuolation in the previtellogenic oocytes leads to the compartmentalization of the nucleolus into granular and fibrous zones. These nucleolar granules first make their appearance in the cytoplasm in the form of membrane-bound vesicles which aggregate beneath the nuclear envelop. Subsequently, they are transferred to the ooplasm by the fusion of the vesicular membranes with the inner nuclear membrane and bleb out from outer nuclear membrane. They can also pass through the nuclear pores. RNA, possibly, mRNA from the nucleus is transferred to the perinuclear ooplasm in the oocytes of other crustaceans such as the crayfish (Kessel and Beams, 1968).

In all higher crustacean orders belonging to Malacostraca, the ovary lacks nurse cells and hence RNA is synthesized within the nucleus of the oocytes and released into the ooplasm, as described above. On the other hand, the ovary of species belonging to the primitive subclass Branchiopoda is similar to meroistic ovarioles of insects in possessing nurse cells, which perform the function of RNA synthesis and transport to adjoining oocytes (see above).

A characteristic feature of anostracan oocyte growth is the formation of accessory nuclei, produced from blebbing of the germinal vesicle membrane. They are arranged at the periphery of the oocytes (Criel and Macrae, 2002). In addition, the so-called yolk nucleus appears in the center of the oocyte. The yolk nucleus is composed of small vesicles as well as multivesicular and dense bodies. Its role in endogenous yolk formation is, however, unknown. Endogenous yolk products include electron-dense protein, yolk platelets, lipid yolk droplets, and clear vesicular yolk bodies, the latter originating as dilated cisternae of the ER. Small yolk platelets, originating in multivesicular bodies, increase in size by fusion with immature granules and dense vesicles. Understandably, endogenous yolk synthesis accounts for the majority of yolk and other deutoplasmic components found in *Artemia* egg, although exogenous yolk sequestration does occur after endogenous yolk synthesis. After the protracted endogenous yolk formation, coated vesicles, a sure sign of endocytosis, are pinched off between and at the bases of newly formed microvilli. These coated vesicles, probably enclosing the endocytosized Vg molecules (van Beek et al., 1987), are transported to the already formed endogenous yolk platelets and fuse with them. Such a kind of integration of exogenously derived yolk with the endogenous yolk materials can be regarded as an initial step in the evolution of heterosynthetic yolk production, found in many advanced decapod species such as crabs. Obviously, exogenous yolk production in *Artemia* egg is kept to a minimum, with more pronounced autosynthesis, producing most of the yolk materials. In the end, four different yolk products are discerned: electron-dense protein yolk, clear lipid yolk droplets, vesicular yolk bodies, surrounded by an undulating membrane, and small intracellular granules (Criel, 1991).

Other ultrastructural studies on decapod crustaceans also emphasized the endogenous yolk synthesis during primary vitellogenesis. In species that are reported to rely predominantly on autosynthesis (eg, *Emerita analoga*, *Pachygrapsus crassipes*, *Cancer* sp., and the lobsters and crayfish), the developing oocytes develop elaborate array of rough ER (Kessel, 1968). Membrane-bound vesicles containing RNA and arising from nuclear membrane as blebs get attached to the ER to give rise to the vesicular type of ER. Gradually, ribosomes become attached to them, and as vitellogenesis advances, rough ER becomes predominant in the ooplasm. These vesicles increase in size and become ultimately the mature yolk granules with granular material inside. In the crayfish *Cambarus virilis* and in the cladoceran *Daphnia pulex*, protein yolk granules appear as dense masses within cisternae of the much branched and anastomosing lamellae of the rough ER (Kessel, 1968). In *Orconectes immunis*, Ganion and Kessel (1972) described the ultrastructural details of intracellular synthesis, transport, and packaging of proteinaceous yolk. In the oocyte, both rough and smooth elements of the ER are frequently associated with the forming face of the Golgi bodies. This association between Golgi apparatus and ER may suggest that newly synthesized proteins in the ER may be conveyed to the Golgi complex for packaging and combining to the carbohydrate moieties, synthesized within the Golgi membranes. Participation of both ER and Golgi apparatus in the synthesis and morphogenesis of protein–carbohydrate yolk bodies have been shown to be the primary feature of predominantly autosynthetic oocytes of other invertebrates, including the coelenterate medusa (Kessel, 1983). In the *O. gammarella* oocytes, the intracisternal glycoprotein granules which appear within the folds of the ER represent only one of the three principal components of the yolk, other than lipovitellin (Zerbib, 1976). The occurrence of other organelles such as free ribosomes, mitochondria, and sometimes annulate lamellae in the cytoplasm of the previtellogenic oocytes may also adduce further evidence that this stage is metabolically active in the production and storage of nutritive substances required for the subsequent growth and development of the embryo.

Support for autosynthetic yolk formation came from in vitro incubation studies using ovaries of crayfish *Procambarus* sp. and of the crab *P. crassipes* (Lui and O'Connor, 1977). When ovarian tissues were cultured in the presence of radioactive amino acids, two principal subunits of lipovitellin of *P. crassipes* and three of *Procambarus* incorporated the label, suggesting that the ovaries of these species are capable of synthesizing proteinaceous yolk intraoocytically. Cycloheximide, which interferes with the transfer of amino acids from aminoacyl-tRNA to polypeptide chains, inhibited leucine-^{3}H incorporation into the yolk protein of the crayfish *O. immunis*. Autosynthesis, however, prevails partially during the early phase of vitellogenesis when ovarian growth proceeds at a relatively slow rate, as in *L. emarginata* (Hinsch and Cone, 1969). Extensive electron microscopic studies on a number of crustacean species have also revealed differences in the relative reliance of intraoocytic synthesis and extraoocytic yolk incorporation during oogenesis.

In the penaeid shrimp, yolk content of the egg is meager, and hence oocytes may be in a position to synthesize most of them, with only a very limited contribution from extraovarian sites. Under in vitro conditions, the ovary of *M. japonicus* incorporated radioactive amino acids into a protein immunologically identical to vitellin (Yano and Chinzei, 1987). However, later immunofluorescence studies with antivitellin IgG revealed that only follicular epithelial cells rather than oocytes synthesized yolk proteins in this shrimp (Tsutsui et al., 2000). In other penaeids such as *P. vannamei* and *Penaeus semisulcatus*, in vitro studies using radiolabeled amino acids showed incorporation into the proteins of the oocytes, suggesting protein synthetic ability of the penaeid shrimp (Quackenbush, 1989; Browdy et al., 1990). Khayat et al. (1994) demonstrated high levels of Vg mRNA in the vitellogenic ovary of *P. semisulcatus*, as evidenced by its ability to direct cell-free synthesis of large amounts of Vg polypeptide. In the penaeid shrimp species such as *M. japonicus*, Northern blot analysis revealed that the Vg gene is expressed in the ovary and hepatopancreas simultaneously (Tsutsui et al., 2000).

8.7 EXOGENOUS YOLK SYNTHESIS

Unlike insects, many crustaceans adopt a biphasic mode of yolk formation, combining both autosynthesis and heterosynthesis, almost in a sequential manner (see Adiyodi and Subramoniam, 1983). As seen above, crustaceans thus far investigated rely, at least in part, on an early intraoocytic synthesis of proteins, comprising primarily of glycoprotein containing yolk, RNA materials, and lipid yolk. The exogenous yolk formation or secondary vitellogenesis is characterized by appearance of micropinocytotic vesicles, suggesting endocytosis of yolk protein from the hemolymph (Kessel, 1968). In many decapods, there is distinction, between the endogenous and exogenous yolk products. Thus, in the crab, *Cancer pagurus*, Eurenius (1973) described two types of yolk precursors, with type I accumulating within the cisternae of granular ER and type II occurring near the periphery of the oocytes in smooth and coated vesicles, both signifying endogenous and exogenous products, respectively.

Further evidence for the exogenous yolk formation in crustaceans comes with the discovery of a female-specific protein fraction (Vg) in the crab *Carcinus maenas* by Frentz (1960), which led to the search for the yolk protein precursor in the hemolymph of other crustaceans. Vg is a lipoglycoprotein conjugated invariably to a carotenoid pigment in decapod crustaceans (Ceccaldi, 1967). Kerr (1969) found the first evidence of extraovarian synthesis of Vg in the hemocytes of vitellogenic females of the crab *C. sapidus*, as inferred from in vitro incorporation studies using labeled leucine. In the brachyuran crab *Scylla serrata*, Ezhilarasi and Subramoniam (1986) provided serological evidence that hemocytes could be a synthetic site of Vg. Alternatively, these proteins could be the sequestered products from the hemolymph, inasmuch as hepatopancreas and subepidermal adipose are responsible for Vg synthesis in *S. serrata* (Rani and Subramoniam, 1997).

Subsequent studies have, however, implicated the fat body, subepidermal adipose tissue, and hepatopancreas as the other extraovarian sites for Vg synthesis in different species. For example, immunofluorescence and tracer studies provided strong evidence for Vg synthesis in the fat body of the terrestrial isopod, *Porcellio dilatatus* (Picaud, 1980) and of the marine isopod *Idotea balthica basteri* (Souty and Picaud, 1981). Similar to these isopods, the fat body of the amphipod *Orchestia gammerella* has been shown to be the principal site of Vg synthesis (Junera and Croissille, 1980; Meusy et al., 1983). While the fat body could be the organ for synthesis of Vg in the peracarids, such as isopods and amphipods, sporadic demonstration of this organ or its equivalent adipose tissues in the synthesis of Vg in disparate crustacean taxa has also been indicated in the literature. For instance, the fat storage cells of thoracopods are shown to be the Vg synthetic site of the anostracan *Artemia* (van Beek et al., 1987). In decapods such as the freshwater prawn *P. serratus* and *Macrobrachium nipponense* (Han et al., 1994), *Parapenaeus longirostris* (Tom et al., 1987), and *S. serrata* (Rani and Subramoniam, 1997), the subepidermal adipose tissue is one of the sites of Vg synthesis.

Notwithstanding, hepatopancreas has been shown to be the main extra-ovarian site for Vg synthesis in many decapod species (Yang et al., 2000). In crustaceans, hepatopancreas is the primary organ of intermediary metabolism, involved in lipid synthesis and carotenoid pigment metabolism. Li et al. (2006) demonstrated that in the Chinese crab *E. sinensis* hepatopancreas is the main site of Vg synthesis, although immunocytochemical studies have suggested a parallel role for ovary. On the other hand, in the red crab, *Charybdis feriatus*, Northern blot analysis revealed that the crab expressed Vg only in the hepatopancreas (Chan et al., 2005). Hepatopancreas of this crab possessed a large 8.0-kb transcript and several smaller Vg-specific transcripts, originating from alternate splicing and alternative use of promoter and/or termination signals. Using quantitative real-time PCR techniques, Zmora et al. (2007) found evidence that Vg is primarily expressed in the hepatopancreas of the vitellogenic females, with only minor expression in the ovary of the blue crab *C. sapidus*. Vg expression in the hepatopancreas of this anecdysic brachyuran crab is correlated with the ovarian maturation, with a remarkable 8000-fold increase in expression from stage III to IV of ovarian development. Ding et al. (2010), however, detected using Northern blots a single 8.6-kb Vg transcript in both hepatopancreas and ovary of the green crab, *C. maenas*. In the ovary, the Vg mRNA levels increased from the previtellogenic stage to reach the maximum in stage II vitellogenesis, and declined drastically in stage III. However, Vg mRNA levels in the hepatopancreas increased 60-fold between previtellogenesis and stage I vitellogenesis, remaining elevated through stages II and III females. Apparently, when the crab reached the vitellogenic stage III, the Vg mRNA levels in the ovary dropped to a low level but remained elevated in the hepatopancreas, providing a continued source of Vg for the most rapid phase of oocyte growth.

In the lobster *H. americanus*, cloning and expression studies adduced further evidence that the hepatopancreas is the primary organ for Vg synthesis (Tiu et al., 2009). The lobster Vg gene, *HaVg1*, mainly expressed in the hepatopancreas, contains 14 introns and 15 exons and its precursor protein is characterized by the lipid-binding lipoprotein domain at the N-terminus, followed by a domain of unknown function in the middle. The von Willebrand factor type-D domain is located at the C-terminus of the precursor.

Further studies have evidenced that Vg is synthesized in the hepatopancreas, as a single large peptide, then cleaved into subunits, and secreted into the hemolymph, to be endocytosized into the oocyte of the caridean shrimp, *P. hypsinotus* and the mud shrimp, *Upogebia major* (Tsutsui et al., 2004; Kang et al., 2008). Perhaps, the freshwater prawn *M. rosenbergii* is the single species which has received considerable attention with respect to hepatopancreas as the sole site of Vg synthesis among decapods crustaceans. Soroka et al. (2000) used cell culture techniques to demonstrate that hepatopancreas of a secondary vitellogenic female released vitellin immuno-cross-reactive polypeptides into the culture medium. Yang et al. (2000), using Northern blot analysis found evidence for the expression of Vg mRNA only in the vitellogenic female *M. rosenbergii* hepatopancreas, and not in any other tissue, including ovary.

A compelling conclusion from these studies is that hepatopancreas is the principal site of Vg synthesis in brachyuran crabs, caridean shrimps, lobsters, and probably other representative species under the suborder Pleocyamata. Conversely, in Dendrobranchiata, including mainly the penaeid shrimp, both the hepatopancreas and ovary impart equal contribution toward Vg synthesis (see Subramoniam, 2011). Furthermore, Vg synthesized at extraovarian sites such as hepatopancreas undergoes several modifications, such as glycosylation and lipid addition, bringing about changes in molecular weight when compared with the final yolk products accumulated within the ovary (Zmora et al., 2007).

Recent molecular studies furthermore suggest that ovary is the potential site for Vg synthesis in penaeiodean shrimps. However, immunocytochemical studies have revealed that follicular epithelial cells surrounding the oocyte are involved in the Vg synthesis in *M. japonicus* (Tsutsui et al., 2000). Apparently, three modes of yolk formation can be distinguished among crustaceans. In the first type, the majority of yolk may be synthesized within the oocytes themselves (eg, *Artemia* and other branchiopods). The second type of yolk formation is found in penaeid species, in which ovarian synthesis of yolk dominates, with the follicle cells participating in yolk precursor synthesis. Additionally, hepatopancreas also takes part in Vg synthesis. In the third type of vitellogenesis, found in the lobsters, crabs, and freshwater prawn, the yolk precursor molecules are exclusively synthesized in the hepatopancreas with little contribution from the ovary. Nevertheless, it may be that no species exclusively follows any one strategy, the mode of yolk formation being determined by the lecithality of the eggs and the embryonic duration.

8.8 VITELLOGENIN

In Crustacea, pioneering studies on Vg were done by simple electrophoretic analysis of the hemolymph proteins during different stages of ovarian maturation. Immunological studies, followed by determination of molecular weight of both Vg and the ovarian yolk proteins (lipovitellins) confirmed the precursor–product relationship between these two molecules in a number of malacostracan crustaceans (Croisille et al., 1974). In the giant freshwater prawn, *M. rosenbergii*, Vg is composed of three subunits, one having a molecular weight of 200 kDa and the remaining two having molecular weights of around 90 and 100 kDa. Vitellin, on the other hand, consists of only two subunits, both of which have identical molecular weights and are immunologically equivalent to the smaller Vg subunits (Wilder et al., 1994; Lee et al., 1997). Understandably, Vg is secreted into the hemolymph as a larger precursor, which is cleaved into two smaller subunits before being taken up by the ovary. However, in the land crab, *Potamon potamios*, Vg and vitellin are identical in terms of molecular weight, both being composed of polypeptides of 115, 105, and 85 kDa (Pateraki and

Stratakis, 1997). Similarity between Vg and lipovitellin of different decapods is also reflected in their amino acid composition. Nevertheless, quantitative differences found in the molar percentages of amino acids of different species may suggest possible differences in the primary structure of their vitellins.

Crustacean Vg is not only a precursor protein to ovarian lipovitellin but also an effective transporter of lipids to the ovary from the hemolymph. Lipovitellin is a high-density glycolipocarotenoprotein, containing as much as 30% lipids. In the blue crab, *C. sapidus*, lipid transport through hemolymph is accomplished by two high-density lipoproteins (HDLs) and a very high-density lipoprotein (VHDL) (Lee and Puppione, 1998). Female-specific Vg is one of the HDLs, with its production being correlated with ovarian development in female crustaceans, whereas the other HDLs and VHDLs are found in both males and females. In the penaeid shrimp, *P. semisulcatus*, the nonsex-specific hemolymph lipoprotein, LP I, consists of one 110-kDa peptide unit, whereas the sex-specific LP II consists of three subunits of 200, 120, and 80 kDa (Lubzens et al., 1997). Interestingly, the same subunits were also present in the lipovitellin of this shrimp. However, the lipid composition of these two HDLs in *P. semisulcatus* differs: LP II (Vg) has a lower lipid content than does LP I, in addition to differences found in lipid classes, linked to the apolipoprotein. Apparently, Vg and lipovitellin have similar protein structures, but show differences in their lipid contents, with the lipovitellin having more percentage lipid acquired through adsorption within the oocytes. LP I is also different from LP II in its protein composition, as the former does not cross-react with antivitellin antiserum.

In the sand crab, *E. asiatica*, Subramoniam and Gunamalai (2003) described three hemolymph lipoproteins: LP 1, LP 2, and LP 3. LP 1 is nonsex-specific but is accumulated into the oocytes along with LP 2, which is the female-specific Vg. LP 3, which appears only during the premolt of male and female crabs, might play a role in the transport of lipids to the epidermis during cuticle formation. In addition to transporting a variety of lipophilic compounds, such as triglycerides and phospholipids, crustacean Vgs are known to transport steroidal hormones like ecdysteroids and vertebrate steroids, including progesterone and estradiol-17β (Subramoniam, 2000).These hormones are stored within the oocytes as conjugates of yolk proteins and serve regulatory functions during embryogenesis (see chapter: Yolk Utilization and Embryonic Nutrition).

8.8.1 Phylogenetic Analysis of Vg

Tata and Smith (1979) pointed out that Vg -like yolk proteins are homologous in all egg-laying animals. Analysis of amino acid sequence in many vertebrate yolk proteins supported this contention among submammalian oviparous families (Babin et al., 2007). Crustacean Vg is a multidomain apolipoprotein that is cleaved into distinct yolk proteins. Multiple alignments of all known crustacean Vg sequences have revealed almost similar cleavage sites. Clustal W alignment of *M. rosenbergii* Vg with that of 17 other crustacean species has shown that the first cleavage site RXRR occurs at amino acid residue 707–710, and the homology for the first segment is high when compared with the rest of the module (Okuno et al., 2002). The results from BLAST searches indicate that the N-terminal region of crustacean Vgs is conserved, as in the vertebrate apolipoproteins that are involved in lipid transport. This property is in accord with the fact that Vg, insect apolipophorin II/I, apoB, and microsomal triglyceride transfer protein (MTP) are members of the same multigene superfamily of large lipid transfer proteins (LLTPs) (Babin et al., 1999). Next to the N-terminal segment, the middle segment is comparable to a lipovitellin domain of unknown function called DUF1943. The C-terminal domain of *M. rosenbergii* Vg harbored a von Willebrand-factor type domain (YGP4) found in mammals. Similarity in the amino acid sequence of the von Willebrand factor at the C-terminal region has also been reported for the insect apolipophorin (Sundermeyer et al., 1996).

A phylogenetic tree constructed based on the alignment of amino acid sequences of 30 crustacean Vgs using the Clustal W program shows six distinct lineage groups: Penaeiodea (a, b, c), Caridea (d), Astacidea (e), Thalassinidea (f), Brachyura (g), and Copepoda and Branchiopoda, forming a separate group (h) (Fig. 8.5, Table 8.3). As seen from the figure, Vgs of the penaeiodean shrimps seem to be homogeneous, with 92% identity in amino acid sequence, except

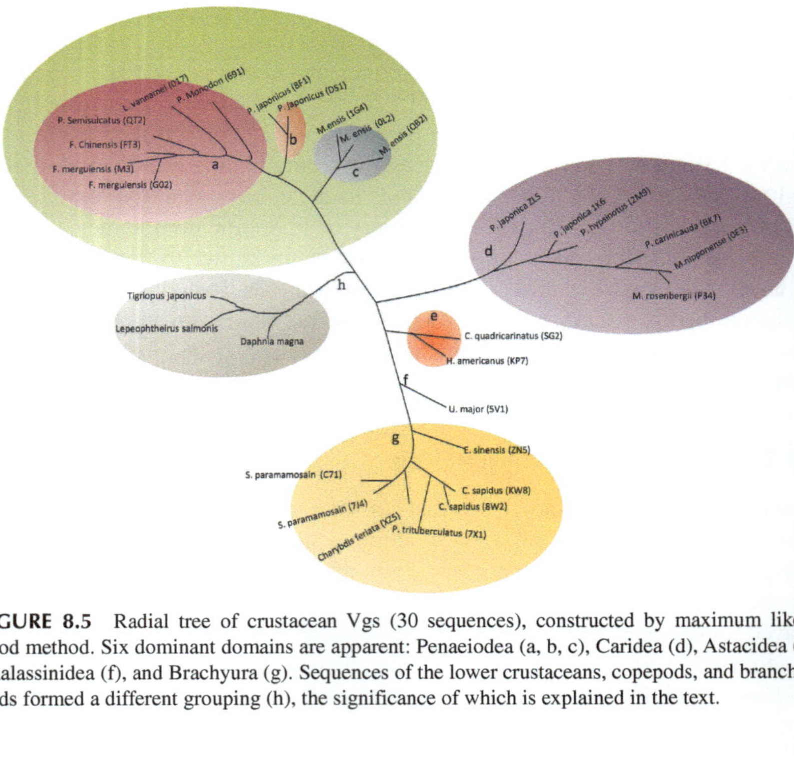

FIGURE 8.5 Radial tree of crustacean Vgs (30 sequences), constructed by maximum likelihood method. Six dominant domains are apparent: Penaeiodea (a, b, c), Caridea (d), Astacidea (e), Thalassinidea (f), and Brachyura (g). Sequences of the lower crustaceans, copepods, and branchiopods formed a different grouping (h), the significance of which is explained in the text.

TABLE 8.3 Vg Accession Numbers of 18 Crustaceans, Along With Their Taxonomic Classifications

S.No.	UID	Species	Systematics	Group
1	A1XFT3	*Fenneropenaeus chinensis*	Malacostraca; Eumalacostraca; Eucarida; Decapoda; Dendrobranchiata; Penaeoidea; Penaeidae; Fenneropenaeus	a
2	D2SSM3	*Fenneropenaeus merguiensis*	Malacostraca; Eumalacostraca; Eucarida; Decapoda; Dendrobranchiata; Penaeoidea; Penaeidae; Fenneropenaeus	a
3	Q6RG02	*Fenneropenaeus merguiensis*	Malacostraca; Eumalacostraca; Eucarida; Decapoda; Dendrobranchiata; Penaeoidea; Penaeidae; Fenneropenaeus	a
4	Q8WQT2	*Penaeus semisulcatus*	Malacostraca; Eumalacostraca; Eucarida; Decapoda; Dendrobranchiata; Penaeoidea; Penaeidae; Penaeus	a
5	Q2I691	*Penaeus monodon*	Malacostraca; Eumalacostraca; Eucarida; Decapoda; Dendrobranchiata; Penaeoidea; Penaeidae; Penaeus	a
6	Q7Z017	*Litopenaeus vannamei*	Malacostraca; Eumalacostraca; Eucarida; Decapoda; Dendrobranchiata; Penaeoidea; Penaeidae; Litopenaeus	a
7	Q4W8F1	*Marsupenaeus japonicus*	Malacostraca; Eumalacostraca; Eucarida; Decapoda; Dendrobranchiata; Penaeoidea; Penaeidae; Marsupenaeus	b

Continued

TABLE 8.3 Vg Accession Numbers of 18 Crustaceans, Along With Their Taxonomic Classifications—cont'd

S.No.	UID	Species	Systematics	Group
8	Q9NDS1	*Marsupenaeus japonicus*	Malacostraca; Eumalacostraca; Eucarida; Decapoda; Dendrobranchiata; Penaeoidea; Penaeidae; Marsupenaeus	b
9	Q7Z1G4	*Metapenaeus ensis*	Malacostraca; Eumalacostraca; Eucarida; Decapoda; Dendrobranchiata; Penaeoidea; Penaeidae; Metapenaeus	c
10	Q8I0L2	*Metapenaeus ensis*	Malacostraca; Eumalacostraca; Eucarida; Decapoda; Dendrobranchiata; Penaeoidea; Penaeidae; Metapenaeus	c
11	Q6QQB2	*Metapenaeus ensis*	Malacostraca; Eumalacostraca; Eucarida; Decapoda; Dendrobranchiata; Penaeoidea; Penaeidae; Metapenaeus	c
12	D3J1K6	*Pandalopsis japonica*	Malacostraca; Eumalacostraca; Eucarida; Decapoda; Pleocyemata; Caridea; Pandaloidea; Pandalidae; Pandalopsis	d
13	Q762M9	*Pandalus hypsinotus*	Malacostraca; Eumalacostraca; Eucarida; Decapoda; Pleocyemata; Caridea; Pandaloidea; Pandalidae; Pandalus	d
14	Q95P34	*Macrobrachium rosenbergii*	Malacostraca; Eumalacostraca; Eucarida; Decapoda; Pleocyemata; Caridea; Palaemonoidea; Palaemonidae; Macrobrachium	d

TABLE 8.3 Vg Accession Numbers of 18 Crustaceans, Along With Their Taxonomic Classifications—cont'd

S.No.	UID	Species	Systematics	Group
15	S4SBK7	*Palaemon carinicauda*	Malacostraca; Eumalacostraca; Eucarida; Decapoda; Pleocyemata; Caridea; Palaemonoidea; Palaemonidae; Palaemon	d
16	V9XZL5	*Pandalopsis japonica*	Malacostraca; Eumalacostraca; Eucarida; Decapoda; Pleocyemata; Caridea; Pandaloidea; Pandalidae; Pandalopsis	d
17	A0A0E3JBN6	*Macrobrachium nipponense*	Malacostraca; Eumalacostraca; Eucarida; Decapoda; Pleocyemata; Caridea; Palaemonoidea; Palaemonidae; Macrobrachium	d
18	Q9GSG2	*Cherax quadricarinatus*	Malacostraca; Eumalacostraca; Eucarida; Decapoda; Pleocyemata; Astacidea; Parastacoidea; Parastacidae; Cherax	e
19	A3RKP7	*Homarus americanus*	Malacostraca; Eumalacostraca; Eucarida; Decapoda; Pleocyemata; Astacidea; Nephropoidea; Nephropidae; Homarus	e
20	A8R5V1	*Upogebia major*	Malacostraca; Eumalacostraca; Eucarida; Decapoda; Pleocyemata; Thalassinidea; Callianassoidea; Upogebiidae; Upogebia	f
21	A1X8W2	*Callinectes sapidus*	Malacostraca; Eumalacostraca; Eucarida; Decapoda; Pleocyemata; Brachyura; Eubrachyura; Portunoidea; Portunidae; Callinectes	g

Continued

TABLE 8.3 Vg Accession Numbers of 18 Crustaceans, Along With Their Taxonomic Classifications—cont'd

S.No.	UID	Species	Systematics	Group
22	C1KC71	*Scylla paramamosain*	Malacostraca; Eumalacostraca; Eucarida; Decapoda; Pleocyemata; Brachyura; Eubrachyura; Portunoidea; Portunidae; Scylla	g
23	M4GKW8	*Callinectes sapidus*	Malacostraca; Eumalacostraca; Eucarida; Decapoda; Pleocyemata; Brachyura; Eubrachyura; Portunoidea; Portunidae; Callinectes	g
24	Q1L7X1	*Portunus trituberculatus*	Malacostraca; Eumalacostraca; Eucarida; Decapoda; Pleocyemata; Brachyura; Eubrachyura; Portunoidea; Portunidae; Portunus	g
25	Q5XXZ5	*Charybdis feriata*	Malacostraca; Eumalacostraca; Eucarida; Decapoda; Pleocyemata; Brachyura; Eubrachyura; Portunoidea; Portunidae; Charybdis	g
26	R9W7J4	*Scylla paramamosain*	Malacostraca; Eumalacostraca; Eucarida; Decapoda; Pleocyemata; Brachyura; Eubrachyura; Portunoidea; Portunidae; Scylla	g
27	R9RZN5	*Eriocheir sinensis*	Malacostraca; Eumalacostraca; Eucarida; Decapoda; Pleocyemata; Brachyura; Eubrachyura; Grapsoidea; Varunidae; Eriocheir	g

TABLE 8.3 Vg Accession Numbers of 18 Crustaceans, Along With Their Taxonomic Classifications—cont'd

S.No.	UID	Species	Systematics	Group
28	BAE94324	*Daphnia magna*	Eukaryota; Metazoa; Ecdysozoa; Arthropoda; Crustacea; Branchiopoda; Diplostraca; Cladocera; Anomopoda; Daphniidae; Daphnia	h
29	ABU41135	*Lepeophtheirus salmonis*	Eukaryota; Metazoa; Ecdysozoa; Arthropoda; Crustacea; Maxillopoda; Copepoda; Siphonostomatoida; Caligidae; Lepeophtheirus	h
30	ABZ91537	*Tigriopus japonicus*	Eukaryota; Metazoa; Ecdysozoa; Arthropoda; Crustacea; Maxillopoda; Copepoda; Harpacticoida; Harpacticidae; Tigriopus	h

in the case of *Metapenaeus ensis*, which has two Vgs (MeVg1 and MeVg2), expressed independently in the ovary and hepatopancreas. Incidentally, these two gene products have only 56% sequence identity between them (Tsang et al., 2003; Kung et al., 2004). Similarly, *Penaeus japonicus* also formed a small subgroup within the penaeiodean shrimps. Vg sequence comparison in different crustacean groups is advantageous in proposing a phylogenetic relationship with other arthropod groups. In the primitive branchiopod, *Daphnia magna*, there are two Vgs, *DmagVg1* and *DmagVg2*. The lipid transport module in the N-terminal region of *DmagVg1* is more closely related to those of insect Vgs than to those of decapod crustaceans (Tokishita et al., 2006). In addition, crustacean Vg shows closer homology in amino acid sequence with coral Vgs, pointing to the emergence of Vg as an egg protein precursor before the cnidarians–bilaterian divergence (Hayakawa et al., 2006). Warrier and Subramoniam (2003) indicated the immunological relatedness between Vg of the crab *S. serrata* and apoB, the major component of low-density lipoprotein (LDL) and very low-density lipoprotein of mammalian serum. Avarre et al. (2003) studied the homology between crustacean Vgs and members of the LDL superfamily of lipoproteins and arrived at the conclusion that crustacean Vgs are closer to mammalian LDL and insectan apolipophorins. The closer relationship between crustacean Vg and apoB indicates not only the high conservancy in the lipid-binding domains of both these proteins but also point to the evolutionary derivation of vertebrate apolipo B proteins from invertebrate (crustacean) Vg (Subramoniam, 2011).

8.8.2 Vg Gene Expression and Yolk Synthesis

Recent spurt in the molecular techniques has enhanced our understanding on the genetic control of yolk protein synthesis in decapod crustaceans. The sequencing of amino acids of crustacean Vgs has enabled the construction of cDNA followed by the identification of their genes together with their expression pattern in different tissues. The first elucidation of the complete primary structure of Vg in Crustacea was made in the kuruma prawn, *M. japonicus* (Tsutsui et al., 2000). Following this study, primary structure of Vg in several prawn/shrimp species as well as other crustaceans has been elucidated. These studies have important implications in the Vg gene expression studies in different synthetic sites such as hepatopancreas and ovary. Nevertheless, hormonal induction of Vg gene expression, as in vertebrates, has not been demonstrated in any decapod crustacean.

In the giant freshwater prawn *M. rosenbergii*, Yang et al. (2000) obtained partial amino acid sequence for the four vitellin fractions (Vn A,B,C/D). Using gene-cloning techniques, full cDNA sequence and the deduced primary structure of Vg was accomplished (Okuno et al., 2002). The cDNA of *M. rosenbergii* Vg is 7800 bp and encodes a protein corresponding to 2537 amino acids containing three R-X-K/R-R consensus cleavage sites. It has a putative signal peptide sequence followed by the mature Vg sequence that contains three consensus cleavage sites, targeted by endoproteases of the subtilisin family. The Vg sequence analysis further revealed that the four vitellin fractions are all connected with one another in the sequential order of Vn-A to D.

With the cloning of full-length cDNAs of Vg precursors from different crustacean species, clarifications on the site of Vg synthesis have been made by gene expression studies in several species. For example, Chen et al. (1999) cloned a cDNA fragment encoding Vg in *M. rosenbergii* and found its expression only in the hepatopancreas of vitellogenic females. Using the cDNA fragments of four vitellin fractions as probes, Yang et al. (2000) found the exclusive expression of Vg mRNAs for these vitellins in the hepatopancreas of this freshwater prawn. Furthermore, Jayasankar et al. (2002) measured the expression levels of Vg mRNA using quantitative real-time PCR techniques in the hepatopancreas of this prawn. Interestingly, this mRNA expression level in the hepatopancreas corresponded with a gradual increase of hemolymph Vg levels. A simultaneous increase in the gonadosomatic index also suggested that hepatopancreas is the sole Vg synthetic site in *M. rosenbergii*. However, in the penaeid shrimps, Vg gene expression occurs both inside the ovary and hepatopancreas. Multiple Vg gene expression is also reported in the ovary and hepatopancreas in the penaeid shrimp, *M. ensis*, in which two Vg genes, MeVg1 and MeVg2, have been identified (Tsang et al., 2003). Whereas MeVg1 is expressed equally in the ovary and hepatopancreas, MeVg2 is expressed only in the hepatopancreas. MeVg2 could also give rise to smaller transcripts, resulting in the production of many smaller subunits.

However, the relative levels of Vg mRNA synthesis in the ovary and hepatopancreas may differ among different shrimp species. In *P. monodon*, expression of PmVtg1mRNA in the hepatopancreas was 25- to 40-fold greater than in ovaries (stages I to III) of intact shrimps (Hiransuchalert et al., 2013). Conversely, in *Fenneropenaeus merguiensis*, the Vg mRNA level in the hepatopancreas is much lower than in the ovary at all stages of ovarian development (Phiriyangkul et al., 2007). Gene expression studies particularly solved earlier controversies regarding the site of synthesis of Vg. In the brachyuran crab, *C. feriatus*, Northern blot analysis revealed that the crab expresses Vg only in the hepatopancreas (Mak et al., 2005).

8.8.3 Vitellogenin Receptor and Yolk Protein Uptake

It is well established now that the last phase of vitellogenesis in crustaceans consists of pinocytotic uptake of Vg. The endocytosis of Vg is invariably mediated by a membrane receptor. In insects, vitellogenin receptor (VgR) is characterized and their mediation in Vg uptake well studied (Sappington et al., 2002). On the other hand, only a few studies have been carried out with reference to crustacean VgRs. Jugan and Soyez (1985) first demonstrated Vg uptake by the oocytes, employing colloidal gold-conjugated vitellin in the freshwater prawn *M. rosenbergii*. The gold labeling was localized in the microvilli, coated pits, and intraooplasmic vesicles. It was further observed that a sinus gland neuropeptide inhibited vitellogenesis by blocking the membrane receptor. Subsequently, Laverdure and Soyez (1988) solubilized the VgR from the oocyte membrane of the lobster *H. americanus*, and characterized it using an enzyme-linked immunosorbent assay. Furthermore, Jugan and Van Herp (1989) found that the solubilized oocyte membrane receptor of *Orconectus limosus* had a molecular weight of 28–30kDa, which binds specifically to the Vg of the same crayfish. In a subsequent detailed study, Warrier and Subramoniam (2002) purified the VgR in the brachyuran crab *S. serrata* using HPLC and determined its molecular weight to be 230kDa, based on size exclusion chromatography. Incidentally, the molecular weight of insect VgR is also in the range of 200–230 (Sappington et al., 2002). In direct-binding studies using ^{125}I-labeled Vg, *S. serrata* VgR was observed to have increased affinity to its ligand in the presence of Ca^{2+} and was inhibited by suramin, a receptor antagonist. The receptor possessed high affinity for crab Vg with a dissociation constant of K_d 0.8×10^{-6} M. Further

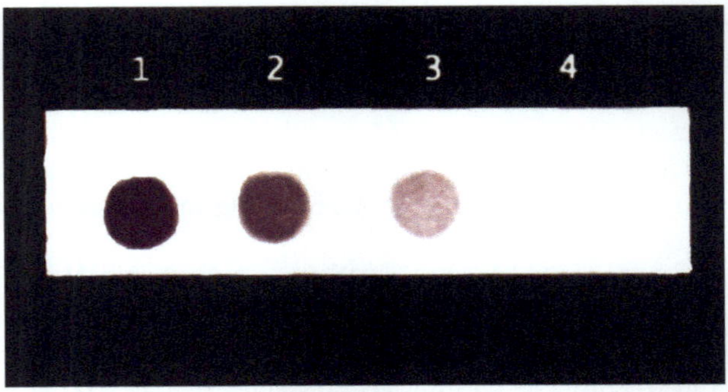

FIGURE 8.6 Dot blot analysis of crab Vg (1), rat LDL (2), VLDL (3), and HDL (4) using anticrab Vg antibodies (dilution 1:2000). Anti-Lv antibodies are seen to react well with Vg, LDL, and VLDL but there was no reaction with HDL. *HDL*, High-density lipoprotein; *LDL*, Low-density lipoprotein; *VLDL*, Very low-density lipoprotein. *Reproduced from Warrier, S.R., Subramoniam, T., 2003. Instability of crab vitellogenin and its immunological relatedness with mammalian atherogenic lipoproteins. Mol. Reprod. Dev. 64 (3), 329–340.*

investigations revealed an interesting immunological relatedness between VgR of *S. serrata* and mammalian low-density lipoprotein receptor (LDLR) by virtue of the ability of VgR to bind rat LDL and VLDL, suggesting the presence of conserved receptor-binding sites in the crab Vg and the mammalian LDL and VLDL (Fig. 8.6) (Warrier and Subramoniam, 2002).

Following this, Tiu et al. (2008) cloned the cDNA encoding a putative vitellogenin receptor (PmVgR) from the shrimp *P. monodon*. The PmVgR cDNA is 6.8 kb, and the deduced protein has 1943 amino acids with a molecular weight of 211 kDa, which is in the close range of *S. serrata* VgR. PmVgR is ovary-specific and it consists of conserved cysteine-rich domains, epidermal growth factor (EGF)-like domains, and Tyr-Trp-Thr-Asp (YWTD) motifs, similar to the mammalian LDL receptor. PmVgR expression in the ovary coincides with the rapid pace of Vg production by the hepatopancreas. PmVgR expression was knocked-down in shrimp after they were injected with PmVgR dsRNA, leading to a decrease in PmVgR protein synthesis in the ovary and at the same time elevating the level of hemolymph Vg. Molecular characterization of VgR has also been reported for the kuruma prawn *M. japonicus* (Mekuchi et al., 2008). Structural analysis of the VgR of this shrimp also reconfirmed its inclusion in the LDLR superfamily.

In a recent study, Roth and Khalaila (2012) identified and sequenced the full-length VgR of the freshwater prawn *M. rosenbergii*. The putative MrVgR showed high sequence similarity to VgRs of other crustaceans studied, insects, and vertebrates. Its structure includes typical elements, such as an extracellular, lipoprotein-binding domain, EGF-like domain, and O-glycosylation domains, a transmembrane domain, and a short C-terminal, cytosolic tail. The

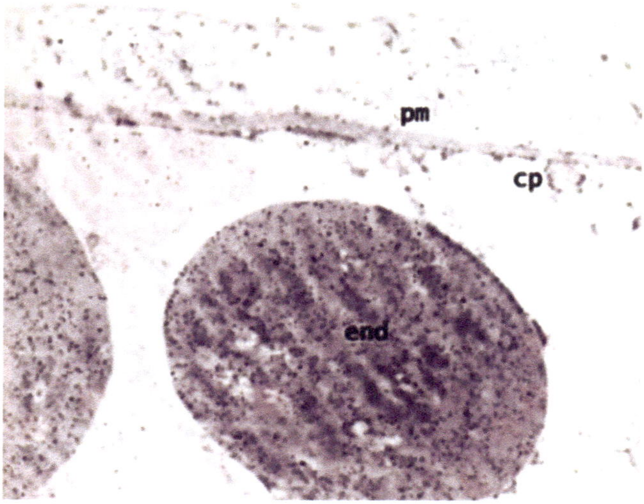

FIGURE 8.7 Immunogold labeling of Vg in ultrathin section of S. serrata oocyte. Plasma membrane (pm) is visible and Vg labeling is seen in coated pits (CP) and is concentrated in the endosome (end). 2900×. *Reproduced from Warrier, S., Subramoniam, T., 2002. Receptor mediated yolk protein uptake in the crab* Scylla serrata*: crustacean vitellogenin receptor recognizes related mammalian serum lipoproteins. Mol. Reprod. Dev. 61 (4), 536–548.*

expression dynamics of MrVgR during vitellogenesis revealed that it was high early in the vitellogenic process, peaked during the mid-vitellogenic stage, and then fell to its lowest level in the late-vitellogenic stage and in females that had just released their eggs. It is not known whether crustacean VgR facilitates the endocytosis of other hemolymph lipoproteins into the ovary, as reported for avian VgRs (Schneider, 1996) and insectan lipophorin receptors (Atella et al., 2006).

Receptor-mediated internalization of Vg into the oocytes of the mud crab *S. serrata* was demonstrated using gold-labeled antiVg as the primary antibody (Warrier and Subramoniam, 2002). Immunogold labeling against Vg antibody was first visualized in the coated pits, found on the plasma membrane of the vitellogenic oocytes (Fig. 8.7). This is followed by their appearance in the pinched-off coated vesicles as well as in early endosomes, which fuse together to form the mature electron-dense late endosomes (Figs. 8.8 and 8.9). In the penaeid shrimp, *P. monodon*, after the binding of Vg with VgR, the complex moves into the oocyte cytoplasm, aided by internalization signals present in VgR (Tiu et al., 2008). The VgR of *P. monodon* has two putative internalization signals, as found in vertebrate VgRs; this redundancy with the internalization signals present in the shrimp oocytes could increase the efficiency of receptor–ligand binding during crustacean vitellogenesis.

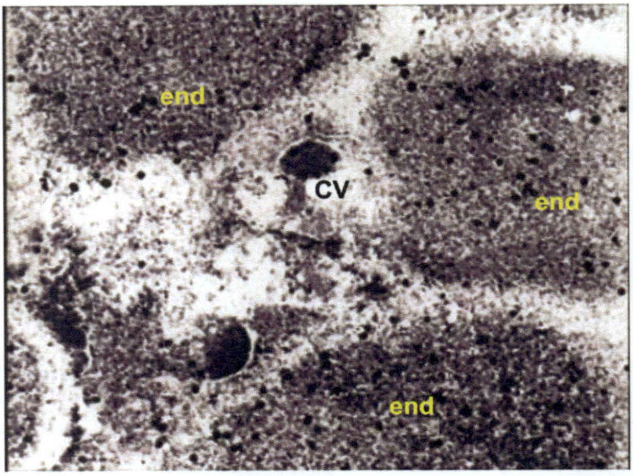

FIGURE 8.8 Vg labeling is seen along the luminal surface of the coated vesicle (cv). Vg particles are seen in endosomes (end) found close to the coated vesicles. Clustering of endosomes filled with Vg are seen. 15,000×. *Reproduced from Warrier, S., Subramoniam, T., 2002. Receptor mediated yolk protein uptake in the crab* Scylla serrata*: crustacean vitellogenin receptor recognizes related mammalian serum lipoproteins. Mol. Reprod. Dev. 61 (4), 536–548.*

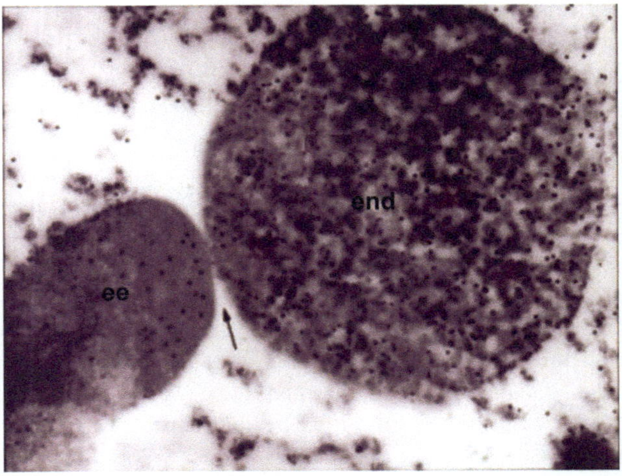

FIGURE 8.9 Fusion of an early endosome (ee) with a mature endosome (end) is indicated by an arrow. Vg is labeled extensively in the mature endosome compared to the early endosome. 8000×. *Reproduced from Warrier, S., Subramoniam, T., 2002. Receptor mediated yolk protein uptake in the crab Scylla serrata: crustacean vitellogenin receptor recognizes related mammalian serum lipoproteins. Mol. Reprod. Dev. 61 (4), 536–548.*

8.9 YOLK PROCESSING

The processing of Vg, since its synthesis in the extraovarian site, into the vitellin product inside the ovary is unclear in crustaceans. In insects, Vg undergoes posttranslational proteolytic cleavage at the site of synthesis (fat body) in insects (Raikhel and Dhadialla, 1992), whereas the processing of Vg occurs after sequestration into the ovary in amphibians (Wall and Patel, 1987). SDS-PAGE analysis of hemolymph and ovary in several decapod crustaceans has shown the occurrence of varying numbers of Vg and vitellin fractions, suggesting that Vgs are already fragmented at the time of endocytotic entry into the ovary. In the terrestrial isopod, *Armadillidium vulgare*, up to four female-specific glycoprotein bands were detected in the hemolymph by SDS-PAGE, and the same bands were also detected in the ovarian extract (Suzuki, 1987). Furthermore, cloning of a cDNA encoding Vg revealed the presence of an amino-terminal sequence identical to those of lipovitellin subunits of the egg (Okuno et al., 2000).

The instability of Vg to undergo proteolytic cleavage may be a general feature, but in a brachyuran crab, *S. serrata*, the Vg itself possesses proteinase activity (Warrier and Subramoniam, 2003). These authors demonstrated that conformational changes in the native Vg could bring about such proteolytic cleavage, as indicated in a study using urea as a destabilizer. Vg showed a spectral change with 8M urea treatment, due to exposure of the hydrophobic core containing aromatic residues (absorption at 274 nm), whereas lipovitellin did not show such spectral shift. Clearly, Vg is a relatively labile lipoprotein, but the ovarian lipovitellin is more stable.

Cloning of the full-length cDNA encoding Vg in the giant freshwater prawn, *M. rosenbergii*, based on the amino acid sequence and partial cDNA sequence of vitellins (Yang et al., 2000), facilitated deduction of the complete primary structure of Vg in this prawn (Okuno et al., 2002). The deduced amino acid sequence possessed three consensus cleavage sites, R-X-R-R, similar to those reported in insects. In *M. rosenbergii*, Vg, after being synthesized as a single precursor protein, undergoes initial cleavage by a subtilisin-like endoprotease to give rise to two subunits, A and pro-B, within the hepatopancreas. After secretion into the hemolymph, subunit A is directly sequestered into the ovary, whereas pro-B is cleaved by another processing enzyme to give rise to subunits B and C/D. They are subsequently taken up by the ovary to give rise to yolk proteins, Vn A, Vn B, and Vn C/D (Fig. 8.10). Examination of subunit composition of Vg in the hemolymph and vitellin in the ovary by SDS-PAGE and Western blotting has also supported the above sequence of Vg processing to vitellin fractions. Furthermore, N-terminal amino acid sequencing of these Vg and Vn fractions as appearing in the hemolymph and ovary has also provided additional support to the scheme of Vg processing in this freshwater prawn. Evidence for secondary cleavage of Vg after its uptake into the ovary has been given in freshwater crayfish, *Ibacus ciliatus* (Komatsu and Ando, 1992).

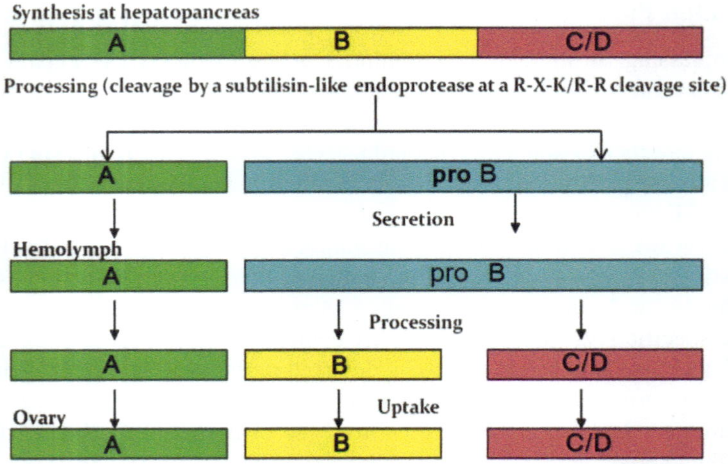

FIGURE 8.10 Schematic representation of synthesis and processing of Vg. In *M. rosenbergii* Vg is synthesized as a single precursor molecule, A-B-C/D in hepatopancreas, which is then cleaved into two subunits, A and pro-B. Subunits A and pro-B are released into the hemolymph and pro-B is cleaved to form two subunits B and C/D. The three processed subunits, A, B, and C/D are incorporated into the ovary. *Adapted from Okuno, A., Yang, W.J., Jayasankar, V., Saido-Sakanaka, H., Huong, D.T.T., Jasmani, S., Atmomarsono, M., Subramoniam, T., Tsutsui, N., Ohira, T. and Kawazoe, I., 2002. Deduced primary structure of vitellogenin in the giant freshwater prawn,* Macrobrachium rosenbergii, *and yolk processing during ovarian maturation. J. Exp. Zool. 292 (5), 417–429.*

8.10 CRUSTACEAN LIPOVITELLIN

Yolk provides the developing embryo with nutrients essential for growth within the egg. In vertebrates, yolk proteins are phosphoglycoproteins which, besides being a rich source of amino acids, phosphate, and carbohydrate, contain lipids, vitamins, and mineral ions necessary for embryonic development (Byrne et al., 1989). Yolk proteins have nonenzymatic action in oviparous animals and hence can be compared to milk proteins in mammals or the storage proteins of plant seeds. The term "vitellogenin" was first adopted by Pan et al. (1969) to denote the female-specific blood proteins of insects that had common functional and compositional characteristics with that of the egg yolk. In the literature, the term "vitellin" is used to denote the egg yolk form of Vg in insects (see Engelmann, 1970). Similar terminologies are also adopted for crustaceans. Wallace et al. (1967) first characterized the egg yolk protein (vitellin) in selected decapod crustaceans as a high-density lipoprotein, by virtue of high lipid content (30%), but low protein phosphates, in this respect differing from those of oviparous vertebrates. In general, it has a molecular weight of around 350 kDa. In view of its differing chemical composition especially of relatively high lipid content, these authors preferred to designate vitellin as crustacean lipovitellin, a term that was adopted by later crustacean

TABLE 8.4 Relative Percentage Composition of Phospholipids of the Lv II and I Stage Egg of *E. asiatica*

	Percentage Composition	
Phospholipids	**Egg (Mean ± SD)**	**Lv II (Mean ± SD)**
Cardiolipin	13.2 ± 1.5	–
Phosphatidyl ethanolamine	14.0 ± 2.3	–
Cardiolipin + phosphatidyl ethanolamine	–	23.3 ± 1.9
Phosphatidyl inositol	6.6 ± 1.0	7.5 ± 1.5
Phosphatidyl serine	27.5 ± 2.1	25.3 ± 1.6
Phosphatidyl choline	33.2 ± 1.5	38.1 ± 2.1
Sphingomyelin	3.7 ± 1.8	3.5 ± 1.6

Reproduced from Tirumalai, R., Subramoniam, T., 1992. Purification and characterization of vitellogenin and lipovitellins of the sand crab Emerita asiatica: molecular aspects of crab yolk proteins. Mol. Reprod. Dev. 58, 54–62.

workers. The lipovitellin of *Procambarus* sp. is immunologically identical to its Vg, having a lipid content of 35% (Fyffe and O'Connor, 1974). The high lipid content of lipovitellin is due to the increased lipid transport ability of Vg which is internalized into the oocytes as the major lipovitellin in many decapod species (Kerr, 1969). Sedimentation coefficients vary from 10.0 to 11.1 Svedbergs in the crustacean vitellins (Lee and Puppione, 1998).

Several studies have reported posttranslational modification of extraovarially synthesized Vg as well as its conjugation with several organic and inorganic components such as lipids, carbohydrates, carotenoids, hormones, and metals (Wilder et al., 2002). In the mole crab, *E. asiatica*, the major egg yolk protein consists of two lipovitellins (Lv I and Lv II) constituting as much as 90% of the total egg proteins (Tirumalai and Subramoniam, 1992, 2001). In SDS-PAGE analysis, Lv I yielded two subunits with molecular weights of approximately 109,000 and 105,000 Da, respectively, whereas Lv II resolved into six subunits with molecular weights of 65,000, 54,000, 50,000, 47,000, 44,000, and 42,000 Da, respectively.

As in other protein synthetic system, glycosylation is an early step in the posttranslational modification of Vg at its synthetic site. The carbohydrate component of the *Emerita* Lv II yolk exists in three forms, namely free carbohydrates, protein-bound carbohydrates, and lipid-bound carbohydrates (Table 8.4). The protein-bound carbohydrates are dominated by hexose, hexosamine, and galactosamine. The Lv II contains the highest amount of *N*-linked oligosaccharides than *O*-linked oligosaccharides, but sialic acid is absent. Abdu et al. (2002) identified 10 putative glycosylation sites in the lipovitellin of the crayfish *Cherax quadricarinatus*. Subsequently, Khalaila et al. (2004) found that the three N-glycosylation sites in the Vg of this crayfish at Asn^{152}, Asn^{160}, and Asn^{2493} were glycosylated with high mannose glycans identified as Man_{5-9} $GlcNAc_2$ (Man = mannose; Glc = glucose; GlcNac = *N*-acetyl glucosamine) species. Glycosylation of Vg may be important in the folding, processing, and recognition of its receptors on the oocyte membrane during yolk accumulation. In addition, O-glycosylation may also render the major yolk protein resistant to proteolytic cleavage during yolk degradation (Berman and Lasky, 1985).

High levels of lipids are characteristic of eggs of marine invertebrates as a source of metabolic energy during egg maturation and embryonic development (Holland, 1978). In *E. asiatica*, purified lipovitellin (Lv II) contained neutral lipids, glycolipids, and phospholipids (Tirumalai and Subramoniam, 1992). Among them, phospholipids are the dominant lipid class with phosphatidyl choline and phosphatidyl serine as the major class (Table 8.5). These phospholipids play an important role in embryogenesis. The neutral lipids of *E. asiatica* Lv II include cholesterol, 1,3-diglyceride, fatty acid methyl esters, and carotenoid pigments. In addition to saturated and unsaturated fatty acids, arachidonic acid is also predominant in the neutral fatty acid fraction of Lv II. The increased percentage of neutral lipids in the eggs may result from the presence of glycerol, free fatty acids, and different carotenoids in *Emerita* (Kour and Subramoniam,

TABLE 8.5 Sugar Composition of Delipidated Lv II of *E. asiatica*

Carbohydrates	Percentage Composition
Hexose	1.375 ± 0.32
Hexosamine	1.460 ± 0.14
Galactosamine	1.020 ± 0.09
Mannose	0.730 ± 0.11
Fucose	ND
Glucose	ND
Sialic acid	1.690 ± 0.11
N-linked oligosaccharides mannose	0.680 ± 0.04
O-linked oligosaccharides	1.045 ± 0.16
O-linked oligosaccharides with *N*-acetyl hexosamine as the terminal residue	0.193 ± 0.03

Reproduced from Tirumalai, R., Subramoniam, T., 1992. Purification and characterization of vitellogenin and lipovitellins of the sand crab Emerita asiatica: molecular aspects of crab yolk proteins. Mol. Reprod. Dev. 58, 54–62.

1992). The presence of both glucose (monoglycosylceramide) and galactose (diglycosylceramide) containing glycolipids in the Vg is important for the recognition of its receptor in the oocyte membrane (Van Berkel et al., 1985). In the penaeid shrimp *P. semisulcatus*, vitellin carried considerable amounts of triacylglycerols and negligible amounts of diacylglycerols in contrast to Vg (Lubzens et al., 1997). Similar differences in lipid composition of vitellin and Vg has also been shown in other crustaceans (Wilder et al., 2002).

Crustacean lipovitellin is unique in its possession of a variety of carotenoid pigments. Carotenoids can either be linked directly to protein chains through amino groups (carotenoproteins) or esterified to the fatty acids of the lipovitellin molecules (Cheesman et al., 1967). Lipovitellin may possess a single carotenoid component or a mixture of all of the egg carotenoids. Therefore, ovary of different decapods has different colors during maturation, although lipovitellin in general is rich in beta-carotene (Wallace et al., 1967). For instance, lobster egg lipovitellin contains astaxanthine and is green colored (Zagalsky, 1985), while lipovitellin from the anostracan crustacean *Branchipus stagnalis* contains blue-colored canthaxanthin (Zagalsky and Gilchrist, 1976). According to Cheesman et al. (1967), carotenoid pigment of lipovitellin might be required for the stabilization of the protein backbone of the major yolk protein. Crustaceans do not synthesize carotenoid pigments but ingest them from their plant food. Hence, interspecific variation in lipovitellin color may arise from differences in their dietary carotenoids. Crustacean vitellin also undergoes characteristic spectral shifts due to the interaction of beta-carotene with the lipid and protein of the lipoprotein complex (Zagalsky and Gilchrist, 1976). Carotenoids provide protection to the ovary and developing eggs from radiation, besides serving a metabolic function based on antioxidant properties and a source of pigmentation during embryonic development (see chapter: Yolk Utilization and Embryonic Nutrition). The photoprotective action of the carotenoids is a function of the number of conjugated double bonds in the molecule and the lipovitellin complex acts as a quencher of single-excited oxygen (Krinsky, 1964).

The immature ovary is colorless and it shows various shades of colors such as yellow, orange, and dark green in different decapods as the vitellogenesis advances. The occurrence of various carotenoid pigments present in the ripe ovaries of decapods crustaceans is listed in Table 8.6. In *M. rosenbergii*, Soroka et al. (2000) showed that both carotenoids and lipids are conjugated to Vg within the hepatopancreas and then transported to ovary through hemolymph. The occurrence of various carotenoid pigments present within the ovary may suggest a possible interconversion of Vg-bound carotenoids into various other forms.

Several steroid hormones, like ecdysteroids and vertebrate steroids, are also bound to crustacean lipovitellin and they play a significant role in the control of embryogenesis. In addition to these organic conjugates, crustacean vitellins also contain significant quantities of metal ions. The yolk protein of *E. asiatica* contains several metal ions such as copper, iron, sodium, calcium, and phosphorous.

TABLE 8.6 Carotenoids Present in the Ovary of Crustaceans

Carotenoids	Species	References
β-carotene	*Cherax quadricarinatus*	Sagi et al. (1995)
Astaxanthin	*Plesionika*	Zagalsky (1985)
	Penaeus esculentus	Dall et al. (1995)
	Cherax quadricarinatus	Sagi et al. (1995)
	Cancer pagurus	Sagi et al. (1995)
	Emerita analoga	Goodwin (1984)
	Homarus americanus	Goodwin (1984)
	Homarus gammarus	Goodwin (1984)
	Palinurus vulgaris	Goodwin (1984)
	Branchipus stagnalis	Goodwin (1984)
	Portunus trituberculatus	Goodwin (1984)
	Chionecetes opilio	Miki et al. (1982)
	Pandalus borealis	Miki et al. (1982)
	Penaeus orientalix	Miki et al. (1982)
Canthaxanthin	*Branchipus stagnalis*	Zagalsky and Gilchrist
	Emerita asiatica	(1976)
	Carcinus maenas	Tirumalai (1996)
	Idothea montereyensis	Gilchrist and Lee (1967)
	Idothea granulose	Goodwin (1984)
	Artemia salina	Goodwin (1984)
		Goodwin (1984)
δ-carotene	*Carcinus maenas*	Gilchrist and Lee, 1967
Echinenone	*Carcinus maenas*	Gilchrist and Lee (1967)
Isocryptoxanthin	*Carcinus maenas*	Gilchrist and Lee (1967)
Zeaxanthin	*Carcinus maenas*	Gilchrist and Lee (1967)
	Chionecetes opilio	
Lutein	*Cherax quadricarinatus*	Sagi et al. (1995)
	Carcinus maenas	Sagi et al. (1995)
Astaxanthin monoester	*Penaeus esculentus*	Dall et al. (1995)
Astaxanthin diester	*Eupagurus bernhardus*	Goodwin (1984)
	Chionecetes opilio	Dall et al. (1995)
	Penaeus esculentus	Miki et al. (1982)
	Portunus trituberculatus	
Doradexanthin	*Portunus trituberculatus*	Miki et al. (1982)
	Pandalus borealis	Miki et al. (1982)
	Penaeus orientalix	Miki et al. (1982)
Tetrol	*Portunus trituberculatus*	Miki et al. (1982)
Unidentified	*Portunus trituberculatus*	Miki et al. (1982)
	Penaeus orientalix	Miki et al. (1982)
	Chionecetes opilio	Miki et al. (1982)

Whereas calcium and copper are bound to lipid in Lv II, iron, phosphorous, and sodium are both lipid and protein bound. These ions constituted as much as 3.5% of the purified major yolk protein in *E. asiatica* (Tirumalai and Subramoniam, 1992). Whereas phosphorous is bound to protein by way of phosphorylation in insect vitellin, in *E. asiatica*, large amounts of phosphorous is linked to the lipid component of the lipovitellin (Tirumalai, 1996). The presence of a meager amount of protein-bound phosphorous in crustacean lipovitellin may result from the *O*-linked glycosylation of serine moieties before phosphorylation (Dhadiall and Raikhel, 1990). Crustacean lipovitellin is apparently a composite protein fraction, the production of which is likely to be a multigene-and multihormone-controlled process.

8.11 CORTICAL ROD FORMATION

After the completion of vitellogenesis by endocytotic accumulation of yolk precursors from the hemolymph, the microvilli on the oocyte periphery disappear, and the oocyte plasma membrane becomes smooth, and the cortical granules appear in the oocyte cytoplasm (Charniaux-Cotton, 1980). In the crab, *C. maenas*, Goudeau (1984) described the formation of cortical granules as the last phase of endogenous yolk synthesis and its composition is mainly of glycoprotein nature. In the penaeid shrimp, completion of yolk accumulation during ovarian development is marked by the formation of cortical rods in the oocytes. In *P. monodon* and *P. vannamei*, cortical rods appear first as spherical bodies near the periphery of the oocyte cytoplasm and elongate toward the center as maturation proceeds to give a radial arrangement around the periphery of the oocyte plasma membrane (Clark et al., 1990). The cortical rods in the shrimp are composed of glycoprotein and contain no lipids (Tan-Fermin and Pudadera, 1989). In general, all endogenously synthesized proteins are glycoproteinous in nature, whereas the yolk precursors sequestered from outside the ovary are all primarily lipoproteins. During the formative stage, the cortical rod peptides are found within vesicles, formed from dilated ER (Rankin and Davis, 1990). In *P. semisulcatus*, Khayat et al. (2001) characterized the cortical rod protein as shrimp ovarian peritrophin, whereas in *M. japonicus*, this protein is called cortical rod protein. In the latter shrimp species, another protein namely, thrombospondin, has also been identified in the cortical rod (Yamano et al., 2004). Messenger RNAs for all the three cortical rod proteins are expressed in the previtellogenic oocytes, but the respective proteins are synthesized in the last phase of vitellogenesis. They are first scattered throughout the cytoplasm but become localized within the cortical rod during its formation (Kim et al., 2004, 2005). Apparently, a single gene generates three transcriptional variants corresponding to the three cortical rod proteins and the transcription, translation of cortical rod proteins, and the formation of the cortical rod structures occurred at different stages of oocyte development (Khayat et al., 2001). The expression of these genes is controlled at the translational level by CHH-family peptides (Okumura, 2006). A cortical rod protein homologue of penaeid shrimp oocyte has also been found in the ovaries of the freshwater prawn *M. rosenbergii*, a species that does not form cortical rod structures in the oocytes (Kim et al., 2007). In the lobster *H. americanus*, the oocytes do not develop cortical rods either, but structures, called bottle brushes, are present as analogs to the cortical rods of penaeid shrimp (Talbot, 1981). In the penaeid shrimps, the contents of the cortical rods are released around the eggs and form into a jelly, on contact with the sea water at spawning (Clark et al., 1990). The jelly coat layer forms a protective layer around eggs immediately after spawning and dissipates following the assembly of the hatching envelop.

To sum up, oocyte differentiation in crustaceans is an extended process, in keeping with the enormous amounts of yolk materials that the oocyte has to accumulate during vitellogenesis. Crustaceans show vast diversity in the morphology and organization of the ovary in different crustacean orders in producing enormous numbers of yolky eggs. A combination of primitive autosynthetic yolk formation and yolk incorporation by sequestration of yolk precursor molecule synthesized elsewhere is the cornerstone in crustacean vitellogenesis. This makes the crustaceans a good model system to investigate the mechanism and the control of vitellogenesis in invertebrates. Nevertheless, the process of yolk synthesis and uptake by the ovary are similar to those of insects and many oviparous vertebrates. Several gene expression studies have revealed the dynamics of Vg synthesis within the hepatopancreas, fat body, and ovary. Employing multiple sites for Vg synthesis is perhaps unique in the evolution of vitellogenic system in crustaceans, but this is one way to augment yolk production involving many reproductive and nonreproductive tissues. No wonder, hormonal regulation of cellular activity concerning Vg synthesis and uptake by the ovary is a fascinating area of research in crustacean reproduction.

Chapter 9

Endocrine Regulation of Vitellogenesis

9.1 INTRODUCTION

Crustaceans have unique egg-laying abilities, by which they lay a large number of yolk-laden eggs at each spawning. Therefore, vitellogenesis, the process of yolk formation, is the central event in the oogenic cycle. The egg stores complex molecules of yolk proteins to meet the energy demands of embryonic development. Although the process of vitellogenesis is similar in crustaceans and insects, hormonal mechanisms controlling this energy-demanding process is widely different, inasmuch as the endocrine systems of these two arthropod groups vary vastly. Unlike insects, crustaceans continue growth and molting even after sexual maturity. Hence, in the adult reproductive females, vitellogenic activities are sandwiched between two consecutive molt cycles. Therefore, hormonal interactions could play a pivotal role in the regulation of nutritive supply to molting and vitellogenic processes. The hormonal controlling mechanisms, enabling the temporal separation of these major energy-demanding processes, involve principally the inhibitory neuropeptides—vitellogenesis-inhibiting hormone (VIH) and molt-inhibiting hormone (MIH)—originating in the optic ganglia of the eyestalk. Conversely, many hormonal factors originating from different sources are implicated in the stimulation of vitellogenesis (Fig. 9.1). Evidently, a coordinated control of molting and vitellogenesis becomes imperative to accomplish continued body growth and increased fecundity (Subramoniam, 2011). Yet another intervention in the coordinated control of molting and reproductive cycles of crustaceans is the egg brooding within the pleopods of several malacostracans.

9.2 EYESTALK INHIBITORY HORMONES

Hanström (1937) first described the anatomy of neurosecretory structures in the optic ganglia of the eyestalk in crustaceans. The X-organ is part of the medulla terminalis found within the eyestalk. Neurosecretory cells from the X-organ extend their axonal ends into the sinus gland, a neurohemal organ, lying in between the medulla externa and medulla interna in the eyestalk (Fig. 9.2). In those species lacking stalked eyes (eg, isopods and amphipods), the X-organ–sinus-gland (XOSG) complex is embedded in the brain. The neurosecretory

Sexual Biology and Reproduction in Crustaceans. http://dx.doi.org/10.1016/B978-0-12-809337-5.00009-5

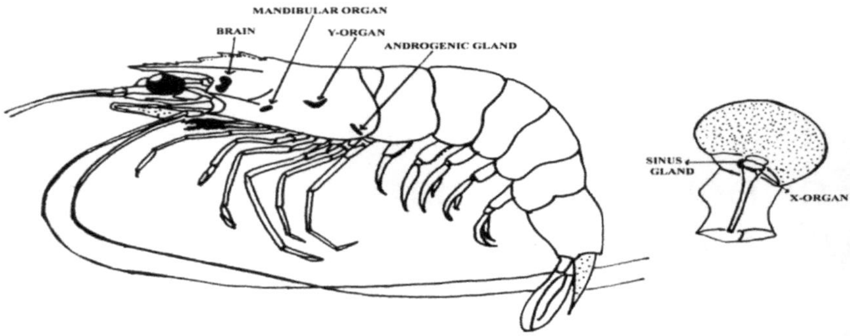

FIGURE 9.1 Various neuroendocrine glands in a penaeid shrimp. *Redrawn from Subramoniam, T., 1999. Endocrine regulation of egg production in economically important crustaceans. Curr. Sci. 76, 350–360.*

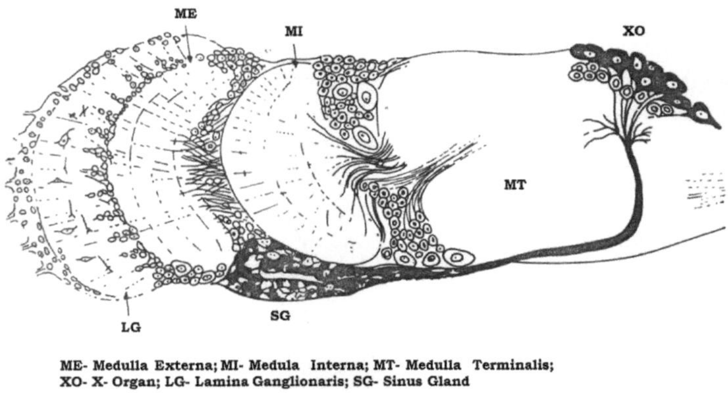

FIGURE 9.2 Eyestalk neurosecretory centers. *Redrawn from Subramoniam, T., 1999. Endocrine regulation of egg production in economically important crustaceans. Curr. Sci. 76, 350–360.*

perikarya of the X-organ produce the "crustacean eyestalk hormones" that comprise, among others, a family of multifunctional peptides, called crustacean hyperglycemic hormone (CHH). Structurally, CHH are highly conserved among Crustacea with typically 72–83 amino acid sequences, including six cysteine residues paired in three disulfide bridges in conserved positions (Keller, 1992). Peptides included in the family, other than the CHH, are (1) ion transport peptides (ITPs), (2) MIH, VIH, and mandibular organ-inhibiting hormone (MOIH). These eyestalk hormones exhibit diversified physiological functions such as carbohydrate metabolism, molting, gonad maturation, as well as ionic and osmotic regulation (Webster et al., 2012).

9.3 VITELLOGENESIS-INHIBITING HORMONE

The existence of a gonad inhibitory principle in the eyestalk of decapod crustaceans was first indicated when eyestalk ablation (ESA) resulted in accelerated maturation of the ovary in the grass shrimp *Leander serratus* (Panouse, 1943). Since then, many studies involving extraction and injection of eyestalk hormones adduced further evidence that GIH originate from XOSG of several crustaceans. Eyestalk removal results in accelerated secondary vitellogenesis due to increased synthesis and uptake of vitellogenin (Vg) (Adiyodi and Subramoniam, 1983). A significant outcome of the discovery of gonad inhibitory principle within the eyestalk is the practical application of ESA technique to induce ovarian maturation and spawning in commercially important penaeid shrimps under captive conditions. Table 9.1 enlists the experimental ESA studies to induce ovarian maturation and spawning in various decapod crustaceans. The regulatory role of the GIH is better understood on vitellogenesis than on reproductive processes in males. By virtue of its specific inhibitory effect on vitellogenesis, GIH is rechristened as VIH.

The first isolation and partial characterization of VIH were made in the crab *Cancer magister* using gel chromatography techniques (Bomirsky et al., 1981). It was found to be a 2 kDa peptide with inhibitory effect on the ovarian growth of the shrimp *Crangon crangon*. Quackenbush and Keely (1988) purified the VIH from the crude extract of the shrimp *Penaeus setiferus* using sephadex G-25 chromatography and determined the molecular weight as 3300 Da. This peptide inhibited the ^{14}C-leucine incorporation into the Vg of the cultured ovary of *Uca pugilator*, while the incorporation of the radioactivity in the other proteins remained unaffected. The VIH of *P. setiferus* exhibited similar inhibitory effect on the yolk protein synthesis in the ovary and hepatopancreas of another penaeid shrimp *Penaeus vannamei* (Quackenbush, 1989). These results incidentally demonstrate the nonspecificity of VIH action in controlling vitellogenesis among different crustacean species.

The complete characterization of VIH was, however, made possible by the microsequencing (Soyez et al., 1991) and cDNA cloning (De Kleijn et al., 1992) in the American lobster *Homarus americanus*. VIH of this lobster consists of a 78 residue peptide (Mr. 9135), with an amidated C-terminus and a free N-terminus. Interestingly, this molecule showed considerable sequence similarity (53%) to *Carcinus maenas* MIH (Webster, 1991). VIH of *H. americanus* exists as two isoforms of identical sequence and mass, with the gonad-inhibiting effect found only in one isoform, when tested with an in vivo heterologous assay developed in the grass shrimp *Palaemonetes varians* (Soyez et al., 1987, 1991). Following this, Aguilar et al. (1992) isolated a peptide from the eyestalk of the Mexican crayfish *Procambarus bouvieri*, which inhibited the growth of oocytes from *P. vannamei*, in vitro. This neuropeptide, however, contains lesser number of amino acid residues (72–74) with a molecular weight of 8388 Da and a blocked N-terminus. Structurally, the VIH of Mexican crayfish is more closely related to its CHHs than to the lobster VIH. Again, its VIH activity in the endogenous Vg synthesis by *P.*

TABLE 9.1 Effect of Eyestalk Ablation on Ovarian Maturation and Spawning in Decapods Crustaceans

Species	Type of Ablation	Effect of Ablation	References
Penaeidae (Shrimps)			
Penaeus monodon	Unilateral ablation	Induction of ovarian maturation	Aquacop (1979)
Penaeus monodon	Bilateral ablation	Higher number of eggs and nauplii	Primavera and Cuballero (1992)
Penaeus monodon	Unilateral	Maturation and spawning induced	Halder (1978)
Penaeus monodon	Unilateral	Spawning with viable nauplii	Primavera (1978)
Penaeus monodon	Unilateral	Spawning with viable eggs	Beard and Wickins (1980)
Penaeus monodon	Unilateral	Rematuration accelerated	Primavera and Borlongan (1978)
Penaeus monodon	Unilateral	Induced gonadal development	Santiago (1977)
	Bilateral	High mortality	
Penaeus monodon	Unilateral	Ovarian maturation induced, bigger oocytes	Tan-Fermin (1991)
Penaeus monodon	Unilateral	Spawning index and egg fertilization rates not affected	Kelemec and Smith (1984)
Penaeus monodon	Unilateral	Ovarian maturation induced	Pudadera and Primavera (1981)
Penaeus monodon	Unilateral	Ovarian maturation induced	Ruangpanit et al. (1981)
Penaeus monodon	Unilateral	Maximum growth with fresh feed	Murugesan et al. (1998)
Penaeus japonicus	Bilateral tying/cutting	Ovarian maturation	Teshima et al. (1988)
		Increased ovarian lipid	
		Decrease in hepatopancreas lipids	
Penaeus japonicus	Bilateral	Lipid transfer from hepatopancreas to ovary	Teshima et al. (1989)

Penaeus notalis and *Penaeus schmitti*	Unilateral	Maturation induced; low hatching rate	Trujillo and Primavera (1986)
Penaeus schmitti	Unilateral	Maturation induced	Nascimento et al. (1991)
Penaeus stylirostris	Unilateral	Ovarian maturation but the spawn not fertilizable	Chamberlain and Gervais (1984)
Penaeus stylirostris	Unilateral	Induced ovarian maturation	Chamberlain and Lawrence (1981)
Penaeus setiferus	Unilateral	Fertile egg produced	Lawrence et al. (1980)
Penaeus setiferus	Unilateral	Egg production increased	Brown et al. (1979)
Penaeus indicus	Unilateral (pinched/tied)	Ovarian maturation induced	Makinouchi and Honculada-Primavera (1987)
Penaeus indicus	Unilateral	Induced ovarian maturation	Emmerson (1980)
Penaeus orientalis	Unilateral	Induced ovarian maturation	Arnstein and Beard (1975)
Penaeus semisculcatus	Unilateral	Increased energy demand; growth decreased; egg quality not affected; spawning induced	Browdy and Samocha (1985a,b)
Penaeus notialis	Unilateral	Ovarian development quickened	Ramos-Trujillo and Gonzales Flores (1983)
Penaeus plebejus	Unilateral	Induced ovarian development	Kelemec and Smith (1984)
Penaeus monodon and *Metapenaeus dobsoni*	Unilateral	Enhanced ovarian maturation and spawning	Muthu and Laxminarayana (1977)
Parapenaeopsis stylifera	Unilateral	Induced ovarian maturation and spawning	Emmerson (1983)
Penaeus monodon		Meiotic maturation	Phinyo et al. (2013)
Marsupenaeus japonicus	Bilateral	Ablation during nonreproductive period results vitellogenesis	Okumura and Sakiyama, 2004

Continued

TABLE 9.1 Effect of Eyestalk Ablation on Ovarian Maturation and Spawning in Decapods Crustaceans—cont'd

Species	Type of Ablation	Effect of Ablation	References
Marsupenaeus japonicus	Bilateral	ESA in immature shrimp induces transcription of Vg mRNA in ovary but not in hepatopancreas	Okumura et al. (2006)
Liptopenaeus vannamei	Unilateral	Increased Vg concentration; more maturation efficiency	Arcos et al. (2003)
Liptopenaeus vannamei	Unilateral	Increased spawning	Palacios et al. (1999 a, b)
Liptopenaeus vannamei	Unilateral	Ablated females matured sooner with high maturation efficiency compared to serotonin-injected animals	Vaca and Alfaro (2000)
Penaeus (Marsupenaeus) japonicus	Unilateral	Increase of Vg mRNA in ovary but not in hepatopancreas	Tsutsui et al. (2000)
Penaeus semisulcatus	Unilateral	ESA results predictable maturation in off-season	Aktaş et al., 2003
Penaeus monodon	Unilateral	ESA increased ovarian maturation, compared with serotonin injection. Serotonin-injected females showed comparable maturation performance with eyestalk-ablated females	Wongprasert et al. (2006)
Penaeus monodon	Unilateral	Higher rate of ovarian maturation in the eyestalk-ablated females administrated with vertebrate-like sex steroid	Merlin et al. (2015)
Penaeus indicus	Unilateral	ESA along with serotonin injection provided higher level of maturation and spawning	Tomy et al. (2016)
Palaemonidae (Caridea)			
Macrobrachium nobilli	Unilateral	Growth and maturation induced ablation alone	Kumari and Pandian (1991)
Macrobrachium rosenbergii	Snipping both eyestalks	Serum Vg appeared, increased, and then decreased	Wilder et al. (1994)
Macrobrachium lanchesteri	Bilateral	High maturity and increase in growth rate	Ponnuchamy et al. (1981)

Macrobrachium malcolmsonii	Unilateral	Increased fecundity	Murugadass et al. (1988)
Macrobrachium nipponese	Unilateral	Ovarian maturation induced	Bingru and Hong (1990)
Palaemon serratus	Uni/bilateral	Induced ovarian development	Panouse (1943)
Macrobrachium rosenbergii	Bilateral	ES in males and females shortened the molt cycle. In females it induced ovarian development	Okumura and Aida (2001)
Macrobrachium rosenbergii	Bilateral	Vg synthesis in ovary is negligible when compared to Vg synthesis in the hepatopancreas	Jayasankar et al. (2002)
Pandalidae (Shrimps)			
Pandalopsis japonica	Bilateral	ESA upregulates JH esterase like carboxylase in the ovary	Lee et al. (2011)
Palinuridae (Lobsters)			
Panulirus argus	Bilateral	Increase in gonadal size	Quackenbush and Herrnkind (1981)
Homarus americanus	Unilateral	Accelerated ovarian maturation and spawning	Waddy et al. (1995)
Panulirus homarus	Unilateral and bilateral	ESA induced precocious ovarian maturation in the immature lobsters weighing less than 100 g body weight	Subramoniam and Kirubakaran (2010)
Brachyura (freshwater crab)			
Paratelphusa hydrodomous	Bilateral	• Low level of organic substances • More oocytes • Lipid:protein ratio in ovary high	Anilkumar and Adiyodi (1980)
Astacidae (European crayfish)			
Astacus astacus	Bilateral	Molting rate increased	Huner and Lindquist (1984)

setiferus ovaries occurred only at a high concentration of 10 sinus gland equivalents. Furthermore, microsequencing revealed the primary structure of VIH in another lobster *H. gammarus* (Ollivaux et al., 2009). Yet another authentic VIH to be identified in another malacostracan is that of the isopod *Armadillidium vulgaris* (Greve et al., 1999).

In general, VIH consists of a signal peptide (20–31 amino acid residues) and mature peptide, having 77–83 amino acid residues. The VIH sequence contains six cysteine residues paired in three disulfide bridges in conserved positions (Böcking et al., 2002). The suppressive effect of VIH on vitellogenesis is evident from numerous experimental studies involving the removal of VIH by way of eyestalk extirpation. Several molecular studies have shown that ESA leads to acceleration of Vg gene expression and increased levels of Vg in the hemolymph as well as shortening of the ovarian cycle (Subramoniam, 2011).

Amino acid sequence homology studies on the VIH of various crustacean species have revealed significant similarities with other CHH family peptides (Keller, 1992). Thus, CHH family peptides are grouped under two subclasses, based on their structural and functional characteristics. The type I peptides include CHHs which possess 72 residues and blocked N-termini (pyroglutamate ring) and amidated C-termini. Type II consists of three other related peptides such as VIH, MIH, and MOIH, having more amino acid residues (78–83) and an unblocked N-terminus and C-terminus. Further support for this classification came from the sequence analysis of hormone precursors. In type I, the CHH peptide consists of three regions: a signal sequence, a short sequence called CHH precursor-related peptide, and the mature hormone with its basic processing site. On the other hand, in the VIH subclass, the mature hormone is directly linked to the signal peptide (Lacombe et al., 1999).

In the penaeid shrimps, a notable exception occurs in that a type II hormone corresponding to VIH appears to be absent. Instead, in the marine shrimp *Penaeus semisulcatus*, Khayat et al. (1998) found that up to seven sinus gland neuropeptides of CHH family (type I) inhibited protein and mRNA synthesis in in vitro incubated ovarian explants of vitellogenic females. Furthermore, in the penaeid shrimps, type I CHH-active peptides exhibit a wide range of overlapping functions on female reproduction and molting. For example, in the kuruma prawn *Marsupenaeus japonicus*, as many as six CHH peptides are shown to have additional activity on ovarian tissues (Tsutsui et al., 2005a). To be specific, fraction number three (Pej-SPG-III) reduced Vg mRNA levels in a dose-dependent manner but did not exhibit any hyperglycaemic activity. Understandably, the existence of Pej-SPG-III in the XOSG complex could account for the significant increase in Vg mRNA levels in the ovaries of eyestalk-ablated shrimps (Tsutsui et al., 2005b). Similarly, in the South African spiny lobster, *Jasus lalandii*, two CHHs, having a C-terminal amide, inhibited Vg synthesis in the incubated ovarian tissues of *P. semisulcatus* (Marco et al., 2002).

These studies emphasize the fact that in penaeid shrimps, VIH activity lies in type I CHH hormone in multiple molecules, contradicting the classification of CHH peptides into CHH and VIH subtypes. The existence of CHHs in the eyestalk in multiple VIH forms has also been reported in other shrimps such as *Penaeus monodon* (Vincent et al., 2001). More recently, Tsutsui et al. (2007) isolated seven type I CHH family peptides, namely Liv-SGP-A, -B, -C, -D, -E, -F, and -G in the eyestalk of the white leg shrimp *Litopenaeus vannamei*. When their VIH activity was tested by in vitro incubation of *M. japonicus* ovarian fragments, all the molecules, except Liv-SGP-D, showed significant inhibitory activity with different efficacies on Vg mRNA synthesis, in a dose-dependent manner. Vitellogenesis-inhibiting ability of these type I peptides in *L. vannamei* may reside in the carboxyamide C-terminal moiety (Wilder et al., 2010). Furthermore, structural analysis showed that Liv-SGP-D is a type II peptide by virtue of the presence of a glycine residue at position 12. Therefore, the absence of a VIH activity by this peptide only adds to the confirmation that in the penaeid shrimps, CHH type II hormones either have weak or no vitellogenesis-inhibiting effects on the ovary (Webster et al., 2012).

Despite the structural relationship existing between the two groups of CHH family, functional distinctiveness occurs only for the VIH peptides. Thus, VIH is rather homogeneous in terms of biological functions, inasmuch as these peptides do not show any cross-function for molt control or glycemic activity. Even though all CHH family peptides possess CHH activity to some extent, none of the other inhibitory peptides (VIH, MIH, and MOIH) exhibited any interfunctional role among them. The structural similarity as well as the physiological overlapping among these super family peptides may suggest the ancestral nature of the CHH peptides, giving origin to the inhibitory peptides such as VIH, MIH, and MOIH by way of gene duplication (Lacombe et al., 1999).

Immunocytochemical localization of VIH in the XOSG complex was made in *H. americanus*, using polyclonal antisera raised against VIH (Meusy et al., 1987). Furthermore, detection of mRNA encoding VIH in the neurosecretory cells of XOSG was made in the lobster using in situ hybridization techniques. Similarly, De Kleijn et al. (1992), by employing complementary RNA probes, colocalized VIH mRNA with CHH mRNA in the same neurosecretory perikarya, although the two neuropeptides originate from two different precursors.

VIH activity has also been observed in other parts of the central nervous system. For example, Edomi et al. (2002) found mRNA expression of VIH in the Norway lobster, *Nephrops norvegicus*, not only in the eyestalk but also in the supraesophageal ganglia. In the penaeid shrimp, *P. monodon*, VIH activity has also been reported in the brain, thoracic ganglia, and abdominal nerve cord, in addition to the eyestalk (Treerattrakool et al., 2008). Posttranscription gene silencing using GIH-specific double-stranded RNA triggers sequence-specific suppression of its mRNA, resulting in increased expression levels of Vg in *P. monodon* (Treerattrakool et al., 2008). Injection of GIH-dsRNA into previtellogenic female *P. monodon* not only inhibited GIH expression but also induced ovarian maturation and eventual spawning, with a comparable effect to ESA in wild shrimps (Treeattrakool et al., 2011). However, these authors found the effect of *P. monodon* VIH-knockout on Vg synthesis only in ovary and not in hepatopancreas.

While the inhibitory neuropeptides MIH and MOIH have their action on their respective targets such as the Y-organ and mandibular organ (MO), the target organ for VIH action has been found to be the site of Vg synthesis. In other words, it has been suggested to act directly on the oocytes by inhibiting Vg synthesis or its uptake by the ovary. Jugan and Soyez (1985) first showed that VIH could inhibit the gold-labeled vitellin binding to oocyte microvilli in incubation medium containing sinus gland extract, thus demonstrating that VIH could bind to the receptor to block the Vg binding site. No supporting evidence could be found in the literature to substantiate this finding.

On the other hand, more evidence has been brought forward to the tenet that VIH inhibited the synthesis of Vg at its synthetic site. Vincent et al. (2001) quantified the hemolymph Vg level using a sensitive sandwich enzyme-linked immunosorbent assay in the penaeid shrimp *P. monodon*. They showed that two of the HPLC-purified XOSG fractions reduced hemolymph Vg concentrations in a time-dependent manner suggesting their probable inhibitory effect on Vg synthetic sites.

It is interesting to find that most of the in vitro studies to demonstrate VIH influence on Vg synthesis have been done using ovarian tissues. Clearly, ovarian tissues are more suitable for culture studies than the hepatopancreas. Tsutsui et al. (2005b) measured the levels of Vg mRNA expression, using a quantitative real-time PCR system, in both ovary and hepatopancreas in the normal and eyestalk-ablated adult shrimp, *M. japonicus*. However, significant increments in Vg mRNA level were observed only in the ovary of the eyestalk-ablated females, and no such rise in Vg mRNA was observed after eyestalk removal in the hepatopancreas. Understandably, hepatopancreas is refractory to VIH influence as far as Vg gene expression is concerned in this shrimp. In the shrimp ovary, Vg mRNA expression is localized only in the follicle cells enveloping the vitellogenic oocytes and the oocyte cytoplasm is positive to vitellin antibody, suggesting that VIH acts at the level of follicle cells to exert its inhibitory control on Vg gene transcription. Further data on the presence of VIH receptor in follicle cells is necessary to confirm the VIH action.

On the contrary, hepatopancreas, the sole site of Vg synthesis in the freshwater prawn *Macrobrachium rosenbergii*, has been shown to respond to the inhibitory action of VIH (Jayasankar et al., 2002). In yet another study on the Pacific white shrimp, *L. vannamei* Chen et al. (2014) revealed the inhibitory effect of VIH also on the hepatopancreas. These authors isolated a potent VIH neuropeptide, Lv VIH, from both eyestalk and brain. This neuropeptide has all characteristics of type II CHH family peptides, in contradistinction with the work of Tsutsui et al. (2007), who considered all the VIH-active sinus gland peptides under type I CHH. Interestingly, recombinant Lv VIH protein was effective in suppressing Vg mRNA expression in the primary hepatopancreatic cell culture and in vivo injection experiments. Hence, in *L. vannamei*, brain and eyestalk (XOSG complex) are both important sites for VIH expression and are possibly coregulators of hepatopancreatic Vg mRNA expression (Chen et al., 2014). However, these authors did not test its effect on ovarian synthesis of Vg. A recent work by Merlin et al. (2015) also found that eyestalk removal resulted in the elevation of Vg mRNA levels both in the ovary and hepatopancreas of *P. monodon*.

In the shrimps, and possibly other decapods, the distribution of VIH is now known to extend throughout the central nervous system, including thoracic and abdominal nerve ganglia. Nevertheless, the neurosecretory control of reproduction is a primitive feature existing in subarthropod invertebrate phyla-like annelids, in which the so-called cerebral hormone induces somatic growth but inhibits oocyte growth (Olive, 1983). In annelids, no ovarian stimulatory hormones have been reported so far, the oogenesis occurring in the absence of cerebral hormone by a process of autodifferentiation.

The isolation, characterization, and in vitro testing of VIH effect on the target organs have unraveled the unique inhibitory pathways to control Vg synthesis in Crustacea. As with other peptide hormones, VIH acts through the second messenger systems. Okumura (2007) found the involvement of cyclic nucleotides, Ca^{2+}, and protein kinase C in the signaling pathways that regulate Vg mRNA synthesis in the ovary and hepatopancreas of *L. vannamei*. Cyclic AMP and GMP probably mediate the action of VIH on Vg synthesis in the follicle cells of the ovary. In *L. vannamei*, the responsiveness of ovary to VIH is high during previtellogenesis, compared to the vitellogenic stage.

Whereas VIH exerts its inhibitory effect directly on Vg synthesis, other type II peptide such as MOIH can inhibit vitellogenesis indirectly. Liu and Laufer (1996) showed that MOIH represses methyl farnesoate (MF) synthesis in MO in the spider crab *Libinia emarginata*. A role for MF to induce vitellogenesis and ovarian maturation has been shown in this crab (Laufer et al., 1986). In the shore crab *Cancer pagurus*, the sensitivity of MO to MOIH is high at the beginning of vitellogenesis and declines drastically during peak vitellogenesis, indicating a stage-specific role for MF on Vg synthesis (Wainwright et al., 1998). Any direct role of MOIH on Vg synthesis in ovary or hepatopancreas is yet to be demonstrated.

9.4 ANDROGENIC HORMONE

A number of malacostracan crustaceans are known to be hermaphroditic, with protandry dominating in caridean shrimps. The sex reversal in crustaceans is mediated by the androgenic gland hormone. This hormone is responsible for the male differentiation and maintenance of testicular activities. With the disappearance of androgenic gland, the male primary and secondary characters transform into those of female. This means that androgenic gland hormone could also suppress the vitellogenic activities until the transformation of male phase into female occurs. Accordingly, the VIH-like effects of androgenic gland hormones are well recognized in the protandric hermaphrodites. Thus, in *Pandalus hypsinotus*, a caridean protandric prawn, Vg is not expressed in males and immature females but becomes detectable from the late male phase associated with the degeneration of the androgenic glands and the appearance of vitellogenic oocytes in testicular tissues (Tsutsui et al., 2004).

In another sexually plastic crayfish, *Cherax quadricarinatus*, Vg is not expressed in intersexual individuals, while transcription of the Vg gene is induced in the hepatopancreas when the androgenic gland is extirpated, suggesting that androgenic hormone exerts its inhibitory effect directly on Vg synthesis (Sagi et al., 2002). Furthermore, androgenic gland itself is under the inhibitory control of GIH from the XOSG, as ESA induced androgenic hormone secretion in the crayfish *C. quadricarinatus*. It is not clear whether the removal of androgenic gland could have a feedback effect on arresting the release of GIH neuropeptide, thereby causing inducement of Vg synthesis in this crayfish.

9.5 GONAD-STIMULATING HORMONES

Numerous studies have shown that crustaceans employ multiple hormonal factors to positively control vitellogenesis. They are presumed to act species-specifically or in synergy with one another, but data on this aspect is still inadequate. Further, the stimulatory hormones are varied in their chemical nature. Though new hormonal factors continue to be discovered in Crustacea, the following are the hormones implicated with the positive control of vitellogenesis: (1) the neurosecretory hormones from the brain/thoracic ganglia, (2) biogenic amines secreted from the central nervous system, (3) Vg-stimulating ovarian hormone, (4) MF, a structural homologue of insect juvenile hormone III (JH III), and farnesoic acid (FA) secreted by the MO, (5) ecdysteroids, originating from Y-organ, and (6) a variety of vertebrate-type sex steroids, including estrogen and progesterone of uncertain origin (Subramoniam, 2000).

9.5.1 Vitellogenesis-Stimulating Hormones

As early as 1963, Otsu provided the first evidence for a gonad/vitellogenesis-stimulating principle in the central nervous system of Crustacea by noticing precocious ovarian development in the crab *Potamon dehaani* after the implantation of thoracic ganglia. This was followed by several attempts to implant brain and thoracic ganglia or to inject their extracts to stimulate vitellogenesis in different crustacean species (Gomez, 1965; Hinsch and Bennett, 1979; Takayanagi et al., 1986; Yano, 1988; Yano and Wyban, 1992). Despite the reported gonad stimulatory effect of this peptide from brain and thoracic ganglia of different decapod crustaceans, its chemical nature as well as the mechanism of action on target organs has not been so far elucidated, although Yano (1998) reported that in the shrimp *L. vannamei*, this is a 10 kDa peptide that is inactivated by trypsin. This active principle, which is presumably secreted by the neurosecretory cells of brain and thoracic ganglia, has been named as gonad stimulatory hormone (GSH). More work is certainly needed to substantiate the gonadotropic role of GSH in stimulating Vg synthesis. The antagonistic actions of GIH and GSH have long been held as the bihormonal theory to control female reproduction in decapod crustaceans (Adiyodi and Adiyodi, 1970; Adiyodi and Subramoniam, 1983).

Although the chemical nature of GSH is not adequately understood and its gonadotropic influence is largely known from ablation or implantation experiments, some information is available on the neurotransmitter control of GSH release from the brain. Several studies indicate that 5-hydroxytryptamine (5-HT) could stimulate ovarian maturation and spawning in crustaceans such as the freshwater shrimp *Paratya compressa* (Mattson and Spazaini, 1985) and the crayfish *Procambarus clarkii* (Bauchau and Mangeot, 1966). Furthermore, Sarojini et al. (1996, 1997) demonstrated in *P. clarkii* that dopamine (DA) inhibits 5-HT-stimulated ovarian maturation by inhibiting the release of GSH from the brain or thoracic ganglia or enhancing the release of VIH from the eyestalk neurosecretory centers (see below).

Despite the fact that GSH is the first reproductive hormone to be discovered in Crustacea, we still do not have any experimental clue on their influencing effect, if any, on the proximate endocrine glands/factors involved in the control of gonad maturation. It is also not clear whether GSH could directly act on ovary or hepatopancreas, as does the VIH.

9.5.2 Stimulatory Action of CHH Peptides

Although several CHH peptides show inhibitory activity on molting and reproduction, in some instances they also exhibit gonad stimulatory action. In the American lobster *H. americanus*, CHH exists in two isoforms, CHH-A and CHH-B (de Kleijn et al., 1995). Measurements of mRNA levels as well as CHH titers in the hemolymph indicate that CHH-B expression peaks during intense vitellogenesis, suggesting a positive influence on vitellogenic activities. Similarly, MIH of several decapod species are also reported to have gonad stimulatory functions. For example, in the penaeid shrimp, *Metapenaeus ensis*, MIH exists in two isoforms, MIH-A and MIH-B. MIH-B is not only expressed in the XOSG, but also in the rest of nervous system, such as ventral nerve cord, thoracic ganglia, and brain during vitellogenic activities in the ovary (Gu et al., 2002). The levels of MIH-B mRNA transcripts in the eyestalk is low during the initial phase of vitellogenesis but increase toward the end of ovarian maturation, indicating a stimulatory role for this MIH neuropeptide in the control of vitellogenesis. Injection of rMIH-B delayed the molting cycle of the maturing female and increased the levels of Vg mRNA expression and Vg synthesis in the ovary and hepatopancreas of this shrimp. Furthermore, injection of MeMIH-B dsRNA into the female shrimp caused a decrease in the MeMIH-B transcript levels in the eyestalk and thoracic ganglia.

Stimulatory activity of MIH on vitellogenesis has been reported for other decapod species too. For instance, Zmora et al. (2009) observed MIH titers of the blue crab *Callinectes sapidus* to increase significantly in the mid-vitellogenic period, compared to early vitellogenic stages. This crab is an anecdysic brachyuran, and hence continued secretion of MIH is a necessity to inhibit any more molting. Nevertheless, the specific elevation of MIH during the vitellogenic period is suggestive of a role in reproduction. In *C. sapidus*, the stimulatory role of MIH in Vg synthesis is further substantiated by the specific binding of MIH to its receptors in the hepatopancreas followed by the modulation of a cAMP pathway involved in the Vg synthesis. In addition, actinomycin D blocked the stimulatory effects of MIH on Vg mRNA and Vg synthesis, while cycloheximide lowers Vg level only, thereby confirming the role of MIH in Vg transcription and translation (Zmora et al., 2009).

Taken together, it may be suggested that in the anecdysic crab, *C. sapidus* and the penaeid shrimp, *M. ensis*, MIH achieves the coordination of somatic growth and female reproduction by stimulating Vg synthesis and, at the same time, extending the intermolt conditions, making it favorable for the rather protracted vitellogenic activities. More evidence from other decapod species is, however, required to confirm this dual role of MIH in both molt inhibition and vitellogenesis induction.

9.5.3 Biogenic Amines and Opiod Peptides

In Crustacea, biogenic amines function as neurotransmitters and neuromodulators, involving themselves in the regulation of a wide array of physiological activities (Table 9.2; Fig. 9.3) (Fingerman, 1997). Some amines also serve as neuroregulators that control the release of crustacean neurohormones. For instance, 5-HT stimulates the release of the crayfish hyperglycemic hormone, red pigment dispersing hormone, neurodepressing hormone, and MIH (Fingerman et al., 1994). The role of 5-HT in the ovarian development was first indicated by its stimulatory effect on the production of GSH in the brain and thoracic ganglia of the fiddler crab, *U. pugilator* (Richardson et al., 1991). Subsequently, Sarojini et al. (1995) demonstrated that 5-HT, under in vitro conditions, stimulated the release of GSH from the brain and thoracic ganglia of the crayfish *P. clarkii*. On the contrary, an endogenous opioid met-enkephalin inhibited ovarian development in the same crayfish by stimulating the GIH release from the XOSG and/or by a concomitant prohibition of the GSH release from the brain or thoracic ganglia.

In the spiny lobster *Panulirus homarus*, immunocytochemical localization of 5-HT in the brain and thoracic ganglia revealed a rise in the synthesis of this biogenic amine, correlated to ovarian maturation (Subramoniam and Kirubagaran, 2010). While the occurrence of 5-HT in the CNS has been reported in many more crustacean species, other forms of biogenic amines are also suggested to take part in the integrative control of ovarian maturation. For example, Sainath and Reddy (2011) found evidence that the freshwater crab *Oziotelphusa senex senex* responded to the injection of melatonin by increasing ovarian index, mean oocyte

TABLE 9.2 Biogenic Amines and Opiods Involved in Vitellogenesis and Their Physiological Properties

Biogenic Amines and Opiods	Species	Mode of Action		Site of Production	Target Organs/ Tissues
		Stimulation	Inhibition		
Serotonin (5-HT)	*Uca pugilator* *Procambarus clarkii* *Homarus americanus* *Libinia emarginata* *Litopenaeus stylirostris* *Liptopenaeus vannamei* *Penaeus vannamei* *Penaeus monodon* *Macrobrachium rosenbergii* *Oziotelphusa senex senex* *Anomalocera patersoni;* *Centropages typicus;* *Hemidiaptomus roubani* *Palaemon elegans*	Ovarian development Agonists – ovarian maturation Testicular development Ovarian maturation Ovarian maturation and spawning Ovarian maturation and spawning Ovarian maturation and spawning VSH production and gonadal maturation Ovarian maturation Release of VSH Release of CHH	Low enhancement of egg-extruding process Synthesis of MF from MO	XO-SG, brain, and TG XO-SG, brain, and TG XO-SG, brain, and TG	Gonads, HP, brain, and TG XO-SG, brain and TG XO-SG, brain, and TG
Melatonine (*N*-acetyl-5-methoxy-tryptamine)	*Oziotelphusa senex senex* *Scylla serrata*	Increased ovarian Vg levels		Hemolymph, eyestalk, and brain	Brain and TG
Dopamine (DA)	*Procambarus clarkii* *Oziotelphusa senex senex* *Macrobrachium rosenbergii*		Ovarian-stimulating action of 5-HT Synthesis of VSH	XO-SG, brain and TG CNS and ovary	Gonads, HP, brain, and TG
Leucine-enkephalin	*Penaeus indicus* *Oziotelphusa senex senex*	Mean ovarian index Mean ovarian index		CNS	Brain, TG, ovary, and HP
Methionine-enkephalin	*Procambarus clarkiii* *Procambarus clarkii*		Decreased mean ovarian index Inhibits the synthesis of VSH	CNS	Brain, TG, ovary, and HP

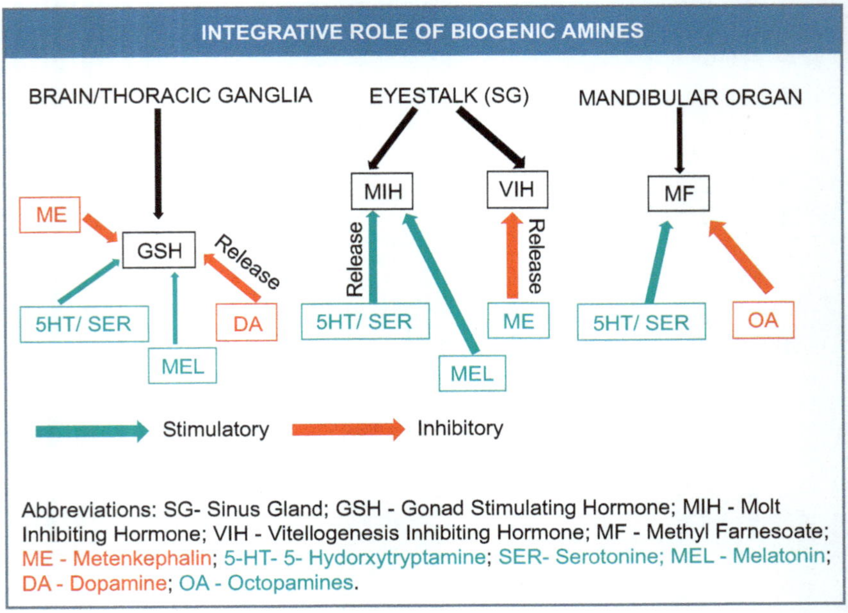

FIGURE 9.3 Integrative role of biogenic amines.

diameter, and ovarian vitellin level. Interestingly, acceleration of ovarian growth by melatonin was similar in both eyestalk-ablated and eyestalk-intact females.

An antagonistic role for the action of 5-HT in the stimulation of ovarian development by DA has been reported in a wide variety of decapod crustacean species. In the red swamp crayfish, Sarojini et al. (1995) demonstrated the in vitro inhibition by DA of 5-HT stimulated ovarian maturation. Using this antagonistic effect between these two amines, as well as their natural occurrence in the central nervous system, Alfaro et al. (2004) induced ovarian maturation and spawning in the penaeid shrimps, *Litopenaeus stylirostris* and *L. vannamei* by a combined treatment with 5-HT and the dopaminergic antagonist, spiperone. Chen et al. (2003) showed that in the giant freshwater river prawn *M. rosenbergii* the site of inhibitory action of DA against 5-HT was at the thoracic ganglia by inhibiting the release of GSH rather than enhancing secretion of VIH from XOSG. Furthermore, Meeratana et al. (2006) found supporting evidence on the inducement of ovarian maturation in *M. rosenbergii* by injecting 5-HT into the female carrying first stage (spent) ovary. Pretreatment with a 5-HT receptor antagonist, cyproheptadine, significantly suppressed the effect of this biogenic amine on ovarian inducement. Administration of 5-HT furthermore enhanced the synthesis and release of Vg into the hemolymph, in addition to increasing the total Vg mRNA in the hepatopancreas of this prawn (Kuo et al., 2009). However, the stimulation of Vg synthesis in hepatopancreas by 5-HT was complete only when it was injected into the eyestalk-ablated females. In a

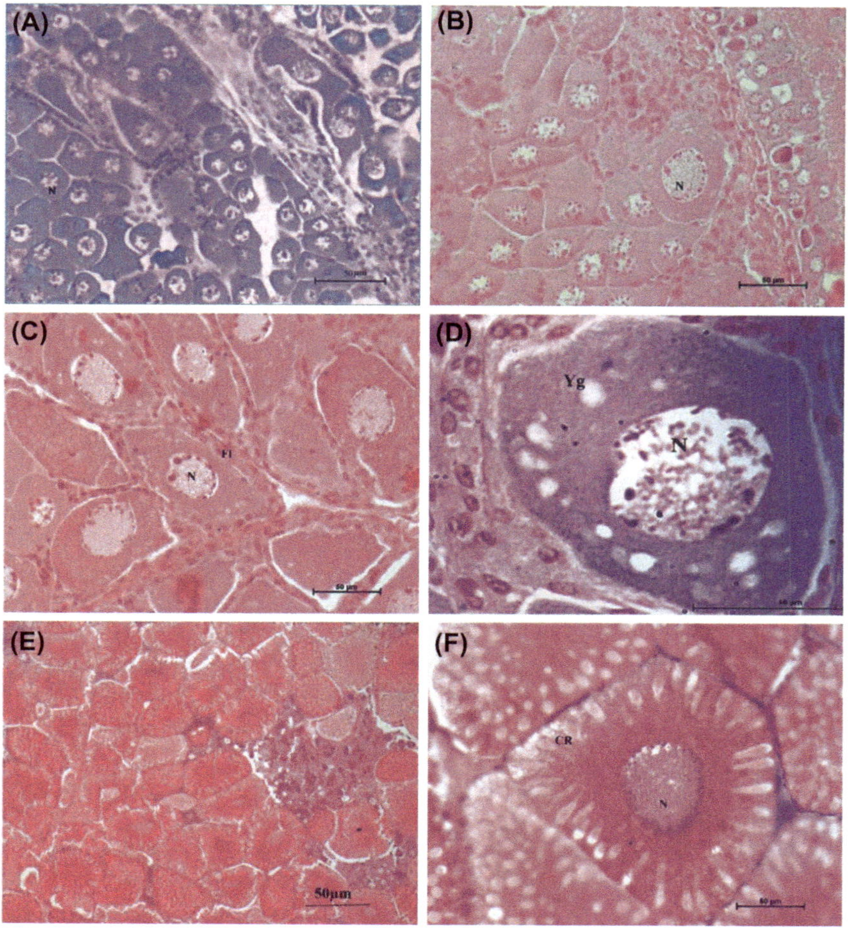

FIGURE 9.4 Histological sections of ovarian tissues of *Fenneropenaeus indicus* on day 14 of treatment and stained with hematoxylin and eosin. Ovarian tissues from (A) control (10×), (B) 5-HT group (40×), (C) ESA group in early vitellogenic stage with numerous eosinophilic yolk granules in the peripheral ooplasm, (D) ESA group with early vitellogenic oocyte (100×), (E) 5-HT+ESA group with oocytes having club-shaped cortical rods extending toward the nucleus (10×), and (F) mature oocyte with cortical rods (40×). *CR*, cortical rod; *Fl*, follicle layer; *N*, nucleus; *Yg*, yolk granules. *From Tomy, S., Saikrithi, P., James, N., Balasubramanian, C.P., Panigrahi, A., Otta, S.K., et al., 2016. Serotonin induced changes in the expression of ovarian gene network in the Indian white shrimp, Penaeus indicus. Aquaculture 452, 239–246. Fig. 1, p. 242; Elsevier, 2015.*

recent study on the Indian white shrimp, *Fenneropenaeus indicus*, Tomy et al. (2016) found differential expression of ovarian genes involved in vitellogenesis (Vg and Vg receptor (VgR)) in serotonin-treated female shrimps. Increase in gene expression was, however, higher in the eyestalk-ablated females, after receiving serotonin treatment. Fig. 9.4 shows the effect of 5-HT and the combined effect of ESA on vitellogenesis in this penaeid shrimp. It is clear from these studies that 5-HT might not act directly on the ovary or other Vg synthetic sites to induce vitellogenesis, but it enhanced the release of gonad-stimulating neuropeptides in order that the latter can stimulate the proximate gonadotropic hormonal sources to initiate vitellogenesis, especially in the absence of VIH. Since the secretory activity of the biogenic amines is correlated to VIH and VSH release, a hierarchical control of biogenic amines on the synthesis and secretion of the regulatory neuropeptides of eyestalk and brain during the ovarian maturation is postulated (Fig. 9.3).

Tinikul et al. (2008) measured the levels of 5-HT and DA in the brain and thoracic ganglia along with the hemolymph Vg levels during the ovarian cycle of the freshwater prawn *M. rosenbergii*. They found a progressive increase of 5-HT in the brain during vitellogenesis, whereas the DA concentration showed a decreasing trend. Injection of 5-HT significantly increased the Vg concentration but the injection of DA and another neurotransmitter octopamine (OA) showed an opposite effect. Interestingly, the same result has also been found on the embryonic development in the brood. Delaying of brood development is also another way of slowing down the succeeding ovarian cycle in *M. rosenbergii*. Unless the brooded embryos are released, the next molting, which is a prerequisite for the ovulation, will not occur.

Tinikul et al. (2009) also observed a gradual rise in 5-HT in the ovary from stage I to IV of *M. rosenbergii*. On the contrary, DA concentration steadily decreased in the oocytes toward maturation of the ovary. It is of interest to note here that Ongvarrasopone et al. (2006) have reported expression of a 5-HT receptor, 5-HTI, in the membrane of mature oocyte in stages III and IV of the ovarian cycle in the tiger shrimp, *P. monodon*. Evidently, 5-HT may play a critical role in regulating meiotic maturation in this shrimp.

Different types of biogenic amines are synthesized in the CNS of crustaceans. In general, these biogenic amines can be categorized into two types with reference to their effect on the release of neuropeptides involved in the control of reproduction. The first category includes 5-HT/serotonin and melatonin having proven positive effect on the release of GSH from the brain and thoracic ganglion as described above. These compounds are derived from the amino acid tryptophan (see Fig. 9.5). The second category includes mainly the DA which shows an antagonistic action against 5-HT in inducing the release of GSH. Incidentally DA is a derivative of another aromatic amino acid namely tyrosine (see Fig. 9.5).

9.5.4 Gonadotropin-Releasing Hormone

Gonadotropin-releasing hormone (GnRH) is a well-studied decapeptide, initiating hormonal induction in the brain-pituitary-gonadal axis of vertebrates (Millar, 2005). In vertebrates, GnRH has a role in stimulating the release of gonadotropins such as follicle-stimulating hormone (FSH) and luteinizing hormone (LH)

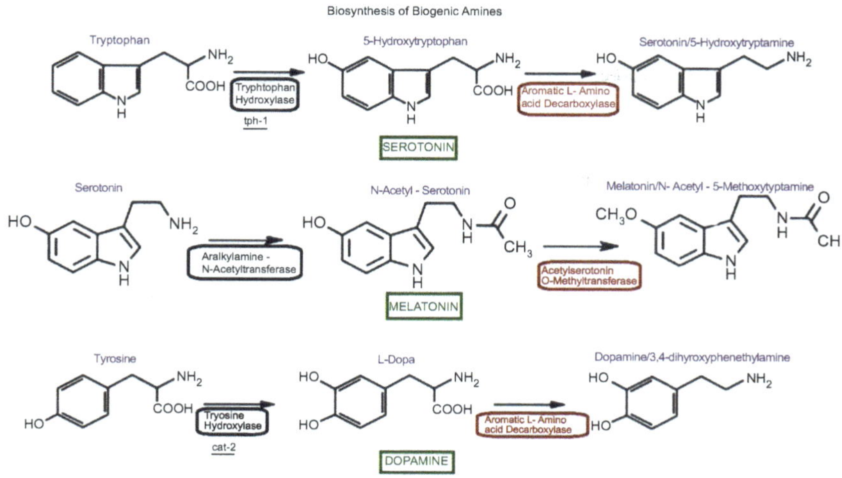

FIGURE 9.5 Biosynthesis of biogenic amines.

from the pituitary. However, several studies have reported on the occurrence of GnRH or GnRH-like peptides in diversified invertebrate phyla ranging from corals to prochordates. Although their functional role in invertebrates is generally unknown, in mollusks, and the prochordate, *Ciona intestinalis*, the presence of GnRH has been related to reproductive activities, particularly stimulating yolk synthesis (Di Cosmo and Di Cristo, 1998; Di Fiore et al., 2000).

Recent studies have revealed the immunocytochemical detection of GnRH peptides in the central nervous system as well as developing oocytes of the freshwater prawn *M. rosenbergii* and the penaeid shrimp, *P. monodon* (Ngernsoungnern et al., 2008a,b). GnRH, existing in different isoforms such as GnRH-I, lGnRH-III, and octGnRH are specifically localized in the follicular epithelial cells of proliferative, vitellogenic, and mature ovary of the penaeid shrimp, *P. monodon*, indicating a role in oogenesis. Furthermore, ir-GnRH in *P. monodon* was more related to octGnRH and lGnRH than to other forms. However, in the blue crab *Portunus pelagicus*, only the isoform, lGnRH-III, has been shown to induce ovarian maturation (Saetan et al., 2013).

While the immunoreactivity of GnRH, found in the neurons and the nerve fibers innervating neurons in brain and thoracic ganglia of both *M. rosenbergii* and *P. monodon* might suggest a stimulatory role in the release of serotonin and gonad stimulatory neuropeptides, the expression of GnRH in the ovary, especially the follicle cells, could indicate a regulatory role in the ovarian synthesis and release of the steroid hormones such as estradiol, into the hemolymph to influence Vg synthesis in the hepatopancreas. Tinikul et al. (2011) also localized two isoforms, octGnRH-ir and tunicate (t) GnRH-I-ir, both in the oocytes and follicular epithelial cells of previtellogenic and vitellogenic ovary. Furthermore, from the ovary of the crayfish, *P. clarkii*, Guan et al. (2013) isolated a novel form of GnRH (pcGnRH), which is structurally different from all known vertebrate and invertebrate GnRH peptides. The synthetic pcGnRH, when injected into the female crayfish, has a profound influence in augmenting the gonad index as well as ovarian maturation. However, the mechanism of GnRH action on the proposed gonadal control in Crustacea is still unknown.

9.5.5 Vertebrate-Type Gonadotropins

In vertebrates, the synthesis of the ovarian steroid hormones is stimulated by the gonadotropic hormones such as FSH, LH, and chorionic gonadotropin (CG) (Weghofer et al., 2007). The expression of GnRH-like peptides in both CNS and the ovary during vitellogenesis in the decapods might suggest the possible occurrence of such vertebrate-type gonadotropin in crustaceans too. It has been indicated in early studies that mammalian gonadotropic hormones have an influencing effect on crustacean oocyte maturation. For example, injection of human gonadotropins (FSH and LH) induced ovarian maturation in the sand shrimp *C. crangon* (Żukowska-Arendarczyk, 1981). Sarojini et al. (1986) found an increase in the number and size of follicle cells as well as accumulation of yolk granules in the oocytes of freshwater prawn *Macrobrachium lamerri*, after injection of mammalian FSH. Human chorionic gonadotropin (HGC) had also a similar influence on gonad maturation and spawning in penaeid shrimps, *P. semisulcatus* (Aktas and Kumlu, 2005), *M. japonicus* (Yano, 1993), and the sand shrimp *C. crangon* (Bomirski and Kelk-Kawinska, 1976). Further, Laufer and Landu (1991) found enhanced ovarian maturation in *Penaeus indicus*, fed with the food containing HGC.

While the above studies indicate a strong response of crustacean ovary to mammalian gonadotropic hormones, the natural occurrence of them has been rarely reported in crustaceans. Nevertheless, Huang et al. (2008) first reported the existence of FSH and LH-like peptides in the brain and thoracic ganglia of the swimming crab, *Portunus trituberculatus*. In addition, an FSH-like peptide was also detected in the hemolymph of *M. japonicus*, and its level increased progressively during vitellogenesis (Ye et al., 2011).

9.5.6 Bursicon

Recent studies in arthropods have furthermore revealed the occurrence of a novel heterodimeric glycopeptide, bursicon, which has structural similarities with vertebrate gonadotropins. Bursicon was first discovered in insects as a developmental hormone having control over cuticle tanning and wing expansion (Fraenkel and Hsiao, 1965; Huang et al., 2007). Subsequent discovery of bursicon in the blue crab *C. sapidus* extended its presence in crustaceans also (Chung et al., 2012). Bursicon is synthesized mainly in thoracic ganglia, playing a pivotal role in cuticle tanning during the molt cycle of crustaceans (Webster et al., 2012). However, Sathapondecha et al. (2015) found that the treatment of primary ovarian cells of the penaeid shrimp, *P. monodon* with bursicon, rbursa $\alpha\beta$, resulted in the elevation of Vg mRNA expression twofold, compared to control. Although the exact mechanism of bursicon action on Vg synthesis has not been determined, it could stimulate steroidogenesis, in much the same way the gonadotropins stimulated gonadal activity in vertebrates (Erickson and Hsueh, 1978). Identification of bursicon-like peptides in crustaceans may also fill the gap in the vertebrate-type endocrine axis in these invertebrates.

9.5.7 Prostaglandins

Prostaglandin (PG), a C_{20} polyunsaturated fatty acid derivative has wide distribution in mammalian tissues with diverse physiological functions. The occurrence of PGs in invertebrates ranges from the gorgonian corals to mollusks (Weinheimer and Spraggins, 1969; Varaskin and Reunova, 1993). In the snail *Helisoma durgi*, the occurrence of PGs has been related to reproductive functions, such as egg production (Kunigelis and Saleuddin, 1986). Among Crustacea, the reproductive role of PG has been unraveled first as a hatching factor in the barnacle *Balanus balanoides* (Holland et al., 1985). Later, Sarojini et al. (1989) indicated its role in the regulation of ovarian maturation in a penaeid shrimp *Metapenaeus affinis*. In the crayfishes *Procambarus paeninsulanus* and *C. quadricarinatus*, the level of prostaglandins such as PGE_2 and PGF_2 in the ovary increased gradually during oocyte maturation, suggesting a role in vitellogenesis (Spaziani et al., 1993; Silkovsky et al., 1998). The mechanism of action by prostaglandin E in the promotion of vitellogenesis may involve elevation of cAMP in the ovary of *M. rosenbergii* (Sagi et al., 1995). Incorporation of prostaglandins into the food pellets stimulated final stages of ovarian maturation in shrimps in vivo (Yano et al., 2000). In a freshwater field crab, *O. senex senex*, Reddy et al. (2004) found an increased uptake of arachidonic acid as well as an augmented activity of the enzyme prostaglandin H synthetase in the isolated mature ovary, indicating its biosynthetic ability of prostaglandin during maturation. Besides, injection of prostaglandin $F_{2\alpha}$ and prostaglandin E_2 significantly increased ovarian index and oocyte diameter in a dose-dependent manner in this crab. Obviously, crustacean ovary and probably other tissues produce prostaglandin endogenously, although its role in stimulating vitellogenic activity remains enigmatic.

9.5.8 Vg-Stimulating Ovarian Hormone

In the amphipod, *Orchestia gammarellus*, a different type of vitellogenic system has been described (Charniaux-Cotton, 1985). The yolk precursor, Vg, is synthesized in the fat body, instead of hepatopancreas. Experimental extirpation of ovary in the mature female results in the arrest of Vg in the fat body, suggesting that a hormonal factor from the ovary is responsible for the control of Vg synthesis in the fat body. Terming this hormone "vitellogenin-stimulating ovarian hormone," Junera et al. (1977) considered that the follicle cells surrounding the vitellogenic oocytes could be the source of this hormone. Meusy et al. (1971) demonstrated that the Vg synthesis in *O. gammarellus* is under the inhibitory control of androgenic hormone. However, removal of androgenic gland in the male amphipod did not elicit Vg appearance in the hemolymph. Nevertheless, Vg appeared in the hemolymph after the grafting of the ovary, confirming the role of an ovarian hormone in the production of Vg in the amphipods. Although the presence of an ovary is a prerequisite for Vg synthesis by the fat body, there is no evidence that the ovary could synthesize Vg by itself. The chemical nature of this ovarian hormone is still elusive, as there was no attempt to characterize this Vg inducing hormone in the amphipods. Fat body exists in peracarids, in addition to hepatopancreas, as the central organ of intermediary metabolism and a site of Vg synthesis in many decapod crustaceans. A homologous subepidermal fat body tissue is also implicated with Vg synthesis in the brachyuran crab, *Scylla serrata* (Rani and Subramoniam, 1997). However, no ovarian hormone has been suggested to be responsible for the synthesis of Vg in the fat body of this crab or any other decapod. Extirpation experiments of ovary in the decapod crustaceans are difficult by virtue of its large size and volume, preventing any attempt to prove the existence of the Vg-stimulating ovarian hormone.

9.5.9 Methyl Farnesoate

MO, first described by Le Roux in 1968, secretes a sesquiterpenoid compound, MF. MF (Fig. 9.6) is the unepoxidated form of the insect JH (JH III), and hence considered to have a gonadotropic role, like JH in the insects (Borst et al., 1987).

A functional role for MF in crustacean reproduction was first indicated by Hinsch (1980) from her observation that the active MO implants stimulated ovarian growth in the immature female spider crab *L. emarginata*. A rise in MF levels in the hemolymph as well as MO during vitellogenesis in this crab further suggested its involvement in the control of Vg synthesis (Laufer et al., 1986). Similarly, in the crab *C. pagurus*, concentrations of MF in the hemolymph varied during ovarian development, with a peak at the beginning of secondary vitellogenesis and subsequent fall to basal level thereafter (Wainwright et al., 1996). Positive effects of MF on oocyte growth were also found in the crayfish *P. clarkii*, when MF was injected alone or in combination with 17β-estradiol (Rodriguez et al., 2002). In in vitro culture of ovary with MF, the level of incorporation of labeled leucine increased significantly, suggesting that MF has a positive effect on vitellogenic activity within the ovary of this crayfish. Since the crayfish ovary is shown to have a structural machinery

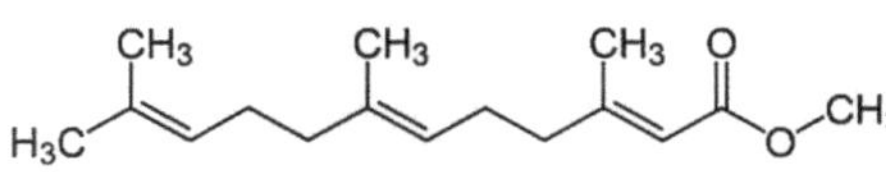

FIGURE 9.6 Methyl farnesoate.

for autosynthesis of vitellin (Kessel, 1968), MF could be considered as the hormone to influence such intraovarial yolk synthesis. Further studies also revealed that MF binds to its receptor RXR in the ovary of *C. maenas*, adducing more evidence that MF could stimulate vitellin synthesis within the ovary of this brachyuran crab (Nagaraju et al., 2011). In this crab, RXR mRNA levels increased in the hepatopancreas and the ovary in a stage-dependent manner during vitellogenesis. Possibly, MF and RXR form a complex that could stimulate Vg synthesis. Expression of RXR gene has also been reported in the vitellogenic ovary of other brachyuran crabs like *Paratelphusa hydrodromous* (Sarika and Anilkumar, 2014). MO's activity seems to be under the control of the eyestalk neuropeptide, MOIH, as evidenced by several ESA studies (Tsukimura and Borst, 1992).

The involvement of cGMP as a second messenger in the inhibitory activity of the sinus gland factor on the MF synthesis was demonstrated by Tsukimura et al. (1993). Complete characterization of the eyestalk neuropeptide inhibiting MF synthesis by MO was first made by Wainwright et al. (1996), who purified two MOIH from the HPLC separation of sinus gland in the crab, *C. pagurus*. Both the peptides have 78 amino acid residues and a molecular weight of 9235 Da with free N- and C-termini. Sequence alignment studies with other CHH family peptides revealed 50–60% similarity with MIH of *C. pagurus*. However, MOIH had only limited activity in the MIH assay. Later on, Liu and Laufer (1996) reported three MOIHs from the eyestalk of the spider crab, *L. emarginata*, one of which is structurally related to CHH.

Similar to MF, its precursor, FA (Fig. 9.7) is also found to have stimulatory action on Vg synthesis in Crustacea. In the red crab *Charybdis feriatus*, Vg gene expression increased in the hepatopancreas, when low dosage of FA was injected (Mak et al., 2005). Conversely, injection of MF together with JH III was effective in inducing Vg expression only in high concentration. Interestingly, FA administered with 20-hydroxyecdysone (20E) synergistically increased Vg gene expression in the lobster *H. americanus* (Tiu et al., 2009).

The mode of action of MF on vitellogenesis is still undetermined. In insects, JH III, the structural homologue of crustacean juvenoid, MF, stimulates Vg synthesis in the fat body and facilitates uptake of Vg into the oocytes through follicular epithelial layer by "patency," a process by which interfollicular space is created by the shrinkage of follicle cells. A rise in the Na/K-ATPase activity in the follicle cells facilitates this process, under the influence of JH (Davey et al. , 1993). In crustaceans, Soroka et al. (2000) observed that injection of MF into the vitellogenic crayfish *C. quadricarinatus* resulted in the elevation of

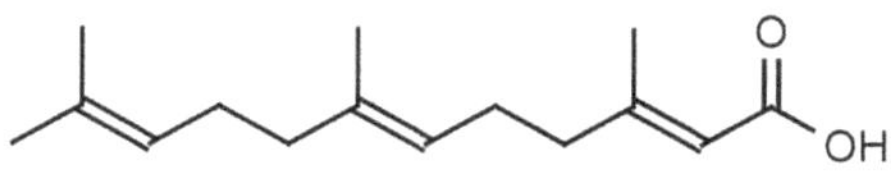

FIGURE 9.7 Farnesoic acid.

protein kinase C, an isoenzyme involved in Vg uptake by oocytes and follicle cells. Although the occurrence of JH III has never been reported in crustaceans, injection of exogenous JH III purportedly induced ovarian development in a freshwater crab *P. hydrodromous* (Sasikala and Subramoniam, 1991). Besides, a JH-responsive element has been reported in the promoter region of the Vg gene of a cladoceran, *Daphnia magna* (Tokishita et al., 2006). When JH agonists, like pyriproxyfen and fenoxycarb were injected into *Daphnia*, there was a strong repression of the Vg gene expression. More evidence is, however, needed to show whether MF or FA acts alone or in synergy with other hormonal factors in controlling Vg gene expression in crustaceans.

9.5.10 Steroidal Control of Vitellogenesis

9.5.10.1 Ecdysteroids

In all oviparous vertebrates, sex steroids are the exclusive activators of Vg gene transcription. In invertebrates, a new class of polyhydroxylated ketosteroids, the ecdysteroids, native to arthropods, serve the primary function of molting hormone (Buckmann, 1989). In crustaceans, ecdysteroids are synthesized by the ecdysial gland, called Y-organ. Y-organ secretes ecdysone, which on release into hemolymph, is converted by the peripheral tissues into active 20E (synonyms: crustecdysone, ecdysterone) (Goodwin et al., 1978). 20E, besides controlling molting, also plays a decisive role in the transcriptional activation of the Vg gene in dipteran insects and certain ticks and mites (Sappingotn et al., 2002). In a similar vein, several workers found correlation between ecdysone and ovarian activities in crustaceans. Arvy et al. (1954) found evidence that there is a rise in hemolymph ecdysteroids coincident with the initial stages of gametogenesis (oogonial and spermatogonial mitoses) of the shore crab *C. maenas*. However, in isopods and amphipods, Steel and Vafopoulou (1998) found hemolymph Vg levels paralleling ecdysteroid titers during the vitellogenic cycle, suggesting a role in Vg synthesis. Similarly, Vg synthesis occurs in the shrimp *Lysmata seticaudata* under a high titer of ecdysone (Charniaux-Cotton and Touir, 1973). A high titer of 20E (Fig. 9.8) also appears to be necessary for the Vg synthesis

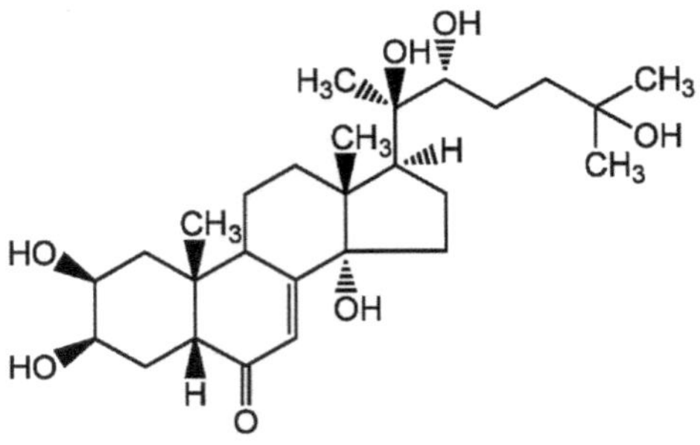

FIGURE 9.8 20-Hydroxyecdysone.

in the amphipod *O. gammarellus*, as Y-organectomy in postecdysis inhibits the onset of Vg synthesis (Blanchet-Tournier, 1982). In addition, Okumura et al. (1992) found a close correlation between hemolymph ecdysteroids titer and ovarian maturation stages in the reproductive molt cycle of the freshwater prawn, *M. rosenbergii*.

That the 20E has a control over the Vg synthesis at transcriptional level was shown in the lobster, *H. americanus* in which injection of 20E alone or in combination with FA stimulated *HaVg1* gene expression (Tiu et al., 2009). Contrastingly, noninvolvement of ecdysteroids in vitellogenesis has been shown in the shore crab, *C. maenas*, as Y-organ removal did not halt this process (Demeusy, 1962). In support of this observation, in the anecdysic oxyrhynchan crab *Acanthonyx lunulatus*, the Y-organ degenerates at the pubertal molt, but there are two more vitellogenic cycles that are completed in the absence of ecdysteroids (Chaix and Reggi, 1982). The above observations have not, however, considered the possibility of ecdysone being synthesized in alternative sites such as the ovary. In general, the levels of ecdysteroid are kept low during ovarian maturation, as its main role is to control molting; keeping a low level of ecdysone in the intermolt is indeed necessary for allowing vitellogenesis to progress, uninterruptedly by any precocious molting event. In the freshwater prawn *M. rosenbergii* as well as the penaeid shrimp *P. monodon*, both hemolymph and ovarian ecdysteroid levels declined from the immature to the late vitellogenic ovarian stages (Young et al., 1993).

Any role played by ecdysteroids in Vg synthesis is rather enigmatic, as it is primarily a molting hormone, and hence their role in reproduction could only be supplementary. Again, the pattern of programming of both growth and reproduction in any given taxonomic group could have a bearing on the timing of ecdysone action for molting and reproduction. In a continuously breeding sand crab *Emerita asiatica*, exhibiting synchronization of molting and oogenic cycles, hemolymph ecdysteroids revealed a biphasic increase during intermolt and premolt, the small peak corresponding to vitellogenic activities in the ovary, and a prominent premolt peak (Gunamalai et al., 2004). In this crab, injection of 20E enhanced protein synthesis in the ovary, hepatopancreas, and integumentary tissues, substantiating a dual role in the control of molting and reproduction. In a fast growing, highly fecund estuarine crab, *Metopograpsus messor*, Sudha and Anilkumar (2007) found a simultaneous acceleration of molting and oogenic activities, when eyestalks were ablated in the early intermolt female crabs.

Obviously, the effects of ecdysteroids (molting hormone) on vitellogenesis in Crustacea are indirect; probably playing a metabolic role in the synthesis of different yolk precursor components. In the freshwater prawn, *M. rosenbergii*, molting alternates with female reproduction and hence, their activities often overlap. With the result, elevated levels of ecdysteroids in the premolt may permit or promote increased yolk synthesis during the extended intermolt and premolt stages (Okumura et al., 1992). Clearly, synergistic programming of molt and breeding activities is found to be characteristic of palaemonid shrimps, wherein premolt growth is closely entrained with yolk deposition, suggesting that yolk protein synthesis could continue to occur under high titers of ecdysteroids (Phillipphen et al., 2000; Tsukimura, 2001).

The question of ecdysteroid control of Vg could be resolved only by molecular studies pertaining to their receptor activities. In insects, the action of ecdysteroids is similar to vertebrate steroid hormones: the transcriptional control being mediated via two nuclear receptor superfamily members, ecdysone receptors (EcR) and its heterodimeric partner, ultraspiracle (USP). Such dimerization of nuclear receptors is vital to induce transcription of Vg gene in the cyclorraphan dipteran flies such as *Aedes aegypti* (Zhu et al., 2003). In crustaceans, EcR has been identified in blastimal tissues of regenerating limbs of *U. pugilator*, but it dimerizes with retinoid X receptor, RXR (Chung et al., 1998). Furthermore, Durica et al. (2002) found coexpression of these two receptors (UpEcR and UpRXR) in the ovary of *U. pugilator* during ovarian cycle, suggesting that the ovary is a potential target for ecdysteroid action. It is not clear whether this receptor activity is related to the ovarian synthesis of yolk. Ecdysteroid-responsive genes such as HHR3 have also been identified in the epidermis and eyestalk tissues of the lobster, *H. americanus* (Haj et al., 1997). In a recent study, Girish et al. (2015) found increased expression of RXR and EcR in the hepatopancreas of the freshwater crab *O. senex senex*, followed by upregulation of ecdysone-responsive gene E75 during vitellogenic stages I and II. Interestingly, Tokishita et al. (2006) reported on the occurrence of an ecdysone-responsive element in the upstream of the Vg gene in a primitive cladoceran, *D. magna*. In addition, other binding sites for E74, E75, and those for GATA factors have also been identified in *D. magna* genome, suggesting the possible transcriptional activation of Vg gene by ecdysteroids, as in dipteran insects (Kokoza et al., 2001).

9.5.10.2 Vertebrate Steroids

Many crustacean species have been shown to possess steroidogenic ability to produce a variety of vertebrate-type steroids (Table 9.3). In Crustacea, the occurrence of estrogenic compounds was first recorded in the berried eggs of the American lobster *H. americanus* and their activity was thought to inhibit precocious molting of the female that carried the berry (Donahue, 1940, 1955). Later work on *H. americanus* revealed the occurrence of 17β-estradiol (Fig. 9.9) and progesterone (Fig. 9.10) in the MO, green gland, kidney, hepatopancreas, ovary and hemolymph, with MO showing the highest concentration (Couch et al., 1987). These authors found a parallelism between the levels of estradiol in the hemolymph and the maturity stages of the ovary in the lobster. While estradiol was found to be present only in females with mature ovary, the level of progesterone in the MO was unrelated to the state of the ovary and was even found in the immature females. In *Nephrops norvegicus*, the eggs and hemolymph contained unconjugated 17β-estradiol, while the ovary and hemolymph

TABLE 9.3 Occurrence of Vertebrate-Type Steroids in Crustacea

Steroids	Species	Detected sites	References
Pregnenolone	*Penaeus japonicus*	Ovary	Cardoso et al. (1997)
	Penaeus monodon	Ovary	Fairs et al. (1990)
	Astacus leptodactylus	Hemolymph	Ollevier et al. (1986)
	Carcinus maenas	Ovary	Hazel (1986)
Progesterone	*Penaeus japonicus*	Ovary	Cardoso et al. (1997)
	Penaeus monodon	Ovary, hepatopancreas, and hemolymph	Fairs et al. (1990) and Quinitio et al. (1994)
	Euphausia superba	Whole-body extract	Nikitima et al. (1977)
	Homarus americanus	Hemolymph, ovary, mandibular organ, green gland, hepatopancreas, testis, and serum	Couch et al. (1987) and Burns et al. (1984)
	Artemia sp.	Whole-body extract	Van Beek and De Loof (1988)
	Pandalus kessleri	Hemolymph	Quinitio et al. (1991)
	Emerita asiatica	Hemolymph, ovary, and hepatopancreas	Gunamalai et al. (2006)
	Macrobrachium rosenbergii	Hemolymph, ovary, and hepatopancreas	Gunamalai et al. (2006)
	Marsupenaeus japonicus	Hemolymph and ovary	Okumura and Sakiyama (2004)
	Metapenaeus ensis	Hepatopancreas and ovary	Tiu et al. (2006)
	Cherax albidus	Hepatopancreas	Coccia et al. (2010)
	Eriocheir sinensis	Ovary	Wu et al. (2014)
	Astacus leptodactylus	Hemolymph	Mirheydari et al. (2014)

Continued

TABLE 9.3 Occurrence of Vertebrate-Type Steroids in Crustacea—cont'd

Steroids	Species	Detected sites	References
17αHydroxyprogesterone	*Penaeus japonicus*	Ovary	Cardoso et al. (1997)
	Macrobrachium rosenbergii	Hemolymph and ovary	Martins et al. (2007)
	Oziotelphusa senex senex	Ovary	Reddy et al. (2006)
	Astacus leptodactylus	Hemolymph, ovary, and hepatopancreas	Malati et al. (2013)
17α-Hydroxypregnenolone	*Astacus leptodactylus*	Hemolymph	Ollevier et al., 1986
6β-Hydroxyprogesterone	*Astacus leptodactylus*	Hemolymph	Ollevier et al., 1986
20α-Hydroxyprogesterone	*Nephrops norvegicus*	Ovary	Fairs et al. (1989)
17β-Estradiol	*Scylla serrata*	Ovary (lipovitellin), hepatopancreas, and hemolymph (Vg)	Warrier et al. (2001)
	Penaeus monodon	Ovary, hepatopancreas, and hemolymph	Fairs et al. (1990) and Yano et al. (2000)
	Homarus americanus	Mandibular organ, green gland, hepatopancreas, ovary, and hemolymph	Couch et al. (1987), Donahue (1948, 1952, 1957) and Lisk (1961)
	Parapenaeus fissures	Ovary	Jeng et al. (1978)
	Pandalus kessleri	Hemolymph	Quinitio et al. (1991)
	Macrobrachium rosenbergii	Ovary, hemolymph, and hepatopancreas	Ghosh and Ray (1993), Gunamalai et al. (2006) and Martins et al. (2007)
	Emerita asiatica	Ovary, hemolymph, and hepatopancreas	Gunamalai et al. (2006)
	Marsupenaeus japonicus	Hemolymph and ovary	Okumura and Sakiyama (2004)
	Metapenaeus ensis	Hepatopancreas and ovary	Tiu et al. (2006)

	Cherax albidus	Hepatopancreas	Coccia et al. (2010)
	Astacus leptodactylus	Hemolymph, ovary, and hepatopancreas	Malati et al. (2013) and Mirheydari et al. (2014)
Estrone	*Penaeus monodon*	Ovary	Fairs et al. (1990)
	Parapenaeus fissures	Ovary	Jeng et al. (1978)
Testosterone	*Scylla serrata*	Ovary (lipovitellin), hepatopancreas, and hemolymph (Vg)	Warrier et al. (2001)
	Penaeus japonicus	Ovary	Cardoso et al. (1997)
	Penaeus monodon	Ovary	Fairs et al. (1990)
	Euphausia superba	Whole-body extract	Nikitima et al. (1977)
	Homarus americanus	Serum and testis	Burns et al. (1984)
	Nephrops norvegicus	Ovary	Fairs et al. (1989)
	Carcinus maenas	Ovary	Hazel (1986)
	Astacus leptodactylus	Hemolymph	Ollevier et al. (1986)
	Artemia sp.	Ovary	Van Beek and De Loof (1988)
	Marsupenaeus japonicus	Hemolymph and ovary	Okumura and Sakiyama (2004)
	Macrobrachium rosenbergii	Hemolymph and ovary	Martins et al. (2007)
	Astacus leptodactylus	Hemolymph	Mirheydari et al. (2014)
Dehydroepi-androsterone	*Penaeus monodon*	Ovary	Fairs et al. (1990)
5α-Dihyrotestosterone	*Artemia* sp.	Ovary	Van Beek and De Loof (1988)
Estriol	*Marsupenaeus japonicus*	Hemolymph and ovary	Okumura and Sakiyama (2004)
11-Ketotestosterone	*Marsupenaeus japonicus*	Hemolymph and ovary	Okumura and Sakiyama (2004)

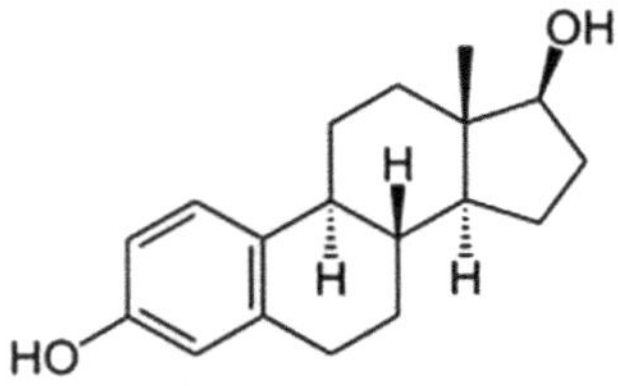

FIGURE 9.9 17β-Estradiol.

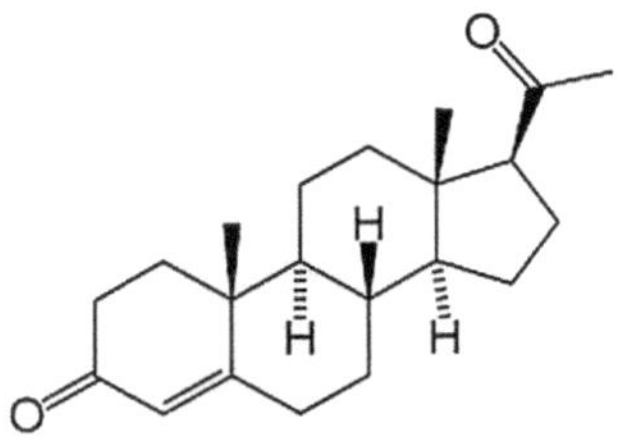

FIGURE 9.10 Progesterone.

contained 5α-dihydrotestosterone. Similarly, in the ovary of *P. monodon*, and the crab *S. serrata*, both conjugated and nonconjugated steroidal compounds have been found to accumulate during the progression of vitellogenesis (Fairs et al., 1990; Warrier et al., 2001).

These studies show that steroidal compounds could either be synthesized or stored within the ovary, but evidence on their role in the control of ovarian maturation has been provided in other studies. For instance, in the deep sea prawn *Pandalus kessleri*, Quinitio et al. (1991) found a correlative fluctuation of hemolymph estrogen and progesterone levels in relation to ovarian development, implying a role in the control of vitellogenesis. Progesterone (P) levels increased at the onset of vitellogenesis and decreased during vitellogenesis, whereas estradiol (E_2) concentrations rose during the peak of vitellogenesis and dropped after the release of mature eggs from the ovary. Furthermore, in the penaeid shrimp, *P. monodon*, 17β-estradiol and P levels in the hemolymph, hepatopancreas, and ovary were found to fluctuate closely with serum Vg levels during ovarian maturation (Quinitio et al., 1994). Similarly, in the mud crab *S. serrata*, Warrier et al. (2001) found that the levels of both E2 and P fluctuated in the hepatopancreas, hemolymph, and ovary during ovarian development, suggesting a correlation between hormone profiles and stages of vitellogenesis. These authors also reported the presence of E2 and P in the lipovitellin and Vg molecules, implying a role for the Vg as steroid carriers. While both E2 and P could be involved in the transcriptional control of Vg gene expression in both hepatopancreas and ovary, progesterone might have additional functions in the meiotic maturation of postvitellogenic oocytes. Correlative fluctuations of these hormones during the ovarian cycle have been observed in other decapods such as the spiny lobster *P. homarus* (Kirubagaran et al., 2005), the anomuran crab *E. asiatica* and the giant freshwater prawn *M. rosenbergii* (Gunamalai et al., 2006). In *M. rosenbergii*, the hormonal fluctuations have been found only in the reproductive molt cycle, whereas during the nonreproductive common molt, characterized by the presence of inactive ovary, the level of estradiol and progesterone in the hemolymph was not detectable at any molt stage. The inactive ovary and the hepatopancreas showed only basal levels of estrogen during the nonreproductive molt, while the progesterone levels were undetectable. Taken together, these studies may suggest that estradiol potentially plays a role in crustacean vitellogenesis either upregulating Vg synthesis as in vertebrates, and/or stimulating related metabolic pathways initiated during vitellogenesis, such as lipogenesis.

Since these hormones appear only during the ovarian cycle, it could mean that their secretion is under the eyestalk inhibitory hormones. In the penaeid shrimp, *P. monodon*, a recent study has indicated that there is stage-specific differences in the appearance of estrogen and progesterone in the hemolymph; the estrogen levels increased significantly in the immature and vitellogenic stages, whereas progesterone levels increased only in late vitellogenic and ripe stages (Merlin et al., 2015). Furthermore, injection of extraneous hormones resulted in the stage-specific increase of Vg mRNA transcripts in the ovary. The injection of progesterone led to increased Vg mRNA in the ovary only in the vitellogenic ovary and not in immature stage. However, both immature and mature females responded to estrogen injection by elevating the Vg mRNA expression in the ovary. These results suggest that estrogen has a role in the stimulation of Vg synthesis, whereas progesterone may accelerate the process of vitellogenesis in this shrimp. More interestingly, ESA increased the hormone levels in the hemolymph significantly, suggesting that the synthesis and the release of the sex hormones are under the inhibitory control of eyestalk hormone, VIH (Merlin et al., 2015).

Further evidence for the influence of vertebrate steroids on crustacean vitellogenesis has been adduced from the injection of exogenous vertebrate sex steroids. In vitro culture of previtellogenic ovary of immature kuruma prawn, *M. japonicus* with 17β-estradiol resulted in the inducement of Vg synthesis into the medium, as well as the appearance of primary vitellogenic oocytes (Yano and Hoshino, 2006). In the freshwater crayfish, *Cherax albidus*, the in vivo treatment with 17β-estradiol and progesterone, alone or in combination, brought about an increase in Vg mRNA in the early vitellogenic females and, although to a lesser extent, in full vitellogenic females. However, in this crayfish, 17β-estradiol seems to be more effective than progesterone on Vg mRNA synthesis in the hepatopancreas (Coccia et al., 2010). They did not show any effect on the nonvitellogenic females, suggesting that these hormones will act only under proper hormonal conditions already set for vitellogenesis. A similar result was also reported in the ridgeback shrimp *Sicyonia ingentis* by Tsukimura (2001) who did not find increased levels of Vg in the sexually quiescent females treated with progesterone, 17α-hydroxyprogesterone and 17β-estradiol. Tsukimura reasoned that the GIHs present in high titer during ovarian inactive state could override the stimulatory activities of the vertebrate sex steroids. These observations indicate the strong antagonistic nature of VIH and the vertebrate steroids in impacting their control over vitellogenesis.

Further work of Yano and Hoshino (2006) on the kuruma prawn *M. japonicus*, suggested that 17β-estradiol could be the actual hormone that stimulated Vg synthesis, using progesterone as the precursor.

In addition, Tiu et al. (2006) showed that 17β-estradiol and progesterone stimulated Vg gene expression, although the former is more effective in increasing the mRNA transcription of Vg in *M. ensis*. In support, Wu and Chu (2008) indicated a strong correlation between the estrogen hormones and the heat shock protein Hsp90 expression in *M. ensis*, a condition similar to vertebrates.

Both these steroids were also shown to increase Vg mRNA expression in ovary and hepatopancreas of the freshwater crayfish *C. albidus* (Coccia et al., 2010). Hormone-injected *C. albidus* females also showed significant changes in the fatty acid profiles in the hepatopancreas, suggesting their specific influence on lipogenesis related to Vg synthesis. Lu et al. (2010), measuring the levels of progesterone in the hepatopancreas, ovary, and hemolymph in different stages of ovarian development in the Chinese mitten crab, *Eriocheir sinensis*, found a correlative fluctuation to suggest a stimulatory role in vitellogenesis.

In vertebrate reproduction, binding of sex steroids with their respective nuclear receptors is a prerequisite for them to regulate gene expression. In recent years, localization of nuclear receptors for both progesterone and estradiol has been reported in several decapod crustaceans. In the freshwater crayfish *Austropotamobius pallipes*, using immunocytochemistry and Western blotting, Paolucci et al. (2002) showed the presence of progesterone receptor (PR) in the ovary and hepatopancreas, and estrogen receptor (ER) solely in the hepatopancreas. Understandably, the absence of ER in the ovary, but the presence of estrogen in it may suggest that this steroid may be synthesized in the ovary and then released into the hemolymph to finally reach hepatopancreas, where the estrogen and progesterone may bind to their respective nuclear receptors to control Vg synthesis. In an earlier study, Ghosh and Ray (1993) found the induction of two lipogenic enzymes, the cytosolic NADP-linked malate dehydrogenase and the glucose-6-phosphate dehydrogenase in the hepatopancreas of *M. rosenbergii*, when injected with estrogen. This finding has relevance to the fact that hepatopancreas produces enormous amounts of lipids, necessary for binding to the Vg synthesized there.

In *P. monodon*, a progestin receptor (*Pgmrc1*) was localized in the membrane of the oocytes and the surrounding follicle cells of the developing ovary (Preechaphol et al., 2010). In the ovaries of eyestalk-intact broodstock females, the *Pgmrc1* was upregulated only in stage IV (mature) ovaries, but on ESA, upregulation of this receptor was noticed in the stage II (vitellogenic) ovaries. This result indicates a negative controlling effect of eyestalk factors on the expression of the PR and hence progesterone action. Furthermore, EST analysis of the cDNA library established from vitellogenic ovary revealed the expression of a PR-related protein P23 (Pm-p23) during vitellogenesis in *P. monodon* (Preechaphol et al., 2010). In this shrimp, in situ hybridization also indicated that Pm-p23 is localized in the ooplasm of previtellogenic oocytes. Immunolocalization of both ER and PR in the maturing ovary of *P. monodon* has also been reported recently (Merlin et al., 2015), indicating control of Vg gene transcription in the ovary of this penaeid shrimp. As the *Pm-p23* gene products peaked during vitellogenesis and decreased thereafter, progesterone is suggested to play the main function during vitellogenesis rather than final maturation of *P. monodon* oocytes. Furthermore, Wu et al. (2014) found the immunolocalization of the PR in the follicle cells of the ovary, hepatopancreas, optic ganglia, brain, and thoracic ganglia during ovarian maturation. In the oocytes, positive PR was found in the cytoplasm only during early ovarian developmental stages (I–III), whereas positivity was noted in the nucleus from stage III to V. However, in the hepatopancreas, PR was localized both in the cytoplasm and nucleus of follicular cells in all stages of ovarian development. The expression of PR in the follicle cells surrounding the oocytes may suggest a role for progesterone in the follicle cell-mediated Vg transport from hemolymph. The high titer of progesterone in the hepatopancreas during early stages of vitellogenesis, together with expression of PR, may provide evidence for the control of Vg synthesis in the hepatopancreas. These authors also suggest that hepatopancreas could be the site of synthesis of progesterone.

Warrier et al. (2001) also found correlative evidence that progesterone could be synthesized in the hepatopancreas and then transported through the hemolymph by binding to the Vg, which will be ultimately sequestered by the oocytes. The expression of progesterone in the neuronal tissues of optic ganglia and the brain and thoracic ganglia of *E. sinensis* and *Scylla paranamosain* could imply a mediating role on the neuroendocrine regulation of ovarian development.

By and large, recent investigations on the hormonal control of vitellogenesis have indicated the occurrence and the correlative fluctuations of both estrogen and progesterone during ovarian maturation in many decapod crustaceans. The expression pattern of their receptors in the Vg synthetic sites such as hepatopancreas and the ovary during vitellogenic processes adduces further evidence on their possible role in the stimulation of Vg synthesis at the transcriptional level. However, direct evidence on the binding of hormone-receptor to the respective response elements on Vg gene in crustaceans is still lacking.

The presence of different vertebrate gonadotropin-like glycopeptides in crustaceans, together with the steroidogenic ability of the ovary lends support to the contention that a steroidal control of vitellogenesis would be operative in the crustaceans too. Pooling information on multifarious molecules purported to be connected with the control of oocyte maturation in crustaceans, a possible networking of interactive hormonal effectors to control the brain–ovary axis, concerned with the regulation of the all important vitellogenic activities in crustaceans is proposed in comparison with the well-studied system in vertebrates (Fig. 9.11). Working of

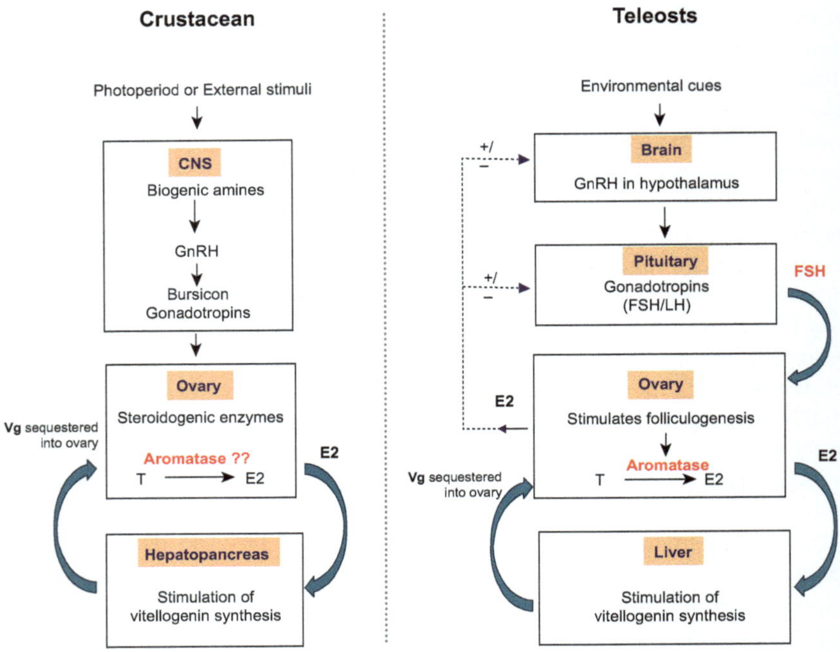

FIGURE 9.11 Schematic graphical diagram of the pathway regulating vitellogenesis in crustaceans and teleosts. *CNS*, central nervous system; *E2*, 17β-Estradiol; *FSH*, follicle-stimulating hormone; *GnRH*, gonadotropin-releasing hormone; *LH*, luteinizing hormone; *T*, testosterone; *Vg*, vitellogenin; ±, feedback.

a linear hormone signaling cascade, starting with the perception of environmental cues by the biogenic amines, ending with the ovary to stimulate steroidogenesis of progesterone and estrogen is possible. These steroids then could control Vg synthesis and associated lipogenesis in the hepatopancreas and ovary of crustaceans. Incidentally, VIH does not seem to interlude in this hormonal signaling pathway, with its direct inhibitory action through second messenger system on the transcriptional activation of Vg synthesis. Among the various vitellogenesis-stimulating hormones, only the synthesis of estrogen and progesterone in the ovary seems to be under the inhibitory control of eyestalk hormones, as revealed by ESA studies. Furthermore, all other hormones implicated with a control over vitellogenesis have only a secondary role, their primary role being concerned with the control of molting. In many fast growing commercially important shrimps and prawns, molting and female reproductive events are overlapping and sometimes, synergistic, and hence, their influence could be incidental and conducive to support vitellogenesis. To conclude, evidence in favor of vertebrate steroids controlling vitellogenesis overweighs the role of other hormones in crustaceans. Nevertheless, a multihormonal influence, in which the vertebrate steroids play a pivotal role in vitellogenesis, could not be overruled.

9.6 FUTURE PERSPECTIVES ON APPLICATION OF ENDOCRINE RESEARCH TO CRUSTACEAN AQUACULTURE

With the discovery that ESA induced ovarian maturation in the grass shrimp, the unique existence of a gonad inhibitory principle in the eyestalk of decapod crustaceans was proposed (Panouse, 1943). The occurrence of a VIH or GIH has been validated in diversified decapod species, in which, depending on the season and physiological status, eyestalk removal results in accelerated secondary vitellogenesis due to increased synthesis and ovarian uptake of Vg. Since then, crustacean endocrinology has made rapid progress, thanks to introduction of modern microanalytical procedures in identifying a variety of molecules with hormonal functions.

While VIH has been found to exist in the XOSG complex, a corresponding gonad stimulatory neurosecretory center has been proposed in the brain and thoracic ganglia to secrete the GSH in several decapods such as the crabs, lobsters, and shrimps. Such a neurosecretory control of reproduction in crustaceans led to the concept that the GIH/VIH of XOSG acts antagonistically against the GSH of brain/thoracic ganglia. Continued growth and reproduction is a characteristic feature of many crustacean species; hence, these two energy demanding physiological processes are believed to be under a bihormonal control mechanism. The discovery of yet another inhibitory hormone (MIH) originating from the XOSG complex and having a negative control over the ecdysteroid synthesis by the Y-organ adduces further evidence that there could be interrelationship and interplay between the inhibitory neuropeptides, in coordinating the molting and reproductive processes (Adiyodi and Adiyodi, 1970). Although the antagonistic action of GIH and GSH is yet be verified, the concept persisted for quite some time, until other GSHs from nonneural sources were discovered. The concept of reproductive endocrinology has also undergone fascinating changes due to discovery of many other factors with putative role in the control of reproduction. These include several of the GSHs such as biogenic amines and opioid peptides, originating from the brain and thoracic ganglia, MF, secreted from the MO, ecdysteroids from Y-organ, vertebrate-type steroids such as 17β-estradiol and progesterone from the ovary, in addition to other ovarian hormones (vitellogenesis-stimulating ovarian hormone, VSOH), identified in the amphipods. Obviously, these multifactorial regulatory molecules may have different roles in the synthesis of Vg and other organic substances like lipids, associated with the formation of the complex yolk proteins. Since the effects of these hormonal factors were described in different crustacean species, generalization on the gonadotropic role of any particular hormone for the entire crustacean taxa is untenable. With the advent of molecular techniques, peptide sequencing of the eyestalk hormones as well as the gene expression studies enabled several insights into the Vg gene action under specific hormonal control. An important outcome of the recent research is that VIH has strong influencing effect on all other stimulatory hormonal factors and hence, all the hormone-induction studies with stimulatory hormones require ESA to achieve complete stimulation.

An attempt has been made to integrate the different hormonal factors involved in the control of ovarian maturation in a typical crustacean (Fig. 9.12). This diversity in the hormonal control of female reproduction has a bearing on the complex vitellogenic system that exists among different crustaceans. As seen from the figure, there is a hierarchical nature of hormone action on the target organs of Vg synthesis and uptake. Environmental signaling is perceived through the neurons of the central nervous system by neurotransmitters and passed on to the next level of neurohormones secreted by the XOSG and the brain/thoracic ganglia. The neurohormones in crustaceans play a dual role, one inhibitory (VIH) and the other stimulatory (GSH). Although several molecular studies have delineated the role of VIH on Vg synthesis, the interplay between different eyestalk neuropeptides such as VIH, MIH, and CHH as well as with the gonad stimulatory peptides of the brain has not been clearly understood. Yet again, recent evidence brings forth a clear existence and functioning of different proximate endocrine glandular secretions on the control of Vg synthesis. Among them, vertebrate steroids, estrogen and progesterone, have been known to have direct influencing effect on Vg synthesis in a variety of decapods species. Expression analysis of their receptors in the yolk-synthesizing organs such as ovary and hepatopancreas, correlated to vitellogenic stages, corroborates further evidence on the control of Vg synthesis. A setback expressed in the ESA technique, commonly adopted in the hatchery operations, is the poor quality of

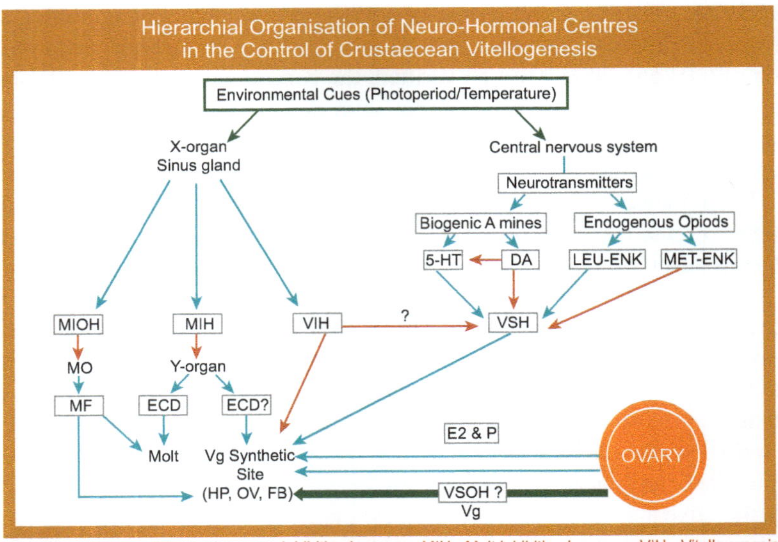

Abbreviation: MIOH - Mandibular organ inhibiting hormone; MIH - Molt inhibiting hormone; VIH - Vitellogenesis inhibiting hormone; MO - Mandibular organ; MF - Methyl farnesoate; ECD - Ecdysteroids; VSH - Vitellogenesis simulating hormone; E2 - Estradiol - 17 B; P - Progestrone; VSOH - Vitellogenesis simulating ovarian hormone; Ovary as endocrine gland; Vg - Vitellogenin; DA - Dopamine LEU - ENK - Leucine Enkephalin; MET - ENK - Methionine Enkephalin.

FIGURE 9.12 Hierarchical organization of neurohormonal factor in the control of vitellogenesis.

seeds. Lipids stored in the eggs are considered to be a determinant of the quality of seeds produced in the shrimp hatcheries. In this regard, the role of vertebrate steroids such as estrogen in inducing lipogenesis during vitellogenesis is of great importance. Continued work is required to recognize the endocrine targets, organized hierarchically, in the control of vitellogenesis of the commercially significant crustacean species.

Chapter 10

Yolk Utilization and Embryonic Nutrition

10.1 INTRODUCTION

Crustaceans show a great diversity of developmental modes, but majority of them incubate their yolky eggs in the pleopodal broods. The quantity and distribution of yolk in the crustacean eggs have a relationship with the embryonic developmental patterns (Anderson, 2013). In most branchiopods and penaeid shrimps, producing oligolecithal eggs, the cleavage is holoblastic, whereas in most of the decapods, producing yolk-laden centrolecithal eggs, the cleavage is superficial. Accordingly, the oligolecithal eggs release the most primitive nauplius larva, and the centrolecithal eggs delay the embryonic development to release the highly advanced zoea or megalopa larva (Scholtz, 2000).

The crustacean yolk protein, accumulated during vitellogenesis, eventually forms the primary source of nutrients for the developing embryo. In most decapods such as crabs, embryogenesis is epimorphic in that the naupliar stage is superseded by a period of extensive embryonic morphogenesis, widely known as a "metanauplius" stage. Nevertheless, the developmental phenotype exhibited at hatching varies widely among crustacean species (Goudeau and Lachaise, 1980). Such abbreviated larval development accompanied by embryonic molting within the egg is evidenced by the occurrence of exuvial layers, subjacent to the hatching envelope, at the time of hatching. For example, in the crab, *Carcinus maenas*, progression from the nauplius to metanauplius is recognized by an additional egg envelope. The postmetanaupliar embryonic development is associated with the formation of a loosely fitting cuticle, which is rapidly shed just after hatching into the prezoea stage. This is followed by another larval molt which sets free the free swimming zoea into the water (Goudeau and Becker, 1982). To meet this elaborate and complex embryonic development, crustaceans have to accumulate both nutritive and morphogenesis-regulative hormonal molecules at the time of vitellogenesis.

10.2 EMBRYONIC DEVELOPMENT

Several studies have been carried out on the embryonic development of crustaceans, using developmental features such as yolk clearance, eyespot formation, heartbeat,

Sexual Biology and Reproduction in Crustaceans. http://dx.doi.org/10.1016/B978-0-12-809337-5.00010-1

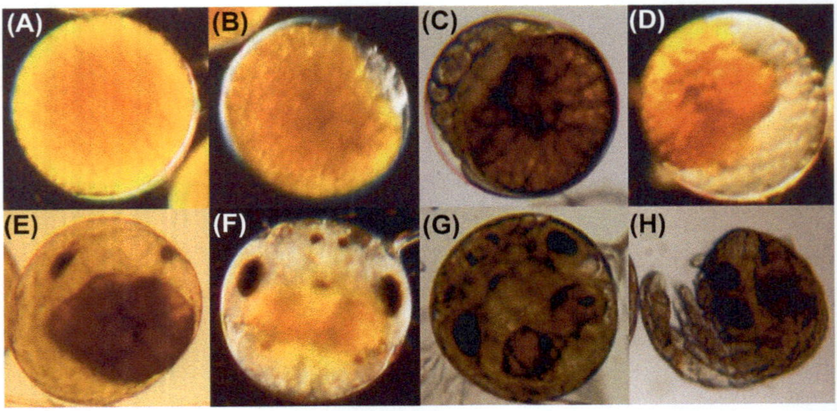

FIGURE 10.1 Embryonic stages of *Emerita asiatica*. (A) Stage I; (B) stage III; (C) stage IV; (D) stage V; (E) stage VI; (F) stage VII; (G) stage VIII; (H) stage X (for description see Table 10.1).

and appendage formation (Müller et al., 2004; Browne et al., 2005; Tracey et al., 2013; Klann and Scholtz, 2014). In the snapping shrimp, *Alpheus angulogus*, Tracey et al. (2013) correlated the rate of yolk clearance with different stages of embryonic development. In *Homarus americanus*, Perkins (1972) used the eye index (defined as half the sum of width and length of the eye of the embryo) to correlate the rate of development in lobster embryos. Still others used the distinct color change of the eggs during embryonic development as the criteria for classifying the developmental stages of the caridean shrimp, *Stenopus hispidus* (Wang et al., 2015). Combining all these features, Subramoniam (1979a, 1991a) classified egg development into 10 stages in the mole crab *Emerita asiatica* (Fig. 10.1 and Table 10.1).

10.3 BIOCHEMICAL COMPOSITION AND ANALYSIS

Crustacean vitellin, comprising the major yolk ingredient of the egg, is a high-density lipoprotein, called lipovitellin, often conjugated to a variety of carotenoid pigments (Wallace et al., 1967). Crustacean lipovitellin differs from that of vertebrates in that it lacks protein phosphate but has a higher content of lipids and carbohydrates. The yolk protein in addition includes important inorganic ions, to serve as micronutrients during embryogenesis. Furthermore, free lipid globules and glycogen droplets are stored in the ooplasm during vitellogenesis to increase the efficiency of embryonic nutrition. Several studies have uncovered the presence of steroidal hormones, present both in free and conjugated form. The stage-specific release of a host of hydrolytic enzymes splits up the complex yolk molecules, which serve as raw organic substrates for tissue buildup and function as regulatory factors during embryogenesis. For example, the regulated release of active ecdysteroids from their conjugates at a specific time during embryogenesis could trigger embryonic cuticle formation and larval molting and egg hatching.

TABLE 10.1 **Developmental Stage of Embryogenesis in *Emerita asiatica* (Classification According to Subramoniam, 1991a)**

Stage	Duration (Days) at 28°C	Description
I	5	Freshly laid egg with dense yolk granules: egg mass bright orange in color
II	5	Cleavage has taken place and blastomeres are seen: egg mass bright orange in color
III	1	A yolk-free white streak (marked as arrow) makes its appearance at the animal pole
IV	1	One-quarter of the yolk cleared; embryo in the late gastrulation stage
V	2	One-third of the yolk is cleared: beginning of organogenesis
VI	2	Two-thirds of the yolk is cleared: yolk is found in the vegetal pole: eyes well developed
VII	1	Yolk is found as a single cluster in center: appendages in developing stage
VIII	1	Yolk is found as two clusters in the centre: well-developed appendages and eyespots: heart beat seen
IX	1	Embryo fully formed: egg mass white in color: no yolk globule seen: colorless yolk in the form of oil globules seen just below the eyes: about to hatch
X	1	Hatched out larvae with little yolk

Number of days calculated based on the direct observation in the laboratory.
Reproduced from Subramoniam, T., Gunamalai, V., 2003. Breeding biology of the Intertidal sand crab, *Emerita* (Decapoda: Anomura). Adv. Mar. Biol. 46, 91–182. Table 15. p. 148; Academic press, 2003.

Biochemical analysis of the egg in different crustaceans has revealed the predominance of lipids and proteins with meager quantities of carbohydrates. Table 10.2 gives the results of the changes in the substrate utilization during embryogenesis of different crustaceans. Variations in the preferential utilization of lipids and proteins are discernible among different crustacean species, as related to the environmental conditions in which they live and reproduce. Particularly, changes in the enormously accumulated lipids could suggest resynthesis of new lipid classes to serve in the physiological control of embryogenesis, besides forming the chief fuel for development.

TABLE 10.2 Biochemical Composition During Embryonic Development in Crustaceans

Embryo Stages	Protein (%)	Lipid (%)	Carbohydrate (%)	References
Emerita asiatica				
Stage I	13.2	18.9	1.2	Subramoniam (1991a)
Stage IX	1.6	4.9	7.2	
Homarus gammarus				
Stage I	47.65	44.12	5.29	Pandian (1970)
Stage V	51.08	25.18	4.32	
Paratelphusa hydrodromous				
Stage I	29.15	61.66	2.265	Pillai and Subramoniam (1985)
Stage IV	40.71	40.07	0.859	
Macrobrachium idella idella				
Stage I	69.32	15.86	3.24	Soundarapandian et al. (2013)
Stage V	52.42	4.82	4.16	
Portunus pelagicus				
Stage I	44.22	10.24	0.74	Muthuvelu et al. (2013)
Stage V	37.61	12.5	0.58	
Thalamita crenata				
Stage I	73.18	20.76	4.00	Kannupandi et al. (1999)
Stage VI	43.18	37.45	5.03	
Chionoecetes opilio				
Stage I	68.88	30.45	3.66	Reppond et al. (2008)
Stage V	69.11	13.81	10.28	
Cherax quadricarinatus				
Stage I	58.97	37.10	3.70	García-Guerrero et al. (2003)
Stage XVII	67.11	27.10	5.87	

In a typical crustacean species, *Macrobrachium borellii*, the freshly laid eggs contain 29.3% lipids (on dry weight basis), followed by proteins (28%), and carbohydrates (0.2%) (Heras et al., 2000). Lipids and proteins are catabolized in equal proportion for energy production until stage VII. From this stage on, a greater proportion of energy appears to have been derived from the catabolism of lipids, while proteins are retained for converting into body components. On the other hand, carbohydrate content was low throughout the development but increased sharply toward the end of embryogenesis.

10.4 LIPID UTILIZATION IN EMBRYOS AND LARVAE

In the freshwater shrimp, *M. borellii*, Heras et al. (2000) studied the lipid composition and its quantitative fluctuation in the vitellus during embryo development. Phospholipids and triacylglycerol (TAG) were the most important lipids in the vitellus, representing more than 89% w/w at stage I of embryogenesis. TAG level showed a slight decline from stages I to IV (55.40–48.99 mg lipid/g egg wet weight); thereupon a sharp decline was noticed in the subsequent stage. This decrease in TAG amounted to 33% of the total lipids in the embryo. However, in the prehatching stage, the level increased to 37.38 mg lipid/g egg wet weight. The phospholipid, phosphatidylcholine changed its concentration from 14% to 25% by weight without any defined pattern during development. However, phosphatidylethanolamine showed significant changes in the vitellus between stages IV and V. Changes observed at stage V yolk were accompanied by an increase of phospholipids in the embryos. TAG lipase activity increased during the first half of the embryonic development, reaching a peak in stage V, and then maintained a high level of activity till hatching. This resulted in an increase in the available free fatty acids for energy use (González-Baró et al., 2000). Thus, in *M. borellii*, maximum TAG and fatty acid degradation occurred at stage V, and then declined toward the hatching stage. In addition, fatty acid β-oxidation, measured in the mitochondrial fraction of developing eggs, also showed the highest level in stage V. However, the enzyme activity declined thereafter. A similar hike in the esterase activity in the mid-embryonic stage (V) and the subsequent decline toward the end of the development has also been reported in the crab, *E. asiatica* (Subramoniam, 1991a). The fact that the highest enzyme activity relating to lipid metabolism occurs in the mid-stage of embryogenesis may indicate both lipid degradation and resynthesis, required for organogenesis. This is consistent with the finding that fatty acid activation reached the highest activity at the subsequent stage VII, suggesting that the fatty acid metabolism becomes more dynamic at the time of hatching. Clearly, there exists a positive correlation between the degradation as well as synthesis of TAG, and the activation and oxidation of fatty acids in *M. borellii*.

In crustaceans, lipid plays a major role in embryonic metabolism and constitutes an important source of energy reserves, providing approximately 60% of the total energy expended during embryo development in decapods (Graeve and Wehrtmann, 2003). In the caridean shrimp, *S. hispidus*, large concentrations of free fatty acids were found throughout embryo development (Wang et al., 2015). Free fatty acids may be associated with the utilization of endogenous lipid resources, particularly for metabolic energetic purposes. In *S. hispidus*, the main lipid in the embryo includes wax ester, TAG, free fatty acid, phosphatidylethanolamine, and phosphatidylcholine (Wang et al., 2015). Among them, free fatty acids and the phospholipids dominated in all stages of development and showed a steady decline from the eye formation to prehatching stage. Even though phospholipids are important membrane components, the high concentration of these lipids in decapod crustaceans suggests that embryos utilize them as energy sources also.

Neutral lipids, mainly in the form of TAG are also major energy sources in aquatic crustaceans (Morais et al., 2002). In crustacean embryos, the utilization of TAG is related to the release of fatty acids (see above). Naturally, a large amount of fatty acids may be utilized during larval development (Nates and McKenney, 2000). Although lipid concentrations and composition decreased with increasing embryonic developmental stages, prehatch embryos contained 31% of the free fatty acids content. This high concentration of lipids in the newly hatched larvae is required as energy source for different physiological processes, including the successful first larval molt.

In two tropical caridean shrimps from Pacific Costa Rica, Wehrtmann and Graeve (1998) investigated the egg size and its relationship with the lipid composition and change during embryogenesis and early larval development. The recently spawned embryo of *Alpheus sexidomus* is 4.5 times larger and contained 4.5 times more lipid than those of *Palaemonetes schmitti*. Like other caridean shrimps discussed above, both TAG and phospholipids are the major classes of lipids in these two species. Total fatty acids are, however, higher in *A. sexidomus* (28.4 μg of whole lipids) than in *P. schmitti* (6.2 μg). Phospholipids also showed variations in between these two caridean shrimps; 48.9% of whole lipids in *A. sexidomus* and 68.6% in *P. schmitti*. TAG level in both the shrimps decreased during the incubation period. Most of the fatty acids in the embryos of both the species are either saturated or monounsaturated, suggesting a high proportion of de novo synthesis of fatty acids as the embryogenesis advances. Interestingly, embryos of *A. sexidomus* at the time of hatching contained 70% of the initially deposited fatty acid content. In this shrimp, the larvae hatch as zoeae, which molt within a few hours into zoea II. Such a rapid first molt, quite common in the alpheid shrimps, requires a high level energy in the form of fatty acids at the time of hatching. Arguably, lower fatty acid utilization and the higher lipid content in eggs close to hatching are related to habitat conditions and life history pattern of these caridean shrimps.

In a similar way, the naupli larvae of the brine shrimp, *Artemia salina* retains considerable quantity of lipids and fatty acids (Benijts et al., 1976). From a total lipid content of 22.7% of dry weight in the first instar, the value declined to 16.7% in II and III instar. Similarly, free fatty acid level also declined from an initial value of 14.7% in the first instar to 10.9% in II and III instar.

In the stone crab, *Lithodes maja*, embryonic yolk utilization continued into the lecithotrophic larval stage, which commences feeding only in the first juvenile crab stage, after metamorphosing from the megalopa (Anger, 1996). Lipid fraction of biomass is the major energy source during both embryonic and lecithotrophic larval development of this crab. The freshly hatched zoea I larvae have enhanced lipid stores, which are utilized as the major metabolic substrate during larval development. However, larval proteins decreased only very little from 55% at hatching to 48% in the megalopa stage. Clearly, lipids, rather than proteins, may be utilized as the principal biochemical substrate for metabolic needs during food-independent development (lecithotropy) in embryo and larval stages.

10.5 YOLK PROTEINS

A comprehensive study of yolk protein in an anomuran crab, *E. asiatica* revealed that it is comprised mainly of two lipovitellins, Lv I and Lv II, constituting as much as 90% of the total egg proteins (Subramoniam, 1991a). In SDS-PAGE analysis, Lv I yielded two subunits with molecular weights of 109,000 and 105,000 Da, whereas, Lv II resolves into six subunits with molecular weights of 65,000, 54,000, 50,000, 47,000, 44,000, and 42,000 Da (Tirumalai and Subramoniam, 1992). Many crabs, including *Emerita*, also contain hemocyanin bound to the yolk protein, providing a source of protein and copper during yolk utilization (Gilchrist and Lee, 1972; Gunamalai, 2002). In addition, Terwilliger (1991) suggested that embryonic accumulation of hemocyanin may help in oxygen transport or diffusion in the developing embryos in the brood, exposed to seawater in the marine benthic decapods.

The carbohydrate component of the yolk exists in three forms, namely free carbohydrate, protein-bound carbohydrate, and lipid-bound carbohydrate. The protein-bound carbohydrates are dominated by hexose, hexosamine, and galactosamine. Further, Lv II contains higher amount of *N*-linked oligosaccharides than *O*-linked oligosaccharides. In addition, three neutral glycolipids such as monoglycosylceramide, diglycosylceramide, and triglycosylceramide are also present in the major yolk protein, Lv II (Tirumalai and Subramoniam, 2001).

In agreement with the defining characteristic of marine eggs, *Emerita* eggs accumulate high levels of lipids, constituting the main source of metabolic energy during embryonic development. Tirumalai and Subramoniam (1992) have calculated the lipid content of major lipoprotein (Lv II) to be 30% in *E. asiatica*. The purified Lv II contained neutral lipids, glycolipids, and phospholipids. Among these, phospholipids are the dominant lipid class with phosphatidylcholine and phosphatidylserine as the major species. In addition to the lipid, bound to the yolk protein, free lipid droplets are also distributed in the ooplasm.

Yolk proteins of *E. asiatica* also contain significant quantities of metal ions such as copper, iron, sodium, calcium, and phosphorus. These ions constituted as much as 3.5% of the purified major yolk protein. Whereas calcium and copper are bound to lipid in Lv II, phosphorus and sodium are bound to lipid and protein (Subramoniam and Gunamalai, 2003). The metalloprotein nature of *Emerita* lipovitellin signifies its metabolic role during embryogenesis.

10.5.1 Breakdown of Lipovitellin

Crustacean lipovitellin is a complex glycolipocarotenoprotein, which requires an enzymatic breakdown to release component substrates for energy utilization and to tissue buildup during embryogenesis. In the electrophoretic separation, the lipovitellin of stage I eggs of *E. asiatica* is resolved into a thin proximal fraction of Lv I, and a thick fraction, Lv II in the slow-moving region of the electropherogram (Subramoniam, 1991a). Both the fractions were of glycolipoproteinous nature, as revealed from their intense staining with amido black, Oil Red O, and PAS. During the progression of embryogenesis, there is a natural cleaving of the lipovitellin fractions with the loss of prosthetic groups such as lipids and carbohydrates. Lv I cleaved apart into two fractions in stage VII, before totally disappearing in stage VIII/IX. Similarly, Lv II also cleaved into three fast migrating fractions. From stage VII onward, the staining intensities of Lv II and its subunits declined and disappeared in stage VIII/IX, suggesting the loss of their lipid and carbohydrate components. An important consequence of lipovitellin breakdown is the release of ecdysteroid conjugates, which are then cleaved by esterase to form hormonally active ecdysone and 20-hydroxyecdysone (20E). In a similar way, the bound vertebrate steroids were also released during the mid-stage of embryonic development (see Section 10.10).

10.5.2 Yolk Utilization in *Emerita asiatica*

Early investigations on crustacean egg development discussed yolk utilization from the energetic aspect and environmental adaptation (Pandian, 1970; Pillai and Subramoniam, 1985). The yolk utilization in crustacean eggs varies according to the percentage composition of different yolk materials as well as ecological conditions in which the animals live (Needham, 1950). In crustaceans, as in other oviparous animals, lipids and proteins are preferentially used, with the carbohydrates, constituting a very low percentage of total yolk, being utilized very quickly.

Biochemical analysis of the freshly laid eggs of *E. asiatica*, revealed that the major nutrients were lipid (4.6 mg/10 mg dry weight), followed by protein (3.2 mg/10 mg dry weight) and carbohydrates (0.48 mg/10 mg dry weight) (Subramoniam, 1991a). Protein content steadily declined from stages I to IX, corresponding to increasing water content of the growing embryo. Conversely, lipid content has increased up to 25% in stage V; thereupon, the value fell precipitously, reaching the lowest of 27% (on dry weight basis) in stage IX. Thus, lipid is preferentially used in the second half of *Emerita* embryogenesis, when the tissue synthesis and organ formation occur rapidly. Nevertheless, a substantial quantity of lipid is retained within the embryo in the form of oil globules, which increase the buoyancy of the hatched zoea larva. The lipid reserve in the larva would also be advantageous in delaying starvation during the protracted larval life of *Emerita* species (Efford, 1970; Sunil et al., 2006). Such tardiness in lipid utilization during initial stages of embryogenesis is reported in the gammarid amphipods (Clarke et al., 1985).

This pattern of lipid utilization is in contrast to the condition found in lecithotrophic eggs of the freshwater field crab, *Paratelphusa hydrodromous*, in which the enormous quantity of lipid stored is used gradually all through the embryonic development (Pillai and Subramoniam, 1985). This crab depends mainly on lipids for metabolic energy in the initial stages of embryogenesis and for making available substrates for new tissue synthesis during later stages of development. Such a pattern of yolk (lipid) utilization is considered to be an adaptive strategy of terrestrialization in this freshwater crab (Pillai and Subramoniam, 1985).

Compared to lipid and protein, the total carbohydrate was low initially in *E. asiatica* but different carbohydrate components exhibited an interesting pattern of fluctuation during embryonic development (Subramoniam, 1991a). For example, the total free carbohydrates on a dry weight basis were 2.9% in stage I, but rose to a high-percentage value of 4 in stage IX. The free glycogen content also exhibited a similar increase during embryonic development. Concurrently, the protein-bound polysaccharides decreased from a high value of 0.178 mg/10 mg of dry weight in stage I to 0.117 mg/10 mg of dry weight in stage IX. During the entire embryonic development, there is steady increase in the water content from an initial value of 59% in stage I to 82% in stage IX. The increased levels of free carbohydrates in the later phase of embryogenesis is significant in that both free and protein-bound sugars are involved in the chitin synthesis of embryonic as well as larval cuticle formation (McWhinnie and Saller, 1960; Bergreen et al., 1961).

10.6 ENERGY UTILIZATION

In oviparous animals, the pattern of yolk utilization varies according to the percentage composition of different yolk materials as well as ecological conditions in which the animal lives (Needham, 1950). Such differences in the nature of yolk composition are reflected in the preferential utilization of protein or lipid substances. Crustaceans have adapted to live and reproduce in the marine, freshwater, and terrestrial habitats. Therefore, the pattern of energy utilization from the stored substrates differs markedly among them. In the freshwater crab, *P. hydrodromous*, around 95% of energy is derived from lipids and the rest from carbohydrates, during stages I and II of embryonic development (Pillai and Subramoniam, 1985). During this period, no loss of protein occurs, instead the total protein registered an increase, suggesting that the progressive lipid utilization may represent its conversion to tissue-building proteins as well as providing energy for embryogenesis. Carbohydrates might be utilized as an additional energy-yielding substrate. Correlated with the depletion of lipids and carbohydrates, the calorific content of the egg is also reduced. It is only during the late stage of embryonic development (stages III and IV), there is a small decline in the total protein, whereas the efficiency of lipid utilization is considerably reduced. That the fat oxidation is the main source of energy for embryonic metabolism has also been demonstrated in marine decapods such as the European lobster, *Homarus gammarus* (Pandian, 1970).

Mobilization of energy from the organic substrates during egg development of *E. asiatica* was calculated by applying energy equivalents, according to Brody (1968). The energy equivalent for total carbohydrates was calculated to be 17.3 kJ/g dry weight, for protein, 23.5 kJ/g, and for lipids 39.5 kJ/g at the initial stage of embryogenesis (Table 10.3) (Subramoniam and Gunamalai, 2003). It is seen from the table that the energy derived from the protein is continuously expended from stage I onward (75.6 J per mg dry weight to 20.9 J in the last stage). On the other hand, carbohydrate-based energy is continuously built up from 8.4 J per mg dry weight in stage I to 73.2 J in stage IX. In the same way, lipid energy increased from stage II to stage V. This period in embryogenesis corresponded to maximum yolk clearance, coupled with faster organogenesis (eye and appendage formation). This is followed by a quick decrease in lipid energy, although considerable amount of lipid energy is still retained in stage IX. Evidently, this remnant lipid energy could be utilized by the hatched zoea larvae, during periods of food scarcity and starvation. The above-mentioned differences in the energy utilization particularly between the field crab, *P. hydrodromous* and the marine crab *E. asiatica* reflect on their ecophysiological conditions and their adaptation to their respective habitats.

TABLE 10.3 Mobilization of Energy During Egg Development in *Emerita asiatica*: Energy Values (J per mg Dry Tissue) Calculated From the Organic Composition Values Given in Table 10.2 and by Applying Energy Equivalents Suggested by Brody (1968)

Egg Stages	Protein	Carbohydrates[a]	Lipids	Total
I	75.6	8.4	182.1	266.1
II	52.3	11.7	151.6	215.6
III	57.2	15.0	189.4	261.6
IV	64.1	20.1	221.4	305.6
V	58.4	21.9	229.3	309.6
VI	55.4	33.7	90.3	179.1
VII	39.8	37.9	79.0	156.7
VIII	29.9	57.0	91.6	178.5
IX	20.9	73.2	107.5	201.6

[a]*Free carbohydrates + glycogen + protein-bound sugars.*

10.7 ENZYME ACTIVITY DURING YOLK PROTEIN DEGRADATION

As seen from the preceding sections, crustaceans accumulate large reserves of complex yolk in the egg to meet the high energy demands of embryogenesis. Therefore, enzymatic digestion of lipovitellin becomes the major biochemical activity, crucial to embryogenesis. During egg maturation, the ovary produces a host of hydrolytic enzymes, which are stored in inactive form, along with the yolk. In *E. asiatica*, Subramoniam (1991a) found a high activity of nonspecific esterase during embryogenesis to dismantle lipovitellin molecule, thereby releasing component substrates in an utilizable form. In addition, other enzymes such as proteases and glycosidases are also active during embryogenesis. Lipid being the major organic reserve of *E. asiatica*, nonspecific esterase activity is intense during embryonic development. However, its activity appears only from stage IV onward, reaching a peak in stage V and declines thereafter. As seen from Fig. 10.1, five fractions of nonspecific esterase are present, although no fraction could be observed until stage IV. In stage IV, the zymogram consisted of a major fraction (E1), a moderately staining E2 fraction, and two other thin fractions (E3 and E4). In stages V and VI, E2 fraction declined in intensity, but another fraction E5 appeared in the fast-moving zone in stage VI but persisted up to stage VII only. E1 fraction was present in all stages of embryonic development, but others decreased in intensity and disappeared in the last stages of development. Similar activity of esterases during mid-embryonic stage was reported in a freshwater prawn, *M. borellii* (Heras et al., 2000).

All esterase fractions of *E. asiatica* are isozymes of the carboxylesterase, as they were inhibited by silver nitrate and malathion and unaffected by pCMP, EDTA, and eserine sulfate. An interesting observation is that the thick fraction, E1 appears to overlap the main lipovitellin in its relative mobility. Obviously, this esterase could bind to the vitellogenin and then get incorporated into the oocyte; thereafter, E1 is kept in an inactive state, and then activated during embryogenesis. All other isozymes (E2–E5) of *E. asiatica* might be synthesized by the embryonic cells with their stage-specific appearance or disappearance indicating changes in cellular differentiation and metabolic pathways. In a parallel instance, a cathepsin-B-like thiol protease is secreted by the fat body as a latent proenzyme and then accumulated by developing oocytes, where it is stored in yolk bodies of the mosquito, *Aedes aegypti*. This inactive protease undergoes further processing at the onset of embryogenesis to an active state, to get involved in vitellin degradation (Liu et al., 1996). Another enzyme that is active at the time of lipovitellin degradation (stages V and VI) in *E. asiatica* is phospholipase C (Ramachandran, 1992). Esterase activity, in addition to releasing the free fatty acids from the conjugated lipids, may also release carotenoids that are esterified to the long chain fatty acids of the lipovitellin.

Contrary to esterase activity, protease activity during embryonic development in *E. asiatica* commenced quite early and increased during mid-stage (V),

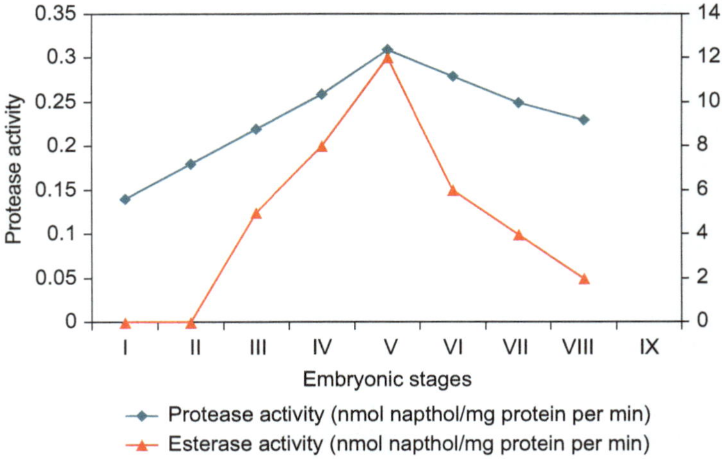

FIGURE 10.2 *Reproduced from Subramoniam, T., 2000. Crustacean ecdysteroids in reproduction and embryogenesis. Comp. Biochem. Physiol. 125, 135–156. Table 2. p. 146; Elsevier, 2000.*

followed by a slight reduction in the end stages, suggesting protein degradation throughout embryogenesis (Fig. 10.2).

Embryonic utilization of yolk protein contained in the yolk granules has been well studied in *A. salina*. In this brine shrimp, the yolk granules found in the early embryonic cells undergo membrane rupture when the dormant dry cysts are hydrated to resume development. This allowed the hydrolytic enzymes to degrade the lipovitellin components of the yolk into simpler units. The proteases that metabolize yolk into amino acids and usable energy include a neutral cytosolic protease and another protease of lysosomal location (Ezquieta and Vallejo, 1985). The cytosolic proteases, localized within the yolk granules, are kept in an inactive condition until hydration, when the release of these enzymes into the cytosol occurs. This protease undergoes further processing to reduce its molecular weight and becomes activated to digest the vitellin molecules. The programmed events of protease activation, due to molecular processing, coincide with the cellular differentiation that restarts after hydration of the *Artemia* cysts.

In accordance with the relatively high carbohydrate content of the crustacean lipovitellin, *Emerita* eggs store different carbohydrases. Gunamalai (1993) found the activity of two forms of α-galactosidases and three forms of β-galactosidases in the developing embryos of *E. asiatica* and found their activity peaking in stage VII/VIII of embryonic development (Fig. 10.3).

These glycosidases may be required to release bound glucose and galactose from the glycolipid and oligosaccharide components of the lipovitellins and to hydrolyze stored glycogen during embryogenesis of *E. asiatica*. A consequent accumulation of free sugars such as glucose toward the final stages of embryogenesis may be required for the synthesis of chitin precursor molecules for the embryonic cuticle synthesis. Interestingly, the activity of all the three enzymes

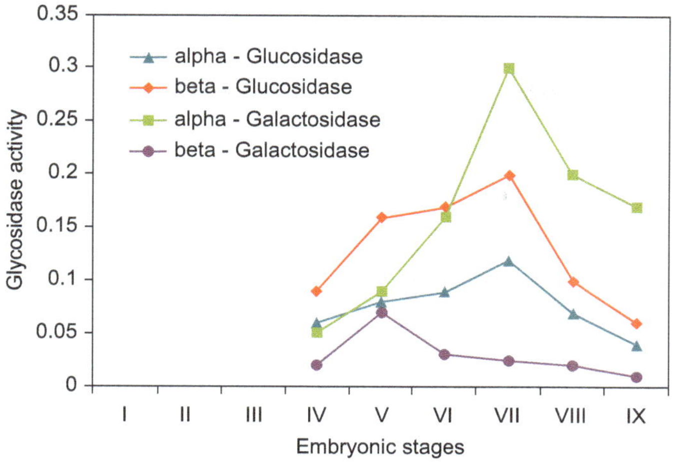

FIGURE 10.3 *Reproduced from Subramoniam, T., 2000. Crustacean ecdysteroids in reproduction and embryogenesis. Comp. Biochem. Physiol. 125, 135–156. Table 2. p. 146; Elsevier, 2000.*

is correlated with breakdown of the complex yolk proteins into simpler utilizable subunits and the onset of organogenesis. Furthermore, small amount of enzyme activity is left in the hatching embryos, suggesting that the remnant yolk present in the released larvae continued to release energy materials for larval survival under adverse conditions.

10.8 CAROTENOID METABOLISM

Decapod crustaceans are well known for accumulating carotenoid pigments in the ovary and eggs in the form of chromoproteins (Zagalsky et al., 1967; Wallace et al., 1967). Carotenoids form two classes by their chemical structure: (1) carotenes that are constituted by carbon and hydrogen; (2) oxycarotenoids or xanthophylls that have carbon, hydrogen, and additionally, oxygen (Isler, 1971).

Astaxanthin is the most commonly identified carotenoid pigment found in conjugation with the lipovitellin, giving it a golden yellow color to the eggs of crabs. Unlike the other carotenoid–protein complexes, in which the carotenoid components are directly linked to protein chains through amino groups, in the eggs of decapod crustaceans, they are dissolved nonstoichiometrically in the lipid prosthetic groups of the lipovitellin (Zagalsky et al., 1967). Additionally, carotenoids can be esterified to the fatty acids of the lipovitellin molecules. The pigments of the eggs are derived from the hemolymph as conjugates of vitellogenin, which is then sequestered into the growing oocytes for final deposition. Changes in the color of the egg mass of decapod crustaceans are an easy index of carotenoid metabolism, as occurring during embryogenesis.

Kour and Subramoniam (1992) investigated the qualitative and quantitative changes in the carotenoids during embryogenesis of a sand crab *E. asiatica*. Table 10.4 shows the variation in the occurrence of different carotenoids in

TABLE 10.4 Carotenoid Content in Different Egg Developmental Stages of *Emerita asiatica* (μg/g Wet Weight)

Carotenoid Species	Stages						
	I	III	V	VII	VIII	IX	X
α-carotene	0.853 ± 0.056	0.921 ± 0.189	1.490 ± 0.026	0.833 ± 0.013	0.960 ± 0.012	0.031 ± 0.002	–
β-carotene	15.560 ± 0.122	16.072 ± 0.141	15.445 ± 0.087	14.320 ± 0.097	12.220 ± 0.034	7.220 ± 0.034	3.700 ± 0.069
Lutein	2.080 ± 0.067	–	–	–	–	–	–
Echinenone	0.846 ± 0.031	1.960 ± 0.036	3.540 ± 0.036	–	–	–	–
Isozeaxanthin	4.373 ± 0.068	1.500 ± 0.019	3.540 ± 0.039	3.380 ± 0.048	0.031 ± 0.010	–	–
Zeaxanthin	4.034 ± 0.045	–	4.510 ± 0.058	4.093 ± 0.248	–	5.971 ± 0.372	–
Canthaxanthin	–	–	–	2.972 ± 0.323	5.806 ± 0.528	4.613 ± 0.264	2.606 ± 0.264
α-doradexanthin	–	–	–	–	–	–	0.666 ± 0.117
Isocryptoxanthin	6.712 ± 0.198	5.100 ± 0.197	3.910 ± 0.153	3.630 ± 0.161	2.540 ± 0.236	2.136 ± 0.142	2.104 ± 0.173
Free astaxanthin	0.600 ± 0.022	0.216 ± 0.016	0.686 ± 0.034	0.608 ± 0.016	1.192 ± 0.055	2.440 ± 0.100	0.848 ± 0.044
Esterified astaxanthin	–	–	–	–	–	–	4.280 ± 0.018

Data from Kour, D.V.R., Subramoniam, T., 1992. Carotenoid metabolism during embryonic development of a marine crab, *Emerita asiatica* (Milne-Edwards). Invertebr. Reprod. Dev. 21 (2), 99–106.

different stages of embryo development. By far, the most abundant form of carotenoid deposited in the developing eggs is β-carotene, with its concentration varying between 15.4 μg/g wet weight and 16.1 μg/g wet weights in the early stages of embryogenesis. After maintaining almost the same level up to stage V, β-carotene concentration declined gradually to reach a low level of 3.7 μg/g wet weight in the newly hatched larvae. α-Carotene also showed a declining trend during embryogenesis. Obviously, these two parent carotenoids of dietary origin undergo bioconversion into more oxidized forms such as hydroxyl and ketocarotenoids. Other pigments such as echinenone and isozeaxanthin occurred mainly in the early stages, whereas zeaxanthin was found in stages I, V, VII, and IX. Canthaxanthin made its appearance only from stage VII onward. While free astaxanthin was observed in all stages, esterified astaxanthin was visualized only in the hatching stage. In spite of these metabolic conversions taking place during embryogenesis, the total carotenoids remained almost same. Based on the initial occurrence of α- and β-carotene, lutein, and astaxanthin, as well as the identification of several intermediate compounds during the course of embryogenesis, the following pathways have been proposed for carotenoid metabolism in crustacean eggs (Fig. 10.4). Apparently, astaxanthin is the final product of β-carotene metabolism in the crustacean embryos. The highest amount of astaxanthin present in embryogenesis is due to the conversion of dietary pigments such as α- and β carotene. Lutein also contributed to the formation of astaxanthin through several intermediary compounds such as α- and β-doradexanthin. Compounds such as zeaxanthin, echinenone, canthaxanthin, isocryptoxanthin, and isozeaxanthin appear during the formation of astaxanthin from β-carotene. Esterification of astaxanthin toward the last stage of embryonic development is associated with the origin of chromatophores and the possible biosynthesis of visual pigments.

In a freshwater crayfish, *Astacus leptodactylus*, Berticat et al. (2000) studied the carotenoid content in the embryos and followed the metabolic conversions of free astaxanthin and lutein, the latter being predominant in freshwater species. In this crayfish, astaxanthin is noncovalently bound to protein in the vitellus, and is freed during yolk utilization. The steady decline of lutein toward the hatching stage is also related to their metabolic conversion into astaxanthin. After giving rise to many intermediate compounds, astaxanthin is esterified at the closing stage of hatching. The appearance of blue and red chromatophores in the later stages of embryonic development is related to the esterified astaxanthin. The appearance of several other compounds such as zeaxanthin, canthaxanthin, echinenone, β-doradexanthin, and α-carotene was detected by HPLC separation. The high amount of astaxanthin in the embryos of *A. leptodactylus* as well as other decapod crustaceans studied is related to metabolic adaptation of the embryos, developing under endotrophic conditions.

The carotenoids in the freshly laid eggs of the freshwater prawn *Macrobrachium olfersii* consist of α-carotene (0.8–6.4 μg/g fresh egg weight), β-carotene (0.13–13.0 μg/g fresh egg weight), and astaxanthin (23.2–38.8 μg/g fresh egg weight), together with a mixture of four unidentified yellow pigments

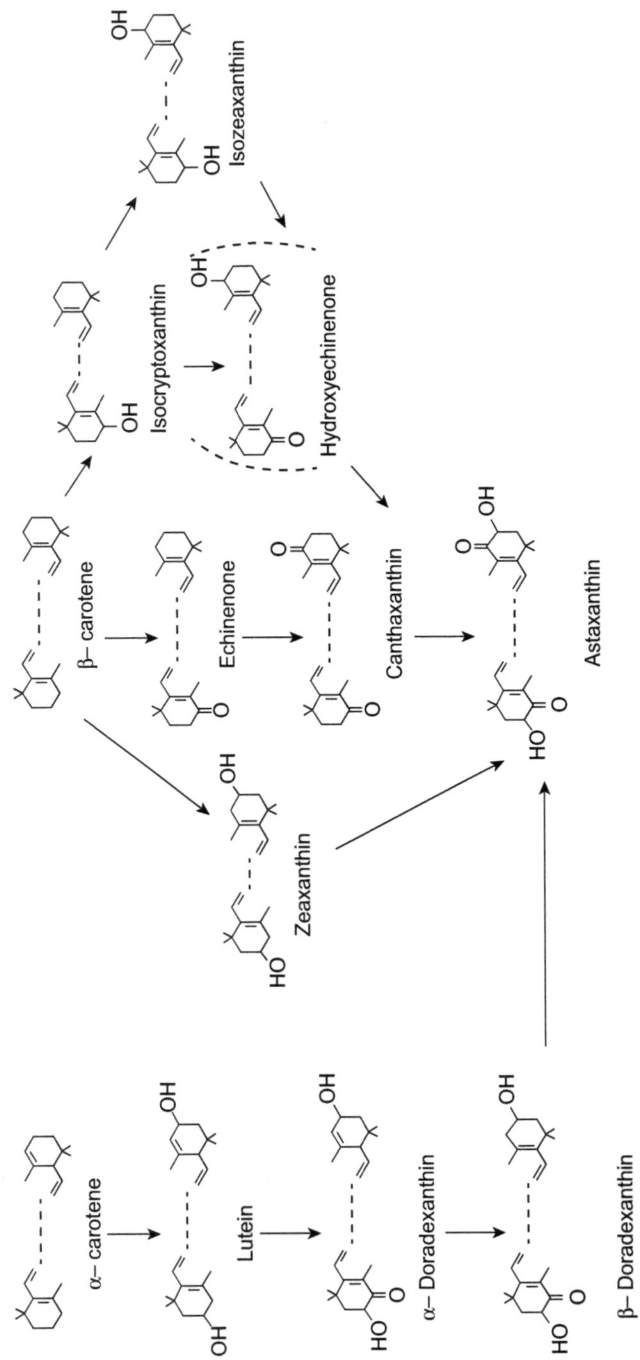

β– carotene
Isocryptoxanthin
Isozeaxanthin
Hydroxyechinenone
Echinenone
Canthaxanthin
Astaxanthin
Zeaxanthin
α– carotene
Lutein
α– Doradexanthin
β– Doradexanthin
OH
HO
O

FIGURE 10.4

(Ribeiro et al., 2001). During stages I, II, and III of embryogenesis, the relative concentrations of α- and β-carotene as well as astaxanthin increased significantly, but their concentration decreased significantly at the latter stages (IV and V).

Zadorozhny et al. (2008) analyzed the qualitative and quantitative changes in the composition of carotenoids at various stages of embryo development in the crab species *Chionoecetes opilio*, *Paralithodes camtschaticus*, and *P. platypus*. Major carotenoids are astaxanthin and β-carotene. The carotenoid content in the early development stages (orange colored) in *C. opilio* was found to be 22.7 ng, and in *P. camtschaticus* and *P. platypus* (the violet egg) it amounted to 49.2 and 23.3 ng, respectively. At the later development stage (the brown egg), the carotenoid content was decreased to 13.1 ng in *C. opilio* and to 20.1 ng in *P. camtschaticus*.

Change in carotenoid titers during embryogenesis is a useful metabolic parameter to monitor the development of embryo. Whether total carotenoids increase, decrease, or remain unchanged may reflect underlying biochemical and physiological events and should help to disclose the role of the carotenoids in crustacean embryonic development. The esterified astaxanthin in the juvenile stages of *A. leptodactylus* could impart protective coloration that could make them less vulnerable to predators (Berticat et al., 2000). In addition, astaxanthin could play a structural role in the stabilization of protein structure and protection from oxidative damage by free radical species (Vershinin and Lukyanova, 1993).

Embryonic metabolism of carotenoid pigments continues into larval stages in decapod crustaceans. Toward the end of embryogenesis, astaxanthin is partly esterified into new compounds and the rest stored as pigmentary source for their larval metabolism. In the shrimp, *Penaeus japonicus*, the hatched out nonfeeding larva relies on astaxanthin precursors for further metabolism (Petit et al., 1991). However, the pigmentary composition changes in the next zoea and mysis stages, when it starts feeding on microalgae and then *Artemia* nauplius larva. In *A. salina*, *cis*-canthaxanthin was found exclusively in the ovaries, and eggs (Nelis et al., 1988). *cis*-Canthaxanthin disappeared progressively during the nauplius stage and was isomerized to the all *trans*-canthaxanthin. When fed to the mysis stage of *P. japonicus*, canthaxanthin is not converted into astaxanthin but distributed to all larval tissues in a storage form. In the postlarval stage, these precursor pigments undergo oxidation and esterification. In the P20 stage, most of the astaxanthin is esterified to mono- and diesters, whereas canthaxanthin is retained as it is.

10.9 EMBRYONIC ECDYSTEROIDS

The ovary of several crustacean species accumulates significant quantities of ecdysteroids during maturation, mainly by sequestration from hemolymph, in contrast to the insects, in which the follicles synthesize them for storage within the ovary. In the embryos, the molting hormone, 20E as well as its precursor, ecdysone, have been reported in many malacostracan crustaceans. In addition, a new ecdysteroid, ponasterone A (25-deoxy-20-hydroxyecdysone)

(PoA) with high activity during embryogenesis has been isolated in crabs such as *C. maenas* (Lachaise and Hoffmann, 1982) and *Gecarcinus lateralis* (McCarthy, 1982).

As in the brachyuran crab, *C. maenas*, the ovary of crustaceans accumulates significant quantities of ecdysteroids throughout intermolt stage (Lachaise and Hoffmann, 1982). The sequestered ecdysteroids are passed onto the eggs for possible elimination as ecdysonic acid and to function as morphogenetic hormones partaking in the control of embryogenesis and early development (Lachaise et al., 1981). Within the embryos, the three major forms of free ecdysteroids, viz., ecdysone, 20E, and PoA are also found as conjugates, which could be esters with phosphoric acid or long chain fatty acids, as in insects (Koolman, 1990). During embryogenesis, these accumulated ecdysteroids within the eggs undergo active metabolism, leading to the formation of inactivation products, including the polar and apolar conjugates. In insects, phosphate and fatty acyl ecdysteroid conjugates in newly laid eggs represent hormonally inactive storage forms of maternal hormones for utilization following enzymatic hydrolysis early in embryogenesis (Lanot and Clédon, 1989).

The conjugated ecdysteroids in the freshwater prawn *Macrobrachium rosenbergii* are mainly represented as high and low polar products (HPP and LPP). The conjugated HPP on enzymatic hydrolysis releases a variety of free ecdysteroids such as ecdysone, 20E, 20,26-hydroxyecdysone, and enokosterone, whereas the LPP contain 2-deoxyecdysone and PoA (Wilder et al., 1995). Apart from being used as recyclable conjugates, catabolism of ecdysteroids could result in other products such as 20,26-dihydroxyecdysone and ecdysonic acids, both forming the inactive end products for elimination through storage excretion (McCarthy and Skinner, 1979; Lachaise and Lafont, 1984).

Lachaise and Hoffman (1982) further observed that PoA alone showed high concentration with two peaks, one during early embryonic development and the second before the secretion of the first embryonic envelope in *C. maenas*, in which the late stage embryonic development is comprised of four stages, viz., nauplius, metanauplius, protozoea, and prezoea that hatches into zoea larva. Radioimmunoassay of the embryonic ecdysteroids revealed that only PoA showed significant rise, corresponding to molting in the four embryonic larval stages. Conversely, ecdysone, 20E, and the ecdysteroid-conjugates maintained minimal levels throughout embryogenesis. Evidently, PoA is the only ecdysteroid taking part in the control of larval cuticular secretion and molting in *C. maenas*.

The PoA control of embryonic cuticle secretion and molting, though unique to the brachyuran crab *C. maenas*, has also been reported for the European lobster *H. gammarus* (Goudeau et al., 1990). In the course of embryogenesis, six new coatings are secreted successively from the ectodermal embryonic cells, beneath the existing outermost fertilization envelope. These cuticular layers will remain intact until hatching, freeing the mysis larva in concentric order without exuviations. Correlated with the formation of these embryonic layers, the total ecdysteroid levels in the embryo also fluctuated in accordance with the secretion of each layer. Yet, in the European lobster, both 20E and PoA appear to exert a sequential but independent control on embryonic envelope secretion. While high titer of 20E was maintained during the onset of embryonic envelope formation, the last phase of envelope secretion proceeded under high titer of PoA. Similar concentration peaks of free ecdysteroids, correlated with deposition of cuticle or embryonic molt, has been reported in insects like *Locusta migratoria* (Lagueux et al., 1981).

Occurrence of embryonic ecdysteroids and their fluctuation, correlated to specific developmental stages, including secretion of embryonic envelopes and cuticle, has been reported for many crustacean species. In the spider crab, *Acanthonyx lunulatus*, Chaix and Reggi (1982) reported the presence of ecdysone, 20E, and HPP and LPP in the ovary and early embryos. However, the polar compounds were not present in the ovary and appeared only in the beginning of embryogenesis. Both ecdysone and 20E were detected in high concentration at the early stage of embryogenesis. The 20E maintained a high concentration during naupliar development, metamerization leading to metanaupliar stage, and at the end of embryonic development when the formation of first exoskeleton occurs. Any role for PoA in the control of embryonic molting is not reported in these crabs.

As seen above, most of the embryonic ecdysteroids are derived from maternal sources, although de novo synthesis by the differentiating Y-organ could further augment the total ecdysteroids. For example, in the crab, *Cancer magister*, embryos displayed a biphasic pattern of ecdysteroid fluctuation during embryonic development; hormone titers decreased until mid-embryogenesis and then increased to give a peak before hatching (Okazaki and Chang, 1991). On the contrary, in *Cancer anthonyi*, with relatively short embryonic development, ecdysteroid titer in the embryo marked a steady decrease from a high initial concentration. Evidently, both the *Cancer* species start using the maternally derived ecdysteroids first; however, in *C. magister*, the Y-organ, formed during mid-embryogenesis, augments a fresh supply of ecdysteroids, resulting in the second peak which occurs before the prehatching stage.

The consistent rise in the free ecdysteroids during embryogenesis with a peak at the prehatching stage in many decapods is indicative of fresh synthesis of ecdysteroids by the embryonic Y-organ. In the caridean prawn *Palaemon serratus*, ecdysone and 20E are the predominant ecdysteoroids undergoing profound changes during embryogenesis. Spindler et al. (1987) correlated changes in embryonic development, including the formation of Y-organ, with the level of embryonic free ecdysteroids. Two such embryonic events are (1) the metamerization beginning at embryonic stage C and (2) the appearance of the Y-organ shortly before embryonic stage C and its completion in embryonic stage D. The peaking of ecdysteroid level during metamerization is common in insects (Lagueux et al., 1984) and also reported in the spider crab *Acanthonyx* (Chaix and De Reggi, 1982). In *Palaemon*, the Y-organ appears at a time when the embryonic ecdysteroids begins to increase (stage C) (Spindler et al., 1987). Yet again, the ratio of 20E to ecdysone increases steadily during embryogenesis suggesting that the embryo possesses an active C_{20}-hydroxylase system to supply the embryo with active 20E continuously. Concurrently, the conjugated HPP also accumulate toward the end of embryonic development, suggesting a mechanism for the storage excretion of ecdysteroids.

In another caridean prawn *M. rosenbergii*, Young et al. (1991) analyzed ecdysteroid composition in the embryos and observed changes in their profiles during embryogenesis. The newly laid eggs contained only free ecdysteroids such as ecdysone and 20E. However, the eggs at early stage of embryogenesis possessed ecdysone and low levels of apolar ecdysteroid conjugates, but their concentration increased significantly in mid-stage embryo. The overall ecdysteroid level reached the maximum in the late stage embryo, indicating cumulative contribution from newly formed Y-organ. In addition to these ecdysteroids, the late embryos contained 26-hydroxyecdysone, ecdysonic acid, and 20-hydroxyecdysonic acid at relatively low levels, and polar conjugates, 2,3-diacetylecdysone 22-phosphate at a high level. Evidently, the embryo not only possessed biosynthetic capacity to produce ecdysteroids but also to convert them into apolar conjugates and inactivation products 26-oic acids and polar conjugates. The authors proposed the following inactivation pathway for the ecdysteroids in *M. rosenbergii* (Fig. 10.5):

On the contrary, in the free-spawning penaeid shrimp *Sycionia ingentis*, with short embryogenesis, the bulk of embryonic ecdysteriod is derived from the differentiating Y-organ (Chang et al., 1992). In this marine shrimp, the freshly spawned eggs contain very little maternally derived, suggesting de novo synthesis from the embryonic Y-organ. In this shrimp, three embryonic envelopes, viz., hatching, embryonic, and exoskeletal layers are formed during embryogenesis, and the peaking of embryonic ecdysteroids coincided with the shedding of the hatching and embryonic envelopes by the prenaupliar larva (Chang et al., 1992).

In the anomuran crab, *E. asiatica*, there is evidence that the ecdysteroids are bound to the vitellogenin in the hemolymph and carried to the maturing ovary,

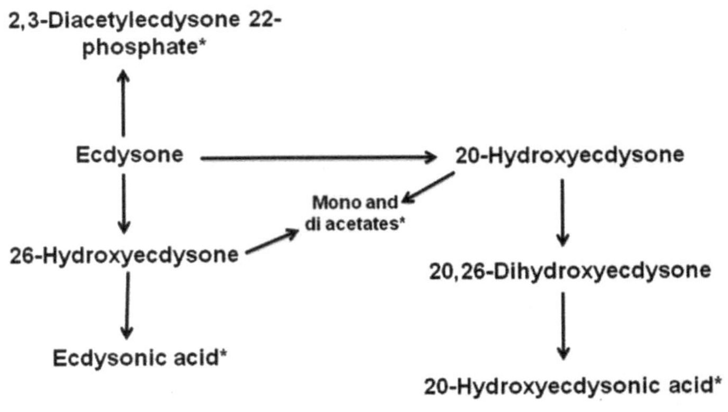

FIGURE 10.5

where they are stored along with lipovitellin in the oocytes. The purified lipovitellin are found to contain both free and conjugated ecdysteroids (Subramoniam et al., 1999). The reverse phase HPLC/RIA of the free ecdysteroids revealed a higher percentage concentration of 20E than ecdysone. However, the free ecdysteroid level in the ovary is drastically declined during previtellogenic stage. This decline in the ovarian ecdysteroids is inversely related to rising hemolymph ecdysteroids, suggesting a release from the ovary (Gunamalai et al., 2004). Nevertheless, significant quantities of free and conjugated ecdysteroids are retained in the spawned eggs. Wilder et al. (1991) found a rise in the ovarian ecdysteroids during gonad development and the accompanying molting stages of the freshwater prawn *M. rosenbergii*, suggesting ovarian sequestration of ecdysone, as a function of oocyte maturation. In *Emerita* and *Macrobrachium*, the ovarian development extends into the premolt stage, when the hemolymph ecdysteroids show a peak.

Subramoniam et al. (1999) reported the occurrence of a complex mixture of free and conjugated ecdysteroids in the developing embryos of *E. asiatica* (Fig. 10.6). These hormonal complexes exhibited multiphasic fluctuation during the course of embryonic development. The first rise in free, active ecdysteroids was noticed in stage III, when the blastoderm extends. The next increase in free ecdysteroids was seen in stage VI, when the embryonic eyes are fully developed and the first limb buds become apparent. The sharp rise of total free ecdysteroids toward the prehatching period may be correlated with the deposition of the embryonic cuticle of the proto- and prezoea larvae.

Separate determination of titers of conjugated and free ecdysteroids during embryonic development of *E. asiatica* showed that the concentration of free

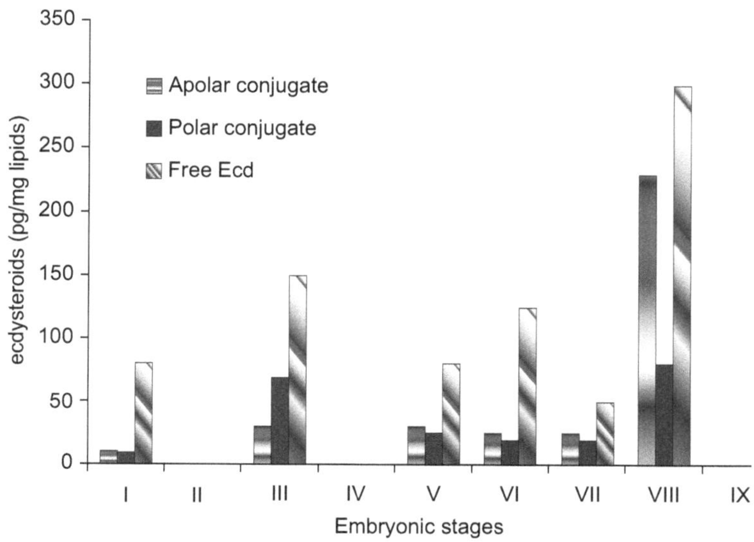

FIGURE 10.6

ecdysteroids always predominated over the conjugated ones in all the developmental stages. That the ecdysteroids have a definitive role in the brood development of egg carrying decapods is clearly evidenced in *Emerita*. In this crab, Gunamalai et al. (2004) reported that hatching of the embryos, attached to the pleopods of the ovigerous females occur under a high titer of hemolymph ecdysteroids. 20E injection at C_3 stage crabs indicated a significant reduction in time duration of embryonic development leading to hatching of zoea larvae (Fig. 10.7). Evidently, the augmented hemolymph ecdysteroid titer helps in the synchronization of embryo

(A)

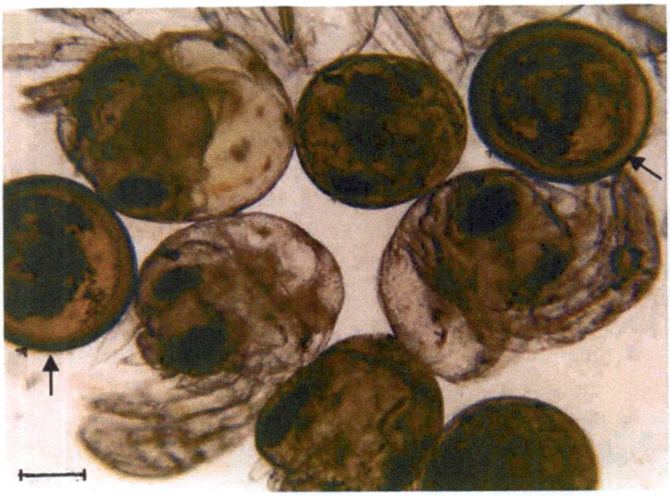

(B)

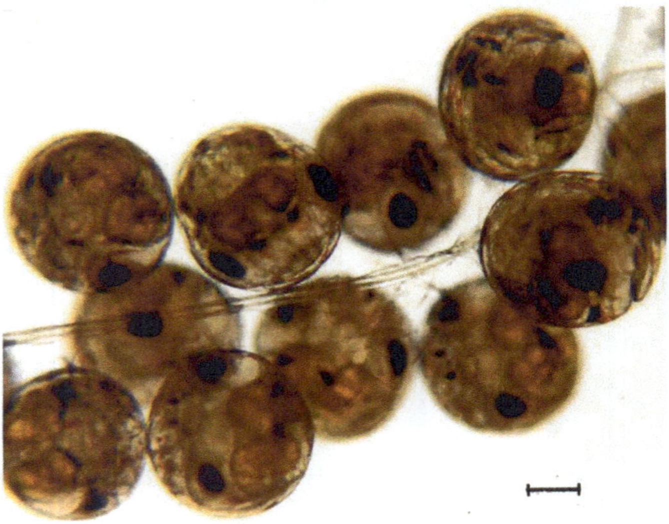

FIGURE 10.7 *Reproduced from Gunamalai, V., Kirubagaran, R., Subramoniam, T., 2004. Hormonal coordination of molting and female reproduction byecdysteroids in the mole crab* Emerita asiatica *(Milne Edwards). Gen. Comp. Endocrinol. 18, 128–138. Figure 5. p. 134; Elsevier, 2004.*

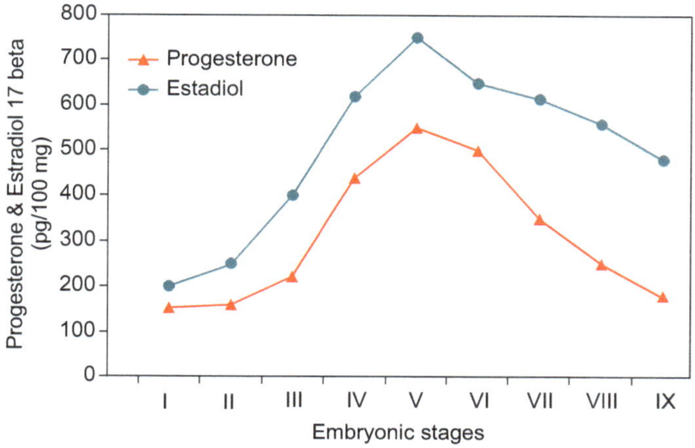

FIGURE 10.8 *Reproduced from Sudha Warrier R., Tirumalai R., Subramoniam T., 2001. Occurrence of vertebrate steroids, estradiol 17-β and progesterone in the reproducing females of the mud crab* Sylla serrata. *Comp. Biochem. Physiol. 130, 283–294. Figures 4 and 5. p. 289–290; Elsevier, 2001.*

hatching and the premolt changes. Taken together, the above data indicate the importance of ecdysteroids in accomplishing molt-related reproductive processes such as brood development and larval hatching in those decapods that incubate the eggs, facilitating normal ecdysis without affecting brood development.

10.10 VERTEBRATE STEROIDS

Several studies have revealed the occurrence of vertebrate steroid hormones such as 17β-estradiol (E_2) and progesterone (PG) in the eggs of decapod crustaceans (see chapter: Endocrine Regulation of Vitellogenesis). Warrier et al. (2001) found that these hormones were conjugated to hemolymph vitellogenin and lipovitellin in the oocytes of *E. asiatica* and *Scylla serrata*. During embryonic development of *E. asiatica*, both E_2 and progesterone levels increased from stages I to V (Fig. 10.8). Previous studies have indicated that in stage V of *E. asiatica* embryogenesis, intense enzyme activity occurs to dismantle lipovitellin molecules, thereby releasing the bound steroid hormones into the general pool of free steroids. It has been suggested that the presence of the steroid hormones in the eggs of the two crabs could possibly potentiate and facilitate the morphogenetic processes in embryonic development. However, evidence to support this supposition is lacking in crustaceans.

Chapter 11

Spermatogenesis

11.1 INTRODUCTION

Gametogenesis is the most conspicuous and central event in the reproductive cycle of all metazoan animals. The formation and maturation of male and female gametes in their respective sexual organs lead to their expulsion into the reproductive tract. Crustaceans are unique among invertebrates in that the male gametes are not only atypical but also aflagellate, entailing special transport mechanisms to meet eggs during fertilization. In Crustacea, the spermatogenesis, comprising a sequence of events by which spermatogonia are transformed into spermatozoa, occurs within the seminiferous tubules of the testis. However, in many species, spermatogenesis is not restricted to testis, and the sperm cells have yet to achieve their definitive characteristics upon entering the testicular lumen and vas deferens. Thus, in many decapods, many later stages of spermiogenesis are completed within the vas deferens to assume complete morphological sperm characteristics. In a few instances, sperm capacitation, the maturational process by which fertilization ability is achieved, occurs after sperm deposition into the female seminal receptacle.

The morphology of the spermatozoa varies widely among crustaceans, almost in a species-specific manner. Sperm cell organization also deviates significantly from the typical flagellate spermatozoa of other animal groups. In well-studied decapods, the spermatozoa are not only atypical but also are incapable of motility. Nevertheless, many primitive groups exhibit motile spermatozoa, although the organization of microtubules (9 + 2 filamentous pattern) is very different from that of the typical motile sperm.

Among Decapoda, cellular organelles such as the acrosome and nucleus show remarkable variability, reflecting phylogenetic relationship. Interestingly, the branchiopod spermatozoa lack all the characteristic features usually associated with the concept "spermatozoon" except that they are used to fertilize the eggs (Wingstrand, 1978). Such diversity in the morphological features of sperm among different crustacean taxa may suggest independent evolution from primitive motile spermatozoa.

11.2 MALE REPRODUCTIVE SYSTEM

The male reproductive system in the Crustacea comprises testis and vas deferens and placed dorsally or dorsolaterally with respect to the underlying

Sexual Biology and Reproduction in Crustaceans. http://dx.doi.org/10.1016/B978-0-12-809337-5.00011-3

hepatopancreas (Mauchline, 1980). The testis is either fused or partially joined at the anterior end. In the red claw crayfish *Cherax quadricarinatus*, the testes are paired structures all along their complete length (López Greco et al., 2007). On the other hand, in another crayfish, *Astacus leptodactylus*, the testis has two anterior lobes and only one posterior lobule, suggesting variability in testes morphology within the same group of crustacean (Erkan et al., 2009). Each genital duct consists of collecting tubules within the testis, the vas deferens, a dilated region called seminal vesicle, and finally the ejaculatory duct.

In most of the brachyuran crabs, the paired genital system consists of the testis, vas deferens, and the ejaculatory duct opening into the penis. The reproductive system, especially the vas deferens, fills much of the body cavity in the adult males. In the swimming crab, *Portunus sanguinolentus*, the paired testes are medially interconnected by a commissure, giving the testes, the shape of "H" (Fig. 11.1). The testes are a pair of convoluted tubular organs with many lobes. The testes lie dorsal to the hepatopancreas. The vas deferens arises from the posterior end of the testis (see Fig. 11.1). The male reproductive tract is divided into the proximal, mid, and distal vas deferens. The vas deferens ends with the penile papilla on the medial surface of the coxa of the last periopod (Ryan, 1967). The bilateral male reproductive system in the form of "H," located in the cephalothoracic cavity seems to be a general feature among brachyuran crabs. In *Callinectes danae*, the paired testes are joined to each other by a transverse commissure near the vas deferens (Zara et al., 2012).

The male genital system opens to the exterior through simple gonopores, elevated genital papillae (in some cases) or elaborates copulatory structures that are used as a taxonomic character (Wilson, 2009). In addition, several accessory sex

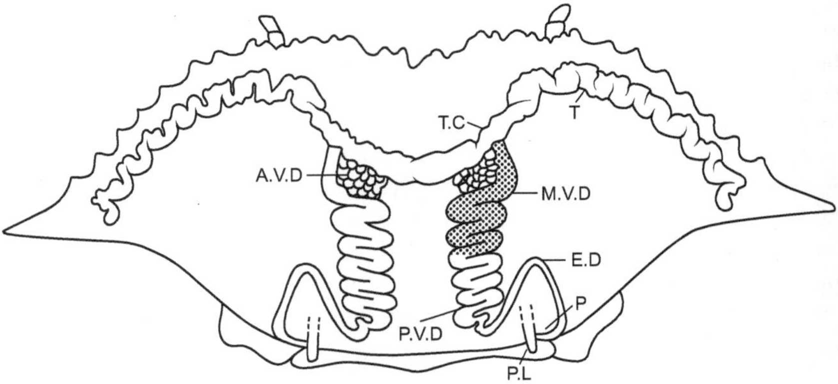

FIGURE 11.1 Male reproductive system in brachyuran crab *Portunus sanguinolentus*. *T*, testis; *T.C.*, testicular commissure; *A.V.D.*, anterior vas deferens; *M.V.D.*, mid vas deferens; *P.V.D.*, posterior vas deferens; *E.D.*, ejaculatory duct; *P*, penis; *P.L.*, first pleopod. *Adapted from Ryan, E.P., 1967. Structure and function of the reproductive system of the crab* Portunus sanguinolentus *(Herbst) (Brachyura: Portunidae). II. The female system. Proc. Symp. Crustacea 2, 522–544.*

glands are also reported to be part of the male reproductive system. Their secretions have obvious functions, such as sperm nutrition, sperm activation, spermatophore dehiscence, and other physiological functions in modifying the behavior of the females after mating (see chapter: Accessory Reproductive Glands).

11.2.1 Morphology of the Testis

The testicular morphology as well as the vas deferens glandular structures varies widely among crustacean taxa, in accordance with the visceral anatomy and the structural complexity of spermatophores. The testes are generally paired structures in most crustacean species, however, it is unpaired in the copepods (Boxshall, 1992), some branchiopods (Martin, 1992), and in Ostracoda (Cohen and Morin, 1990). In peracarides, the male genital organs are paired tubelike structures with testis at the anterior end (Johnson et al., 2001). Furthermore, in Cirripedia, the paired testes look like diffused organs, scattered throughout the connective tissue of the prosoma (Walker, 1992). More variations in the testis morphology are present in other lower crustacean forms (see Lopez-Greco, 2013). In the malacostracan crustaceans, the testes are generally elongated, but some degree of medial fusion is found in decapods. The presence of testicular lobes is also evidenced in penaeiodean shrimps and many crabs. In some decapods, the testes are V-shaped and in others it is H-shaped like the ovary (Kronenberger et al., 2004). For example, in the crayfishes, the H-shaped testes are present in Parastacoidea, whereas in Astacoidea it is Y-shaped (Hobbs et al., 2007).

In the cancrid crab, *Cancer borealis*, the testis, positioned dorsal to the hepatopancreas and under the hypodermis of the carapace, is composed of seminiferous ducts surrounded by numerous large lobes lying in the anterodorsal region (Moriyasu et al., 2002). Each seminiferous tubule is covered by a two-layered wall. The inner cellular wall is thin, whereas the outer layer is composed of thick connective tissue. The germinal epithelium is found on a ridge of the connective tissue. The spermatogonia develop from the germinal layer of the testis and give rise to spermatocytes and spermatids that fill the seminiferous tubule lumen. In all brachyuran crabs, the mature spermatozoa is pushed into the vas deferens, where they are clustered together to form the spermatophore, with their wall secreted by the glandular epithelial cells. In the penaeid shrimp *Trachypenaeus similis*, the testis on each side is composed of five to six lobes, the first with an anterior elongation and a posterior extension inserted among the coils of the median vas deferens (Fig. 11.2). Each lobe is composed of a single, highly convoluted seminiferous tubule (Bauer and Min, 1993). In *Sicyonia ingentis,* the paired testes are lobated and fused medially with each other. The highly coiled proximal vas deferens takes origin from the posterior region of the testis (Shigekawa and Clark., 1986).

Observations of atomic force microscopy revealed that the testicular lobules in the mud crab *Scylla serrata* are found in a radial manner (Mahalingam

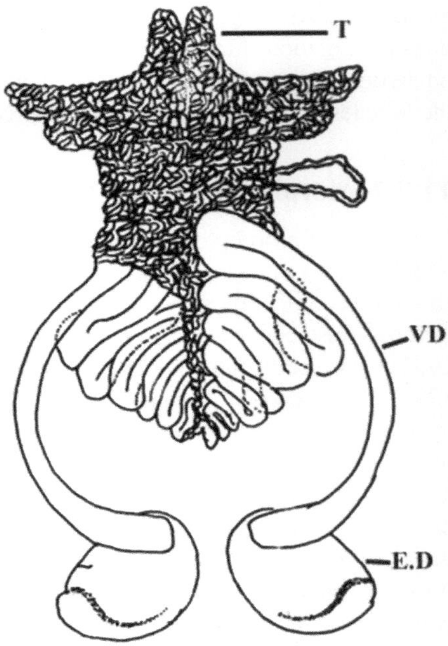

FIGURE 11.2 Male reproductive system in penaeid shrimp *Trachypenaeus similis*. *T*, testis; *V.D.*, vas deferens; *E.D.*, ejaculatory duct. *Adapted from Bauer, R.T., Min, L.J., 1993. Spermatophores and plug substance of the marine shrimp* Trachypenaeus similis *(Crustacea: Decapoda: Penaeidae): formation in the male reproductive tract and disposition in the inseminated female. Biol. Bull. 185 (2), 174–185.*

et al., 2013). These radially branching basal tubules are of different sizes and join together to form a collecting duct (Fig. 11.3). The centrally located tubules serve as a common collecting duct, which then join together in a convoluted manner and run into the proximal vas deferens.

In a typical decapod testis, the seminiferous tubule is a tubelike structure in which the lumen might be visible, depending on the gametogenic stage. From the outer region of the seminiferous tubule toward the interior, the immature stages are found near the base and more advanced stages are found near the center. The lumen of the seminiferous tubule leads into the vas deferens. Within the testicular tubules, besides germ cells, other cell types, variously known as accessory, nurse, Sertoli, Sertoli-like, follicular, intercalary epithelial, or sustentacular, are also reported in different crustaceans (Hinsch, 1993). These somatic cells are reported from lower crustacean orders such as Branchiopoda, Copepoda, Cirripedia, Peracarida, as well as Decapoda. However, their functional role in spermatogenesis is not fully understood. Nevertheless, they have been implicated with functions such as maintenance of sperm numbers by phagocytizing excess spermatids, producing flocculent materials that surround the sperm inside the spermatophore and eliminating excess cytoplasm during spermatogenesis (Hinsch, 1993). In the Cephalocarida, the epithelial cells extend centripetally to form thin septae between clusters of sperm cells (Hessler et al., 1995). It is not known whether the Sertoli cells have any endocrine functions, as that of the vertebrate testis.

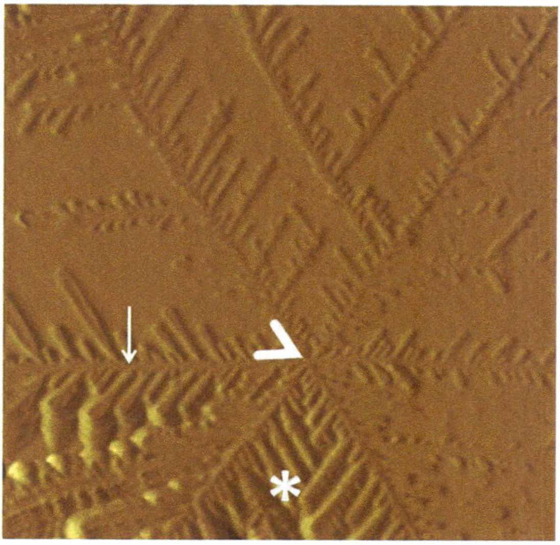

FIGURE 11.3 Arrangement of radially branching testicular lobules in the brachyuran crab *Scylla serrata*. *Arrow* and *arrowhead* indicate a lateral basal tubule and a central collecting duct, respectively. *indicates a typical lobule formation. *Reproduced from Mahalingam, A., Kirubagaran, R., Dharani, G., Subramoniam, T., 2013. Probing the architecture of testis and morphology of male germinal cells in the mud crab* Scylla serrata *with the atomic force microscopy. Turk. J. Biol. 37 (5), 507–513. Permission from* The Scientific and Technological Research Council of Turkey, *2013.*

11.3 SPERMATOGENESIS

Differentiation and maturation of male and female gametes make them competent to take part in fertilization, although their respective role in fertilization is different. Whereas egg is a passive partner, sperm actively moves toward the egg and fertilizes it. Therefore, various structures present in the sperm cells have evolved in such a way as to accomplish their role during fertilization with the egg. Essentially, the function of spermatozoa is to transmit the paternal genome to the egg, in addition to activating it. Obviously, in various animal forms with varied reproductive modes, sperm structure as well as the mechanisms by which the male gametes fertilize the egg is enormously variable. The diversity in spermatozoa morphology and the extent of deviation from the basic pattern prescribed for typical sperm have reached great heights in crustaceans, thanks to the existence of diversified reproductive morphology and phenomena in different taxa.

In all sexually reproducing animals, spermatogenesis occurs in three main phases. The first phase involves the production of spermatogonia, which proliferate, renew, or differentiate to produce spermatocytes; in the second phase primary and secondary spermatocytes undergo meiotic divisions, yielding haploid cells, the spermatids, and the last phase is spermiogenesis, in which transformation of spermatid to mature sperm takes place. These three phases correspond to multiplication phase, meiotic phase, and maturation phase. In Crustacea, the male gametic cells undergo spectacular cellular metamorphosis by way of structural changes, especially during spermiogenesis (Pochon-Masson, 1983).

In Crustacea, spermatogenesis proceeds in a manner similar to that in vertebrates. It begins with the proliferation of spermatogonia and formation of primary spermatocytes in the testis. These then undergo meiosis to become secondary spermatocytes, which differentiates into spermatids, and through spermiogenesis, finally to mature sperm. Mature sperms are then transported to the vas deferens and become encapsulated into spermatophores (see Subramoniam, 1993).

11.3.1 Spermatogenesis in Brachyuran Crabs

In the blue crab *Portunus pelagicus*, Stewart et al. (2010) described spermatogenesis from light and electron microscopic studies. The testis is composed of anterior and posterior lobes. The anterior lobe is highly coiled and contains primarily the germinal zone, while the posterior lobe is less coiled, comprising of germinal, transformational, and evacuation zones in equal proportion. Further, the central region of right and left testes is connected by a commissure, forming an H-shaped paired testicular complex. The two posterior testes are connected with the spermatic duct, which is divisible into three regions, anterior, median, and posterior vas deferens. An ejaculatory duct connects the posterior vas deferens and the spermiductal opening at the base of the last pair of walking legs.

The anterior lobe of the testis is encased in a thin connective tissue capsule and is composed of many lobules. Each lobule contains highly convoluted seminiferous tubules, surrounded by a thin basal lamina. The posterior testis also consists of numerous lobules containing seminiferous tubules and the proximal part of the collecting ducts. The epithelium of a seminiferous lobule is divided into three zones: the germinal zone at one end of the tubule; the transformation zone, filling the central region; and the evacuation zone or collecting tubule that contains mostly spermatozoa. The collecting tubules join together and merge into the proximal vas deferens. The germinal zone contains spermatogonia and accessory nurse cells. Clusters of spermatogonia are located near the basal lamina of the seminiferous tubule. The accessory nurse cells are small and irregular in shape with the nucleus containing highly condensed heterochromatin. In another crab *C. borealis*, Langreth (1969) found the nurse cell to surround several sperm cells and suggested a role in phagocytizing and degradation of residual spermatid cytoplasm. Nurse cells have been described in the testis of other decapod crustaceans, where they are variously designated as intercalary and somatic mesodermal cells, and equated with Sertoli cells of the mammalian testis (Lopez-Greco, 2013).

Regionalized spermatogenic events, starting with the proliferation of spermatogonia in the convex zone at the periphery of the seminiferous tubule may be a characteristic feature of brachyuran testis. Clusters of spermatogonia are located near the basal lamina surrounding the seminiferous tubules. The early stage spermatogonia possess a thin cytoplasm and a large spherical nucleus, which contains small heterochromatin clumps, aggregating at the nuclear envelope. However, in the advanced stage spermatogonia, the nucleus contains only evenly dispersed euchromatin. Within the transformation zone, six stages of primary spermatocytes, including leptotene, zygotene, pachytene, diplotene, diakinesis, and metaphase; a secondary spermatocyte stage, and three stages of spermatids, could be distinguished. The primary spermatocytes are spherical cells and have basophilic cytoplasm, due to the abundant occurrence of ribosomes and mitochondria. Secondary spermatocytes are relatively smaller than primary spermatocytes. Most of the chromatin has become dispersed throughout the nucleus, but some heterochromatin clumps are found bordering the nuclear envelope (Stewart et al., 2010; Zara et al., 2012).

11.4 SPERMIOGENESIS

The spermatogenesis up to the stage of spermatid formation is similar in many crustaceans and, for that matter, does not differ much from other animal species. However, spermiogenesis, the differentiation of spermatid into mature sperm, differs vastly among crustacean taxa, inasmuch as their mature sperm cells exhibit individual variations almost species-specifically (Tudge, 2009; Poljaroen et al., 2010). In typical spermatozoa, spermiogenesis starts with the formation of acrosome, nuclear condensation, and the formation of mitochondrial midpiece and the filamentous tail (Afzelius, 1971). In Crustacea, the acrosome is not only prominent but also shows variation in shape and structure. The nucleus of decapod crustaceans is mainly decondensed, and there is a total lacking of motile tail.

In the crab *P. pelagicus*, the differentiation of spermatids into the mature sperm takes place in three stages (Stewart et al., 2010). The spermatid stage 1 is characterized by the heterochromatin which begins to decondense, while several small clusters of heterochromatin remain close to the nuclear membrane. The second stage is marked by the formation of acrosomal vesicle from the Golgi complex. In stage three, the nucleus is completely decondensed, and there has been formation of a granular belt as the acrosome takes shape.

Fig. 11.4 illustrates the formation of acrosome and changes in nucleus, leading to mature spermatozoa of *P. pelagicus*. The initiation of spermiogenesis is marked by a displacement of nucleus to the posterior pole of the developing sperm as the proacrosomal vesicle grows into the acrosomal vesicle, containing a prominent granule. The sperm nucleus now assumes a cup shape with nuclear expansions into nuclear arms. The nucleus also starts to fold over the acrosomal vesicle. Marked changes occur in the acrosomal matrix, which is divided into a

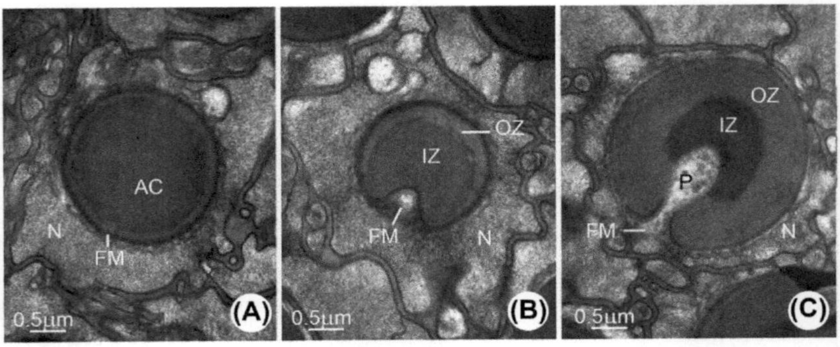

FIGURE 11.4 Maturation of spermatozoa in *Portunus pelagicus*. (A) Invagination of the acrosome vesicle (AC) from the nuclear side, and the space within the invaginated pouch is infiltrated by a loose fibrous material (FM) which originates from the cytoplasm. The nucleus (N) surrounds much of the acrosome which contains a homogenous dense material; (B) invagination extends further into the acrosome and the acrosomal material starts to divide into a dense inner zone (IZ) and a less dense outer zone (OZ); (C) the invagination continues to grow inward to form a tubelike structure that is filled with FM and this structure forms the innermost core of the acrosome, which is equivalent to a subacrosomal space of perforatorium (P). The IZ is now reduced in size but its matrix becomes denser as the OZ. *Reproduced from Stewart, M. J., Stewart, P., Soonklang, N., Linthong, V., Hanna, P. J., Duan, W., Sobhon, P., 2010. Spermatogenesis in the blue swimming crab,* Portunus pelagicus, *and evidence for histones in mature sperm nuclei. Tissue Cell 42 (3), 137–150.*

dense crescent-shaped inner zone, surrounded by a less dense horseshoe-shaped outer zone. The inner zone stretches anteriorly into the acrosomal cap. In the next stage, invagination occurs at the basal region of the acrosomal membrane; as the invagination progresses further inward, it is filled with fibrous materials, leading to the formation of an inner core, called perforatorium. As in other brachyuran crab spermatogenesis reported earlier (Langreth, 1969), decondensation of the chromatin is the result of histone depletion as only traces of sperm nuclear basic proteins remain in the mature sperm. It has been theorized that decondensation of the crustacean sperm chromatin is related to the fact that the sperm do not need to travel far to fertilize the egg; hence there is no need to tightly pack the chromatin to save it from being damaged (Polijaroen, 2010).

The mature sperm of the crab *Uca tangeri* has the typical structure of the brachyuran sperm, with a globular acrosome, cupped by the nucleus, and a thin cytoplasmic band intervening between the former two (Medina and Rodriguez, 1992). At the commencement of spermiogenesis, the nucleus of spermatids has granular chromatin and the cytoplasm contains a proacrosomal vesicle, which is in close approximation with membrane lamellae. In the mid spermatids, a posterior invagination in the acrosomal vesicle membrane gives rise to the formation of the perforatorium, which is a spindle-shaped tubule containing membranous structures. A pair of centrioles located at the base of the acrosome is not directly involved in perforatorial differentiation. The acrosomal vesicle shows a heterogeneous content composed of the operculum, the thickened ring, and three layers of different materials concentrically arranged around the perforatorium. During the last phase of spermatid differentiation, the nuclear profile differentiates numerous slender arms and the chromatin arranges into fibrils. At the same time, membranous tubules from the cytoplasm become incorporated into the perforatorium.

Spermatogenesis in penaeiodean shrimp primarily involves the sequential changes that occur in the morphology of the nucleus. In *Parapenaeus longirostris* the spermatid nucleus is surrounded by the nuclear envelope and consists of condensed chromatin. In the mid spermatid stage, decondensation of the chromatin occurs giving the nucleus a granular appearance. The nuclear envelope is also ruptured. In the late spermatid stage, the chromatin is highly decondensed. In the mature spermatozoa, the chromatin presents a fibrous pattern typical of unistellate spermatozoa. Lipid droplets have also been observed within the nuclear region of several dendrobranchiate spermatozoa (Alfaro et al., 2007). Anterior to the nucleus is the acrosomal complex zoned in subacrosomal region and membrane-bound acrosomal vesicle that is divided into an acrosomal cap and a spike (Medina, 1994, 1995). The acrosomal cap is internally concave and is composed of membrane pouches and anterior granule. The subacrosomal region is located between nucleus and acrosomal vesicle and consists of flocculent material or electron-lucent filaments.

In the marine shrimp, *S. ingentis*, by the completion of meiosis, the early spermatid nuclei are roughly spherical and its chromatin is finely granular, with their heterochromatin scattered uniformly throughout the nucleoplasm (Shigekawa and Clark, 1986). At this first stage of spermiogenesis, the spermatid is highly vesiculated. The vesicles, filled with flocculent materials, originate as outpocketings of the outer membrane of the nuclear envelope. Stage 2 spermatid is characterized by the proliferation of rough endoplasmic reticulum (RER), also formed from the outer membrane of the nuclear envelope. As a result, membrane lamellar bodies (MLBs) are formed and associated with the surface of the nucleus. The membranous vesicles also fuse with one another to form large vesicles. The nucleus is literally surrounded by a layer of convoluted membranes, formed from RER, and the MLB are interspersed in the RER. The mitochondria become segregated and separated from the developing spermatid nucleus. The continued fusion of the vesicles creates extensive membrane-bound regions along the cell periphery which are filled with flocculent materials that in turn give rise to granular materials. The membrane-bound areas continue to coalesce, and the granular materials merge to form the large oval anterior granule.

11.4.1 Acrosome Formation in the Shrimp *Sicyonia ingentis*

Acrosome formation is associated with the Golgi body in many organisms, which produce flagellate spermatozoa (Afzelius, 1972). Acrosomal vesicle formation is the central step in the spermiogenesis of the crustacean sperm. Early cytological investigations on decapod spermatogenesis have attributed acrosome formation to either Golgi complex or mitochondria (Nath, 1965). On the other hand, ultrastructural studies have implicated the involvement of cytoplasmic vacuoles in the formation of acrosomal vesicles. Pochon-Masson (1963) found evidence that these cytoplasmic vacuoles could originate from membranes derived from endoplasmic reticulum as well as nuclear envelopes. In *S. ingentis*, Shigekawa and Clark (1986) observed that blebbing of the outer nuclear membrane gives rise to the proacrosomal vesicle. In addition, the nucleoplasm may be the source of the flocculent material which fills the proacrosomal vesicles. In this marine shrimp, the cellular polarity of the spermatid is also established with the formation of the anterior granule. With the fusion of other membrane-bound vesicles with the anterior granule-containing large vesicle, the acrosomal vesicle is established.

At this stage, the nucleus with its peripheral RER is relegated to the posterior pole of the spermatid. There occurs a decondensation of chromatin to a fibrillar state. The posterior acrosomal membrane merges with the anterior portion of the fibrillar nucleus. The flocculent nucleus, on other sides, is surrounded by a band of cytoplasm consisting of degenerate endoplasmic reticulum. The apical and lateral portions of the acrosomal vesicle membrane become closely opposed to the spermatid plasma membrane, thus forming the unique pentalaminar membrane observed in mature sperm (Kleve et al., 1980). The anterior granule becomes more homogeneous in appearance. The unistellate spike primordia condense in the localized region immediately overlying the anterior granule. At this time, the anterior granule abuts the posterior acrosomal vesicle membrane. The acrosomal vesicle in the mature sperm contains special membrane pouches.

After the completion of the acrosomal vesicle, the subacrosomal region is developed by the formation of an indentation into the posterior surface of the acrosomal vesicle, directly posterior to the anterior granule. In *S. ingentis*, the subacrosomal materials are produced from the perinuclear RER which condense into a granular layer, to finally mature into various subacrosomal components such as the crystalline lattice. These components play a major role in acrosome reaction during sperm activation and fertilization (Clark and Griffin, 1988).

A nuclear plate develops at the posterior subacrosomal region as a narrow layer of dense granular material. The nuclear plate segregates the chromatin from the granular core of the subacrosome. Anteriorly, the subacrosome is separated from the components of the acrosomal vesicle by the inner acrosomal membrane.

Cytoplasmic organelles such as Golgi bodies, mitochondria, and centrioles, organelles typically associated with spermatogenesis, are not universally present in decapod spermatids during their development. During spermatogenesis in *S. ingentis*, mitochondria and Golgi complex are only found in the extracellular matrix of the syncytial spermatids and not found in mature sperm, suggesting that these organelles are expelled from the sperm cell by way of sloughing

(Shigekawa and Clark, 1986). The cytoplasm surrounding the nucleus is segregated from the contents of the acrosomal vesicle and is distinct in appearance. The dense perinuclear cytoplasm is largely occupied by the convoluted membranes of the endoplasmic reticulum.

In *S. ingentis*, the later part of the spermiogenesis is completed when the sperm pass through the vas deferens. The elongation of the spike primordia into the fully polymerized spike takes place inside the vas deferens. Subramoniam (1995) suggested that electron-dense materials present in the mid vas deferens may have a role in sperm maturation (Fig. 11.5). The condensation of the subacrosomal material into structures like crystalline lattice is also completed when the sperm is in the vas deferens. More significantly, the sperm of *S. ingentis* acquires a kind of capacitation only after their transfer to the female seminal receptacle (Clark et al., 1986). Only the "female sperm" is capable of undergoing acrosomal reaction spontaneously in response to treatment with egg water (Griffin et al., 1987).

11.5 SPERM MORPHOLOGY

A typical sperm is a highly polarized cell with its organelles arranged in an anterior–posterior direction. Accordingly, the sperm is comprised of head, midpiece, and tail. The head contains the anteriorly placed acrosome, the organelle

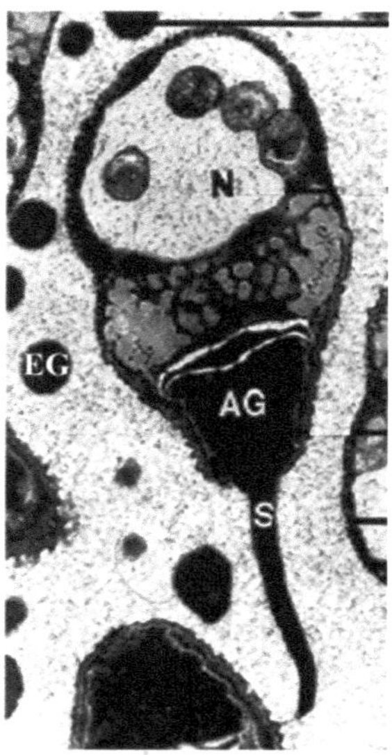

FIGURE 11.5 Electron micrograph of mature spermatozoa and electron-dense granules present in the lumen of the mid vas deferens of *Sicyonia ingentis*. *N*, nucleus; *AG*, acrosomal granule; *S*, spike; *EG*, electron-dense granule. *Reproduced from Subramoniam, T., 1995. Light and electron microscopic studies on the seminal secretions and the vas deferens of the penaeiodean shrimp,* Sicyonia ingentis. *J. Biosci. 20 (5), 691–706. Permission from* Indian Academy of Sciences, *1995.*

unique to sperm cells alone, and an equivalent not found in any somatic cell, and the nucleus having a condensed chromatin material bound by a thick nonporous nuclear membrane. The head is followed by a midpiece characterized by the presence of a well-developed mitochondria ensheathing an inner region containing the centriolar apparatus followed posteriorly by an axonemal complex. The tail contains the 9+2 axial filamental complex, facilitating the lashing movement of spermatozoa. All the important components of the sperm distinctly function in achieving fertilization: acrosome in sperm activation and penetration into the egg; nucleus with its ability to form male pronucleus inside the egg and final fusion with female pronucleus; mitochondrial midpiece providing kinetic energy for sperm motility; and tail acting as the motile apparatus to approach the egg (Fawcett, 1961).

Crustacean spermatozoa are very diverse in their morphology and hence it is difficult to designate sperm features that characterize the entire class (Jamieson, 1991). Nevertheless, these variations in sperm morphology are very useful in determining relationships between crustacean taxa. Crustacean spermatozoa are generally characterized as atypical, nonmotile gametes, with bizarre shapes, not found in any other animal group. Particularly, all decapod crustacean sperm lack a flagellum and a mitochondrial middle piece. Only in primitive groups such as the Ostracoda, members of Maxillopoda, such as Mystacocarida, Branchiura, and Cirripedia, the spermatozoa are motile (Pochon-Masson and Gharagozlou-van-Ginneken, 1978). However, many of the motile spermatozoa described in these groups do not conform to the typical flagellate pattern. Loss of motility of decapod spermatozoa is accompanied by structural modification within the sperm cells, especially with reference to the absence of a midpiece with mitochondria.

11.5.1 Decapod Spermatozoa

Because of their uniqueness and nonmotile characteristics, decapod crustaceans have received considerable attention from spermatologists. In general, the decapod crustacean sperm are comprised of a main body, which houses an uncondensed nucleus, and a highly complex acrosome. From the main body arise variable numbers of stellate processes that project outside. Only in *Stenopus hispidus* (infraorder Stenopodidea), the elliptical sperm cell lacks appendages (Burkenroad, 1981).

The following features characterize the decapod spermatozoa:

- All decapod spermatozoa are aflagellate; lack the 9+2 axoneme complex and hence are immotile.
- The large concentrically zoned acrosomal vesicle constitutes the apical portion of the sperm cell; it is derived from endoplasmic reticular elements, rather than the Golgi complex. It is, however, PAS-positive.
- The sperm nucleus is composed of diffuse, filamentous, heterogeneous chromatin fibers rather than being dense; it has an irregular and often amoeboid

outline, always found at the posterior end of the cell. The absence of nuclear histones and protamine justify the uncondensed condition of chromatin in the nucleus.

- Nucleus composed of diffuse, filamentous, heterogeneous chromatin fibers.
- The decapod sperm nuclear protein is unique and does not belong to any of the sperm-specific basic proteins found in other sperm types; a unique protein decapodin is present in the acrosomal vesicle.
- The appendages or spikes present in the collar region are of nuclear or cytoplasmic origin. The nuclear arms are often packed with microtubules, derived from the spermatid centrioles in some brachyurans, but serving no motile functions. The arms merely serve to attach the sperm to the oocyte surface before fertilization (Brown, 1966).
- The mitochondria in mature sperm are degenerate, nonchristate, and nonfunctional, or absent in mature spermatozoa.
- Positioning of centrioles, if present, is always anterior to the nucleus; they do not show any continuity of structures.
- Decapod spermatozoa have a remarkable abundance of polysaccharides and proteins.

In spite of these common characteristics found in the decapod sperm, further variations do occur almost in a species-specific manner. The sperm of decapods have been divided into two basic morphological groups: (1) reptantian type (crab, lobster and crayfish) characterized by a membrane-limited nucleus with uncondensed chromatin, nuclear spikes containing microtubules, a complex acrosome, and a membrane lamellar region; and (2) natantian type (shrimp and prawns) with only one spike that is thought to be originating from the acrosome, and a nucleus which is not membrane bound (Dudenhausen and Talbot, 1982).

11.5.2 Reptantian Spermatozoa

The reptantian sperm is multistellate, with radiating appendages (also called spikes or arms), extending from the main body. However, these stellate appendages are not homologous to the unistellate spike of the natantian decapods. The origin, morphology, and number of radial arms vary among different decapod species. For example, in some species of brachyuran crabs, the arms appear to originate as extensions of the nucleus, whereas those of other species appear to be extensions of the cytoplasm (Hinsch, 1986). Similarly, microtubules that give structural support to the arms are also present in some species, whereas others are devoid of it. In the spider crab, *Libinia emarginata*, the radiating appendages contain both microtubules and nuclear extension, whereas in the lobsters such as *Homarus americanus* the arms contain microtubules but lack chromatin. A converse condition exists in the crab *Ovalipes ocellatus*, in which the arms lack microtubules but contain chromatin (Hinsch, 1986; Talbot and Summers, 1978). The microtubules when present in the spikes are not arranged in the 9+2 axoneme characteristic of typical flagellated sperm. The spikes are stationary and the sperm is immotile. The number of spikes can also vary from 20 or more in *Procambarus leonensis* to a few (three) robust arms in *Iliacantha subglobosa* (Felgenhauer and Abele, 1991). In addition to the presence of lateral arms, a posterior median process is also present in Majidae and in *Ranina* (Jamieson, 1989).

Reptantian spermatozoa have a complex, but prominent acrosome and a membrane lamellar complex. Usually surrounded by the nucleus, the acrosome consists of a highly refringent, electron-dense acrosomal vesicle that houses the inner acrosomal tubule. The acrosomal vesicle stains positive with the periodic acid-Schiff (PAS) reagent. The acrosomal vesicle is derived mainly from the fusion of RER cisternae as well as some nuclear envelope derivatives. During spermiogenesis, acrosome begins as an osmiophilic granule, which becomes apical in the developing spermatozoan. The acrosomal complex is extremely uniform among the brachyuran crabs. In Brachyura, the acrosomal vesicle has a characteristic bilayered appearance and the layers have been termed the inner and outer regions of the vesicle. The acrosomal tubule (or perforatorium) is located within the acrosomal vesicle and contains filaments or microtubules depending on the species. While this is the basic plan of acrosomal complex in brachyuran crabs, it differs significantly in the astacoid crayfish (Dudenhausen and Talbot, 1983).

Thus, in all freshwater crayfish species, acrosomal vesicle is horseshoe-shaped. The vesicle has a distinct crystalline nature produced by periodically arranged parallel laminae. In addition, no acrosomal tubule or perforatorium within the astacoid spermatozoa has been described. However, Felgenhauer and Abele (1991) have discerned a prominent acrosomal tubule just below the acrosomal cap in the anterior end of the sperm. Interestingly, cell membrane of astacoid spermatozoan is extremely thick and forms what has been termed a cell capsule.

11.5.2.1 Ultrastructure of Crab Spermatozoa

The sperm of *Trapezia coerulea* contains anteriorly a subspheroidal acrosome and an extensive nucleus that surrounds the entire acrosomal vesicle, except the anterior tip which is enveloped by a conical, caplike operculum (Fig. 11.6). The conical operculum has a subopercular region underneath it. The apical operculum in *T. coerulea* extends sideways and looks imperforate, although in other crab species such as *Ranina* and *Menaetheus monoceros,* it is not only perforate but also weakened (Jamieson, 1991). The acrosomal vesicle is divisible into an electron-dense inner zones and outer zone, containing fibrillar materials. In addition, there is an acrosomal ray zone at the periphery of the inner acrosome in the anterior end. The acrosome is invested by an acrosomal membrane, which invaginates at the posterior pole to cover a columnar subacrosomal chamber that contains the perforatorium, a structure equivalent to "acrosomal tubule" (Hinsch, 1973), "axial rod" (Chevaillier, 1967), and percutor organ (Pochon-Masson, 1983) in other brachyuran crab sperm. The perforatorium in *Trapezia* is a stout cylinder, tapering anteriorly, and with a broad stalk toward the posterior

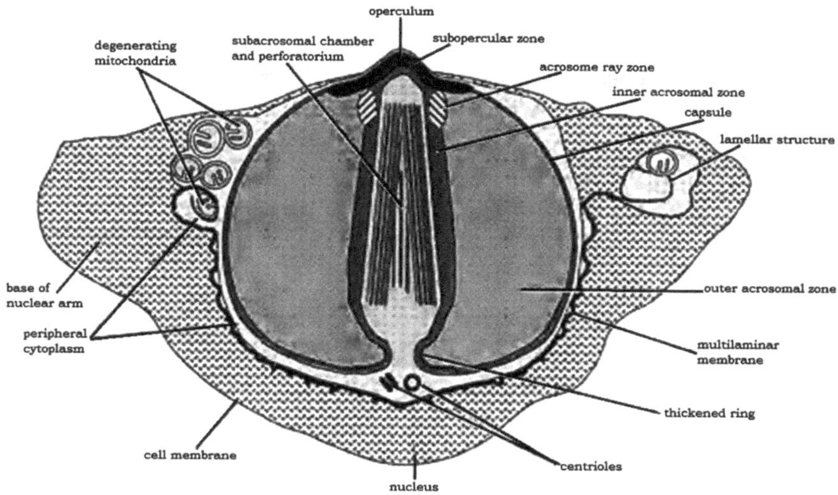

FIGURE 11.6 Diagrammatic representation of a longitudinal sagittal section of the spermatozoan of *Trapezia coerulea*. *Adapted from Jamieson, B.G.M., 1991. Ultrastructure and phylogeny of crustacean spermatozoa. Mem. Queensland Mus. 31, 109–142.*

end. The perforatorium is seen to contain microtubules of varying sizes. A thin layer of cytoplasm surrounds the acrosomal capsule. There are three lateral projections or arms that contain nuclear chromatin materials.

In another brachyuran crab, *Metopograpsus messor*, the acrosomal vesicle is made of three zones which are concentrically arranged around the inner perforatorium. The outer acrosomal zone is made of concentric "onion ring" lamellation, whereas the middle zone is relatively dense. The inner zone encloses the basally open subacrosomal chamber, which in turn contains the perforatorial material (Anilkumar et al., 1996). The perforatorium extends up to the anterior end of the subacrosomal chamber in many brachyuran crabs. In *P. pelagicus*, the subacrosomal material consists of a stout perforatorium which extends from the posterior perforation in the acrosomal capsule to the vicinity of the operculum at the apex of the acrosome (Jamieson, 1989). Although the crab perforatorium looks similar to the acrosomal process of flagellate typical sperm, it originates as an invagination of the acrosomal vesicle membrane during spermiogenesis in the crab *U. tangeri* (Medina and Rodriguez, 1992). The mature sperm of *U. tangeri* is lens-shaped, with a voluminous acrosome, and a cup-shaped nucleus which contains fibrous chromatin and numerous radial arms wrapped around the cell at the equatorial plane. The cytoplasmic region is much reduced, although some membrane materials, including mitochondria are retained anteriorly. The lower acrosome surface is cupped by the nucleus, but its anterior face is free. The perforatorium consists of membrane tubules followed basally by filamentous or granular material, which is in continuity with the cytoplasmic subacrosomal band. Centrioles are absent in the subacrosomal region.

In brachyurans, the cytoplasm of the sperm is restricted to the periacrosomal region, as a thin layer, and surrounds the acrosomal capsule. However, there is extension of the cytoplasm into the basal region of the nuclear arms in the sperm of *Ranina ranina* (Jamieson, 1989). The cytoplasm is enlarged anterolaterally on one side of the acrosomal capsule and contains several subspheroidal and degenerating mitochondria with sparse cristae. The postacrosomal cytoplasm in *Trapezia*, which is typical of brachyurans, is well developed and contains two centrioles. Centrioles are absent in species belonging to xanthids and thoracotremes (Jamieson, 1991). Nuclear material, consisting of uncondensed chromatin, is located in the lateral arms and their branches, and seen around the cup-shaped acrosome and its cytoplasmic sheath. The chromatin also extends around the operculum, leaving only the raised central area. The chromatin consists of electron-dense filaments, which are digested by DNAse and histochemically, they are Feulgen-positive (Yasuzumi, 1960). At the anterior-most portion of the acrosome, there is a conical operculum, with a subopercular region underneath it. Immediately below the subopercular region, and anterior to the acrosome vesicle, there is a dense accessory opercular ring with horizontal orientation. From the base of the acrosome, along its central core, and extending to the subopercular region, there is a longitudinal groove, referred to as the "perforatorium." The lumen of the perforatorium (also called subacrosomal chamber by Jamieson, 1993) has tubular formations. Medina and Rodriguez (1992) suggested that in *U. tangeri*, these perforatorial tubes could serve as a membrane source for male pronuclear formation during fertilization.

Jamieson (1994) investigated spermatozoa belonging to different decapod taxa to investigate the structural basis of the brachyuran phylogeny. In *Neodorippe astute* belonging to Heterotremata the spermatozoal structure is similar to *T. coerulea*. However, the acrosomal ray zone envelops two-thirds of perforatium length. In addition, the bulk of the nucleus is disposed posterior to the acrosome rather than the lateral sides of the sperm. In this respect, the crenulated nuclear extension is more like that of anomuran spermatozoa. The equatorial cytoplasmic band separates the anterior acrosome from the nucleus, as in *T. coerulea* sperm. In addition, the bulk of the nucleus is disposed posterior to the acrosome rather than the lateral sides of the sperm. In this respect, the crenulated nuclear extension is more like anomuran spermatozoa. However, in yet another Portunid crab, *P. pelagicus*, the nucleus envelops the acrosome in all directions, except the anterior-most operculum. Jamieson (1994) used the spermatozoal ultrastructure to suggest that Brachyura is a monophyletic taxon relative to the outgroup, three paguroids.

11.5.2.2 Anomuran Sperm

The species-specific morphological variation in the crustacean spermatozoa has reached a remarkable level in the infraorder Anomura. Despite having its own characteristic features, the sperm morphology of different representative species of its four major superfamilies, Paguroidea, Lomoidea, Galatheoidea, and

Hippoidea, exhibits distinctive variation from the general anomuran plan. The basic sperm morphology among anomurans is similar to that found in Brachyura. Differences that exist between these two groups are mainly related to the shape of the acrosomal vesicle, which is spherical in Brachyura and elongated in anomuran sperm; in addition to opercular shape and acrosome zonation in anomurans. Nevertheless, different taxa belonging to Anomura share a common sperm form consisting of an often large acrosomal vesicle, a posterior nucleus of variable density, intervening cytoplasm containing mitochondria, microtubules, lamellar structures, and sometimes centrioles, and a variable number of arms or spikes (Tudge, 1995).

The sperm structure in the family Coenobitidae combines the following characters: (1) a large cylindrical to oblong-ovoid acrosomal vesicle with a conical operculum and a deeply invaginated columnar perforatorium, which is not divided into a posterior bulb and anterior projection, (2) the posterior wall of the perforatorium extends into the lumen as long, microvillar projections, (3) the subopercular zone is divided into two distinct regions with the posterior-most region extending down the center of the acrosomal vesicle to abut the inner acrosome zone, (4) three long microtubular arms enter the cytoplasmic region, midway between the nucleus and the acrosomal vesicle, and join around the base of the vesicle in a triad, and (5) mitochondria and lamellae are prominent components of the cytoplasm (Tudge and Jamieson, 1991).

In *Coenobita spinosus*, the sperm is composed chiefly of a long cylindrical acrosomal vesicle, a thin cytoplasmic cup which invests the posterior end of the acrosomal vesicle (Fig. 11.7A). Posteriorly, the globular nucleus is limited by a double membrane, termed nucleoplasma membrane and a cytoplasmic region in between. In addition to the concentrically zoned acrosome vesicle, there is a conspicuous acrosome ray zone that forms a thick cylinder extending from the open end of the perforatorial chamber to the operculum. The acrosomal vesicle is penetrated by a columnar perforatorial chamber with microtubular projections inside the lumen. The anterior pole of the acrosome vesicle is capped by a conical operculum, below which occurs a divided subopercular zone.

In *Birgus latro*, the spermatozoan differs from that of *Coenobita* in that the acrosomal vesicle is more oblong-ovoid than cylindrical, and the nuclear material is drawn out into filamentous extensions or arms, in addition to microtubular arms (Fig. 11.7B).

The family Diogenidae is known for its varied sperm features, with reference to the shape of acrosomal vesicle. For example, in *Calcinus* sp., the acrosome vesicle is spherical, whereas in *Cancellus*, it is cylindrical in shape with a conspicuous acrosome ray zone, which is penetrated by a large tapering perforatorial chamber. On the other hand, different species of the genus *Clibanarius* have ovoid acrosome vesicles, penetrated posteriorly by a perforatorial chamber, having a bulbous posterior region. In *Diogenes* species, the perforatorial chamber has the posterior bulb and anterior projection, but there is a prominent acrosome ray zone, as in coenobitids and

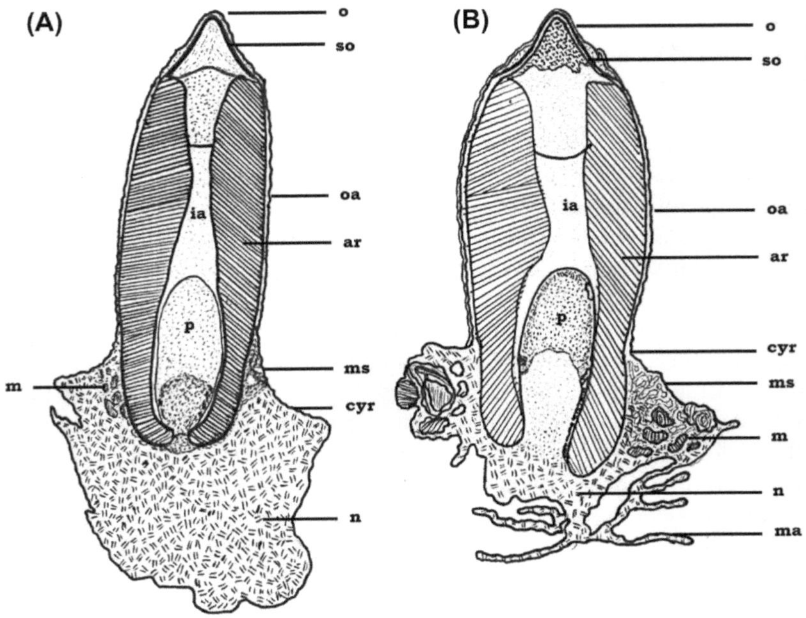

FIGURE 11.7 Semidiagrammatic longitudinal section of spermatozoa in (A) *Coenobita spinosus*, and (B) *Birgus latro. ar*, acrosome ray zone; *cyr*, cytoplasmic region; *ia*, inner acrosome region; *m*, mitochondrion; *ms*, membrane system; *n*, nuclear material; *o*, operculum; *p*, perforatorial column; *so*, subopercular zone; *oa*, outer acrosome zone; *ma*, microtubular arm. *Adapted from Tudge, C.C., Jamieson, B.G.M., 1991. Ultrastructure of the mature spermatozoon of the coconut crab* Birgus latro *(Coenobitidae: Paguroidea: Decapoda). Mar. Biol. 108 (3), 395–402. Permission from Springer.*

dardanids. The spherical acrosome of *Clibanarius virescens* is capped by a thin operculum and features a well-developed perforatorial chamber (Fig. 11.8). The spermatozoa of *Dardanus crassimanus* have also an ovoid acrosome vesicle (Fig. 11.9).

In the family Paguridae, the sperm has an ovoid acrosomal vesicle with one or more reticulated acrosome zones. While other characteristics are similar to that of Diogenidae, the spermatozoa of *Xylopagurus* have the microtubular arms emerging from the cytoplasm in the anterior part of the sperm cell and not posteriorly; this genus also possesses a large cylindrical acrosomal vesicle with unusual zonation. The presence of a reticulated acrosome is also shared with the spermatozoa of the family Parapaguridae. However, the latter differs from other pagurids in having the acrosome vesicle extended anteriorly to more than three times the length of any pagurid acrosome.

In the hermit crab, *Pylocheles* sp., the spermatozoa are large cells with a circular, depressed-ovoid acrosome vesicle, occupying the apical pole of the cell (Tudge et al., 2001). This "depressed shape" of the acrosomal vesicle

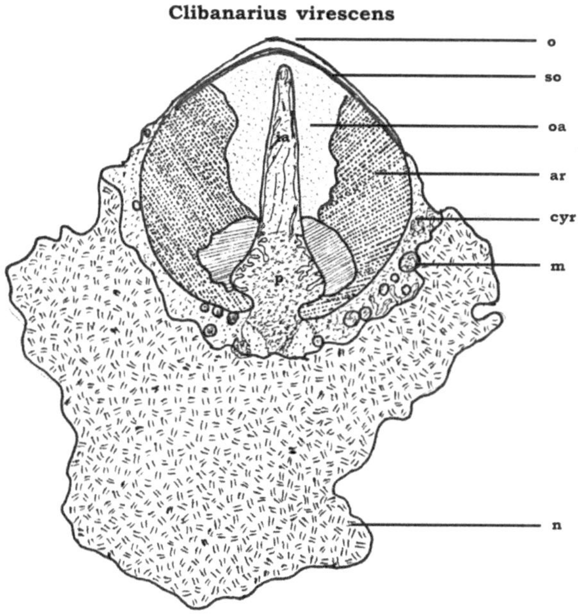

FIGURE 11.8 Semidiagrammatic longitudinal section of spermatozoa in *Clibanarius virescens*. *ar*, acrosome ray zone; *cyr*, cytoplasmic region; *ia*, inner acrosome region; *m*, mitochondrion; *ms*, membrane system; *n*, nuclear material; *o*, operculum; *p*, perforatorial column; *so*, subopercular zone; *oa*, outer acrosome zone. *Adapted from Tudge, C. C., 1992. Comparative ultrastructure of hermit crab spermatozoa (Decapoda: Anomura: Paguroidea). J. Crust. Biol. 12 (3), 397–409.*

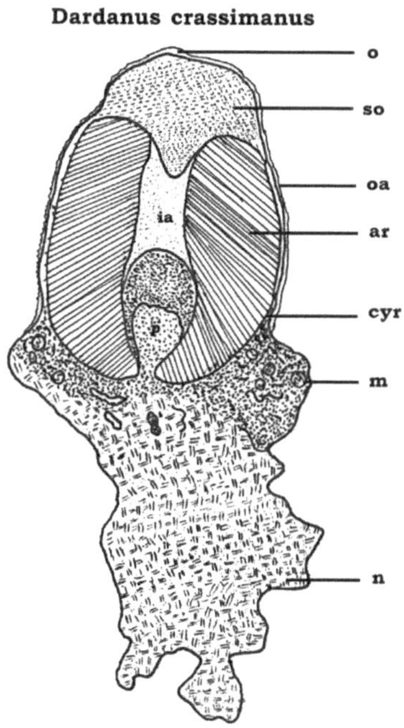

FIGURE 11.9 Semidiagrammatic longitudinal section of spermatozoa in *Dardanus crassimanus*. *ar*, acrosome ray zone; *cyr*, cytoplasmic region; *ia*, inner acrosome region; *m*, mitochondrion; *ms*, membrane system; *n*, nuclear material; *o*, operculum; *p*, perforatorial column; *so*, subopercular zone; *oa*, outer acrosome zone. *Reproduced from Tudge, C. C., 1992. Comparative ultrastructure of hermit crab spermatozoa (Decapoda: Anomura: Paguroidea). J. Crust. Biol. 12 (3), 397–409.*

is unique among anomurans, but closer in shape to podotreme brachyurans and palinurid lobsters and crayfish described by Jamieson (1991). The contents of the acrosomal vesicle are arranged in three concentric zones: an inner, an outer, and a peripheral acrosome zone. The peripheral acrosome zone extends around the periphery of the acrosome vesicle, exterior to the inner and outer acrosome zones, from the perforate operculum to the base of the perforatorial chamber, which is a large but shallow invagination into the posterior region of the acrosomal vesicle. The extensive cytoplasm forms an enveloping layer around the acrosome vesicle except for the region surrounding the apical operculum. The cytoplasm includes many spherical mitochondria and a variety of membrane systems. The arms originate from the cytoplasm and extend away from the main body, but surrounded by nuclear membrane. The contents of the arms appear to be microtubule-like material. The nucleus, occupying the posterior pole of the sperm cell, is a plume of electron-lucent material, enveloping the cytoplasm laterally, as a thin surface layer.

Variation in the shape of the anomuran acrosomal vesicle is encountered in other families also. For example, in Lomidae, the acrosomal vesicle is in the shape of an inverted cup, penetrated by a perforatorial chamber and completely embedded in the cytoplasm. In Galatheidae and Porcellanidae, the acrosomal vesicle is elongate and generally cylindrical like other anomurans, but the inner walls of the perforatorial chamber are folded into septa (Tudge, 1995). In the family Hippidae also, a membranous septum divides the perforatorial chamber longitudinally (Tudge et al., 1999).

Anomuran spermatology continues to give bizarre forms of the male gamete. In the only freshwater anomuran *Aegla longirostri*, there is a distinctive suite of spermatozoa characters. The sperm is divided into two hemispheres, the upper half containing cytoplasm and a very small spherical to ovoid acrosome vesicle embedded in the cytoplasm. The lower half contains the uncondensed nucleus. However, the outer dense region of the acrosome covers the apex in *Lomis*, whereas it is open in *Aegla*. The microtubular bundles are present in both the cytoplasm and nucleus in *Lomis*, while being restricted to the nucleus in *Aegla* (Tudge and Scheltinga, 2002). The spermatozoa of the hairy stone crab, *Lomis hirta*, belonging to Lomoidea shows distinctive variations but possess microtubular arms and a concentrically arranged acrosome vesicle posteriorly penetrated by a perforatorial chamber, which justify its position in the Anomura (Tudge, 1997). However, the small acrosome vesicle completely embedded in the cytoplasm is unique only to this anomuran family, although such a condition approximates thalassinoids such as *Axius* and *Callianassa* (Tudge, 1995). Another characteristic feature in *Lomis* is that the microtubular bundles, which is concentrated in the cytoplasm, extends into the nucleus. The discontinuous operculum surrounds the entire acrosome vesicle, and not just the apical pole, as in other anomurans. Presumably, the anomuran spermatozoa appears to have originated from that of Lomidae by undergoing the following modifications: (1) increasing the size and complexity of the acrosomal vesicle, (2) restricting the operculum to the apical pole, (3) shifting the acrosome vesicle to a superior position on the cytoplasm, and (4) establishing three, equidistant microtubular arms (Tudge and Sccheltinga, 2002).

Ultrastructural aspects of sperm morphology in Lithodidae have unraveled its phylogenetic relationship with the superfamily Paguroidea, by sharing many features, common to the two anomuran taxa (Tudge et al., 1998). Each spermatozoon contains a subspherical, concentrically zoned acrosomal vesicle, capped by a centrally perforate operculum and penetrated from pole to pole by a wide chamber. Relatively thin cytoplasmic region envelops the acrosomal vesicle at it base and contains mitochondria, membrane systems, and lamellae, and the bases of three microtubular arms. The nucleus, with a crenulated external boundary, envelops the acrosomal vesicle and cytoplasm (Tudge et al., 1998).

11.5.2.3 Lobster Spermatozoa

The spermatozoa of lobsters in general conform to the basic reptantian pattern in that the nucleus is uncondensed and membrane-limited, multiple nuclear spikes contain microtubules, the acrosomal region is structurally complex, and a lamellar region and centrioles are present in the subacrosomal region. In the American lobster, *H. americanus*, the sperm consist of an acrosome, subacrosomal region, collar, nucleus, and three spikes which are extensions of the nucleus (Talbot and Chanmanon, 1980a). As shown in Fig. 11.10A, the acrosome is in the form of a cylinder, projecting from the sperm's anterior pole. It is composed of inner and outer acrosomal materials and is limited by a continuous, single membrane. The anteriorly situated cytoplasmic zone has mitochondria and conspicuous lamellar bodies. The apical cap, located at the anterior-most portion of the acrosome contains four zones. The outer three of these are organized concentrically into a ring, while the fourth partially fills the center of this ring. The apical cap has a deep depression in the center and gives a weak reaction to PAS suggesting that the apical cap is not a part of acrosome.

At the base of the acrosome is a large fossa, which is filled with subacrosomal material of finely granular and moderately electron-dense substance. This material is continuous with the chromatin of the nucleus. Immediately below the subacrosomal zone is the "collar" which includes the centrioles and degenerate mitochondria, with poorly defined cristae. The nucleus contains uncondensed chromatin which is finely fibrillar or granular and moderately electron dense. The nucleus extends around the acrosome's lateral edges, forming a small cuff containing chromatin at the acrosome's base. The nuclear membrane between the base of the acrosome and the nucleus appears to be degenerated and the membrane limiting the body of the nucleus is thrown into many folds. *H. americanus* sperm have three spikes, which appear to be lined by nuclear envelope fused with the plasma membrane. Each spike's compartment is continuous with the nucleus, and all spikes are packed with microtubules surrounded by sheets

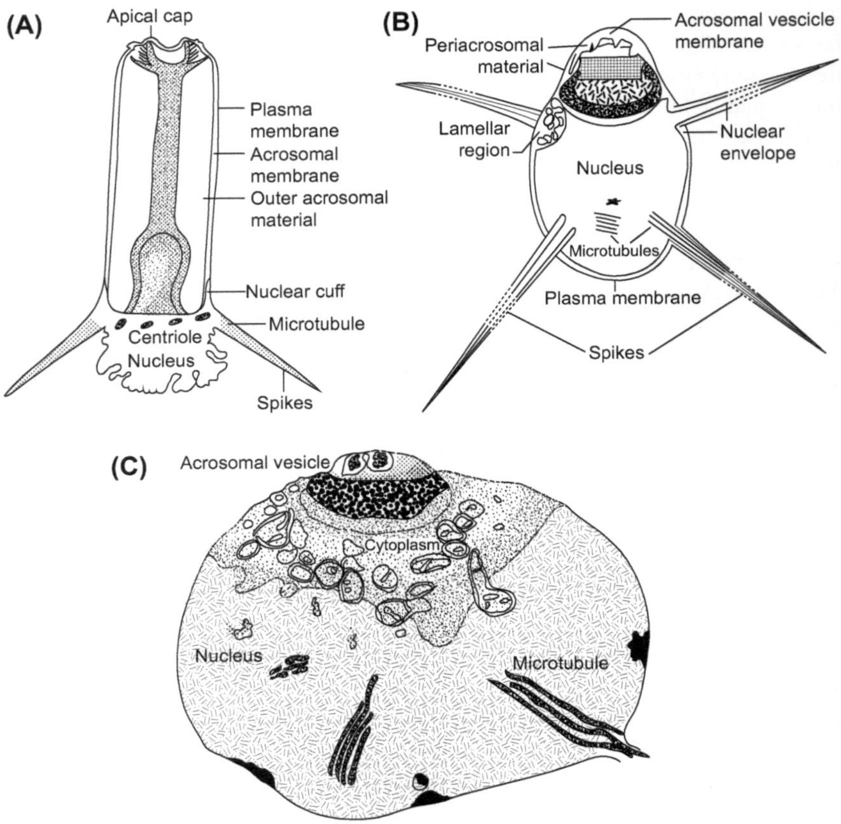

FIGURE 11.10 (A) Sperm structure of American lobster *Homarus americanus (Adapted from Talbot, P., Chanmanon, P., 1980a. The structure of sperm from the lobster, Homarus americanus. J. Ultrastruc. Res. 70, 275–286.)*. The acrosome includes the acrosomal material, outer acrosomal material, and apical cap; (B) sperm structure of spiny lobster *Palinurus argus (Adapted from Talbot, P., Summers, R.G., 1978. The structure of sperm from Panulirus, the spiny lobster, with special regard to the acrosome. J. Ultrastruc. Res. 64, 341–351.)*; (C) semidiagrammatic longitudinal section of spermatozoa of lobster Jasus novaehollandiae *(Redrawn from Tudge, C.C. Scheltinga, D.M., 2002. Spermatozoa morphology of the freshwater anomuran Aegla longirostri Bond-Buckup & Buckup, 1994 (Crustacea: Decapoda: Aeglidae) from South America. Proc. Biol. Soc. Washington 115, 118–128.)*.

of membrane which appear pentalaminar. Obviously, the spikes join each other in the collar region by the continuity of microtubules within the spikes.

The posteriorly placed Feulgen-positive nucleus contains uncondensed finely fibrillar or granular chromatin, which is continuous with the membrane limiting the sperm, but does not project out into the spikes. The nucleus extends around the acrosome's lateral edges, forming a small cuff containing chromatin at the base of the acrosome. The posterior border of the uncondensed nucleus has an irregular and often amoeboid outline.

The structure of the sperm from the spiny lobster *Panulirus* shows considerable variation from that of the American lobster *H. americanus* (Talbot and Summers, 1978). In the Nomarski interference contrast micrograph, the sperm of *Palinurus argus* appears spherical with a refractile lens-shaped acrosomal vesicle at one pole. A variable number of spikes radiate from the sphere except in the region of the PAS-positive acrosomal body. At the fine structural level, each sperm comprises a nucleus, lamellar region, and acrosome (Fig. 11.10B). The nucleus contains uncondensed chromatin and is limited by a nuclear envelope. A variable number of spikes extend from the nucleus. The spikes are lined by evaginated nuclear envelope and are packed with microtubules, which span the nucleus proper. The lamellar region lies outside the nuclear membrane and is situated at one side of the acrosome. The lamellar body is composed of residual cytoplasm, numerous stacks of membrane, and small mitochondrial bodies. The acrosomal vesicle is lens-shaped and limited entirely by the acrosomal membrane. It is structurally complex, comprised by four morphologically discrete regions. The posterior-most homogeneous region forms a narrow cup at the lower periphery of the acrosome and extends laterally to surround the lateral and basal surfaces of the anterior regions of the acrosome. The acrosomal vesicle membrane surrounds the anterior tip of the vesicle and laterally joins with the homogeneous region of the acrosome. There is no acrosomal cap found in this lobster. The periacrosomal material surrounds the acrosomal vesicle anterolaterally. In *Homarus,* the periacrosomal material extends into a shallow cavity in the center of the base of the acrosomal vesicle. However, in *Panulirus*, the acrosome lacks a central depression at the base of the acrosome, and therefore there is no formation of a percutor organ.

In another palinurid spiny lobster, *Jasus novaehollandiae*, the spermatozoa are globular cells with a dense acrosomal vesicle at one pole and three or four microtubular arms (Tudge et al., 1998). The acrosomal vesicle is a membrane-bound, oval-shaped and electron-dense structure divided into three regions which, posterior to anterior, are a homogeneous region, a scroll region, and a flocculent region (Fig. 11.10C). Two-thirds of the acrosomal vesicle is enclosed by periacrosomal material. Posterior to the subacrosomal layer is a substantial cytoplasmic region containing partially cristate mitochondria, irregularly scrolled lamellae, and unidentified dense bodies. The cytoplasm is contiguous with the nuclear material as the intervening membrane is incomplete. The nucleus constitutes the bulk of the sperm cell and appears as a moderately electron-dense matrix of coarse granules and fibrils. The nucleus is interrupted by cytoplasmic protrusions on its anterior border. Bundles of hollow microtubules occur irregularly inside the nucleus; outside the cell, the bundles project as discrete arms or spines. Both microtubular nuclear arms and apical acrosomal vesicles appear typical of this lobster family.

11.5. 3 Natantian Sperm

Natantian decapods include the dendrobranchiate shrimp and caridean prawn. Natantian sperms are unistellate and possess a single appendage or spike and a main body consisting of the nucleus and the acrosome. However, morphological variations do occur between the caridean and dendrobranchiate unistellate spermatozoa with reference to the shape of the main body and the complexity of the acrosomal region. Thus, dendrobranchiates have more complex acrosomal regions than that of carideans, and fertilization involves a visible acrosome reaction, which is not observed in carideans (see chapter: Fertilization). Briefly, the unistellate spermatozoa have two morphological types: the spermatozoa with inverted cup-shaped main body found in carideans, and the dendrobranchiate sperm with spherical, bulged, or elongate main body. A third group of natantian spermatozoa is also reported. They are the dendrobranchiate spikeless spermatozoa found in Aristeidae and Sergestidae (Braga et al., 2013).

11.5.3.1 Dendrobranchiate Spermatozoa

In the pink shrimp, *Farfantepenaeus paulensis*, the spermatozoa are unistellate and are composed of a main body and single spike (Braga et al., 2013). The spheroid main body comprises a perinuclear cytoplasmic band, nucleus, acrosomal cap, and a subacrosomal region. The cytoplasmic band is found at the cell edge, posterolaterally surrounding the nuclear region. The centralized nucleus has a content of granular network, interspersed with filaments. The acrosomal apparatus consists of an acrosomal vesicle and a subacrosomal region. The acrosomal vesicle in turn is composed of an acrosomal cap and a spike. The acrosomal cap is internally concave with its peripheral region filled with electron-dense granular material, whereas the central region of the cap consists of dense homogeneous material, which is aligned with the spike base, thus establishing continuity with the spike material. The subacrosomal region occupies the space between the nucleus and the acrosomal vesicle.

11.5.3.2 Caridean Shrimp Sperm

The caridean sperm cell has a "thumb tack" shape and consists of a single anterior acrosomal spike and a rounded or concave posterior, nuclear region. The spike is longer and larger than that found in dendrobranchiate sperm and may or may not be cross striated. The spike of the caridean sperm is not analogous to the spikes of reptantian sperm, since *Macrobrachium rosenbergii* sperm appendages contain neither nuclear material nor extensive microtubular complexes (Lynn and Clark, 1983a). The nuclear region is homogeneous among most species and corresponds to the general unistellate spermatozoa pattern by occupying almost the entire main body with uncondensed chromatin, which provides a flocculent appearance. A thin cytoplasmic band separates the anterior acrosomal cap region from the nucleus. The acrosomal structures are not membrane-limited. They surround the anterior part of nucleus and extend into the spike. A collar region, composed of lamellae, also exists between the nucleus and the base of the spike. Mitochondria generally occur lateral to the nucleus or adjacent to the subacrosomal cup region.

Lynn and Clark (1983a) described the general sperm morphology of *M. rosenbergii* as an example of unistellate sperm of caridean prawns. The mature sperm as isolated from the spermatophore resembles an inverted umbrella. The main body of the sperm is slightly cupped and a single appendage projects from the convex surface of the base (also called cap region). From the periphery of the base, 14 to 20 radial fibrils converge at the center of the base and anastomose into the spike. These radial fibrils comprise the major substructure of the cap and are continuous with the spike. The cap region, which contains a thin band of cytoplasm, is partially separated from the nucleus by a membrane. In the cytoplasmic band of the cap, a pair of centrioles is housed acentrically in relation to the base of the spike. Each centriole consists of nine doublet microtubules with a pair of dynein-like arms attached to each doublet. Flocculent material occupies the central core of the centrioles and appears to have a spokelike arrangement attached to each of the doublet. However, the spike does not show any motility. The uncondensed nucleus gives a Feulgen-positive reaction and exhibits a green fluorescence when stained with acridine orange. The nucleus contains basic proteins, but the sperm-specific protamine is absent. The nucleus is not limited by a nuclear envelope but rather by the plasmalemma on one side and the membrane-bound region of the cap on the other.

Caridean sperm also takes other forms. For example, in *Thor* spp. (Hippolytidae), the sperm is bullet-shaped. Sperm of the rock shrimp *Rhynchocinetes* are quite unusual, consisting of a long spike with a main body shaped like an elaborate 11-pointed crown instead of a cup (Barros et al., 1986). In *Rhynchocinetes typus*, the radial arms which radiate from radial fibrils remain closed when they are extruded from the vasa deferentia (Bauer and Thiel, 2011). But the radial arms open when the spermatozoa contact sodium ions in seawater (Fig. 11.11). Radial arms, in addition to the acrosomal spikes, have also been observed in another caridean shrimp *Hippolyte obliquimanus* (Terrossi et al., 2011). There are as many as 20 radiating arms emanating from the nuclear rim of the main body. These arms are continuous to nuclear material in the cell body, giving the sperm an appearance of "cnidarians medusa." Unlike the rock shrimp spermatozoa, cross striations are absent in the spike of *H. obliquimanus* sperm, but the spike is composed of granular material.

11.6 SPERM MORPHOLOGY IN THALASSINIDEA

In this infraorder of Decapoda, the sperm cells are spherical with prominent microtubular arms radiating from the equatorial region. The sperm is divided into two hemispheres. The upper hemisphere contains the cytoplasmic organelles and an electron-dense acrosome vesicle with operculum, while the lower hemisphere is composed of nucleus. The microtubular spines originate in the nucleus and appear to pass completely through it. In *Axius glyptocercus*, the perforatorial chamber penetrates the sperm cell from the posterior or lower pole to a subterminal position below the operculum. In the spermatozoa of *Thalassina squamifera*, the large

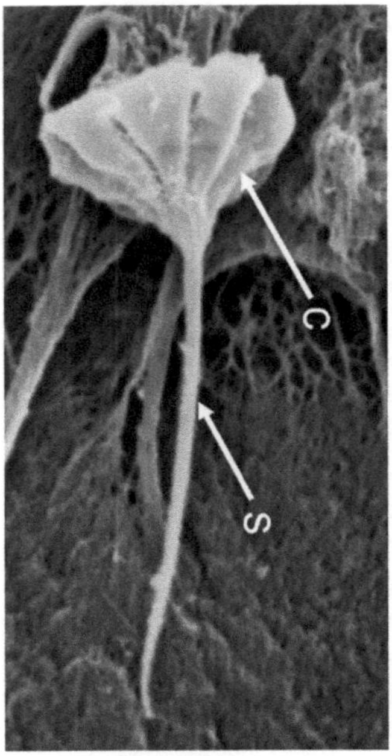

FIGURE 11.11 Spermatozoa of *Rhynchocinetes typus*. *S*, spike; *C*, cap. *Adapted from Bauer, R. T., Thiel, M., 2011. First description of a pure-search mating system and protandry in the shrimp* Rhynchocinetes uritai *(Decapoda: Caridea). J. Crust. Biol. 31, 286–295.*

acrosome is ovoid and has a concentrically zoned acrosomal vesicle, capped by a tri-layered operculum and posteriorly embedded in a ring of cytoplasm and nucleus.

11.7 FLAGELLATE SPERMATOZOA OF CIRRIPEDES

Motile spermatozoa with structural components similar to that of a typical spermatozoon are found only in cirripedes among Crustacea. The spermatozoa are filiform, motile, and flagellate. The head is the longer part and contains the distal centriole and a large part of the flagellum, although the middle piece is not well developed (Pochon-Masson, 1983). In *Balanus balanus*, Munn and Barnes (1970) described the axial filament complex which virtually extends the entire length of the spermatozoa. The complex has the usual 9+2 organization with spokes, central sheath, and connections between the outer fibers, similar to other flagellate spermatozoa. At the anterior end of the sperm, there is a short conical acrosomal region. The axial filament arises immediately behind the acrosome and extends up to the posterior end of the sperm. Lying adjacent to the axial filament complex is a reticulum of narrow tubules which extends about three-quarters of the length of the spermatozoon. The nucleus is filiform and lies in a membranous compartment separate from the reticulum and acrosomal complex. A mitochondrion lies directly posterior to the nucleus.

11.8 MOTILE SPERMATOZOA OF THE OSTRACOD *CYPRIDOPSIS*

Mature spermatozoa of the ostracod, *Cypridopsis* are filiform, spirally patterned cells surrounded by an outer sheath, as found in the male genital duct. Sperm removed in this state from male duct is not capable of motility; however, sperm removed from the female genital ducts are devoid of this outer sheath and are motile. However, the motile spermatozoa are not morphologically dissimilar to the nonmotile sperm removed from the male duct, suggesting that some sort of capacitation occurs in the female reproductive tract. The filiform, spirally patterned ostracod sperm has a head end tapering toward the tail. The head end terminates as a corkscrew-like tip, whereas the tail end is rounded. The pitch of the spiral of the spermatozoa increases from the head end to the tail end, thus deviating from the external morphology of the typical sperm. Ultrastructural studies of Reger and Florendo (1969) showed that the structural plan of *Cypridopsis* sperm is quite unlike any other filiform spermatozoa studied. Particularly, the 9+2 tubule arrangement found in the elongating, differentiating spermatids of *Cypridopsis* sp. are not present in the mature spermatozoa. Possibly, the organelles underlying the motility in the ostracod spermatozoa could be the spirally oriented mitochondria, which extend three-fourths the length of mature spermatozoa. While this orientation of mitochondria is similar to those found in many filiform spermatozoa [ie, modified type of spermatozoa in the nemertine worm *Malacobdell grossa* (Afzelius, 1971)], the unique occurrence of filaments inside the tubule-shaped cristae of the mitochondria in *Cypridopsis* sp. sperm is indicative of its probable role as a mechanotransducer element underlying sperm motility (Reger and Florendo, 1969). Understandably, motility in crustacean spermatozoa has been lost secondarily. Therefore, it is possible to see many intermediary conditions in motility loss, accompanied by ultrastructural alterations in the motile apparatus, namely the 9+2 tubule arrangement, which is basic to motility. Incidentally, the alternate structures found in the form of microtubules and microfilaments in the tail region of these crustaceans may account for sperm motility.

11.9 NONMOTILE SPERMATOZOA OF *ARTEMIA*

The branchiopods spermatozoa lack a flagellum and acrosome but have a well-defined nucleus with a typical double envelope and an unmodified mitochondria (Wingstrand, 1978). Thus the branchiopod spermatozoa lack all the characteristic features associated with a typical sperm except that they are used to fertilize the egg. In *Artemia*, spermatozoa are morphologically unspecialized, as they have changed little from the spermatid stage. Hence, there is no comparison between mature sperm of *Artemia* and other crustaceans (Brown, 1970).

In *Artemia*, spermatozoa are spherical structures with the same diameter as spermatids (Criel and Macrae, 2002). The sperm cells are covered with a

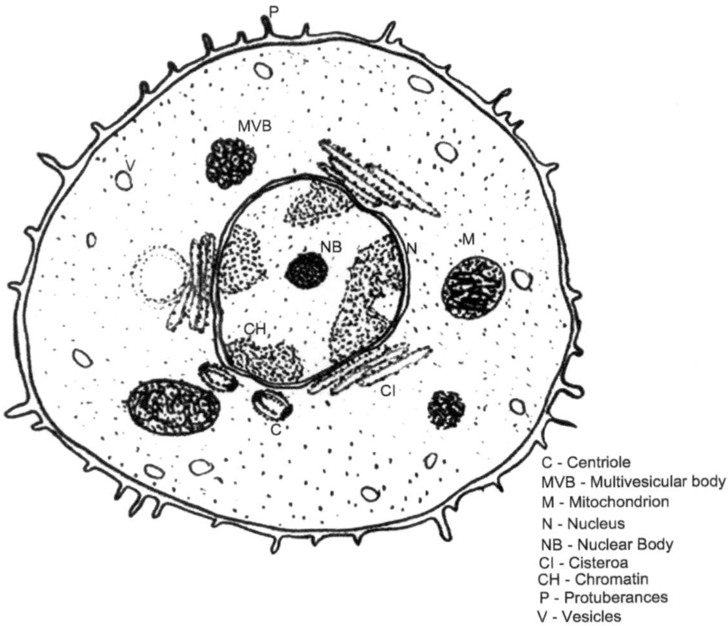

FIGURE 11.12 Spermatozoa of *Artemia. Adapted from Criel, G.R., Macrae, T.H., 2002. Reproductive biology ofArtemia. In: Artemia: basic and applied biology. Springer, The Netherlands, pp. 39–128.*

thin extracellular coat and often show small protuberances (Fig. 11.12). The centrally located nucleus is somewhat smaller than the spermatid nucleus. The chromatin in *Artemia* sperm nucleus is almost uncondensed, as in other crustacean sperm. Patches of chromatin and a characteristic nuclear body are found in every nucleus, which retains the double-walled envelope in the mature sperm. Two centrioles are located close to the nuclear membrane. In the cytoplasm, tightly packed cisternae are found to surround the nucleus, giving the nuclear membrane a pentalaminar appearance. These cisternae obviously arise from the membranous system of *Artemia* sperm. Small flattened vesicles are located close to the plasma membrane in the mature sperm cell. Golgi complexes are absent, but clusters of small vesicles and multivesicular bodies are found in the cytoplasm. Mitochondria do not undergo any structural modification from the spermatic stage. Interestingly, no clearly defined acrosome is present in *Artemia* sperm cells, giving it a nonpolarized appearance. However, the cluster of small vesicles observed in *Artemia* sperm may represent a multiple acrosome system as in the case of some cnidarians (Hinsch and Clark, 1973).

Artemia sperm cells emit pseudopodia-like structures upon release from the male genital tract into the surrounding medium (Brown, 1966). These thread-like extensions that emerge from the sperm cell body also could be seen in spermatozoa that are stored in the lateral pouches of the female reproductive tract. These pseudopodial structures are extensions of the outer nuclear envelope and they are filled with cytoplasm and microfilaments. Thus pseudopodial extensions are analogous to the radial arms of some natantian sperm cells (*M. rosenbergii*) that undergo a characteristic umbrella-like unfolding. They are not known to exhibit any pseudopodial movement.

11.10 SPERM STRUCTURE AND SPERMATOGENESIS IN COPEPODA

As in other groups of crustaceans, spermatozoa of copepods also show structural variations from one group to another. In general, the spermatozoid is ovoid in calanoids or fusiform and highly elongated in the harpacticoid copepods. In both the cases, the gametes are grouped inside a spermatophore during their transmission to the females. In *Calanus finmarchus*, the spermatozoa lie along the wall, while secretory substances are accumulated in the core. In the harpacticoid, *Tisbe holothuriae*, the spermatid is spheroid in nature and gets transformed into fusiform spermatozoa in the spermatophoric sac of the male reproductive tract (Pochon-Masson and Gharagozlou-van-Ginneken, 1977). In *T. holothuriae*, a single testis lies within the cephalothorax, and the proximal vas deferens leaves the testis at its anterior pole. The posterior end of testis shows the young spermatocytes, while the anterior end contains the spermatids, which enter the vas deferens. At this stage of spermiogenesis, the spermatid is round in shape, and the chromatin inside the nucleus soon exhibits an anterior polarity by clustering into a homogeneous osmiophilic body at one pole of the nucleus. The nuclear envelope at the anterior pole exhibits prominent perinuclear cisternae, while posteriorly, the nucleus contains numerous pores. The cytoplasm behind the nucleus contains ribosomes, a few mitochondria and microtubules, a Golgi apparatus with dilated saccules, and a large cisternae lying parallel to the cell membrane. The developing spermatids are surrounded by nurse cells. During the next stage, the sperm cell elongates and becomes narrower while the nucleus tends to be displaced toward the anterior end of the cell. The peripheral cisternae under the cell membrane extend backward in the form of two sheets. Posterior to the nucleus, two long mitochondria with a few irregular cisternae are disposed symmetrically. A few microtubules also occur adjoining the mitochondria. The perinuclear cisternae extend out into the cytoplasm. Anterior to the nucleus, the cytoplasm extends to form a pointed structure which contains osmiophilic lamina. This structure possibly replaces the acrosome. The last step in spermiogenesis is characterized by a marked elongation of the cell and elimination of the posterior mass except the two mitochondria. The nucleus contains chromatin fibrils, clustered in bundles running longitudinally. Centrioles are absent in the neck region, but it contains two long mitochondria, with a few longitudinal cristae. However, no flagellar structures could be observed. The sperm is nonmotile. At this stage, the sperm look ellipsoidal and enter the vas deferens. In the proximal vas deferens, the sperm lie around a central core of secretory materials, as in other copepods. When the sperm enters the spermatophoric sac, the sperm acquires a new coat which is longitudinally striated with dense bands. Secretions from the epithelial cells lining the spermatophore sac give rise to such extracellular coat material, which surrounds the sperm head all over except the mitochondria which are surrounded only by a thick cell wall. As in other copepods, the secretions of the vas deferens contribute to the formation of spermatophores as well as the extracellular coat of the mature sperm.

11.11 REMIPEDE SPERM

Remipedes are the most primitive crustaceans, discovered by Yager in 1981 from anchialine caves in northern Bahamas. They are simultaneous hermaphrodites, having spermatogenesis and oogenesis in the same individuals. The spermatozoa have an ovoid nucleus and a flagellum with 9+2 microtubular arrangement. The mature sperm of *Speleonectes benjamini* has an ovoid nucleus, cup-shaped acrosome, and flagellum with a 9+2 microtubular arrangement (Yager, 1989). The nucleus contains the chromatin which is highly condensed and fibrillar in texture. The cup-shaped acrosome has granular bands in the subacrosomal space. An acrosomal filament or rod extends through the length of the nucleus. Several mitochondria are present in the sperm cytoplasm, but they do not organize into a midpiece, as in typical spermatozoa. However, the flagellum appears to loop around the nucleus before extending out of the distal end of the sperm. Ultrastructural features of the remipede sperm suggest that they are capable of movement. The sperm are packaged into spermatophores with a distinct shape. The sperm nuclei are in the proximal end of the spermatophore, and the flagella extend into the distal portion. The mature sperm of *S. benjamini* seem to resemble the basic sperm model with ovoid nucleus, acrosome, and simple flagellum. This generalized sperm morphology of remipedes is in agreement with those of the arthropods but diverge much from the other crustacean sperm model. Incidentally, the sperm of *S. benjamini* closely resemble that of the horseshoe crab *Limulus*. Although other groups such as Branchiura, Cirripedia, and Mystacocarida have flagellate sperm, they deviate in some way from the conventional sperm model. According to Schram and Jenner (2001), insects, malacostracans, and remipedes form a monophyletic clade, with remipedes and malacostracans, as sister groups. Obviously, motile sperm may be the generalized structure in the primitive crustacean condition, but the motility is secondarily lost in advanced crustaceans such as the decapods.

11.12 ENDOCRINE REGULATION OF SPERM PRODUCTION

Remarkable progress has been made on the endocrine control of female reproduction with special reference to vitellogenesis in crustaceans, thanks to the necessity to control egg production under hatchery conditions. Conversely, our knowledge on the hormonal regulation of spermatogenesis is very little. Although the cost of sperm production is lesser, compared to egg production, many crustaceans produce copious quantities of seminal substances to facilitate sperm transport to females during mating. Whereas the spermatogenesis is completed in the testis, the highly secretory male reproductive tract produces the spermatophores, and the associated seminal substances. Therefore, both testis and vas deferens could be the target organs for endogenous control of male reproduction. Mobilization of resources from the hepatopancreas to the testis during spermatogenesis and spermatophore production would also suggest an endocrine control over these male reproductive activities (Zara et al., 2012).

In the penaeiodean shrimps, Alfaro-Montoya (2010) proposed a model for male maturation within the male reproductive system: (1) spermatogenesis in the testes, (2) sperm maturation in the proximal vas deferens, (3) spermatophore formation in the distal part of vas deferens, and (4) capacitation of sperm within the female thelycum. Therefore, the testis and the vas deferens form the main targets for the male hormones to control male gamete production and maturation. As in the females, the central nervous system consisting of brain, thoracic ganglia, and the neural ganglia, housed in the eyestalk of the decapod crustaceans could provide the apex neuroendocrine regulatory center for male gamete maturation (see chapter: Endocrine Control of Vitellogenesis). However, we have information available only on the X-organ-sinus gland control of spermatogenesis in penaeiodean shrimps and brachyuran crabs (Alfaro-Montoya, 2010). In *Litopenaeus vannamei*, Leung-Trujillo and Lawrence (1985) demonstrated that unilateral eyestalk ablation improved sperm count, testis weight, as well as spermatophore production. Similar results have been reported in various shrimp species, including *L. vannamei*, *Penaeus stylirostris*, *Penaeus indicus*, and *Penaeus monodon* (Leung-Trujillo and Lawrence, 1985; Alfaro-Montoya and Lozano, 1993; Salvador et al., 1988; Muthuraman and Adiyodi, 1998; Babu et al., 2013). In these studies, the eyestalk hormone responsible for restraining testicular activities has not been identified, although the classical gonad-inhibitory hormone (GIH) is implicated in the control of male reproduction. However, in a recent study on GIH silencing by long hairpin RNA resulted in significant increase in testicular activity as well as the secretory activity of the epithelial cells of terminal ampoule in *P. monodon* (Das et al., 2015). The epithelial cells of terminal ampoule are known to produce several sperm maturation peptides called terminal ampoule protein in *M. rosenbergii* (Ma et al., 2010). These peptides also have a role in inducing acrosome reaction and fertilization by degrading vitelline coat (Sroyraya et al., 2013).

In addition to the effect of GIH on spermatogenesis and spermatophore synthesis, studies have shown that eyestalk ablation leads to hypertrophy and increased secretory activity of the androgenic gland in the crayfish, *C. quadricarinatus* (Khalaila et al., 2002). Furthermore, injection of the androgenic gland hormone of this crayfish (*Cq-IAG*) stimulated spermatogenesis in the testis. Similarly, in the freshwater prawn, *M. rosenbergii*, the insulin-like androgenic gland peptide (Mr-IAG) also had the same effect on testis. In vivo silencing of the *Mr-IAG* gene, by injecting the prawns with *Mr-IAG* double-stranded RNA resulted in the arresting of testicular spermatogenesis and spermatophore development in the terminal ampoule of the sperm duct (Khalaila et al., 2001; Kim et al., 2002). Evidently, spermatogenic activities are under the direct control of androgenic gland hormone.

Biogenic amines such as 5-hydroxytryptamine (5-HT) have also been shown to have a controlling effect on testicular maturation by stimulating the release of gonad-stimulatory hormone (GSH) from the brain, which in turn activates the synthesis of androgenic gland hormone (Fingerman, 1997). The effect of 5-HT on testicular maturation was, however, reversed when the animal was injected with dopamine, by blocking the release of GSH from the brain or stimulating the release of GIH (Fingerman, 1997).

There is some evidence to show that methyl farnesoate stimulates testicular maturation in decapods such the freshwater crab, *Oziotelphusa senex senex* (Reddy and Ramamurthy, 1998; Kalavathy et al., 1999); but the results are inconclusive. However, methyl farnesoate has a clear regulatory role in male reproductive morphology and behavior in decapod crustaceans (Laufer and Biggers, 2001). The nonreproductive male morphotype of crayfish lacks aggressive characteristics, due to low levels of methyl farnesoate in the hemolymph (Laufer et al., 2005).

Certainly, our current understanding on the hormonal regulation of spermatogenesis is inadequate, when compared to the information available on the control of female reproduction. Continued work on the control of spermatogenesis and spermatophore formation in the commercially important shrimp species is especially required for brood stock development in the hatcheries. Although eyestalk ablation is a handy technique to promote male reproductive functions, the role of other proximate endocrine factors needs to be investigated to gain a better control over male reproduction under culture conditions. Some early reports have suggested that crustacean testis is a center for steroidogenesis of vertebrate steroids such as estrogen and testosterone (Burns et al., 1984; Couch et al., 1987). It is therefore worthwhile to find out any role for testosterone in the control of spermatogenesis, in addition to the established role of androgenic gland in crustacean reproduction.

Chapter 12

Spermatophore and Sperm Transfer Mechanisms

12.1 INTRODUCTION

Although reproductively unspecialized marine invertebrates such as the corals, annelids, and echinoderms discharge their gametes into the medium, where broadcast fertilization occurs, such primitive type of external fertilization has never been witnessed in crustaceans, which have diversified representation in many marine ecosystems. On the other hand, the sperm transfer in the majority of the crustaceans occurs through spermatophore with the possible exception of the cirripedes, which transfer sperm in a fluid medium of seminal plasm (see Subramoniam, 1993, for review). Interestingly, a marine spionid polychaete, *Polydora ligni*, releases the pelagic spermatophores into the seawater, to be picked up by the females for internal fertilization (Rice, 1978). Though seawater medium is conducive for such indirect sperm transfer mechanism, on no occasion the crustacean spermatophores are released into the medium. They are either deposited onto the exoskeleton of the females or stored in the seminal receptacle of the females. Interestingly, an extant marine primitive arthropod group, Xiphosura (horseshoe crabs) still practices external fertilization with free, motile spermatozoa (Afzelius, 1971; Clapper and Brown, 1980).

The use of spermatophores as the main mode of sperm transfer by crustaceans is important in two respects. First, most of the crustacean spermatozoa are aflagellate and nonmotile (Felgenhauer and Abele, 1991) and hence need a vehicle for their transfer to females in the absence of a copulatory organ. Secondly, crustaceans have employed spermatophores as an effective method for sperm transfer in aquatic media inasmuch as the cryptozoic arthropods used them in land habitat (Schaller, 1980). In the marine condition, the evolution of spermatophoric method of sperm transfer is to minimize the sperm loss that occurs through broadcast fertilization. Although majority of crustaceans are aquatic, some exhibit insemination tendencies suited to terrestrial conditions, as exemplified by certain crabs and wood lice, which possess specialized gonopods for sperm/spermatophore transmission. The occurrence of spermatophores is widespread among different crustacean taxa.

Sexual Biology and Reproduction in Crustaceans. http://dx.doi.org/10.1016/B978-0-12-809337-5.00012-5

12.2 SPERMATOPHORE MORPHOLOGY

The spermatophore morphology is variable in Crustacea in accordance with the type of sperm transfer mechanism as well as fertilization. In general, the spermatophores used in the internal fertilization are simple, sometimes degenerate structures. On the other hand, in the externally fertilizing crustacean species, the spermatophores assume complex structures in order that they could be deposited onto the female body and stored for a long time, pending fertilization. The external fertilization of eggs, using the attached spermatophores is unique to crustacean forms and could be termed as epizoic fertilization. Spermatophore size and morphology are relatively constant within a species and vary between them, making these male reproductive structures taxonomically useful for differentiating closely related species. The occurrence of spermatophores in different groups of crustaceans is listed in Table 12.1.

TABLE 12.1 List of Crustaceans in Which Spermatophores Are Reported

Higher Taxa and Family	Species
Class Remipedia, Yager	
Family Speleonectidae, Yager	*Speleonectus lucayesis* Yager 1981 Speleonectus benjamini Yager 1987
Class Maxillopoda, Pennak and Zinn	
Subclass Mystacocarida, Sars	*Derocheilocaris typicus* Pennak and Zinn 1943
Subclass Copepoda, Milne-Edwards Order Calanoidea, Sars	*Acartia tonsa* Dana 1849 *Calanus finmarchicus* Gunnerus 1770 *Candacia armata* Dana 1846 *Centropages furcatus* Dana 1849 *Centropages typicus* Kroyer 1849 *Euchaeta norvegica* Boeck 1872 *Labidocera aestiva* Wheeler 1900 *Labidocera jollae* Esterly 1906 *Pseudodiaptomus coronatus* Williams 1906 *Diaptomus (Cyclops) castor* Jurine 1820
Order Harpacticoida, Sars	*Diarthrodes cystoecus* Thompson 1882 *Lepeophtherius pectoralis* Müller 1776 *Tisbe holothuriae* Humes 1957
Order Cyclopoida, Bumeister	*Cyclops americanus* Marsh 1892 *Pachypygus gibber* Thorell 1859
Class Malacostraca, Latreille	
Subclass Hoplocarida, Calman	
Order Stomatopoda, Latreille	Squilla holoschista Kemp 1911
Subclass Eumalacostraca, Grobben	
Order Mysidacea, Boas	

TABLE 12.1 List of Crustaceans in Which Spermatophores Are Reported—cont'd

Higher Taxa and Family	Species
Family Mysideae, Dana	*Leptomysis lingvura* Sars 1868 *Mysis relicta* Loven 1862 *Neomysis integer* Leach 1814 *Praunus flexuosus* Müller 1776
Order Euphausiacea, Dana	
Family Euphausiidae, Dana	*Thysanopoda aequalis* Hansen 1905 *Thysanopoda orientalis* Hansen 1910 *Thysanopoda tricuspidata* Milne-Edwards 1837 *Stylocheiron longicorne* Sars 1883
Order Decapoda, Latreille	
Family Penaeidae, Rafinesque	*Penaeus aztecus* Ives 1891 *Penaeus duodarum* Burkenroad 1939 *Penaeus japonicus* Bate 1888 *Penaeus indicus* Milne-Edwards 1837 *Penaeus kerathurus* Forskål 1775 *Penaeus monodon* Fabricius 1798 *Penaeus orientalis* Kishinouye 1900 *Penaeus penicillatus* Alcock 1905 *Penaeus schmitti* Burkenroad 1934 *Penaeus setiferus* Linnaeus 1761 *Penaeus stylirostris* Stimpson 1871 *Penaeus vannamei* Boone 1931 *Metapenaeus monoceros* Fabricius 1798
Superorder Dendrobranchiata, Bate 1888	
Family Sergestidae	*Peisos petrunkevitchi* Burkenroad 1945 *Sergestes arcticus* Kroyer 1855
Family Penaeides	*Artemesia longinaris* Bate 1888 *Penaeus aztecus* Ives 1891 *Penaeus penicillatus* Alcock 1905 *Penaeus vannamei* Boone 1931 *Penaeus satiferus* Linnaeus 1767 *Penaeus japonicas* Bate 1888 *Penaeus kerathurus* Forskal 1775 *Parapenaeus longirostris* Lucas 1846 *Penaeopsis serrata* Bate 1881 *Trachypenaeus similis* Smith 1885 *Sicyonia brevirostris* Stimpson 1871 *Sicyonia carinata* Brunnich 1768 *Sicyonia ingentis* Burkebroad 1938 *Aristaeomorpha foliacea* Rissi 1827 *Aristaeopsis edwardsiana* Johnson 1867 *Aristaeus antennatus* Risso 1816 *Aristaeus varisens* Holthuis 1952 *Pleoticus muelleri* Bate 1888 *Solenocera africana* Stebbing 1917 *Solenocera membranacea* Risso 1816

Continued

TABLE 12.1 List of Crustaceans in Which Spermatophores Are Reported—cont'd

Higher Taxa and Family	Species
Family Sicyoniidae, Ortmann	*Sicyonia ingentis* Burkenroad 1938
Family Palaemonidae, Rafinesque	*Palaemonetes* sp. Heller 1869 *Palaemonetes pugio* Holthuis 1949 *Palaemonetes vulgaris* Say 1818 *Macrobrachium acanthurus* Wiegmann 1836 *Macrobrachium asperulum* Von Martens 1868 *Macrobrachium carcinus* Linnaeus 1758 *Macrobrachium formosense* Bate 1868 *Macrobrachium nipponsense* de Hann 1849 *Macrobrachium rosenbergii* de Man 1879 *Macrobrachium shokitai* Fugino and Baba 1973
Family *Crangonidae*, Haworth	*Crangon crangon* Linnaeus (1758)
Family Nephropidae, Dana	*Enoplometopus occidentalis* Randall 1840
Family Polychelidae, Wood–Mason	*Homarus americanus* Milne-Edwards 1837 *Homarus gammarus* Linnaeus 1758
Family Palinuridae, Latreille	*Jasus lalandii* Milne-Edwards 1837 *Jasus novaehollandiae* Holthuis 1963 *Palinurus gilchristi* Stebbing 1898 *Panulirus homarus* Linnaeus 1758 *Panulirus interruptus* Randall 1839 *Panulirus penicillatus* Oliver 1791 *Panulirus angulatus* Bate 1888 *Potamobius* Fabricius 1819 *Linuparus trigonus* Siebold 1824
Family Scyllaridae, Latreille	*Thenus orientalis* Lund 1793 *Parribacus antarcticus* Lund 1793
Infraorder Anomura, Milne-Edwards	
Family Coenobitidae, Dana	*Coenobita rugosa* Milne-Edwards 1837
Family Diogenidae, Dana	*Clibanarius longitarsus* de Hann 1849 *Clibanarius erythropus* Latreille 1818 *Clibanarius vittatus* Bosc 1802 *Diogenes pugilator* Roux 1828 *Calcinus tubularis* Linnaeus 1767 *Calcinus laevigatus* Randall 1840 *Loxopagurus loxochelis* Moreira 1901

TABLE 12.1 List of Crustaceans in Which Spermatophores Are Reported—cont'd

Higher Taxa and Family	Species
Family Paguridae, Latreille	*Anapagurus hyndmani* Thompson 1843 *Birgus latro* Linnaeus 1767 *Cheiroplatea pumicicola* Forest 1987 *Dardanus asper* de Hann 1839 *Dardanus punctulatus* Hilgendorf 1869 *Eupagurus bernhardus* Linnaeus 1758 *Pagurus novae-zealandiae* Dana 1852 *Pagurus prideauxi* Leach 1865 *Pagurus* bernhardus Linnaeus 1865 *Pagurus excavatus* Herbst 1802 *Pagurus pollicaris* Say 1817 *Pagurus maclaughlinae* Garcia-Gomez 1982 *Pagurus longicarpus* Say 1817 *Pagurus hirtimanus* White 1847 *Pagurus bernhardus* Linnaeus 1758 *Pagurus chevreuxi* Bouvier 1896 *Nematopagurus* Forest and de Saint Laurent 1968 *Solitariopagurus caledonicus* Türkay 1986 *Xylopagurus caledonicus* Milne-Edwards 1880
Family Galatheidae, Samouelle	*Pleuroncodes planipes* Stimpson 1860
Family Hippidae, Latreille	*Albunea symnista* Linnaeus 1758 *Emeriti asiatica* Milne-Edwards 1837 *Hippa pacifica* Fabricius 1787
Family Aeglidae, Dana	*Aegla rostrata* Jara 1977 *Aegla longirostri* Bond-Buckup and Buckup 1994
Family Lomisidae, Bouvier	*Lomis hirta* Lamarck 1818
Family Lithodidae, Samouelle	*Lithodes maja* Linnaeus 1758
Family Porcellanidae, Haworth	*Petrolisthes lamarckii* Leach 1820 *Petrolisthes armatus* Gibbes 1850 *Pisidia longicornis* Linnaeus 1767 *Polyonyx transverses* Haswell 1882
Infraorder, Brachyura, Latreille	
Family Dromiidae, de Hann	*Dromia personata* Linnaeus 1756
Family Homolidae, de Hann	*Homola barbata* Fabricius 1793
Family Raninidae, de Hann	*Ranina ranina* Linnaeus 1758 *Lyreidus tridentatus* de Hann 1839

Continued

TABLE 12.1 List of Crustaceans in Which Spermatophores are Reported—cont'd

Higher Taxa and Family	Species
Family Majidae, Samouelle	*Chionoecetes opilio* Fabricius 1788 *Inachus phalangium* Fabricius 1775 *Libinia emarginata* Leach 1815 *Macrocoeloma trispinosum* Rathbun 1900
Family Portunidae, Rafinesque	*Callinectes sapidus* Rathbun 1896 *Carcinus maenas* Linnaeus 1758 *Portunus pelagicus* Linnaeus 1758 *Ovalipes ocellatus* Herbst 1799 *Portunus sanguinolentus* Herbst 1796 *Scylla serrata* Forskål 1775
Family Goneplacidae, MacLeay	*Geryon fenner* Manning and Holthuis 1984 *Geryon quinquedens* Smith 1879 *Paratelphusa hydrodromus* Herbst 1794
Family Xanthidae, MacLeay	*Menippe mercenaria* Say 1818 *Rithropanopaeus harrisii* Gould 1841
Family Ocypodidae, Rafinesque	*Ocypoda platytarsis* Milne-Edwards 1852 *Uca lacteal* de Hann 1835

The taxonomic arrangement is taken from Bowman, T.E., Abele, L.G., 1982. Classification of the recent Crustacea. Biol. Crust. 1, 1–27.; Reproduced from Subramoniam, T., 1993. Spermatophores and sperm transfer in marine crustaceans. Adv. Mar. Biol. 29, 129–214.

12.2.1 Spermatophore Morphology and Diversity in Decapods

The spermatophores exhibit gross morphological variations among different groups of decapods. The decapod spermatophores could be classified into three general types, depending on their morphology. The simplest type is the small, spherical or ellipsoidal spermatophore found in the brachyuran crabs. The spermatophores are suspended in seminal fluid and transferred to the female by true copulation for storage in the spermatheca. The second type is the pedunculate spermatophore, found in the anomuran crabs. It consists of a sperm-containing ampulla, raised on a stalk or peduncle that is attached onto a gelatinous pedestal to fix the spermatophore on the sternal region of the female. They are usually transferred to the female as a long continuous ribbon to effect epizoic fertilization. The third type, which is produced by most macrurans, is generally complex masses consisting mainly of spermatophoric tubes embedded in a protective gelatinous matrix. Such spermatophoric mass is found in lobsters and crayfish. In addition, spermatophores may contain accessory structures such as attachment wings, as in penaeoid shrimps (Subramoniam, 1993). They are in general stored externally on the ventral sternum or inside special accessory structures, like thelycum.

12.2.1.1 Brachyuran Crabs

Brachyurans represented by true crabs produce the simplest form of spermatophores found among crustaceans. The spermatophores are ovoid, ellipsoidal, or round and contain varying numbers of sperm. In general, these spermatophores are carried in a fluid medium of granulated seminal plasma. Early microscopic studies on the brachyuran spermatophores have revealed the presence of a single layer enveloping the inner sperm mass, as reported in *Carcinus maenas* (Spalding, 1942), *Callinectes sapidus* (Cronin, 1947), and *Portunus sanguinolentus* (Ryan, 1967). Besides, a distinctive spermatophoric layer could be lacking in some crabs, in which cases sperm aggregates retain their integrity merely by differential viscosity of the surrounding seminal fluid (*Ranina ranina*, Ryan, 1984; *Ocypoda platytarsus*, Varadharajan, 1982). Evidently, brachyuran spermatophores are degenerate structures which have lost their original function of protection, serving merely to keep the sperm together during their transmission to the female. The simplicity of spermatophoric structure in the brachyuran crabs could also be due to their transport through seminal plasma to facilitate internal fertilization.

The spermatophores of brachyurans are microscopic; in the mud crab *Scylla serrata*, they measure 3–4 μm in diameter. The spermatozoa within the spermatophores are embedded in a viscous fluid with their multistellate arms folded on the surface of the main body (Uma and Subramoniam, 1979). In this crab the spermatophore layers consist of an outer thick and an inner thin layer, the latter being confluent with the inner sperm mass. A two-layered spermatophore has been reported in the golden crab, *Geryon fenneri* (Hinsch, 1988) and the blue crab, *Callinectes danae* (Zara et al., 2012). In the Oxyrhynchan, *Macrocoeloma trispinosum*, Hinsch (1973) described a spermatophore containing only one sperm with its nucleated arms projecting through the thin spermatophore wall. While the vesiculate spermatophores, bathed in the seminal plasma, is characteristic of the advanced brachyurans, in two primitive families, Gymnopleura and Dromiacea, only a single sperm mass, directly surrounded by viscous seminal secretion is produced (Hartnoll, 1975). A distinctive spermatophore layer is, however, lacking.

In the true crabs, the shape and size of the spermatophores vary considerably in an individual, and hence the number of spermatophores enclosed within them also differs. In the mud crab, *S. serrata*, Jeyalectumie and Subramoniam (1991) reported an increase in the size of the spermatophore during the maturation of male crabs. Interestingly, the size increase is directly correlated to an increase in the organic components of the spermatophore.

12.2.1.1.1 Origin of Spermatophores

In the brachyuran crabs, maturity of female depends on the production of accessory materials to form spermatophores, as well as seminal plasma. In insects, a variety of male accessory sex glands has been reported to be involved in the synthesis of seminal substances with different functions related to gamete maturation, activation as well as transport (Avila et al., 2011). However, in the decapod crustaceans, the inner epithelial cells lining the vas deferens (VD) are highly glandular and secrete a variety of substances that make up the spermatophore and seminal plasma.

In the mud crab, *S. serrata*, the anterior portion of the proximal vas deferens (PVD) secretes substance "A," which agglutinates the sperm, released from testis, and forms the spermatophore layer (Uma and Subramoniam, 1984). The distal part of the PVD secretes a granular substance "B," which constitutes the medium for storing the completed spermatophore. In this crab, the middle VD is membranous and acts as the storage region for the seminal products. The distal VD becomes secretory again with its cuboidal epithelial cells. Its fluid-like secretion "C" is not mixed with the rest of the semen stored in the mid-VD until the time of spermiation. By virtue of its mucoid nature, this secretion may possibly form the material for sperm plug in the mated female as reported in several other brachyuran crabs (Spalding, 1942; Ryan, 1967; Diesel, 1991). Following the distal VD is a short ejaculatory duct, filled with a secretion similar to that found in the distal VD. In the distal part of the VD, a distinct dilated portion in the form of a seminal vesicle is also found in some crabs for storage of seminal substances (Lopez-Greco, 2013).

In the swimming crab, *Portunus pelagicus*, sperm masses are compacted into irregular triangular shapes, surrounded by two secretions from columnar epithelial cells of anterior vas deferens (AVD) (Stewart et al., 2010). However, the spermatophores, found in the anterior and mid-VD, contain immature spermatozoa. In the posterior VD, the spermatophores are ovoid in shape and surrounded by a uniformly thick homogeneous matrix. They also contain mature spermatozoa, suggesting that sperm maturation continues in the VD. Two types of epithelial secretions involved in the production of spermatophore have been described in many other brachyuran crabs, such as *Chionoecetes opilio* (Beninger et al., 1988; Sainte-Marie and Sainte-Marie, 1999) and *Maja brachydactyla* (Simeó et al., 2009). In all of them, the VD secretions are shown to be composed of glycoprotein and polysaccharide chains. In the snow crab, *C. opilio*, these mucopolysaccharide substances have a role in the sperm maturation that occurs within the spermatophores (Sainte-Marie and Sainte-Marie, 1999). Furthermore, the spermatophores stored in the inseminated females contain immature spermatids, mature spermatids, or a mixture of both.

In the blue crab, *C. danae*, the spermatophore formation begins at the anterior part of the AVD. The sperm masses are separated and compacted in small packets by a basophilic and alcianophilic secretion (Zara et al., 2012). More secretions containing protein and neutral polysaccharides are also added to the sperm mass to initiate spermatophore wall formation. In the posterior part of AVD, the mature spermatophores are found to possess a thick glycoproteinacious wall, surrounded by acid mucopolysaccharides. The spermatophores, found in the mid-VD, are bathed in the granular seminal plasm, containing glycoprotein.

In general, the assembly of decapod spermatophores takes place during sperm transit from the testis to the distal part of the VD. In the spider crab, *Libinia emarginata*, the spermatozoa enter the AVD individually and become surrounded by secretion products. They are then compartmentalized into spermatophores of varying sizes. The cytoplasm of the glandular cells lining the VD contains vast arrays of rough endoplasmic reticulum and Golgi complex with few mitochondria (Hinsch and Walker, 1974). The middle VD secretes the seminal fluid surrounding the spermatophores. The posterior VD functions primarily as a storage center for spermatophores until they are released at the time of ejaculation. The cells of this region are also secretory, contributing further to the seminal plasma. Interestingly, in the snow crab, *C. opilio*, Benhalima and Moriyasu (2000) reported that the posterior VD secretes and stores a highly acidic seminal fluid that destroy and resorb the excess spermatophores and sperm.

In a primitive brachyuran crab, *Dromia personata*, belonging to Dromioidea, there is only one spermatophore produced at a time. In this crab, the VD is divisible into three zones. In the first zone, the sperm are aggregated into a white mass, staining blue with Mallory's triple stain. This secretion also forms a layer around the central core of sperm. In the second zone, the central core of sperm is again surrounded by a thick layer of vacuolated blue-staining secretion. The remainder of the VD, comprising the third zone, is filled with a translucent secretion, which is devoid of sperm. The ejaculatory duct also contains a secretion similar to that found in the third zone. A single sperm mass spermatophore is also reported in another primitive gymnopleuran brachyuran, *Homola barbata*. Evidently, in these two primitive crab species, the spermatophore formation is similar to those of macruran Astacidea (Erkan et al., 2009), where the spermatophore is a single mass of sperm surrounded by various secretions, lacking an enveloping layer (Hartnoll, 1975).

12.2.1.1.2 Chemical Composition of Spermatophores and Seminal Plasma

Uma and Subramoniam (1979) investigated the histochemical nature of the spermatophore layers of the crab, *S. serrata*. In this crab, each spermatophore is enveloped by two layers, an outer thick and an inner thin layer, the latter being confluent with the inner sperm mass. Most importantly, this study revealed the presence of a structural polysaccharide, chitin, in the outer wall. The chitinous nature of the spermatophore wall has also been indicated in another crab, *C. maenas* (Spalding, 1942), and in the harpacticoid copepod *Tisbe holothuriae* (Pochon-Masson and Gharagozlou-Van Ginneken, 1977). In addition to chitin, the outer layer of *S. serrata* spermatophore is rich in sulfated mucopolysaccharide. The inner layer also contains acid mucopolysaccharide, rich in carboxyl groups but lacks chitin. Protein components of the two layers appear to be rich in tryptophanyl reactive sites, but other phenolic substances are wanting. Hence, protein stabilization by way of phenolic tanning in these layers may be ruled out, as they fail to give positive reaction to the enzyme phenol oxidase, which is essential for the oxidation of phenolic substances into cross-linking quinines (Richards, 1951). The mucopolysaccharide nature of spermatophore layers has been reported in other crab species also (Sainte-Marie and Sainte-Marie, 1999; Simeó et al., 2009). In the blue crab, *C. danae*, Zara et al. (2012) histochemically detected neutral polysaccharides and protein in the spermatophore wall. Histochemical studies on *S. serrata* have further revealed the presence of acidic mucopolysaccharides in the sperm mass substances inside the spermatophores. By virtue of its liquid imbibing property, the acid mucopolysaccharides of sperm mass absorb low molecular weight substances from the spermathecal fluid, resulting in the dehiscence of the spermatophores at the time of spawning (Uma and Subramoniam, 1979; Ezhilarasi and Subramoniam, 1982). Beninger et al. (1988) suggested that hydration during insemination could cause spermatophore dehiscence in the snow crab *C. opilio*.

Brachyuran crabs secrete copious quantities of fluid semen to carry the spermatophores. Biochemical studies on the spermatophores and seminal plasma of the field crab, *Paratelphusa hydrodromus* revealed that they are rich in protein, free carbohydrates, and lipids (Jeyalectumie and Subramoniam, 1987). A quantitative assay of the enzyme lactate dehydrogenase (LDH) showed its high activity within the spermatophore. Electrophoretic analysis of LDH isoenzyme revealed six fractions, and the most conspicuous of them is a fraction homologous to the LDH_x of mammalian spermatozoa, having an electrophoretic mobility between LDH_3 and LDH_4. Furthermore, LDH_x of the spermatophore seems to be more related to M_4 (LDH_5) (muscle type) than to H_4 (LDH_1) (heart type). In mammals, the LDH_x fraction is associated with the commencement of spermatogenesis (Goldberg and Hawtrey, 1967). It may be inferred from this result that the spermatozoa within the spermatophores may undergo anaerobic metabolism using glycogen-derived glycosyl units or free sugars. Several studies carried out on the biochemistry of seminal plasma of the crab *S. serrata* have revealed that it is rich in acidic mucopolysaccharides, organic substrates, proteolytic enzymes which showed antimicrobial activities (Jeylectumie and Subramoniam, 1991; Jayasankar and Subramoniam, 1997).

In addition, high activity of glycosidase enzymes has been observed in the seminal plasma of the mud crab *S. serrata* (Jayasankar, 1996). Among the glycosidases assayed in crab seminal plasma, α-mannosidase and β-*N*-acetyl hexosaminidases were the most active, conforming rather closely to the pattern present in the seminal secretions of most mammals. The other glycosidases identified showing moderate levels of activity were α-glucosidase, α-L-fucosidase, and β-galactosidase. α-Galactosidase and β-glucosidase, on the other hand, exhibited very weak activity. All these glycosidases were active at highly acidic pH ranges (Table 12.2). Glycosidases have been known to catalyze degradation of heterogeneous polymeric material such as mucoproteins in a variety of tissues. The spermatophore layers of *S. serrata* have been shown to be rich in acid mucopolysaccharides and protein, and the nature of glycosidase enzymes detected in the seminal plasma tends to favor the view that sperm release may

TABLE 12.2 Activity of Glycosidases in Seminal Plasma of the Mud Crab *Scylla serrata*

Glycosidase	Activity (µmoles/ml/min)	Optimum pH Range	Optimum Temperature Range (°C)
α-galactosidase	0.615±0.01	3.5–5.5	40–50
β-galactosidase	0.799±0.02	3.5–6.5	40–60
α-glucosidase	1.257±0.02	3.5–6.5	40–60
β-glucosidase	0.203±0.01	4.5–5.5	40–50
α-mannosidase	1.974±0.01	4.5–5.5	50–60
α-L- fucosidase	1.175±0.02	3.5–5.5	60–80
β-*N*-acetyl glucosaminidase	1.327±0.004	3.5–5.5	30–60
β-*N*-acetyl galactosaminidase	1.576±0.01	4.5–5.5	50–60

Data based on Jayasankar, V., 1996. Biochemical studies on the seminal secretions of the mud crab *Scylla serrata* (Forskal) (Decapoda: Portunidae), with special reference to protease and glycosidase activities. Ph.D thesis submitted to the University of Madras, India.

be aided by the degradation of the mucoprotein and chitin of the spermatophore walls by these enzymes.

In addition to its role in sperm metabolism, the seminal secretion of several brachyuran crabs delivered during mating undergoes a kind of solidification inside the spermatheca, giving rise to a structure called "sperm plug." In some species, the sperm plug extends into the vagina and even protrudes outside from the vulva (Hartnoll, 1969). In *C. maenas*, the sperm plug fits the spermatheca very closely so that its shape is roughly spherical. Various functions have been attributed to sperm plug, but it has been commonly assumed that it helps to prevent the loss of sperm after copulation. In *Cancer borealis*, the spermatophore wall is lysed in the distal VD, and hence only free spermatozoans carried in the liquid seminal plasm is transferred to spermatheca. Therefore, sperm plug formation is a necessity to prevent the loss of sperm after copulation (Moriyasu et al., 2002). However, Diesel (1991) contends that sperm plug may act as a paternity assurance device that prevents subsequent insemination by other male crabs.

12.2.1.1.3 Sperm Transport in Decapods

In the Brachyura, the sperm transfer and storage mechanisms have a bearing on the type of fertilization. In all the advanced brachyuran families, fertilization is internal, whereas in the primitive families, it is external. In both the types, however, there occurs true copulation using the penes as well as the accessory copulatory organs such as the modified pleopods. In the brachyuran crabs, the right and left gonopods (modified I and II pairs of pleopods) and their respective penes form a complete insemination unit. In the mature males, the vasa deferentia open on the coxae or sternum of the last thoracic segment by short muscular penes (Hartnoll, 1975). In the majid crab, *C. opilio*, the first gonopod consists of a basal protopodite and an elongated endopodite, in which the cuticle is infolded the entire length, forming a tube which functions as an ejaculatory canal (Beninger et al., 1991). The second pleopod fits into this cuticular fold, while the penes fit into a slit on the lateral side at the base of the endopodite. During coitus, seminal substances are ejected by muscular action through the penis into the ejaculatory canal of the first pleopod. The ejaculate is further pushed down the canal by the pumping action of the piston-like second pleopod. There is a general similarity in the functional morphology of the first and second gonopod among the advanced brachyurans (Elner et al., 1985).

However, the primitive brachyuran crabs show considerable variation in the insemination morphology. These variations include (1) the size of the penis, (2) length of the first and second gonopods, (3) prominence of the appendix masculina, and (4) number of segment in the pleopod (Hartnoll, 1975). Such differences are bound to reflect on insemination efficiency as seen from the duration of mating in different brachyuran forms. For example, in the primitive forms mating lasts for only a few minutes, whereas in advanced forms it may last for several hours.

Seminal substances, along with spermathecal secretions could provide nutrients for sperm maintenance during their prolonged storage within the spermatheca of the brachyuran crabs. In the mud crab *S. serrata*, the stored sperm use only carbohydrates during their storage inside the spermatheca (Jeyalectumie and Subramoniam, 1991). Furthermore, there is very low activity of the enzyme succinic dehydrogenase and a moderate level of LDH activity in the spermathecal content. However, the fumarate reductase activity is exceedingly high, suggesting anaerobic activity of sperm by using carbohydrates during their storage in the female. Evidently, the true crabs possess excellent sperm storage mechanism to sustain viable sperm for extended periods inside the spermatheca.

12.2.2 Anomura

12.2.2.1 Morphological Diversity

Among the decapods, anomurans show the highest order of variability and complexity in spermatophore morphology. An interesting exception is that the anomuran genus *Aegla* lacks spermatophore in all the species studied so far. In *Aegla platensis*, males transfer unbound spermatozoa, instead of spermatophores to females during mating (Sokolowicz et al., 2007). Anomuran spermatophores are pedunculate and structurally species-specific. Invariably, the spermatophores are attached to the female body to effect epizoic fertilization. Spermatophoric structure in anomurans can be broadly categorized into two major types. The superfamily Paguroidea, represented by the hermit crabs, produce a tripartite, pedunculate type of spermatophore, consisting of a sperm-bearing distal ampulla, raised on a stalk or peduncle, which is fixed on a gelatinous pedestal that is glued to the sternal region of the female during copulation (Tudge, 1999). Exceptions were observed in some species of the genus *Clibanarius* (family Diogenidae), where the stalk and the pedestal are absent (Uma and Subramoniam, 1984; Hess and Bauer, 2002).

The next type is found in Hippidae represented by mole crabs in which the spermatophore is ribbon-like, bearing resemblance to the macruran type of spermatophoric mass (Subramoniam, 1984; Tudge et al., 1999). Among the pedunculate type of anomuran spermatophores, the size and external morphology are relatively constant within a species, but vary between them, making these reproductive structures taxonomically useful for differentiating between closely related species. Phylogenetic relationship apart, variations found in the spermatophore morphology of anomuran crabs and the associated modes of sperm transfer mechanisms may be influenced by the habitat conditions in which these reproductive activities are accomplished (Subramoniam, 1993).

12.2.2.1.1 Paguroidea

Although all paguroidean hermit crabs possess a typical, tripartite, pedunculate spermatophore, hermit crabs belonging to different families exhibit considerable variation in spermatophore morphology that are often species-specific, indicating phylogenetic relationship (Tudge, 1997). The morphological diversity pertains to the number of spermatophores in the ribbon, shape and size of the ampoule, and the length of the peduncle (Mouchet, 1931).

Tudge (1991) studied the morphological variability in the spermatophores of three families, viz., Paguridae, Coenobitidae, and Diogenidae, belonging to the superfamily, Paguroidea. In all the three families, the characteristic features of the spermatophore are distinctively different from one another. Intraspecific differences have also been noticed with respect to size and shape of the spermatophores.

12.2.2.1.1.1 Paguridae Within the Paguridae, the ampullae are elongate, cylindrical, or rhomboid, but the spermatophore sizes vary greatly. Tudge (1999) described 11 species of pagurid hermit crabs with common characteristics of a lateral ridge in the ampulla, and a short peduncle. Lateral ridge is a thickening of the ampullar wall that runs along the elongate spermatophore. The suture found in the lateral ridge gives way to release sperm during dehiscence. Both spermatophore wall and the lateral ridge contain fibrillar layers. In addition, many species possessed an accessory ampulla found at the base of the main sperm-filled ampulla. In many Pagurid species, the peduncle is either short or lack noticeable stalk, as in *Sympagurus* spp. The variation also includes the number of spermatophoric units attached to a single basal plate. Figs. 12.1–12.2 illustrate some of the Pagurian spermatophores showing the morphological diversity in structure and size.

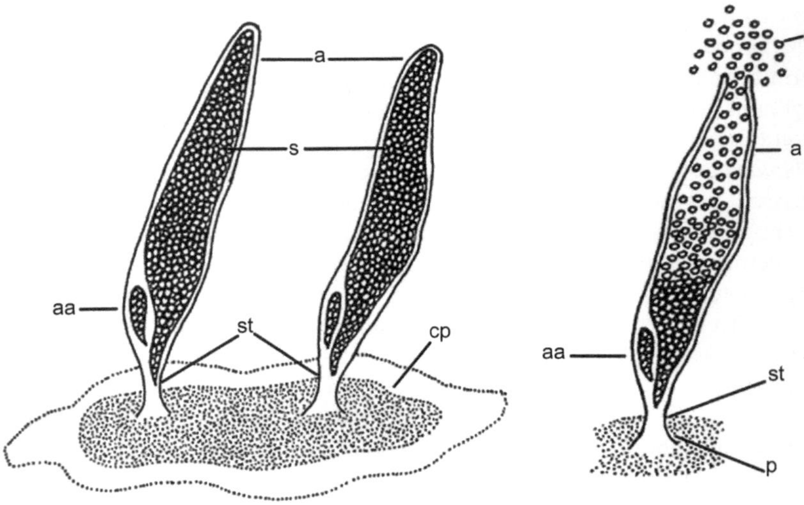

FIGURE 12.1 Before and after sperm release of *Eupagurus bernhardus (After Bloch, F., 1935. Contribution a l'etude des gametes et de la fecondation chez les crustaces decapodes. Travaux de la Station Zoologique de Wimereux 12, 185–279; Adapted from Subramoniam, T., 1993. Spermatophores and sperm transfer in marine crustaceans. Adv. Mar. Biol. 29, 129–214.).* *a*, ampoule; *aa*, accessory ampulla; *cp*, columnar pedestal; *P*, pedestal; *S*, spermatophore; *St*, stalk.

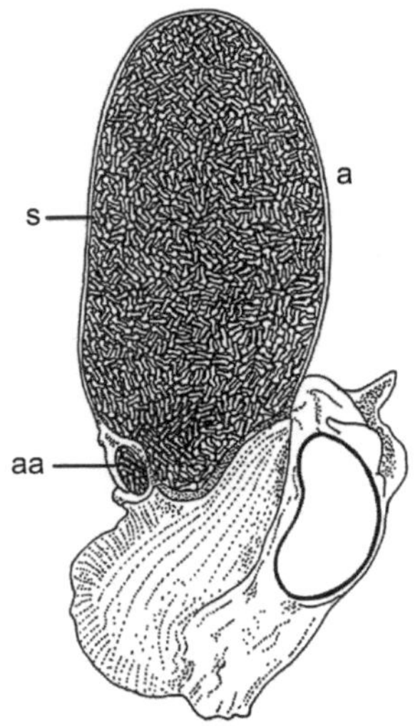

FIGURE 12.2 Spermatophore of hermit crab, *Pagurus hirtimanus* (Tudge, 1991). *a*, ampoule; *aa*, accessory ampulla; *S*, spermatophore. *Permission from The Biological Bulletin.*

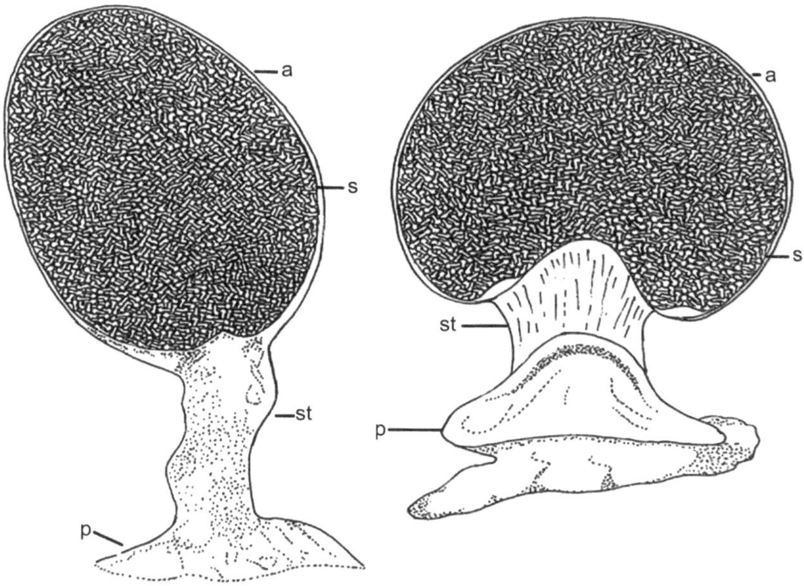

FIGURE 12.3 Spermatophore structure of hermit crab, *Birgus latro* and *Coenobita perlatus* (*Adapted from Tudge, C.C., 1991. Spermatophore diversity within and among the hermit crab families, Coenobitidae, Diogenidae, and Paguridae (Paguroidea, Anomura, Decapoda). Biol. Bull. 181 (2), 238–247.*). *a*, ampoule; *P*, pedestal; *S*, spermatophore; *St*, stalk. *Permission from The Biological Bulletin.*

12.2.2.1.1.2 Coenobitidae The spermatophores in the family, Coenobitidae are characterized by large spherical to ovoid ampullae with short thick stalks, giving them a generally robust appearance. In *Birgus latro*, Tudge and Jamieson (1991) described the spermatophore consisting of a laterally compressed heart-shaped ampulla, a thick stalk, and a broad pedestal (Fig. 12.3). Such robust nature of the spermatophoric mass is especially advantageous for the land hermit crabs, such as the coconut crab, *B. latro* as well as the semiterrestrial hermit crab, *Coenobita perlatus*, which are known to copulate and transfer the sticky spermatophoric mass in a land environment (Fig. 12.3) (Helfman, 1977).

12.2.2.1.1.3 Diogenidae The hermit crabs belonging to family Diogenidae show a great variety of spermatophore shapes but exhibit a characteristic familial morphology with small spherical ampullae and a long thin stalk. The diogenid spermatophores show a great diversity of forms, allowing Tudge (1991) to propose a practical branching key to classify the hermit crabs into their respective families based on the structure of their spermatophores. Furthermore, Scelzo et al. (2004) used spermatophore morphometry as a criterion to distinguish species of diogenid hermit crabs. In *Loxopagurus loxochelis* and *Clibanarius corallinus*, the columnar peduncle penetrates half way into the base of the ampulla, a characteristic feature of this species (Fig. 12.4; Scelzo et al., 2004; Tudge, 1991). The circular flat ampulla is composed of two halves, which meet at the lateral ridge, as in Pagurid hermits. The deposited spermatophore of this species forms a coiled line united by the semicircular-shaped pedestal. In *Clibanarius erythropus*, the ampulla is cap-shaped with its ventral side showing an involution, into which the distal end of the slender stalk fits in. Although a prominent stalk is a characteristic feature of diogenids, two species, *Clibanarius longitarsus* and *Clibanarius virescens*, produce stalkless spermatophores (Uma and Subramoniam, 1984; Matos et al., 1993). In *C. longitarsus*, the spermatophores are kidney-shaped and arranged on both sides of a gelatinous connecting cord which is further embedded in a gelatinous matrix (Fig. 12.5). In yet another species, *Clibanarius vittatus*, Hess and Bauer (2002) described a nonpedunculate spermatophore composed of two halves meeting at the lateral ridge, which also serves as the line of dehiscence for sperm release. Below the ampulla is a basal region that acts as a connecting cord attaching adjacent spermatophores (Fig. 12.6). In *Dardanus asper*, the spermatophores are arranged in a single file with a pear-shaped ampoule kept aloft by the long stalks whose other ends are attached to the pedestal. The whole file of spermatophore is sheathed in a nonsticky veil (Fig. 12.7) (Matthews, 1956). In *Diogenes gardineri*, the ampulla is penetrated ventrally by a thin projection which

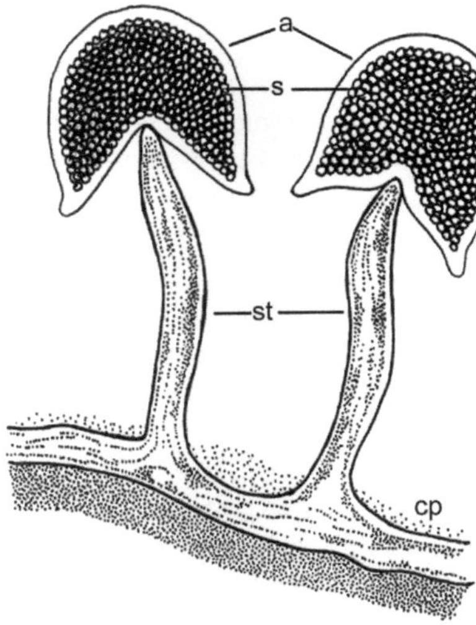

FIGURE 12.4 Spermatophore structure of *Loxopagurus loxochelis (Redrawn from Scelzo, M.A., Mantelatto, F.L., Tudge, C.C., 2004. Spermatophore morphology of the endemic hermit crab Loxopagurus loxochelis (Anomura, Diogenidae) from the southwestern Atlantic—Brazil and Argentina. Inv. Rep. Dev. 46 (1), 1–9.). a*, ampoule; *cp*, columnar pedestal; *P*, pedestal; *S*, spermatophore; *St*, stalk.

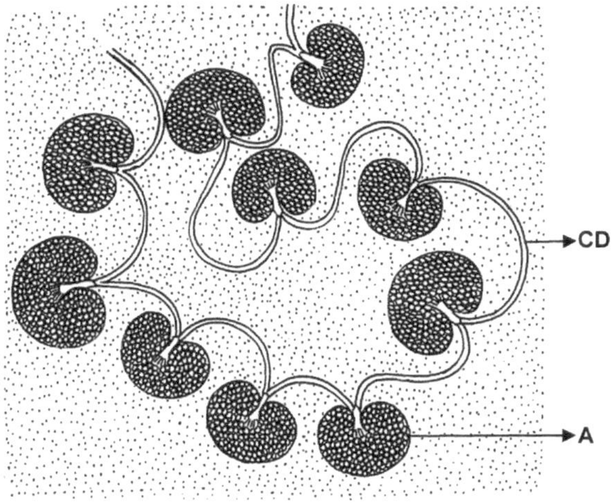

FIGURE 12.5 Spermatophores present within the vas deferens of *Clibanarius longitarsus (After Uma, K., Subramoniam, T. 1984. A comparative study of the spermatophore in Scylla serrata (Forskal) (Decapoda Brachyura) and Clibanarius longitarsus (De Haan) (Decapoda Anomura). J. Mar. Biol. Assoc. India 26, 103–108; Adapted from Subramoniam, T., 1993. Spermatophores and sperm transfer in marine crustaceans. Adv. Mar. Biol. 29, 129–214.)*. *A*, ampulla; *CD*, connecting cord.

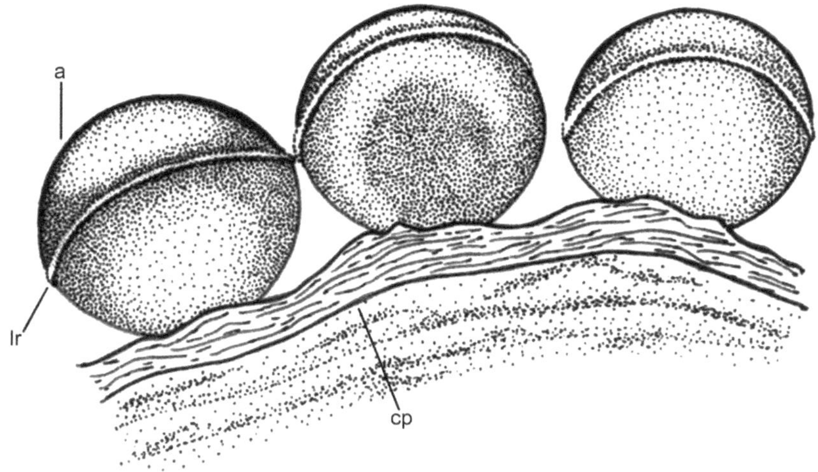

FIGURE 12.6 Nonpedunculate spermatophore of *Clibanarius vittatus*. a, ampoule; cp, columnar pedestal; lr, lateral ridge *(Reconstructed diagrammatic representation from SEM image of Hess, G.S., Bauer, R.T., 2002. Spermatophore transfer in the hermit crab Clibanarius vittatus (Crustacea, Anomura, Diogenidae). J. Morphol. 253 (2), 166–175.). Permission from John Wiley and Sons.*

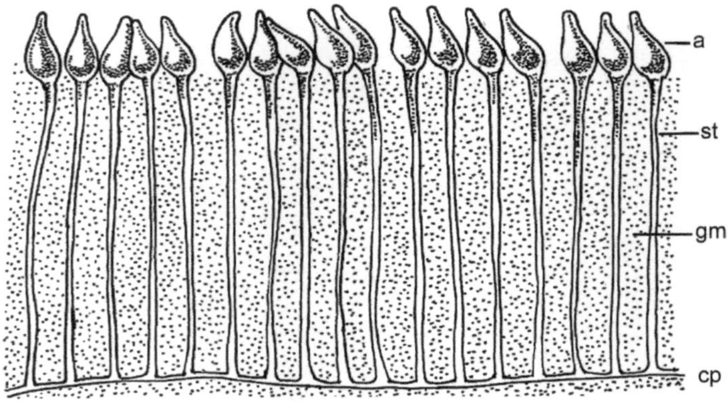

FIGURE 12.7 Spermatophore within the vas deferens of *Dardanus asper (After Matthews, D.C., 1953. The development of the pedunculate spermatophore of a hermit crab, Dardanus asper (De Haan). Pacific Sci. 8, 28–33; Adapted from Subramoniam, T., 1993. Spermatophores and sperm transfer in marine crustaceans. Adv. Mar. Biol. 29, 129–214.). a*, ampoule; *cp*, columnar pedestal; *gm*, gelatinous matrix; *St*, Stalk.

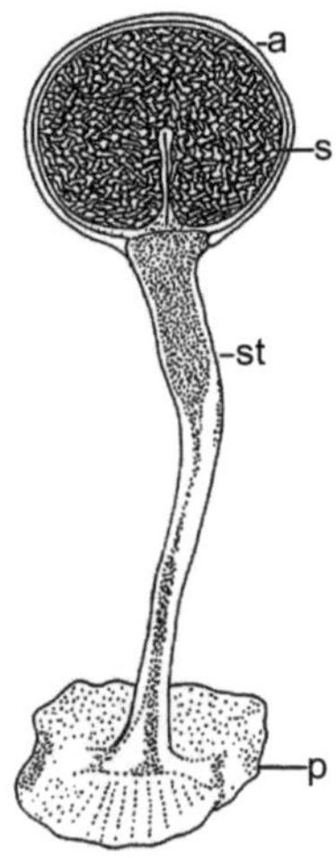

FIGURE 12.8 Spermatophore structure of *Diogenes gardineri (Adapted from Tudge, C.C., 1991. Spermatophore diversity within and among the hermit crab families, Coenobitidae, Diogenidae, and Paguridae (Paguroidea, Anomura, Decapoda). Biol. Bull. 181 (2), 238–247.). a*, ampoule; *P*, pedestal; *S*, spermatophore; *St*, stalk. *Permission from The Biological Bulletin.*

originates at the distal end of the stalk (Fig. 12.8). Interestingly, in *Dardanus lagopodes*, there occurs a double-headed spermatophore on a single peduncle (Fig. 12.9). The majority of Diogenidae are aquatic; hence mating and fertilization occur in water. Spermatophores are placed externally either on the carapace of the females or in the inner side of the gastropod shell the female inhabits.

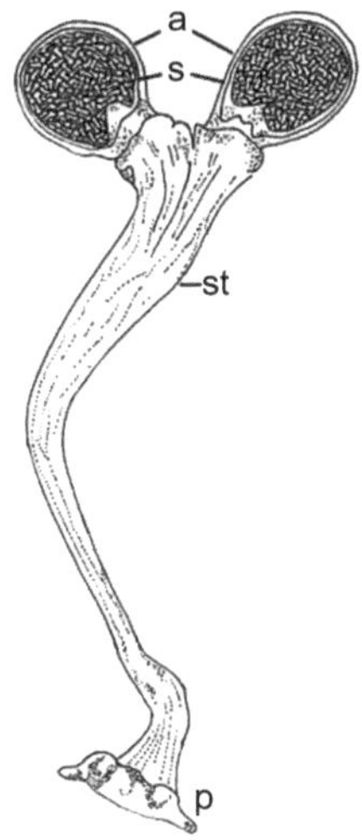

FIGURE 12.9 Spermatophore structure of *Dardanus lagopodes* (Tudge, 1991). *Permission from The Biological Bulletin.*

The morphological variations in anomuran crabs continue with other families also. The spermatophores in the family Zoenovitidae are characterized by large spherical to ovoid ampullae with short thick stalks, attached to the gelatinous pedestal. Tudge (1991) described four species belonging to the genus *Zoenovita* but found similar overall morphology with only size differences distinguishing them. In *Aegla rostrata*, representing Galetheoidia, the spermatophores are either spherical or oblong, enclosing sperm cells bound by a thin translucent membrane (Tudge, 2003). Certain characteristics found in the spermatophores of anomurans also serve to confirm phylogenetic linkages between sister groups. For example, in king crabs (or stone crabs) belonging to Lithodidae, the spermatophore contains the accessory ampulla very similar to pagurid hermit crabs, thus establishing a lithodid–pagurid relationship (Tudge et al., 1998).

In yet another family of Porcellanidae, the spermatophore morphology shows unique variation. In the genus, *Petrolisthes*, the spermatophore is composed of an elongate ampulla, which has a swollen posterior region and a thin apical tubular projection, and is attached basally to a short broad stalk and pedestal (Fig. 12.10) (Tudge and Jamieson, 1996). The spermatozoa are concentrated in the posterior bulbous region of the ampulla and the anterior extension is devoid of spermatozoa. However, in the genuses *Aliaporcellana* and *Pisidia*, the pedunculate spermatophores are small and spherical to ovoid, in contrast to the species of *Petrolisthes*, which are devoid of the tubular extensions. The spermatophores of *Pisidia longicornis*, a lateral ridge, where the ampulla splits to release the spermatozoa, is shown to exist (Mouchet, 1931). The mechanism

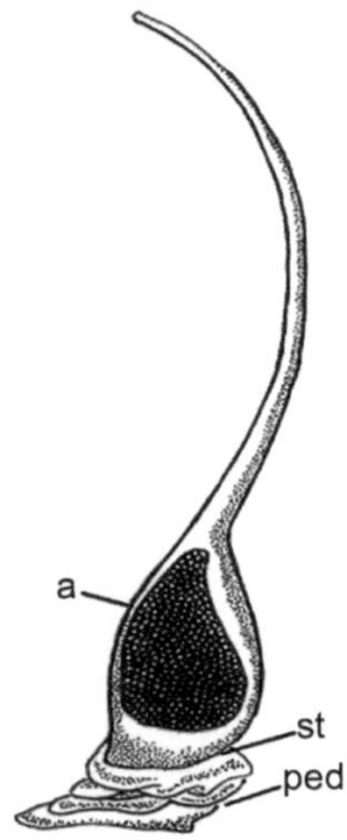

FIGURE 12.10 Spermatophores of *Petrolistheis lamarckii (Adapted from Tudge, C.C., Jamieson, B.G., 1996. Spermatophore and spermatozoal morphology in the Porcellanidae. II. The genera Petrolisthes and Polyonyx (Decapoda: Anomura). J. Crust. Biol. 16 (3), 535–546.). a*, ampoule; *ped*, pedestal; *St*, stalk.

of sperm release in the genus *Petrolisthes* with extended tubule is not known. Extensive morphological variability encountered in the anomuran species truly reflects the polyphyletic assemblage of this infraorder (Tudge, 1997).

12.2.2.1.2 Spermatophores of Hippidae

Although the basic plan of anomuran hermit crab spermatophores is the tripartite pedunculate type, deviation occurs in Hippidae family, consisting of false crabs. In *Hippa pacifica*, a sand crab inhabiting the intertidal region, Matthews (1956) described a spermatophore, intermediate in morphology between the nonpedunculate macruran and the pedunculate anomuran spermatophores. The spermatophore of *H. pacifica* is a continuous, highly convoluted tube raised by a continuous, ribbon-like stalk, attached to a broad foot. Similarly, in another hippid mole crab genus, *Emerita*, several authors have reported deposition of a macruran-type of spermatophoric ribbon on the ventral sternum of the females (MacGinitie, 1938; Wharton, 1942). *Emerita asiatica*, unlike *H. pacifica*, produces pedunculate spermatophore, characteristic of anomurans, albeit with certain peculiarities (Subramoniam, 1984). As in pagurids, the spermatophore consists of three distinct parts: the sperm-containing ampoule, the peduncle, and the gelatinous pedestal. However, the spermatophores are dimorphic in nature: one in the form of a truncated cone and the other in the form of a tumbler. These two types of spermatophores are arranged almost alternatively in a single file (Figs. 12.11 and 12.12). Interestingly, the lateral ridge, so characteristic of paguroidan ampullae, is absent in the spermatophores of *E. asiatica*. The lower ends of the spermatophores possess peduncles, which join with a continuous

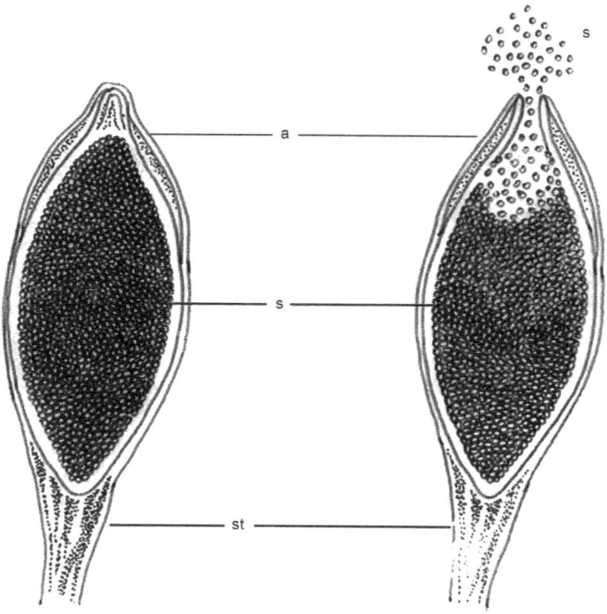

FIGURE 12.11 Truncated cone-shaped spermatophore of *Emerita asiatica* before and during sperm release *(Adapted from Subramoniam, T., 1993. Spermatophores and sperm transfer in marine crustaceans. Adv. Mar. Biol. 29, 129–214.). a*, ampoule; *S*, spermatophore; *St*, stalk.

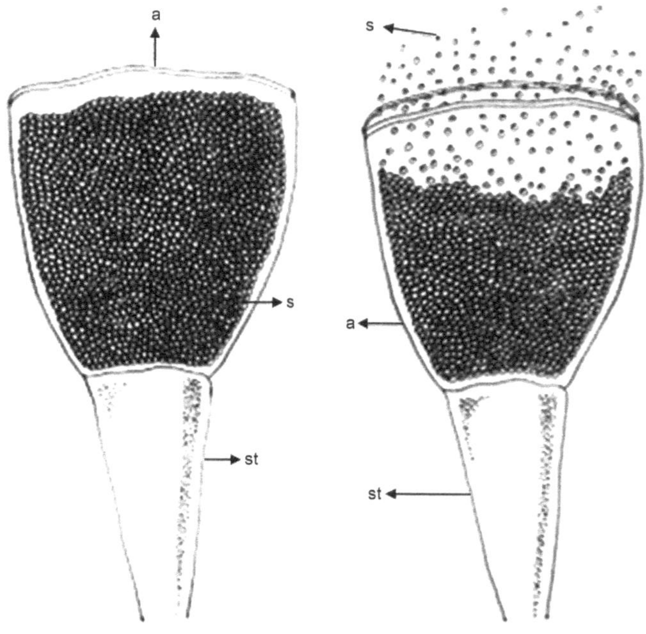

FIGURE 12.12 Tumbler-shaped spermatophore of *Emerita asiatica* before and during sperm release *(Adapted from Subramoniam, T., 1993. Spermatophores and sperm transfer in marine crustaceans. Adv. Mar. Biol. 29, 129–214.). a*, ampoule; *S*, spermatophore; *St*, stalk.

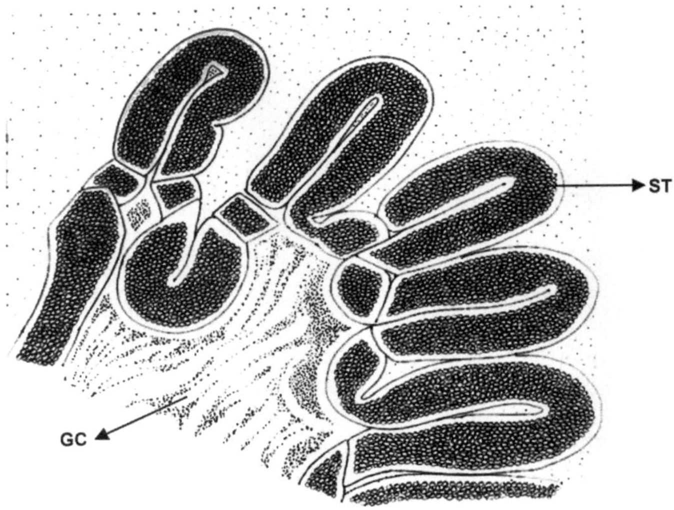

FIGURE 12.13 Freshly extruded spermatophoric mass of *Albunea symnista* showing the loop-like convolution of the spermatophoric tube *(Redrawn from phase contrast micrograph of Subramoniam, T., 1984. Spermatophore formation in two intertidal crabs Albunea symnista and Emerita asiatica (Decapoda: Anomura). Biol. Bull. 166 (1), 78–95.). GC*, gelatinous cord; *ST*, spermatophoric tube). *Permission from The Biological Bulletin.*

gelatinous ribbon. The whole spermatophoric mass is embedded in a protective jelly-like matrix. In this respect, *E. asiatica* differs from other anomuran crabs, such as *Diogenes pugilator* and *Pagurus bernhardus*, wherein spermatophores are attached to the gelatinous base singly or in groups of two or three (Bloch, 1935). The extruded spermatophore has a thick double-layered refractile covering. The spermatozoa are glued together by a viscous fluid and packed closely but irregularly inside the spermatophore.

In another anomuran sand crab, *Albunea symnista*, coexisting with *E. asiatica*, the spermatophore is nonpedunculate and comprises a highly convoluted tube with firm membrane forming a cordlike mass. The spermatophoric tube of *A. symnista*, however, shows nodelike constrictions giving rise to internal discontinuities (Fig. 12.13). The whole spermatophoric ribbon is embedded in a gelatinous matrix (Subramoniam, 1984). Such a breaking up of a continuous spermatophoric tube by constrictions (*Albunea*) and distinct spermatophoric ampullae with drawn-out peduncles set on a basal filamentous pedestal (*Emerita*) suggests that these anomuran sand crabs may be midway forms in the evolution of discrete pedunculate spermatophores of other anomurans from the tubular spermatophores of Macrura (Subramoniam, 1993).

12.2.2.1.2.1 Origin of Spermatophores Despite morphological diversity found in the spermatophores of anomuran crabs, there is a basic similarity in the morphology of the VD in the various species of hermit crabs (Mouchet,

1931). Essentially, the VD of the pagurids exhibits several helical twists in the form of the right- and left-handed coils that determine the production of different spermatophoric components inside the lumen of respective regions. The spiraling VD of hermit crabs is structurally complex, having as many as nine functional regions (Mouchet, 1931).

In general, the VD is made up of an outer muscular and inner glandular layer. In *Eupagurus bernhardus*, the outer muscular layer is extremely thin and the thickness of the inner glandular layer varies significantly in different regions of the duct. Corresponding to the changes in the thickness of the inner secretary layer, the contour of the lumen also changes. In the proximal spiral, the epithelial cells are cuboidal and line a cylindrical canal. The secretary substance produced by these cells is used to agglutinate the sperm released from the testis into a sperm mass. The inner shape of the lumen, however, changes into an elliptical condition by the diminution of epithelial cells at the two diametrically opposite points together with the enlargement of the lateral cells. At the opposite points of cellular depletion are found two cavities filled with a new basophilic secretion, which surrounds the incoming inner sperm mass, forming an envelope as well as a stalk that is connected to a secretary material found in the ventral groove of the VD lumen. Another secretion, emanating from the epithelial cells of the ventral groove spread out to form the basal layer, which is folded laterally on it. Yet another mucoid secretion accumulates in the space between the basal layer and the spermatophore ampoule in the upper axial lumen. All these secretions exhibit different tinctorial properties.

At the end of the second spiral, the fragmentation of this continuous spermatic cord begins. The internal lumen becomes enlarged due to thinning of the epithelial cells. The spermatophores take a helical turn to place themselves at 90° to their initial position. In the terminal region, each of the basal layers, which carry many spermatophores, is placed in the transverse plane. Thus, different regions of the VD implicated with secretions of various spermatophoric components have been found to be remarkably similar in different hermit crabs. Specialized muscular activity as well as the modified lumen shape is thought to be the causative agents for the spermatophoric fragmentation from the continuous sperm sheath in *D. asper* and *Pagurus novae-zealandiae*, respectively (Matthews, 1953; Greenwood, 1972).

Among Anomura, the galatheids, pagurids, and porcellanids are known to elaborate complex spermatophores in the above manner, by which the continuous, convoluted sperm tube is broken into distinct ampullae and elevated on peduncles attached to a footlike base. Conversely, the hippid mole crab, *H. pacifica*, possesses a spermatophore in the form of a continuous convoluted tube raised by a continuous ribbon-like stalk attached to a broad foot. As in pagurids, the changes in the lumen shape and the differential secretary activity of the inner glandular epithelial cells lining the VD could bring about the spermatophore production. However, there is no constriction of the VD lumen and specific muscular activity to bring about fragmentation of the sperm sheath into spermatophore units in *H. pacifica* (Matthews, 1956).

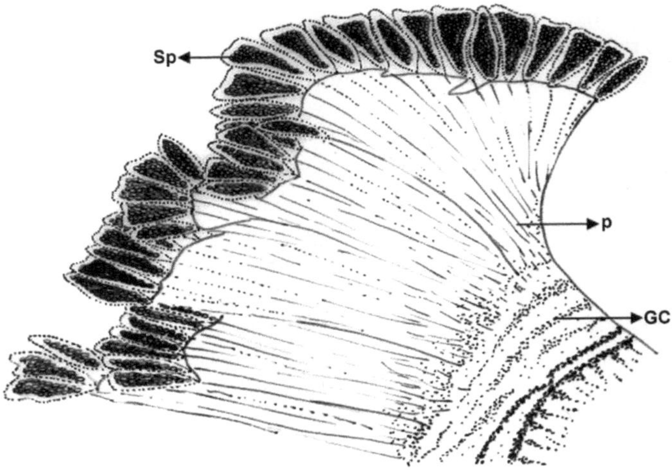

FIGURE 12.14 Spermatophoric ribbon of *E. asiatica* showing the arrangement of the spermatophores (Sp) in the form of a ribbon with the peduncle (P) connected to the connecting gelatinous cords (GC). *Adapted from Subramoniam, T., 1984. Spermatophore formation in two intertidal crabs* Albunea symnista *and* Emerita asiatica *(Decapoda: Anomura). Biol. Bull. 166 (1), 78–95; Permission from The Biological Bulletin.*

In another hippid species, *A. symnista*, VD is a simple straight tube, without much coiling, thus differing from those of hermit crabs, producing typical pedunculate spermatophore. The entire VD is divisible into a short proximal and a dilated distal portion. The spermatozoa as released from the testis are clumped in the anterior part of the VD. This sperm mass is ensheathed in a thin wall by condensation of the secretary material originating in the PVD. As the tubular spermatophore enters the first part of the distal VD, it twists and lies opposed to the ventral epithelial wall. This twisted tube as it extends distally becomes folded into loops and is set on a firm basal membrane. Between the loops, the spermatophoric tube tends to constrict toward the ventral region with nodelike formations, apparently interrupting the continuity of the spermatophoric tube (Fig. 12.14). At the same time, the basal membrane becomes thickened into a gelatinous cord which connects the highly convoluted spermatophoric ribbon ventrally. In the distal VD, two typhlosole-like structures made up of epithelial cells in the dorsal region produce a secretion which fills up the entire VD lumen, constituting the protective matrix. The basal gelatinous cord is produced by the ventral epithelial cells (Subramoniam, 1984).

Contrary to conditions found in *A. symnista*, a continuous sperm sheath is not evident in the PVD of the mole crab *E. asiatica* (Subramoniam, 1984). In the AVD, the rod-shaped mature spermatozoa are agglutinated into many clusters. Each cluster of sperm is covered by a gelatinous membrane originating from the columnar cells lining the VD. The circular lumen changes gradually to an elliptical shape toward the distal part of PVD. In the distal VD, the ventral epithelial cells secrete a thick gelatinous layer that condenses into a thick cord, which gives off branches, connecting to individual spermatophore ampoules, placed on the dorsolateral periphery of the lumen. This gelatinous material also envelops the ampoules, to give rise to a secondary layer. In the dorsal region of the distal VD, the inner epithelial cells produce a typhlosole-like projection, which secretes a frothy substance, constituting the protective matrix material of the extruded spermatophoric ribbon. Typhlosole-like inpushings in the distal VD are characteristic structures of paguridean anomurans such as *Eupagurus prideauxi* (Mouchet, 1931) and the spiny lobster, *Panulirus homarus* (Berry and Heydorn, 1970; Radha and Subramoniam, 1985). Obviously, these glandular formations increase the surface area of the mucus secreting epithelial cells in the VD. It is of interest to note here that the pedunculate spermatophores of *E. asiatica* differ from those of other anomuran crabs in that they are not produced from a continuous sperm cord but are formed from separate sperm clusters, found in the AVD.

12.2.2.1.2.2 Chemical Composition of Spermatophores Detailed histochemical analysis on two sand crabs, *A. symnista* and *E. asiatica* (Subramoniam, 1984) has revealed that mucopolysaccharides, complexed with proteins, form the main components of the spermatophores. The mucoid nature of the spermatophoric components is advantageous for its transference in the aquatic medium where desiccation to spermatophoric layers will not be a limiting factor. However, different parts of the spermatophore contain different types of mucopolysaccharides. In *A. symnista* the spermatophore wall is composed mainly of a neutral mucopolysaccharide and is negative to test for chitin. The sperm mass substance within the spermatophoric tube stains metachromatically with toluidine blue, suggesting the presence of strongly sulfated acid mucopolysaccharides. The sperm cells are PAS positive suggesting the presence of glycogen. The basal gelatinous chord contains a neutral mucopolysaccharide which is conjugated to a protein rich in basic and aromatic amino acids. In contrast, protective matrix contains acid mucopolysaccharide rich in carboxyl groups. Histochemical reactions of mucopolysaccharides in different parts of spermatophores in *A. symnista* and *E. asiatica* are summarized in Table 12.3.

12.2.2.1.2.3 Spermatophore Transfer and Dehiscence In general, anomuran crabs lack intromittent organs for sperm transmission to female genital tract; spermatophores are deposited directly onto the ventral sternum of the female crab. Therefore, the development of a well-defined peduncle in the anomuran spermatophore is a special adaptation for attachment. The mode of attachment, however, differs markedly within anomuran species. In *D. pugilator*, only a single spermatophore is attached with its gelatinous base to the sternum of the females (Bloch, 1935) (Fig. 12.15). On the other hand, in *P. bernhardus*, four to five spermatophores are placed on each gelatinous strip.

In the paguroidean hermit crabs, the lateral ridge, representing the joining line of the two halves of the ampulla, is the region of spermatophore dehiscence.

TABLE 12.3 Histochemical Characteristics of Spermatophores in the Sand Crabs *Emerita* and *Albunea*

Spermatophoric Components	Chemical Components	Origin
Albunea symnista		
1. Sperm mass substance	Sulfated AMP	PVD
2. Spermatophore wall	Neutral MP	PVD
3. Gelatinous cord	Neutral MP	DVD, ventral epithelium
4. Gelatinous matrix	Carboxylated AMP	DVD, dorsal epithelium especially typhlosole
Emerita asiatica		
1. Sperm mass substance	Sulfated AMP	PVD
2. Spermatophore wall		PVD
Inner layer	Carboxylated AMP	DVD, ventral epithelium
Outer layer	Neutral MP	DVD, ventral epithelium
3. Peduncle/gelatinous cord	Neutral MP	DVD, dorsal epithelium
4. Gelatinous matrix	Periodate reactive AMP	especially typhlosole

Data based on Subramoniam, T., 1984. Spermatophore formation in two intertidal crabs *Albunea symnista* and *Emerita asiatica* (Decapoda: Anomura). Biol. Bull. 166 (1), 78–95.

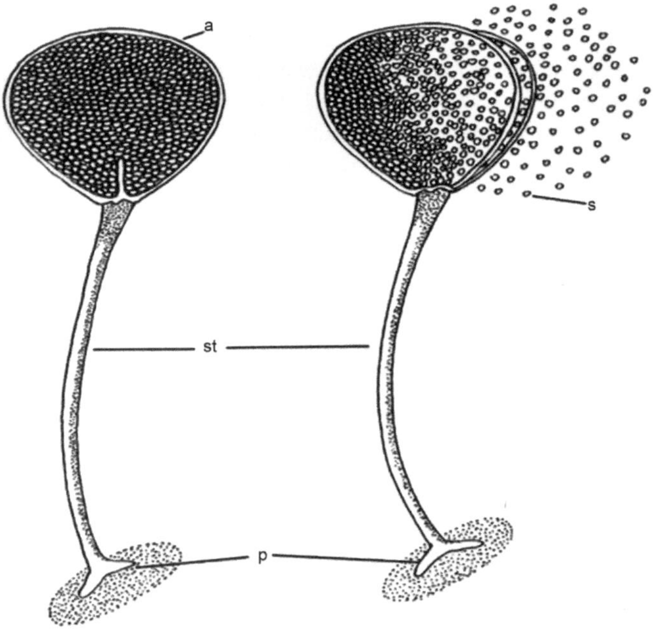

FIGURE 12.15 Before and after sperm release of *Diogenes pugilator (After Bloch, F., 1935. Contribution a l'etude des gametes et de la fecondation chez les crustaces decapodes. Travaux de la Station Zoologique de Wimereux 12, 185–279; Adapted from Subramoniam, T., 1993. Spermatophores and sperm transfer in marine crustaceans. Adv. Mar. Biol. 29, 129–214.). a*, ampoule; *P*, pedestal; *S*, spermatophore; *St*, stalk.

Hamon (1937) theorized that changes in the internal pressure due to mechanical or osmotic forces could be the mechanism underlying spermatophore dehiscence. In addition, chemical substances could also dissolve the suture that unites the two halves of the ampulla. Tudge (1991) showed by electron microscopic studies the fibrillar nature of the spermatophore wall in *Coenobita spinosus*. He also showed that there is a gap in the fibrous matrix in the lateral ridge region, making it a weak point for the release of sperm.

In *E. asiatica*, spermatophore extrusion occurs through a muscular genital papilla, situated at the base of the fifth thoracic leg (Subramoniam, 1977b). In this anomuran crab, the spermatophore deposition occurs only in the fresh molt condition and the spermatophore ribbon sticks to the female pleopodal region very firmly. The sperm release occurs always through a definitive opening. In the truncated conelike spermatophores, the opening is made through the nipple-like projection found at the opposite end of the peduncle. In the other, larger type, the wider region is rimmed by a well-defined lip which is firmly closed before sperm release. Streaming of spermatozoa is first observed in the gaps formed at the corners of the wider end and then in several sites of the center, resulting in the complete opening of the lips (Figs. 12.11 and 12.12). The fact that the spermatozoa release occurs only after contact with the eggs suggests that an oviductal secretion may be responsible for the digestion of the cementing material closing the lip of the spermatophore (Subramoniam, 1977b).

By far, hermit crab species belonging to several subfamilies exhibit complicated spermatophore morphology with basic tripartite plan. On the other hand, sand crab species belonging to Hippidae, show spermatophores which are intermediate between Macrura and true Anomura, establishing a phylogenetic link between these groups of Decapoda. The spermatophore structure has a bearing on the mode of sperm release and the type of fertilization. Table 12.4 shows the salient features characterizing each anomuran group, together with the mode of attachment, sperm release, as well as the habitat conditions in which sperm transfer and fertilization take place.

12.2.3 Macrura

12.2.3.1 Lobsters

Macrurans, including lobsters and crayfish, are typical spermatophore-producing decapod crustaceans. Their spermatophores are generally complex masses consisting mainly of three components: a spermatophoric tube, a basal adhesive matrix, and a protective gelatinous matrix (Berry and Heydorn, 1970). The spermatophores are soft and mucoid when freshly extruded, but undergo "hardening" on exposure to sea water in species employing external fertilization (e.g., spiny lobster of the genus *Panulirus*). In general, the spermatophores are either deposited externally to the ventral sternum of the females (spiny lobsters and rock lobsters) or they are stored within the seminal receptacles (thelycum) of the female homarid and nephropsid lobsters (Aiken and Waddy, 1980).

TABLE 12.4 Types of Spermatophores and Sperm Transfer Modes Within Anomura

Species	Habitat	Spermatophoric Characteristics	Transfer Mechanism
Coenobitidae	Terrestrial; tropical and subtropical. There are also ocean species	The spermatophore tripartite; consisting of large, spherical to oval ampullae with short thick stalks connected to a broad pedestal. Lateral ridge in the ampullae is prominent	Spermatophore transfer occurs in the terrestrial environment for terrestrial species. Sperm release occurs through lateral ridge to sternum of the female
Paguridae	Marine hermit crabs found in coastal and deep waters	Tripartite spermatophore with elongated ampullae, accessory ampullae present at the base of main ampullae. Very short peduncle, often incorporated into the pedestal to give a nonpedunculate appearance	Spermatophore transfer occurs directly to female ventral sternum. Egg release and externally fertilized soon after spermatophore deposition
Diogenidae	Mostly tropical marine hermit crabs and some are terrestrial	Spermatophore consists of small spherical ampullae and a long thin stalk ending with a broad pedestal. The ampullae contain the lateral ridge. Some species produce double- headed spermatophores. A few species of *Clibanarius* produce stalkless spermatophores	Spermatophore transfer occurs in marine conditions. In some cases, spermatophore is released through lateral ridge. Fertilization external. Spermatophore placed externally on the carapace of the females or inside the gastropod shell the female inhabits
Porcellanidae	Intertidal, found in rock crevices	The spermatophore composed of elongated ampullae and a thin epical tubular projection. The base of the ampullae is connected to the pedestal by a short broad stalk. Lateral ridge is present in the ampullae	Sperm release through lateral ridge, fertilized externally
Hippidae	Intertidal region, burrowing crab	**1.** Spermatophore, a continuous highly convoluted tube raised on continuous ribbon like stalk attached to a broad foot—*Hippa pacifica*. **2.** Dimorphic spermatophores with a long drawn-out peduncle joining with a continuous gelatinous ribbonous base—*Emerita asiatica* **3.** Nonpedunculate tubular spermatophore, ventrally connected to a gelatinous cord and embedded in gelatinous matrix—*Albunea symnista*	Spermatophores attached onto the pleopodal region of females. Sperm release through the tip of ampullae. In tubular spermatophore, sperm release could be mechanical. Fertilization external
Lithodidae	Mostly found in cold seas	Spermatophore tripartite; contain accessory ampullae at the base of the main ampullae	Mating occurs after female's molting and face to face between pairs. Release of sperm from spermatophores and the release of eggs occur simultaneously, facilitating external fertilization on abdomen of female
Galatheidae	Marine	Spermatophores are lobulated, spherical to oblong shaped. The sperms inside are bound by a thin translucent membrane	Not known
Aeglidae	Fresh water	Thin walled, spherical to oblong lobes containing sperm cells and were bound by a thin translucent membrane	Not known

In *P. homarus* three distinct horizontal matrix layers could be recognized in the deposited spermatophores. They are an outer crust like layer, termed the protective matrix, a middle layer of highly convoluted spermatophoric tube, containing sperm mass, and a basal spongy layer, termed the adhesive matrix (Berry, 1970). The protective matrix is strongly eosinophilic containing calcium granules, while the underlying spermatophoric matrix differs from the former only in its consistency. The basal adhesive matrix has a striated appearance with strong eosinophilic reaction. The spermatophoric mass of other *Panulirus* species have this basic pattern, although a few deviations, such as the absence of a distinct spermatophoric tube wall, as in *Panulirus angulatus* and *Linuparus trigonus*, do occur (Berry and Heydon, 1970).

A characteristic feature of lobster spermatophores is their structural transformation that occurs in the deposited spermatophoric layers. Martin et al. (1987) investigated the structural transformation and chemical changes after spermatophore attachment in *Panulirus interruptus*. In this species, the spermatophore is composed of a highly coiled sperm tube embedded in an acellular matrix. The extruded spermatophore is white, soft, and sticky on all surfaces. The sperm tube is seen near the surface of the foot of the spermatophore, which is the site of attachment to the female body. The opposite surface, the cap, hardens and darkens after the exposure to sea water. In the homarid, *Homarus americanus*, the spermatophore does not undergo hardening on exposure to sea water and transference to the female, but the stickiness of the freshly extruded spermatophore ensures the successful transfer to the female seminal receptacle (Kooda-Cisco and Talbot, 1982). In this lobster, the inner tubular sperm mass is surrounded by three investment layers: a primary spermatophoric layer, which is amorphous and PAS positive; an intermediate layer containing PAS-positive granules, and an outer bounding layer comprising small filaments and a flocculent material. In a histochemical study on *P. homarus*, Radha and Subramoniam (1985) showed that the wall of the spermatophore tube contains neutral mucopolysaccharide, whereas the sperm mass and the gelatinous matrix are rich in acidic mucopolysaccharides. Further electrophoretic studies revealed that chondroitin sulfate, complexed with protein, is the principal component of the acidic mucopolysaccharides. Chondroitin sulfate could provide elasticity as well as resistance to compression, thus keeping the deposited spermatophore from the risk of desiccation. Understandably, chondroitin sulfate has a dominating role in the "hardening" of the spermatophores as well as maintaining a microenvironment for the delicate spermatozoa within the spermatophoric tube during their prolonged storage in the female lobster.

12.2.3.1.1 The Origin of Spermatophores

In the spiny lobster *P. homarus*, the male reproductive system consists of a paired testis and the VD which is divisible into a convoluted proximal VD and a dilated distal part, the latter leading into a narrow ejaculatory duct. The columnar epithelial cells lining the PVD is highly secretory. Clusters of spermatozoa mixed in eosinophilic secretion are present in the lumen. The columnar cells of PVD secrete a strongly eosinophilic granular matrix, secreted by the columnar

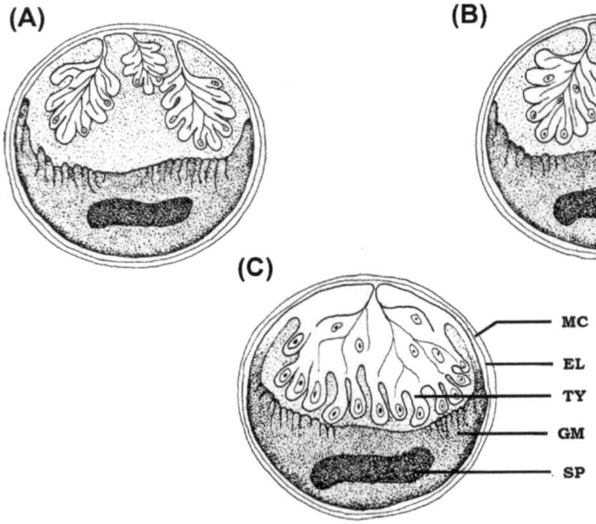

FIGURE 12.16 Diagrams representing the typhlosole in the distal vas deferens of *Panulirus homarus*. (A) Trilobed typhlosole with two large lateral lobes and a small middle lobe. (B) Bilobed typhlosole. (C) Single-lobed typhlosole. *EL*, epithelial layer; *GM*, gelatinous matrix; *MC*, muscle and connective tissue layer; *SP*, spermatophoric mass; *TY*, typhlosole. *Redrawn from Radha, T., Subramoniam, T., 1985. Origin and nature of spermatophoric mass of the spiny lobster* Panulirus homarus. *Mar. Biol. 86, 1, 13–19; Adapted from Subramoniam, T., 1993. Spermatophores and sperm transfer in marine crustaceans. Adv. Mar. Biol. 29, 129–214.*

epithelial cells of the VD surrounds an inner core of tightly packed sperm, forming a layer over it. This forms the spermatophoric tube or spermatophore proper (Radha and Subramoniam, 1985).

In the distal VD, the dorsal epithelium, by proliferation, has given rise to a leaflike typhlosole which runs along the dorsal side up to the terminal end. As shown in Fig. 12.16, the typhlosole is trilobed with two large foliaceous outer lobes and a small middle lobe in the anterior part of the DVD of *P. homarus* (Radha and Subramoniam, 1985). The typhlosole secretes a less viscous eosinophilic granular matrix within which the spermatophore is arranged in regular convolutions and is pushed to the ventral periphery by the profuse production of matrix material from the typhlosole. The ventral epithelium simultaneously secretes a thin matrix consisting of large strongly eosinophilic granules below the spermatophore, its function is to cement the extruded spermatophoric mass to the sternum of the female (Matthews, 1951).

12.2.3.1.2 Spermatophore Composition and the Mode of Attachment

Despite the basic similarities found in the spermatophoric structure, differences do occur in the composition of the mucoid substances and their behavior in the sea water on deposition to the females (Table 12.5). The differences in

TABLE 12.5 Spermatophoric Composition in Lobsters

Species	Spermatophoric Tube	Adhesive Matrix	Protective Matrix	Behavior in Sea Water
Panulirus homarus (Berry, 1970; Berry and Heydorn, 1970)	Highly convoluted cylindrical tube, spermatophore wall distinct, granular	Present, eosinophilic and globular matrix	Granular with a putty-like consistency	Becomes harder and darkens
Palinurus gilchristi (Berry and Heydorn, 1970)	Spermatophore triangular in cross-section, wall composed of a homogeneous eosinophilic substance, spermatophoric wall thickness not uniform	Globular matrix layer eosinophilic	Gelatinous matrix with dense areas of agglutination	Remains insoluble in water
Panulirus angulatus (Berry and Heydorn, 1970)	Consists of large tightly compacted core of spermatozoa embedded in an eosinophilic mucus, but lacks a spermatophore wall	Globular matrix	Gelatinous but without dense areas of agglutination. Resembles the mucus covering spermatozoa	Insoluble in sea water
Linuparus trigonus (Berry and Heydorn, 1970)	Consists of large tightly compacted core of spermatozoa embedded in an eosinophilic mucus but lacks a spermatophore wall	Globular matrix	Gelatinous but without dense areas of agglutination. Resembles the mucus covering spermatozoa	Insoluble in sea water
Jasus lalandii (Fielder, 1964)	Spermatophore a thin continuous thread of spermatozoa with no surrounding wall. Convoluted randomly in the matrix	Absent	Homogeneous gelatinous matrix	Disintegrates in sea water. It is reasonably fluidy and sticky

Continued

TABLE 12.5 Spermatophoric Composition in Lobsters—cont'd

Species	Spermatophoric Tube	Adhesive Matrix	Protective Matrix	Behavior in Sea Water
Penaeus penicillatus (Matthews, 1951)	Crystalline granular spermatophoric wall. Highly convoluted tube	Not distinguishable	Present	Hardens
Enoplometopus occidentalis (Matthews, 1954; Haley, 1984)	Tubular sperm mass. Possesses distinct PAS positive wall	A thin outer bounding layer on sides of spermatophore adheres to the substratum after ejaculation	Gelatinous homogeneous matrix, PAS negative	Not known
Panulirus interruptus (Martin et al., 1987)	Consists of tightly packed spherical cavities in an acellular material within which the sperm lie	Foot of the spermatophore characterized by vertical striations after attaching on the female	Composed of granules embedded in a loose weave of filaments	The cap of the spermatophore will harden and darken after exposure to sea water
Thenus orientalis (Silas and Subramoniam, Unpublished observation)	Sperm are enclosed in oval-shaped vesicles which are connected to one another by a spermatophoric rope. Sperm capsule wall PAS positive	No distinguishable spermatophoric mass attached to the female by any part of the surface	Fibrillar network, contains elastin	Remains insoluble in sea water

Reproduced from Subramoniam, T., 1993. Spermatophores and sperm transfer in marine crustaceans. Adv. Mar. Biol. 29, 129–214.

the spermatophoric components of various lobster species also reflect on the mode of attachment and fertilization process. As an example, in *Jasus lalandii*, the gelatinous matrix disintegrates in sea water, releasing the threadlike spermatophore into the ambient medium, thus necessitating immediate oviposition after mating. On the other hand, in *Palinurus gilchristi* and *P. angulatus*, the protective matrix is gelatinous with a putty-like consistency, which hardens in the sea water (see Table 12.5). In the nephropsid lobster, *Enoplometopus occidentalis*, the tubular spermatophore consists of a highly coiled continuous spermatophoric tube ensheathed by an outer matrix with a narrow sticky region for adhesion to the substratum (Matthews, 1954). The spermatophore of this lobster becomes hard when exposed to sea water, although the female receives and stores the spermatophore within the seminal receptacle. However, structural transformation occurs as a result of spermatophore hardening in those species in which they are fastened to the ventral sternum of the females. Thus, Panulirid spermatophores undergo hardening and blackening after attachment to the female sternum. In the *P. interruptus*, the spermatophore is composed of a highly coiled sperm tube embedded in an acellular matrix. The extruded spermatophore is white, soft, and sticky on all surfaces. The sperm tube is seen near the surface of the foot of the spermatophore. The opposite surface, the cap, hardens and darkens after exposure to sea water (Martin et al., 1987). The sperm release from the hardened spermatophores is achieved by scratching of the outer hardened layer by the powerful chelae of the lobsters (Berry, 1970).

12.2.3.2 Crayfish Spermatophore

In general, the spermatophores of crayfish are similar to that of palinurid lobsters. In the freshwater crayfish, *Pacifastacus leniusculus*, the wall of the unextruded spermatophore is composed of three concentric layers: a thin, primary spermatophoric layer that surrounds the inner sperm mass, a thick middle layer composed mainly of electron-dense spherical granules, and a thick outer globular layer (Dudenhausen and Talbot, 1983). After extrusion, the sticky outer globular layer transforms into a fibrillar thickened ridge, anchoring the oval end of the spermatophore to the female body. At the same time, the outer portion of the middle layer is modified into a reticulated structure, thereby protecting the inner sperm mass during the prolonged storage. The structural organization of the spermatophore envelope in *P. leniusculus* is similar to two other Australian crayfish species, *Cherax tenuimanus* and *Cherax albidus* (Beach and Talbot, 1987). However, in another fresh water crayfish, *Samastacus spinifrons*, belonging to Parastacidae, the sperm mass was embedded in a dense matrix, without any investment layer over it (Rudolph, 2002).

Erkan et al. (2009) studied the formation of a two-layered spermatophore in *Astacus leptodactylus*, belonging to Astacidae. The spermatophore of *A. leptodactylus* is a tubular uninterrupted column, consisting of a sperm mass covered with primary and secondary layers. The spermatozoa, after release from the testis, are accumulated in the collecting tubules and transferred to VD. In the PVD, the spermatozoa are embedded in a PAS-positive secretion, and only in the mid-VD (MVD), the primary layer that surrounds the sperm mass is observed. This layer stained deeply with bromophenol blue, indicating its proteinaceous nature. In the distal part of MVD, a very thin layer surrounds the primary layer, constituting the secondary layer. This layer stains with Alcian blue and aldehyde fuchsin, indicating acidic mucopolysaccharides, containing hyaluronic acid. The primary and secondary layers are thick and prominent in the distal VD (DVD). The histochemical property of spermatophore layers of *A. leptodactylus* is similar to those of the spiny lobster *P. homarus* (Radha and Subramoniam, 1985). The highly acidic mucopolysaccharides present in the spermatophores of this crayfish, by virtue of antimicrobial activity, protect the spermatozoa within the spermatophore after deposition to the sternum of the female. The presence of hyaluronic acid in the outer layer also aids in the hardening process of the deposited spermatophore.

A two-layered spermatophore has also been found in the red claw crayfish, *Cherax quadricarinatus* (López Greco et al., 2007). In the PVD, a PAS-positive material corresponding to the primary layer of spermatophore of *A. leptodactylus* surrounds the sperm mass. In the MVD, a secondary layer, which is a heterogeneous sheath containing cytoplasmic droplets, is formed over the primary layer. The pattern of spermatophore formation in *Cherax quadricarinatus* is similar to that of *C. albidus* with primary and secondary layers being added in successive parts of the VD.

The hardening of spermatophore takes place after its extrusion and deposition onto the female. The structural and chemical changes that occur after transference to the female is described by López Greco and Nostro (2008). The chemical transformation taking place in the secondary layer of the spermatophore produces a change from a soft to a hard consistency and could be related to processes such as phenolic tanning or calcification. Within 24–48 h postmating, the spermatophore acquires a solid intense white color and is tightly attached to the ventral surface of the female between the coxae of periopods 4 and 5. After 72 h post mating, when the eggs are attached to the pleopods, the spermatophore has already started disintegrating. The secondary layer loses the cytoplasmic droplets and changes from a liquid to a sticky condition. Sperm release from the spermatophore is suggested to be achieved by mechanical means using the fifth periopods of the female.

12.2.4 Penaeoid Shrimps

Penaeoidean shrimps show considerable complexity in the spermatophore morphology as well as the associated materials transferred from the male to the female during insemination. The variability found in the morphology of spermatophores is related to the nature of thelycum, which is the external modification of the female's posterior cephalothoracic sternites and/or coxae that are used in sperm receipt and storage (Bauer, 1986b). Females with open thelyca generally receive a spermatophore with complicated sperm-free accessory structures such as wings, whereas in the closed thelycal forms the males produce simpler spermatophoric masses (Burkenroad, 1936). In the latter, the deposited spermatophore is stored and well protected within a seminal receptacle, concealed under the thelycum, until the time of spawning or succeeding molt (Perez Farfante, 1975).

The open thelycal species deposit a compound spermatophore which is pod-shaped after its assembly from two halves, expelled from the paired terminal ampullae, upon mating (King, 1948). In *Penaeus setiferus*, the compound spermatophores consist of the main trunk or germinate body, containing sperm, a pair of wings at the anterior portion, and glutinous material clinging to both sides of the trunk (Chow et al., 1991). The paired spermatophores upon extrusion and deposition onto the female sternum are joined along the medial adhesive layer. The germinate body is surrounded by thick sheaths of gelatinous materials which undergo hardening. Obviously, the anterolateral wings, posterolateral glutinous material as well as the lateral flaps and the dorsal plate serve as the attachment structures of the spermatophores in various *Litopenaeus* species (Perez Farfante, 1975; Fig. 12.17). In *P. setiferus*, the VD consists of two parallel epithelium-lined ducts, referred to as the spermatophoric

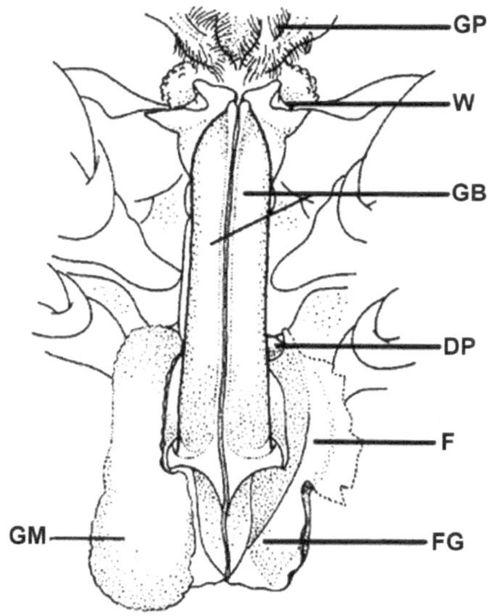

FIGURE 12.17 Ventral view of a compound spermatophore of *Penaeus* (Litopenaeus) *setiferus* attached to female. *DP*, dorsal plate; *F*, flap; *FG*, flange; *GB*, germinate body with inner sperm mass; *GM*, glutinous material; *GP*, gonopore; *W*, wing. *Redrawn from Perez Farfante, I. 1975. Spermatophores and thelyca of the America white shrimps, genus* Penaeus, *subgenus* Litopenaeus. *U.S. Nat. Mar. Fish Ser. Fish. Bull. 73, 463–486; Adapted from Subramoniam, T., 1993. Spermatophores and sperm transfer in marine crustaceans. Adv. Mar. Biol. 29, 129–214.*

and accessory ducts. The germinate body containing the sperm mass is produced and stored in the spermatophoric duct, whereas the wings and other accessory adhesive materials are produced in the accessory duct. Thus, the wing and germinate body are separate structures, formed separately in different chambers of the terminal ampullae. In the open thelycal type, the spermatophores are deposited in the sternal plates, XII to XIV, posterior to the females' ovipores. In *Penaeus occidentalis*, the sperm is released through the anterior opening of the sperm sac, whereas in *Penaeus stylirostris*, *Penaeus schmitti*, and *P. setiferus*, the compound spermatophore splits longitudinally into two parts, exposing the inner sperm mass to the surrounding water (Perez Farfante, 1975). In the open thelycal species, the spermatophores are easily dislodged and hence, the time interval between mating and spawning is usually short (Primavera, 1985).

In comparison, the spermatophores produced by the closed thelycal species are less complex, consisting mainly of two basic divisions, namely the main body and the wing, or the appendage. In *Penaeus indicus*, the main body is composed of a bulky and viscous sperm mass surrounded by a thick envelope, the sperm sac. The membranous wing, attached to the sperm sac, is slimy and comparatively thinner in texture and more transparent than the opaque sperm sac (Sasikala and Subramoniam, 1987). Simplification in the spermatophore structure is found in *Penaeus monodon* (Motoh, 1981) and *Metapenaeus monoceros* (Sasikala and Subramoniam, 1987). In *M. monoceros*, the extruded spermatophore has two morphologically distinguishable parts: a highly viscous sperm mass devoid of sperm sac and four to five adhesive, milky white, small, grain-shaped crystalline structures. In the closed thelycal species, the wings produce a characteristic reaction when placed in sea water. In *P. indicus*, the spermatophore is extruded in the form of a cord which unfolds and spreads like an umbrella on contact with sea water, forming a large membranous foliaceous structure (Sasikala and Subramoniam, 1987). Similar reactions have also been reported in *Penaeus aztecus* and *Penaeus duorarum*, in which the wing material is composed of delicate anastomosing sheaths that delaminate or unfolds upon exposure to sea water (Bauer and Cash, 1991).

In the sicyonid shrimps, with closed thelycum, a typical spermatophore is not produced. Instead, the sperm carried in the seminal plasma is transferred to the females and stored in a pair of baglike seminal receptacles, concealed under the thelycum (Clark et al., 1986). In contrast, other penaeoid shrimps such as *Trachypenaeus* spp. show intermediate degrees of spermatophore complexity and of internalization of sperm storage. For example, in *Trachypenaeus similis*, the sperm cells are contained in spherical spermatophores, enveloped in a pellicle, akin to brachyuran spermatophores (Bauer and Min, 1993). However, the vesiculate type of *T. similis* is transferred to the females as a cord that interconnects the individual spermatophores. More interestingly, a sperm plug substance is produced by the male to form a sperm plug, to seal the openings of the paired seminal receptacles, so as to prevent the backflow of the seminal substances soon after the insemination. In brachyuran crabs, the sperm plug material disintegrates slowly after copulation, but in *T. similis*, the plug material remains in the median pocket of the thelycum from the time of insemination until the next molt, at which time it is cast off with the exuviae. In another penaeoid shrimp, *Xiphocaris kryoyeri*, similar type of simple spermatophores, accompanied by sperm plug substance has also been observed during insemination (Burkenroad, 1934). Sperm plug substances have not been described for Sicyonid species to seal the seminal vesicle openings (Clark et al., 1986).

Penaeoid shrimp have morphologically complex external male and female genitalia to help in the transfer of spermatophore. Male genitalia consist of two structures: the petasma, or modified endopods of the male's first abdominal appendages (pleopods), and the appendix masculina and (in some) the appendix interna, processes on the endopod of the second male pleopods. Female genitalia consist of the thelycum and/or seminal receptacles, thoracic sterna structures associated with spermatophore attachment or storage. In penaeid shrimps, the petasma transfers spermatophores to the females during mating (Bauer, 1991).

As described above, penaeoid spermatophores exhibit considerable variation: spermatophores deposited within some sort of seminal receptacle are morphologically less complex than those deposited externally on the female. For example, in *P. setiferus*, which typifies the externally deposited complex spermatophore, during mating spermatophores of each side are emitted from the male and the main bodies of each adhere to form a compound spermatophore. The deposited spermatophore undergoes a swelling and delamination after several minute's exposure to sea water to release sperm and fertilization should occur without much time gap after the insemination (Chow et al., 1991).

12.2.5 Caridean Shrimp

Caridean shrimps produce relatively simple spermatophores, which are deposited externally onto the females. In the freshwater prawn *Macrobrachium rosenbergii*, Dougherty et al. (1986) described the pod-shaped spermatophores consisting of a lateral sperm mass, a median mucous mass, and a noncellular capsule that sticks to the surface of the sternum of the female during mating. Unlike the penaeid shrimps, where the spermatophoric components are formed in a sequential manner within the VD, the spermatophore is not preformed in the carideans, but the secretions are poured into the distal ampullar region. The spermatophore proper is formed only upon extrusion. In *Heptacarpus sitchensis*, twin spermatophores, one from each male gonopore, are stuck to the underside of the female during copulation. At the ultrastructural level, the capsule material consists of tightly packed fine fibrils that gave positive reactions for glycoproteins and acidic mucopolysaccharides. The matrix material of the sperm mass contains threadlike materials that help in the orientation of the unistellate spermatozoa within the spermatophore (Dougherty, 1987). An interesting deviation from this mode of spermatophore deposition occurs in a marine shrimp *Crangon crangon* (Boddeke et al., 1991). In this species, a tubular spermatophore is formed within the VD; but during mating, only the sperm mass is injected into the oviduct, leaving behind the spermatophoric covering in the VD, where it subsequently undergoes disintegration. The spermatophore of *C. crangon* does not possess any of the adhesive structures; however, no seminal receptacle has been found inside the oviduct. The sperm cells are simply observed along the inner wall of the oviduct.

In the palaemonid shrimp *Macrobrachium* spp., Chow et al. (1982) described spermatophore formation within the VD and its placement on the female. When the sperm mass enters the VD, two secretions are added around it. One is basophilic, staining with hematoxylin, while the other stains with eosin. The latter is adhesive when first exposed to seawater, and this material glues the spermatophore to female exoskeleton during copulation. The nonadhesive basophilic substance protects the inner sperm mass from the external environment until the spermatophore is broken up just before spawning.

12.2.6 Copepoda

The copepods invariably employ spermatophores in sperm transfer. In these predominantly pelagic crustaceans, the males attach the spermatophores directly onto the female genital orifice so that discharge of the sperm allows internal fertilization. The copepod spermatophore is club-shaped or flask-shaped with its proximal end tapering into a tubelike neck, through which sperm flow into the female opening (Fig. 12.18). An adhesive material deposited around the neck region helps in fastening the spermatophore onto the female body. This general structure of copepod spermatophore is found in most of the calanoid copepods such as *Euchaeta norvegica* (Hopkins, 1978). However, in two families, Centropagidae and Pontellidae, a special coupling device is found to replace the

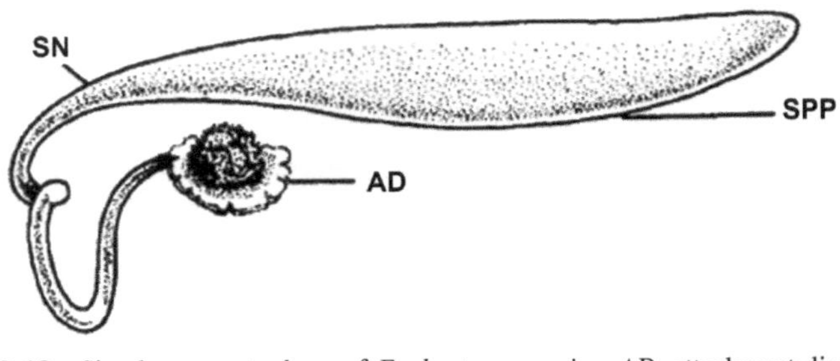

FIGURE 12.18 Simple spermatophore of *Euchaeta norvegica*. *AD*, attachment disc; *SN*, spermatophore neck; *SPP*, spermatophore proper. *Redrawn from a scanning electron micrograph of Blades-Eckelbarger, P.I., 1991. Functional morphology of spermatophores and sperm transfer in calanoid copepods. In: Bauer, R.T., Martin, W.J. (Eds.), Crustacean Sexual Biology, 246–270. Columbia University Press, New York; Adapted from Subramoniam, T., 1993. Spermatophores and sperm transfer in marine crustaceans. Adv. Mar. Biol. 29, 129–214.*

adhesive body in the neck region to effect attachment to the female body. The coupling device consists of a complex group of chitin-like plates, collectively known as coupling apparatus. These two-part coupling devices in *Centropages typicus* hold tightly the tip of the spermatophore stalk to the second urosome segment, which is otherwise called subgenital segment or ancillary genital segment (Lee, 1972). In *Centropages furcatus*, the stalk of the spermatophore is short, bell-shaped, and continuous with the ventral plate of the coupler (Fig. 12.19), which contains a small sac, called spermathecal sac. A granular material, secreted from the neck of the spermatophore, distends the spermathecal sac and forms a plug in the genital opening of the female. The spermatozoal material is ejaculated into a thin-walled tube which coils randomly through the granular material terminating at the plug from where the spermatozoa are eventually absorbed into the female (Fig. 12.19). Thus, the coupler sac acts as a receptacle for the spermatozoa pending their use in fertilization (Lee, 1972).

12.2.6.1 Origin of Spermatophore

In the calanoid copepod, *E. norvegica*, the single pear-shaped testis lies in the anterior region of the cephalosome (Hopkins, 1978). Spermatogenic stages are found in sequential order, starting with spermatogonia at the posterior end with mature spermatozoa found at the anterior end. The VD arises from the left anterior end of the testis. The highly glandular proximal part of the VD produces two secretion products, namely the membrane bound alpha granules and filamentous materials in which the sperm released from testis and the alpha granules lie. The alpha granules are produced in the vicinity of well-developed Golgi apparatus and an elaborate system of rough endoplasmic reticulum. In

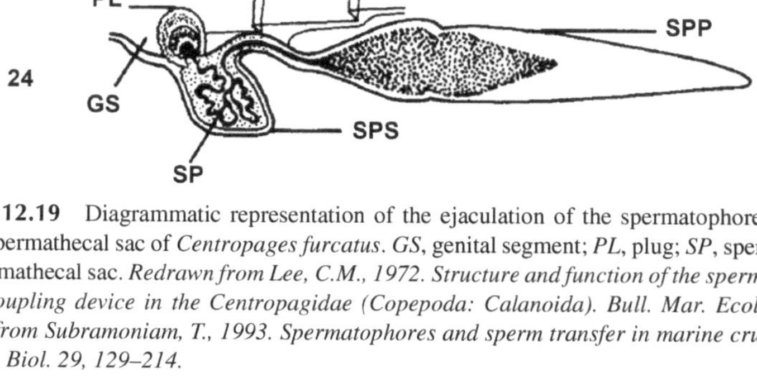

FIGURE 12.19 Diagrammatic representation of the ejaculation of the spermatophore contents into the spermathecal sac of *Centropages furcatus*. *GS*, genital segment; *PL*, plug; *SP*, spermatozoa; *SPS*, spermathecal sac. *Redrawn from Lee, C.M., 1972. Structure and function of the spermatophore and its coupling device in the Centropagidae (Copepoda: Calanoida). Bull. Mar. Ecol 8, 1–20; Adapted from Subramoniam, T., 1993. Spermatophores and sperm transfer in marine crustaceans. Adv. Mar. Biol. 29, 129–214.*

the posterior VD, the epithelial cells secrete the materials for the wall of the presumptive spermatophore, along with another granular material called beta granule. Eventually, the lumen in the posterior VD contains the central core of alpha granules set in a dense matrix around which is arranged a single layer of sparsely distributed spermatozoa. Outside the zone of the spermatozoa, the beta granules are found closely applied to the inner surface of the presumptive spermatophore wall. The seminal vesicle, which follows the posterior VD, also adds more secretions containing globules of fibrillar nature to the spermatophore wall. Such a cylinder of secretions, within the seminal vesicles, is cut off into spermatophores in the short constricted region between the seminal vesicle and dilated spermatophore sac, which bears the completed spermatophore until release. The epithelial cells lining the spermatophore sac produce globular and granular materials which are added onto the spermatophore wall as a new layer on the outside of the spermatophore flask. These secretory products also function as lubricants, aiding ejaculation of the spermatophore. Yet another secretion consisting of birefringent globules fills the cavity of the spermatophore sac and forms special concentrations around the neck, as the "adhesive body."

12.2.6.2 Chemical Composition of Spermatophore

The copepod spermatophores possess well-defined spermatophore layers and a variety of secretory substances that enable spermatophore attachment and sperm expulsion during mating. In *Calanus finmarchicus*, the outer covering of the spermatophore consists of seven concentric layers of amorphous material of medium electron density (Raymont et al., 1974). These layers are positive to histochemical tests for protein, carbohydrates, and lipids. In the spermatophoric layers of the harpacticoid copepod *T. holothuriae*, Pochon-Masson and Gharagozlou-van-Ginneken (1977) found a chitin–protein lamellar pattern, similar to arthropod cuticle. The cuticular envelope extends up to the thin duct, equivalent to the neck region of other copepod spermatophore. The ductal region with its chitinous wall continues with a spherule, which helps in its adhesion to the segmentary fold of the female genital opening. In calanoid copepods, only a sticky adhesive body has been found on the spermatophore neck enabling spermatophore attachment. The neck leads into a tube, through which the sperm passes into the vulva.

In copepods, the spermatophore is an alternative to an intromittent organ inasmuch as it delivers the male gametes directly into the female opening. Hence, the sperm delivery mechanism from the spermatophore to the female orifice is highly complicated in copepods. In certain calanoid copepods such as *Candacia armata*, there is a unique presence of dimorphic spermatozoa inside the spermatophore (Heberer, 1932). They include the Q-spermatozoa (quell = swell) which provide, by swelling, the propulsive force inside the spermatophore to expel the B (Befruchtungs = fertilization) spermatozoa. With the admission of water, during spermatophore transfer, the Q-sperm swell up and develop a strong pressure inside, causing expulsion of the functional B-sperm.

In the absence of Q-sperm, the propulsive force needed to expel the peripherally placed spermatozoa is provided by the centrally placed vesicular foam bodies as well as the alpha granules, as found in *E. norvegica* (Hopkins, 1978). The presence of these core substances to force the expulsion of the spermatozoa to the exterior seems to be a characteristic feature of copepod spermatophores. In the harpacticoid copepod, *Diarthrodes cystoecus*, Fahrenbach (1962) named the inner vesicular bodies (=core substance) responsible for the ejection of Q-bodies by analogy to Heberer's Q-spermatozoa. In another harpacticoid species, *T. holothuriae*, there are several secretory granules that constitute the "core" of the spermatophore (Gharagozlou-van-Ginneken and Pochon-Masson, 1979). Of these, granule A, rich in mucopolysaccharides, corresponds to the Q-bodies in their capacity to increase in volume by water absorption for the expulsion of the gametes from the spermatophore ampulla. Other secretion granules (B, C, and D), mainly proteinaceous and polysaccharidic in nature, may contribute to the attachment of the spermatophore to the female, by forming the spherule lodged in the vulval folds.

12.2.7 Peracarids

Peracarids, such as isopods, amphipods, and mysids, produce filiform spermatozoa which are bundled together by extracellular tubules, secreted by the cells surrounding the developing spermatids within the testis. A matrix material is also used to align the long nonmotile spermatozoa along their long axis into a sperm bundle (Reger and Fain-Maurel, 1973). In all peracarids, the extracellular tubules are considered to function as spermatophores. However, only the sperm mass has been observed to be implanted within the female marsupium in the mysid, *Praunus flexuosus*, by well-developed penes (Nouvel, 1937). Reportedly, the extracellular tubules at the periphery of the spermatophores undergo disaggregation at the lower level of VD in the mysid, *Mysis relicta* (Reger, 1970). Possible involvement of sex pheromones in the attraction of the males to deposit the tubelike spermatophores was suggested for the mysidacean, *Leptomysis lingvura* (Wittmann, 1982).

In the land isopod *Armadillidium vulgare*, each sperm bundle contains 15–30 filamentous spermatozoa, organized with spermatozoan tails aligned parallel to the long axis of the bundle. Each sperm bundle is enclosed by an extracellular investment, consisting of a cone-shaped assembly of longitudinally aligned extracellular tubules and matrix material (Itaya, 1979). In this isopod and probably all other peracarids, testicular follicle cells produce the extracellular tubular and matrix components of the future spermatophores. The synthesis and assembly of extracellular tubules occur within Golgi-derived vesicles, which also contain the matrix-like material. Both tubules and matrix materials are released into the extracellular channel in which the elongation of maturing sperm occurs. Inside the extracellular channels, the tubules become aligned parallel to the long axis of the sperm. Obviously, the assembly of extracellular tubules and surrounding matrix material maintain the shape of the elongated nuclei as well as the arrangement of spermatozoa within the bundle. It is interesting to note that in another isopod, *Porcellio laevis*, Cotelli et al. (1976) could not find the extracellular tubules in the female genital tract.

12.2.8 Spermatophores in Other Crustaceans

Bundles of tube-shaped spermatophores have also been recorded in the VD of a primitive crustacean, *Derocheilocaris typicus* (Mystcocarida) (Brown and Metz, 1967). Each spermatophore contains two spermatozoa which are oriented parallel to each other but face in opposite directions with the acrosomal region at the end of the spermatophore. Interestingly, the mystacocarid spermatozoa are flagellate and motile but become immobilized within the spermatophore. Flagellated sperm enclosed within spermatophore have also been observed in marine cave-dwelling primitive Remipedia (Yager, 1981). The spermatophores are transferred to the female gonopore directly without the help of any copulatory organ.

Variations in the spermatophore morphology occur yet again in Stomatopoda. Unlike several malacostracans, stomatopods do not package their spermatozoa into typical spermatophores but concentrate them into a sperm cord to be transferred to the female during copulation. In *Squilla holoschista*, Deecaraman and Subramoniam (1980) observed a sperm cord that is ejaculated into the female gonopore during an elaborate mating process. These stomatopods possess a pair of male accessory sex organs whose secretions digest the sperm cord inside the female vaginal pouch to liberate active spermatozoa.

12.3 FUNCTIONAL ATTRIBUTES AND EVOLUTIONARY PERSPECTIVES

Spermatophores are male reproductive structures that package sperm cells to aid in their transmission to females during mating in a variety of invertebrate animals (Mann, 1984). The origin of spermatophores in arthropods has been occasioned with the emergence of aquatic annelid-like arthropod ancestors to terrestrial habitat conditions (Schaller, 1971) Hence, the use of spermatophores in sperm transmission to female during mating is more prevalent in primitive terrestrial arthropods. During the transition from the aquatic to terrestrial life, the formation of spermatophores was particularly necessary for animals which had not developed special copulatory or intromittent organs. Additionally, the spermatophores could also protect the delicate sperm from drying during transmission to females in terrestrial environs. Furthermore, in the terrestrial arthropods, sperm transfer by way of spermatophore is through indirect means, as found in apterygote insects and most cryptozoic arachnids. Although Schaller (1980) argues that the production of spermatophores is an adaptation to life on dry land, sperm packaging into spermatophoric structures has evolved many times in aquatic situations, in which prevention from desiccation of sperm was not a selective factor (Subramoniam, 1993).

Crustacean spermatophores are transmitted directly to the females, in contrast to many terrestrial arthropods with indirect transfer mechanisms (Schaller, 1971). Furthermore, spermatophores are employed for sperm transfer mainly in aquatic conditions in crustaceans, thereby precluding any role in protecting the enclosed sperm cells from desiccation. Comparatively, spermatophores of Crustacea are simpler than those found in Chelicerata and primitive insects, but the structure still varies according to the method of sperm transfer. The spermatophores are simple and degenerate structures, when they are transmitted to the females for storage within seminal receptacles for internal fertilization (Brachyura). On the contrary, when the spermatophores are deposited outside of the female body, they assume structural complexity which could be attributed to the presence of multilayered mucoid secretions (lobsters). Careful perusal of spermatophoric structures and their behavior in the sea water may provide insights into their adaptive value to suit the environmental conditions in which the male has to transfer the spermatophore to the females for varying periods of storage, pending fertilization. As an example, in spiny lobsters, the spermatophoric mass quickly adheres to the female body on deposition and undergoes hardening to protect the internal delicate spermatozoa from environmental hazards. At the same time, the acidic mucopolysaccharides present in the inner sperm mass maintain a microenvironment for sperm survival within the stored spermatophore (Radha and Subramoniam, 1985). Again, in the intertidal crabs such as *Emerita* and *Albunia*, the mucoid nature of spermatophores is helpful in the firm adhesion to the females during mating in the turbulent intertidal environment (Subramoniam, 1991b).

Spermatophore production is habitat related. In the marine media, they appear to have evolved to minimize the sperm loss in broadcast fertilization. In crustaceans, the absence of broadcast fertilization and the consequent loss of sperm motility could have been the selective forces for the evolution of spermatophore use in sperm transport. The variability in the spermatophore morphology among different crustaceans reflects on differences in the sperm transport mechanisms as well as the timing of sperm release for fertilization. Crustaceans adopt various mating strategies to transfer spermatophores, often without copulatory organs. Hence, the ubiquitous mucoid nature of spermatophores could be responsible for their easy attachment to the female body. Several mechanical devices adopted by the planktonic copepods in the spermatophore transfer lend support to this tenet. In copepods, the spermatophore may be placed precisely on the female genital apparatus with the sperm mass streaming into the oviduct through the neck or fertilization tube formed from neck materials (Blades-Eckelbarger, 1991). This mechanism of sperm release in copepods parallels the conditions obtained in the spermatophores of cephalopod mollusks (Mann, 1984). Apparently, such analogies in spermatophore use only reflect parallelism in sperm-transporting mechanisms and do not show any phylogenetic relationship between the invertebrate groups.

Evolution of spermatophoric sperm transfer mechanism seems to have occurred independently among crustaceans. Thus, the adaptive advantage of spermatophore use in sperm transmission is exploited even in advanced decapods such as brachyurans, which transfer sperm directly to females by copulatory organs. Interestingly, the primitive orders included in the Euphyllopoda (Anostraca, Cladocera, Notostraca, and Ostracoda) do not produce spermatophores at all; the amoeboid but motile sperm cells are transmitted in fluid semen by penes into the female gonopores for internal fertilization. The terrestrial isopods also have true copulation, but many of them use a primitive type of sperm bundle, equivalent to spermatophores. By analogy, in the more advanced insects, there is a tendency for spermatophores to be replaced by liquid semen, transferred by well-formed penes (Alexander, 1964).

Clearly, spermatophore production is a speciality of the marine crustaceans for sperm transfer in the aquatic medium. There must be a parallel evolution in the use of spermatophores in marine crustaceans, together their counterpart arthropods on land. The common occurrence of spermatophores in other marine invertebrates such as mollusks, annelids, and chaetognaths lends further evidence to the suitability of spermatophores in direct and indirect sperm transfer in sea water (Subramoniam, 1993, for a discussion).

Chapter 13

Accessory Reproductive Glands

13.1 INTRODUCTION

Accessory sex glands (ASG) in higher animals occur as secondary structures that primarily ensure the reproductive success, either by facilitating the transfer of or protecting the gametes. In insects, the secretions of the ASG serve multifunctional roles such as gamete transport, sperm activation, modification of mating behavior, providing nutrients to gametes, and enhancing fecundity (Leopald, 1976; Avila et al., 2011). Crustaceans, on the other hand, do not possess that many accessory glands to perform these multitudes of reproductive functions. Invariably, the accessory glands of crustaceans are attached to the terminal region of the gonoducts, releasing their contents into the lumen of the reproductive tract. In many cases, they may also represent specialized secretory or storage regions of the genital ducts, performing the role of typical accessory gland. In general, majority of decapods are devoid of any male accessory glands, but the vas deferens, consisting of secretory, storage, and ejaculatory regions, produces semen and spermatophores and transfers them to females, during mating (Subramoniam, 1993). Although previous reviewers (eg, Adiyodi and Adiyodi, 1975; Adiyodi and Anilkumar, 1988) classified them based on form, function, and relative location to the primary gonads, most of them could be categorized into male and female gonoductal glands.

13.2 FEMALE DUCTAL GLANDS

Ductal glands have a pivotal role in gamete transport, protection, and nutrition. In the female, they are found as enlarged regions of the oviduct and secrete extra jelly substances to envelop the eggs. In some cases, the accessory gland produces egg cases, as in cirripedes, to amass the eggs into a pod, before releasing them outside and in this case into the mantle cavity (Barnes and Blackstock, 1977b). In the brachyuran crabs, the spermathecae receive semen containing spermatophores during copulation and store them up until ovulation takes place. In general, the secretory activity of the ductal glands synchronizes with the gametogenic activities, implying common regulatory mechanisms for gonadal and accessory gland activities.

Sexual Biology and Reproduction in Crustaceans. http://dx.doi.org/10.1016/B978-0-12-809337-5.00013-7

13.3 OVIDUCTAL GLANDS OF CIRRIPEDES

The oviductal gland is a characteristic female ASG in all the thoracican cirripedes investigated so far (Barnes and Barnes, 1977). They are situated at the terminal part of the oviduct. At the reproductive season, the oviductal gland secretes an elastic sac, into which ova released from the ovary are collected and stored. The ova are fertilized within the sac, and the egg mass, termed as egg sac or egg lamellae, are ultimately released into the mantle cavity of the adult (Fig. 13.1). At the time the embryos are developed, the egg sac undergoes disintegration, facilitating the release of stage I nauplius larvae into the mantle cavity from where the larvae escapes into the surrounding seawater.

The structure of the oviductal gland in the cross-fertilizing hermaphroditic cirripede, *Semibalanus* (*Balanus*) *balanoides* is described by Walley (1965) and Walker (1980). The oviductal gland consists of three distinct regions: (1) the main chamber, (2) the exit canal, and (3) the proximal chamber (see Fig. 13.1). Whereas the first two regions are of ectodermal origin, the third is mesodermal. The ovisac is secreted by the highly secretory columnar epithelial cells of the main chamber. The oviductal gland is differentiated as a small cluster of cells from an epidermal invagination, before the differentiation of the oviduct itself. Before the onset of breeding season, the oviductal gland gives the appearance of

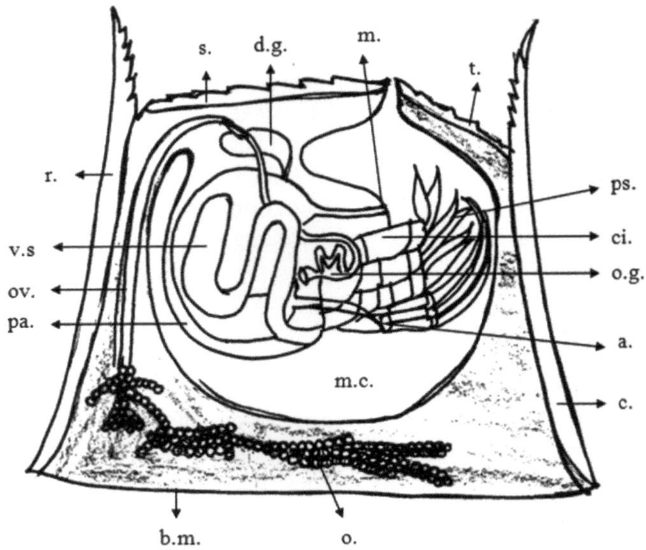

FIGURE 13.1 Diagram to show the female reproductive system and the position of the oviducal gland in the cirrus, of the barnacle *Balanus balanoides*. *a.*, anus; *b.m.*, basal membrane; *c.*, carinal plate; *ci.*, first cirrus (appendages); *d.g.*, digestive gland; *m.*, mouth; *m.c.*, mantle cavity; *o.*, ovary; *o.g.*, oviducal gland; *ov.*, oviduct; *pa.*, prosoma; *ps.*, penis; *r.*, rostral plate; *s.*, scutal plate; *t.*, tergal plate; *v.s.*, seminal vesicle. *Modified from Walley, L.J., 1965. The development and function of the oviducal gland in* Balanus balanoides. *J. Mar. Biol. Assoc. U.K. 45, 115–128.*

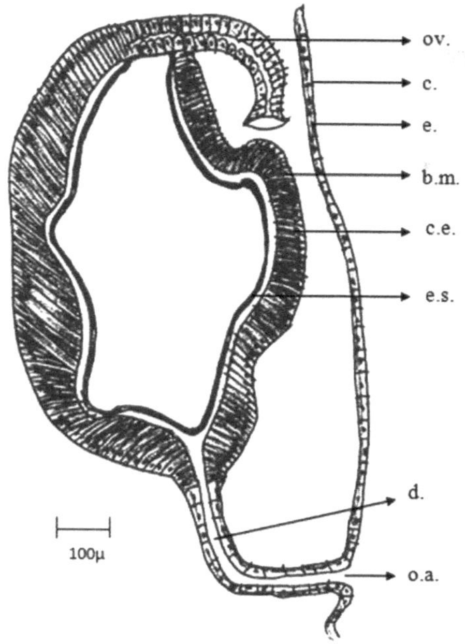

FIGURE 13.2 Diagrammatic longitudinal section through the oviducal gland of *Balanus balanoides* during the late secretory phase. *b.m.*, basement membrane; *c.*, cuticle; *c.e.*, columnar epithelium; *d.*, external duct; *e.*, epidermis; *e.s.*, elastic sac; *o.a.*, external aperture of oviducal gland; *ov.*, oviduct. *Modified from Walley, L.J., 1965. The development and function of the oviducal gland in* Balanus balanoides. *J. Mar. Biol. Assoc. U.K. 45, 115–128.*

a small pouch that opens into the mantle cavity through a short exit duct, lined with a cuticular layer. The gland grows larger, as a result of increased mitotic activity and cellular movement in the epithelium, as it enters the secretory phase during breeding season. At the time of ovisac secretion, the height of the secretory columnar epithelium reaches the maximum (Fig. 13.2).

13.4 NATURE OF OVISAC SECRETION

At the beginning of the secretory phase of the ovisac gland, the cells at the mouth of the oviduct are full of globules which are secreted into the lumen of the developing sac. However, these globular materials do not take part in the development of the ovisac but are thought to be involved in the activation of spermatozoa, present in the mantle cavity of *B. balanoides* (Walley et al., 1971). At this time, each cell of the columnar epithelium is drawn into a fine cytoplasmic thread, around which a cone of secretion is formed. This cytoplasmic thread extends into the lumen and fuses with those secreted by other epithelial cells, giving rise to a wall of 15–20 μm thickness. After the formation of the sac wall, epidermal cells shrink to a small cuboidal element.

The outer part of the sac wall is composed of electron-dense reticulum, whereas the inner region facing the sac lumen contains cytoplasmic filaments, made up of microtubules and small vesicles that coalesce with one another. The cytoplasmic filaments ultimately separate from the underlying cells and fully integrate within the sac wall (Fig. 13.3).

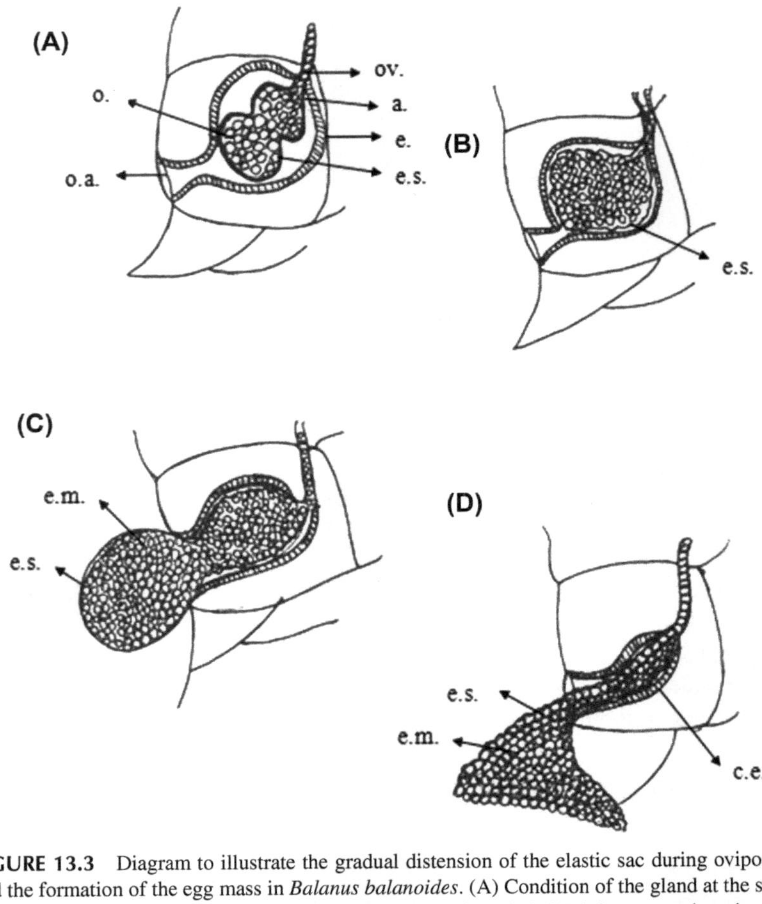

FIGURE 13.3 Diagram to illustrate the gradual distension of the elastic sac during oviposition, and the formation of the egg mass in *Balanus balanoides*. (A) Condition of the gland at the start of oviposition. The eggs fill the elastic sac but it is not yet distended. (B) A later stage in oviposition. The sac has been distended and now fills the oviducal gland. (C) As the number of eggs inside it increases, the sac is forced out through the external duct into the mantle cavity. (D) Oviposition is nearly completed. The oviducal gland has collapsed and the columnar shape of the epithelial cells has been restored. The egg mass is still attached at the mouth of the oviduct as the last few eggs pass into it. *a.*, attachment of the elastic sac at the mouth of the oviduct; *c.e.*, columnar epithelium; *e.*, epithelium; *e.m.*, egg mass; *e.s.*, elastic sac; *o.*, ovum; *o.a.*, external aperture of oviducal gland; *ov.*, oviduct. *Adapted from Walley, L.J., 1965. The development and function of the oviducal gland in* Balanus balanoides. *J. Mar. Biol. Assoc. U.K. 45, 115–128.*

The outer surface of the sac wall has a pitted appearance, the pits lying at the center of surface sculpturing polygons. In the SEM images of the ovisac in the cirrepede *Pollicipes cornucopia*, the pores on the surface are clearly seen to pass directly through the wall of the sac (Klepal and Barnes, 1977). In addition, longitudinal sections of the sac wall revealed the presence of striations, corresponding to the cytoplasmic filaments that run from the pits to the interior of the wall. The pore canals of ovisac wall could be likened to the pore canals of insect cuticle. Klepal and Barnes (1977) compared the formation of ovisac wall with procuticle formation in arthropods (Locke, 1961). In arthropods, the procuticle is formed as a continuous sheet which solidifies around the cytoplasmic projections from the epidermis in the form of pore canals. In a similar way, the cones of secretion from the epithelial cells of oviductal gland solidify around the numerous "pore canals" of the ovisac wall of *B. balanus* (Klepal and Barnes, 1977). The only difference is that arthropod cuticle is made of chitin–protein complex, whereas the sac wall is made up of protein only.

Interestingly, the extrusion of ova from the ovisac is under the influence of mating. Barnes et al. (1977) experimentally proved in *B. balanoides* that the large quantity of seminal plasma poured into the mantle cavity contains a protein that specifically stimulates contraction of the ovisac smooth muscle, enabling extrusion of ova.

In addition to the pores, there are numerous fine holes, which represent irregular spaces where the secretion forming this part of the wall did not become completely confluent with secretions of adjoining cell. The presence of the pores and their canals facilitates the entry of sperm to reach the egg mass stored inside the sac. In cirripedes, semen is deposited by the functional male in the mantle cavity of the functional female. These sperm are stored in an inactive form and are activated by an egg sac secretion, upon the release of egg lamellae into the mantle cavity.

13.5 DISINTEGRATION OF SAC WALL

Toward the end of incubation of egg masses within the egg lamellae, the sac disintegrates first to release the stage I nauplius larvae. The larva then ruptures the egg case to escape into the surrounding water in the mantle cavity. Since the time of breakdown of the oviducal sac synchronizes with the stage of embryonic development, the developing embryo within the lamellae are suggested to release a protease enzyme that will digest the proteinaceous wall of the egg sac (Barnes and Barnes, 1977). The composition of sac wall is always proteinaceous, although its property may vary from species to species. Accordingly, there is a degree of specificity in membrane–enzyme relations existing among different cirripede species. The embryo begins to secrete the enzyme at the time limbs are formed.

Barnes and Blackstock (1977a) eluted a protease from the ripe lamellae with pepsin-like activity at a very low pH of 3.0 from *B. balanoides*. This enzyme is termed as ovisacase in this cirripede. Late stage embryos of *B. balanoides* secrete another protease enzyme, involved in the hatching of the nauplius larva from the hatching envelope. The ovisacase is different from the hatching enzyme in that the former acts at the optimum pH of 3–0, whereas the hatching enzyme acts at a higher pH of 8.3. Furthermore, there is no causal relationship between the hatching factor and the agent causing breakdown of the oviductal sac membrane. The release of hatching substance is initiated at a time when the sac membrane is already disintegrated.

13.6 SPERMATHECA

Crustaceans have evolved various mating systems in an attempt to safely deposit the nonmotile spermatozoa either on the ventral sternum of the female or into specialized seminal receptacle to facilitate internal fertilization. Despite the existing debate on the terminology used for sperm storage organs in female Crustacea (see McLay and Greco, 2011 for a discussion), several workers agree to use the term spermatheca to denote the differentiated region of the oviduct. Spermatheca is also used to represent the depression in the ventral sternal region, morphologically separated from the genital ducts (Hartnoll, 1975, 2006). The term "seminal receptacle" is being used for all the structures involved in sperm/spermatophore storage, regardless of their position, embryonic origin, or relation with the female genital duct. Nevertheless, both "spermatheca" and "seminal vesicle" are terms indicative of storage of seminal substances received from males during mating. Yet, female decapods show tremendous variation in the duration and capacity to store sperm, reflecting differences in their fertilization efficiency.

In a primitive condition, the males simply attach spermatophores on to the abdominal sternites in forms such as many shrimps, some crayfish, lobsters, and all anomuran crabs (Subramoniam, 1993). The epizoic attachment of spermatophores in this way is termed as "ephemeral," as the period of storage is up to the next molt, when the attached seminal substances would be removed along with the old cuticle, requiring fresh insemination for the next oviposition. In other penaeiodean shrimps, crayfishes, lobsters, and more particularly in the primitive brachyuran crabs, females have partly or completely internalized chitinous storage areas.

13.6.1 Spermatheca of Primitive Brachyurans

In the primitive superfamilies of Brachyura, Dromiacea, and Gymnopleura, Hartnoll (1979) described such external spermatheca, formed as invaginations of the integument but having no internal connection with oviduct. Whereas in the Dromiacea, there are paired spermathecae, opening independently on the line of sternal suture between the seventh and the eighth thoracic segments (Hartnoll, 1975), in the Gymnopleura, there is only a single median spermatheca opening on the anterior portion of the seventh thoracic segment. Paired oval spermathecal openings are also found in a raminid, *Lyreidus tridentatus* (Hartnoll, 1979).

The existence of a single median spermatheca in Gymnopleura is akin to that found in the ancestral group, Astacura, which also possesses a single median spermatheca, suggesting an evolutionary trend in the development of spermatheca of these primitive brachyuran crabs. In these spermathecae, the chitin lining over the epithelium prohibits entry of any secretory material into the spermathecal cavity. In addition, because of the chitin lining, these structures have little elasticity, thereby limiting the holding capacity of the deposited spermatophore, and are shed along with their contents with each molt. Therefore, the sperm storage duration is possible only between molts.

13.6.2 Spermatheca of Higher Brachyuran Crabs

It is only in the higher brachyuran crabs, spermatheca with an elaborate structure to store sperm for a longer period is present. In all advanced brachyurans, the spermathecae are dilatations of the oviduct. The spermatheca is spatially divisible into two chambers, the dorsal sperm storage chamber and the ventral insemination chamber (Diesel, 1989). The dorsal chamber is extensible and lined internally with a multilayered glandular epithelium, derived from the proximal region of the oviduct, which connects to the ovary proper. The mesodermally derived dorsal part is devoid of chitin lining, a condition favoring transmolt retention of sperm. Transmolt storage of sperm in spermatheca has been reported in many crabs such as *Menippe mercinaria* (Cheung, 1968), *Portunus sanguinolentus* (Ryan, 1967), *Paratelphusa hydrodromous* (Adiyodi and Anilkumar, 1988), and the snow crab, *Chionoecetes opilio* (Sainte-Marie and Sainte-Marie, 1998).

The ventral part of the spermatheca, derived from ectodermal epithelium, is lined with chitin and leads to the exterior via vagina. The position of spermatheca in relation to the oviduct, however, varies in different species. In the portunid crab, *P. sanguinolentus*, the ovary consists of a central hollow shaft and lateral lobes. The oviduct takes origin from the lateral lobes of the ovary as a tiny portion folded in the wall of the spermatheca, which opens into the vulva (Ryan, 1967). There is no visible oviduct in between spermatheca and the ovary. In the Majoidea crab, *Leurocyclus tuberculosus*, the medially fused lateral ovarian lobes directly open into the spermatheca, and in the distal end, the spermathecal opening leads to the vagina and then opens outside through the vulva (González-Pisani et al., 2012). In the snow crab, *C. opilio*, the female reproductive tract consists of a short oviduct, an epithelium-lined spermatheca, a cuticle-lined intermediate chamber, and the vagina. The oviduct crosses the thick terminal segment of the spermathecal epithelium and opens into the vagina in such a way that it delivers its content directly into the intermediate chamber (Sainte-Marie and Sainte-Marie, 1998). Mixing of ova and the sperm released from spermatheca takes place in the intermediate chamber, facilitating sperm–egg interaction.

Depending upon the location where the oviduct opens into the spermatheca, the spermatheca could be classified into two types. The dorsal type has the oviduct opening in a relatively dorsal position, with the vagina lying ventrally far from the oviduct. In the ventral type, both the oviducal and the vaginal openings are close to each other and are positioned ventrally (Diesel, 1991). In the swimming crab *Callinectes sapidus*, the oviduct enters dorsally (Johnson, 1980), while in the spider crab *Inachus phalangium* (Diesel, 1989) and *C. opilio*, it enters ventrally (Beninger et al., 1988).

The morphology of the female reproductive system in the pea crab, *Pinnotheres pisum* includes the ovary, oviduct, spermatheca, and vagina (Fig. 13.4). Becker et al. (2011) studied the ultrastructural features of the spermatheca and related to their functional role in sperm storage and internal fertilization (Fig. 13.5). The spermatheca consists of two distinct areas, the dorsal sperm storage area, consisting of apocrine glandular epithelium and the ventral part, which includes the fertilization area. The glandular cells are large with lobulated nuclei. Numerous vesicles, formed in the main body, accumulate distally and merge into large secretory bodies. These secretory bodies are emptied into the spermathecal lumen by dissolving the apical part of the cell, a mechanism termed as apocrine. The ventral part is lined with a cuticular layer except for the part where the oviduct enters the spermatheca by the holocrine transfer tissue (see Fig. 13.5). The cuticular layer extends beyond this point, lining the vagina and the mobile operculum. The oviduct is a two-layered epithelium of columnar cells, covered by a thin layer of connective tissue. At its confluence with the spermatheca, the distal part of the oviduct opens into a special tissue, which opens into the lumen of the spermatheca. This tissue secretes a homogeneous substance by holocrine secretory mechanism and hence called "holocrine

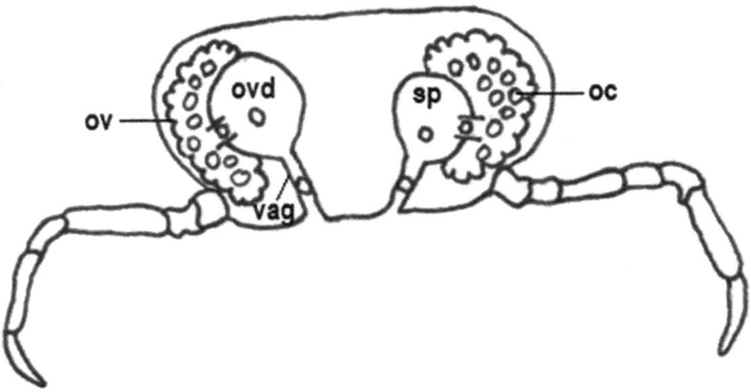

FIGURE 13.4 Female reproductive system of the brachyuran crab *Pinnotheres pisum*, to show the location-wise relationship between the spermatheca and ovary. *Modified from Becker, C., Brandis, D., Storch, V., 2011. Morphology of the female reproductive system of European Pea Crabs (Crustacea, Decapoda, Brachyura, Pinnotheridae). J. Morphol. 272, 12–26, permission from John Wiley and Sons.*

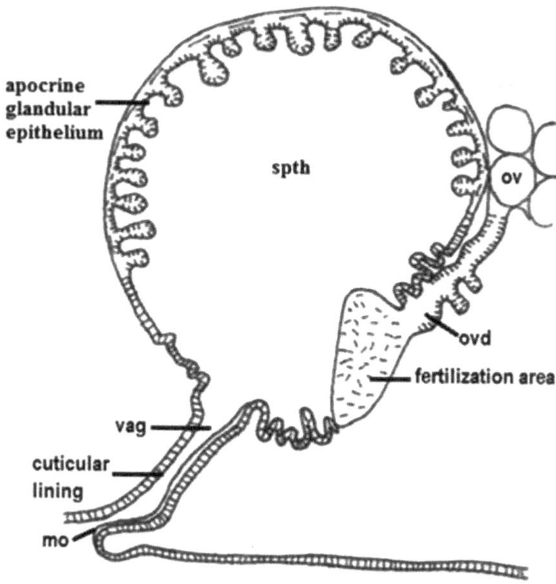

FIGURE 13.5 Diagrammatic representation of spermatheca in the brachyuran crab *Pinnotheres pisum*, the interior lining in the ventral fertilization area and the vagina are lined with cuticle. The oviduct enters the spermatheca by a holocrine transfer tissue in the fertilization area. The dorsal sperm storage area is lined by an apocrine glandular epithelium. *mo*, mobile operculum; *mus*, musculature; *ov*, ovary; *ovd*, oviduct; *spth*, spermatheca; *vag*, vagina. *Modified from Becker, C., Brandis, D., Storch, V., 2011. Morphology of the female reproductive system of European Pea Crabs (Crustacea, Decapoda, Brachyura, Pinnotheridae). J. Morphol. 272, 12–26, permission from John Wiley and Sons.*

transfer tissue" (Becker et al., 2011). The chemical nature of the secretions from dorsal and ventral part of the spermatheca is not fully understood. Nevertheless, it has been suggested by many investigators that they have a role in spermatophore dehiscence, sperm maintenance, and providing antimicrobial protection (Benhalima and Moriyasu, 2001).

The efficiency of sperm storage is very high in the spermatheca of the higher brachyuran crabs. In *Menippe mercenaria*, the spermatozoa stored during one insemination are sufficient to fertilize 10 or more broods within the same intermolt period (Cheung, 1968). The spermathecal secretions, along with seminal plasma derived from the males during mating, form the nutritive materials for the sustenance of sperm during their extended storage.

13.6.3 Secretory Activity of Spermatheca

In the field crab, *P. hydrodromous*, as the female approaches puberty, the columnar epithelial cells, lining the spermatheca, becomes taller and flakes of basophilic substances mixed with acidophilic secretions are evident in the lumen

(Krishnakumar, 1985). With the onset of proecdysis, there is a spurt in the activity of the spermathecal epithelium, leading to the production of substantial quantities of proteinaceous material, rich in tyrosine. During the late premolt stage, the spermathecal epithelium shows villi-like formation toward the lumen. There is also an increase in the basophilic secretion, which contains numerous droplets of varying sizes. This proteinaceous secretion is rich in free amino acids, polysaccharides, and oligosaccharides, but poor in lipids, revealing their mucoidal nature. Mating normally takes place during the postmolt in *P. hydrodromous*. Following molting, the flabby spermatheca gets hardened and its water content reduced to 55%. In the mated females, a second peak of synthesis starts during the prebreeding season (September–October). After vitellogenesis is completed, the spermatheca gets hydrated as a means to develop sufficient hydrostatic pressure to facilitate sperm ejection during oviposition (Adiyodi and Anilkumar, 1988). The development and secretory activity of the spermatheca are closely correlated to the ovarian cycle in *P. hydrodromous*. In the mud crab, *Scylla serrata*, mating in the puberal female influences the secretory activity of the spermatheca, which in turn initiates vitellogenesis in the ovary (Ezhilarasi and Subramoniam, 1982). This observation may indicate a common endocrine mechanism, which is set off by the mating process, regulating spermathecal as well as ovarian activity, at least in the first ovarian cycle.

In the swimming crab, *Arenaeus cribrarius*, Zara et al. (2014) found that the spermatheca of the juvenile crabs contained a secretion rich in acidic mucopolysaccharides. In the mated females with rudimentary ovary, the spermatheca was filled with a glycoproteinous secretion including the sperm plug. However, the sperm plug gets dissolved by the time the ovary reaches maturation. The spermatophore dehiscence has also occurred concurrent to the dissolution of sperm plug. Evidently, there is a synchrony between ovarian maturation, sperm plug dissolution, and spermatophore dehiscence in this crab. More secretions from the dorsal epithelium pushed the free sperm into the ventral part of the spermatheca to facilitate fertilization. In *A. cribrarius*, fertilization could occur both in the dorsal and ventral regions of the spermatheca.

In the snow crab, *C. opilio*, Beninger et al. (1993) studied the ultrastructure of spermatheca and the histochemical nature of its glandular secretion. The spermatheca consists of a dorsal glandular and a ventral chitinous region, which is termed as intermediate chamber (Sainte-Marie and Sainte-Marie, 1998). As in other higher brachyurans, the spermatheca of *C. opilio* is morphologically suited for a secretory and sperm storage function. The dorsal region, representing the pouchlike glandular part, consists of an outer layer of connective tissue, and an inner glandular epithelium. These two layers are separated by collagenous fibers. The glandular epithelium is further divided into an outer germinal zone having many mitotic figures, a middle glandular layer, and the inner stratified squamous epithelial layer. The glandular cells proximal to the spermathecal lumen are filled with globules of electron-dense secretory products, often adhering to the cell membrane. The

Golgi bodies are active, and the nuclei contain condensed heterochromatin, which becomes irregular in cells closest to the squamous layer. The glandular layer contains several membrane-bound collecting sites, which are filled with electron-dense secretions, as they approach the spermathecal lumen. Finally, the cells in the squamous layer as well as the collecting sites degenerate and slough off, liberating their contents in to the lumen. The spermatophores in the dorsal lumen, however, had intact pellicles of spermatophores, but degenerate contents.

13.6.4 Spermatheca in Nonmalacostracans

Among the nonmalacostracans, the structure and location of spermatheca show even more complexity and variation. For example, in the branchiuran form *Dolops ranarum*, the spermathecae are ectodermal in origin and not directly connected with the female reproductive tract. They are located at the base of the abdomen as paired ovoid sacs opening out through a coiled duct, called spermathecal duct. The wall of the spermathecal duct is glandular and thick. This duct terminates in a sharp, sclerotized, and perforated spine, termed spermathecal spine (Fryer, 1960). The inner surface of the spermatheca is lined with a cuticular layer, which is cast off partially at each molt, possibly by digestion and breakage along the longitudinal axis. The remaining undigested part is joined by a new cuticle; the latter goes through the same process at the succeeding molt, resulting in the formation of cuticular annuli, in as many numbers as the number of molts, the animal has passed through.

13.7 SHELL GLANDS IN ANOSTRACA

In anostracans, the paired oviducts originate from the lateral part of the ovary and in *Artemia salina*, the oviducts are short and pouch-like structures appearing temporarily at the time of ovulation (Criel, 1980). After ovulation, the eggs are stored in the dilated lateral pouches of the oviduct; from there, the eggs descend down into the median ovisac or uterus. Fertilization occurs within the uterus. When the eggs are retained in the uterus, they acquire thick shells and develop into late gastrula stage, before being laid. While this kind of embryonated cysts are produced in the parthenogenetic races, in the bisexual forms, *Artemia* eggs in the uterus are enveloped by a thin, membranous shell and develop into nauplii that are born ovoviviparously (Dutrieu, 1960). The shells are formed as tertiary envelope around the eggs, and the materials are derived from the shell gland secretions.

There are three pairs of shell glands, located posterolateral, dorsomedial, and anterolateral to the ovisac or uterus (Anderson et al., 1970). The shell glands of *A. salina* are composed of clusters of shell gland units. Each unit consists of two oblong cells separated by a central, slitlike lumen, which opens into the uterus by a duct, formed by a neck cell and a small number of duct cells

(Fig. 13.6). The two gland cells are joined together by septate desmosomes. At ultrastructural level, the cytoplasmic matrix of the gland cells is shown to contain glycogen, lipids, mitochondria, a few microtubules, Golgi complexes, and large amounts of rough endoplasmic reticulum (Anderson et al., 1970). During the secretory phase of the gland, numerous membrane-bound lipoprotein granules are seen in the cytoplasm. Before their liberation into the lumen, their membrane fuses with the plasmalemma adjacent to the lumen between the gland cells. The lumen of the shell gland unit gets enlarged by the accumulation of these granules that tend to fuse with one another. The shell gland ducts carrying the granular secretions are finally poured into the uterus. The shell gland secretions produce the tertiary envelope of the eggs within the uterus. The tertiary envelope is made up of lipoprotein, rich in tyrosine. Although Anderson et al. (1970) did not find any diphenols and the enzyme phenoloxidase in the egg envelopes to suggest quinone tanning, as in cuticle hardening; they proposed a process of self-tanning involving spontaneous oxidation of the tyrosine residues to quinines. The embryonated *Artemia* eggs are known to undergo dormancy, termed as cryptobiosis. The thick egg shells formed from the shell gland secretions protect the embryo during desiccation and other adverse conditions.

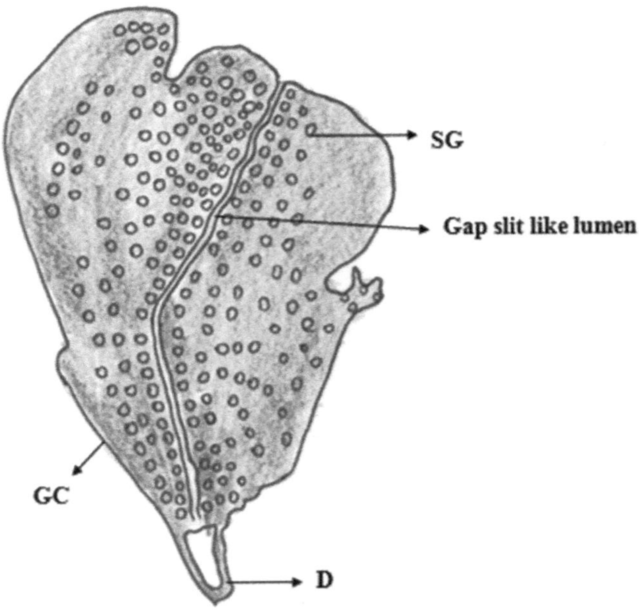

FIGURE 13.6 Diagrammatic representation of shell gland unit in *Artemia*. SG, secretory granules; D, duct; GC, shell gland. *Adapted from Anderson, E., Lochhead, J.H., Lochhead, M.S., Huebner, E., 1970. The origin and structure of the tertiary envelope in thick-shelled eggs of the brine shrimp,* Artemia. *J. Ultra Struct. Res. 32, 497–525.*

Shell gland activity and the egg shell formation have been studied in other freshwater anostracans like *Chirocephalus bundyi* (Linder, 1959). Munuswamy and Subramoniam (1985) made a detailed study on the shell gland activity during breeding season in the fairy shrimp, *Streptocephalus dichotomus*, in which the female reproductive system is morphologically and functionally similar to *Artemia*. After ovulation, the eggs of *S. dichotomus* enter the oviducts where they are stored temporarily, bathing in the oviductal secretions. This secretory phase of the oviducal epithelium is cyclical with the ovulation cycle. During ovulation, the oviduct extends and the glandular cells become more active in secreting granular material into the lumen. Soon after ovulation, and the evacuation of oocytes into the ovisac, the extended oviduct shrinks to its normal size. Within 15–20 h of ovulation, the fine secretory granules reappear in the lumen for the next batch of eggs.

In *S. dichotomus*, copulation initiates the passage of ova into the median ovisac (Fig. 13.7). In the virgin females, reared without males, no shell gland secretions were released into the ovisac, and hence no egg shell formation occurred. The ripe oocytes were retained in the oviduct and the lateral pouch. Obviously, mating induced the coordinated release of shell gland secretions as well as eggs from the lateral pouch into the median uterus.

The synchronization of ovarian and shell gland activities suggests common controlling mechanisms. Further studies also indicated the involvement of brain neurosecretory factors in the control of oocyte development and shell gland activities (Munuswamy and Subramoniam, 1985). Synchronous neuroendocrine control over the ovary and the female accessory glands (cement gland) has also been reported in the stomatopod, *Squilla holoschista* (Deecaraman and Subramoniam, 1983b).

13.7.1 Egg Shell Formation in *Streptocephalus dichotomus*

Detailed investigations on the origin, structure, chemical composition, and mode of hardening of the egg shell in *S. dichotomus* are reported by Munuswamy and Subramoniam (1984) and Munuswamy (1986). The egg shell comprises mainly of two layers, an outer thick cortex and an inner alveolar layer, the latter containing numerous lipid droplets. In addition, a third innermost cuticulin layer, secreted by the embryo surrounds the embryo as a thin membrane. Scanning electron microscopic observations revealed the characteristic ridges on the egg surface with pores. A subcortical space, formed by a split in the tertiary shell between the outer cortex and inner alveolar layer, is present in the region of ridges. In the anostracan eggs, the surface pores open directly into the subcortical space through micropyles, facilitating the entry of air into the alveolar layer for respiratory purposes as well as for insulation of the cryptobiotic embryos (Gilchrist, 1978). The entry of air through the pores of the anostracan egg is equivalent to the so-called "plastron respiration" of insect eggs (Hinton, 1977).

Evidently, the precursors for the egg shell formation are derived from the shell gland secretions in *S. dichotomus*. Both the shell gland and egg shell

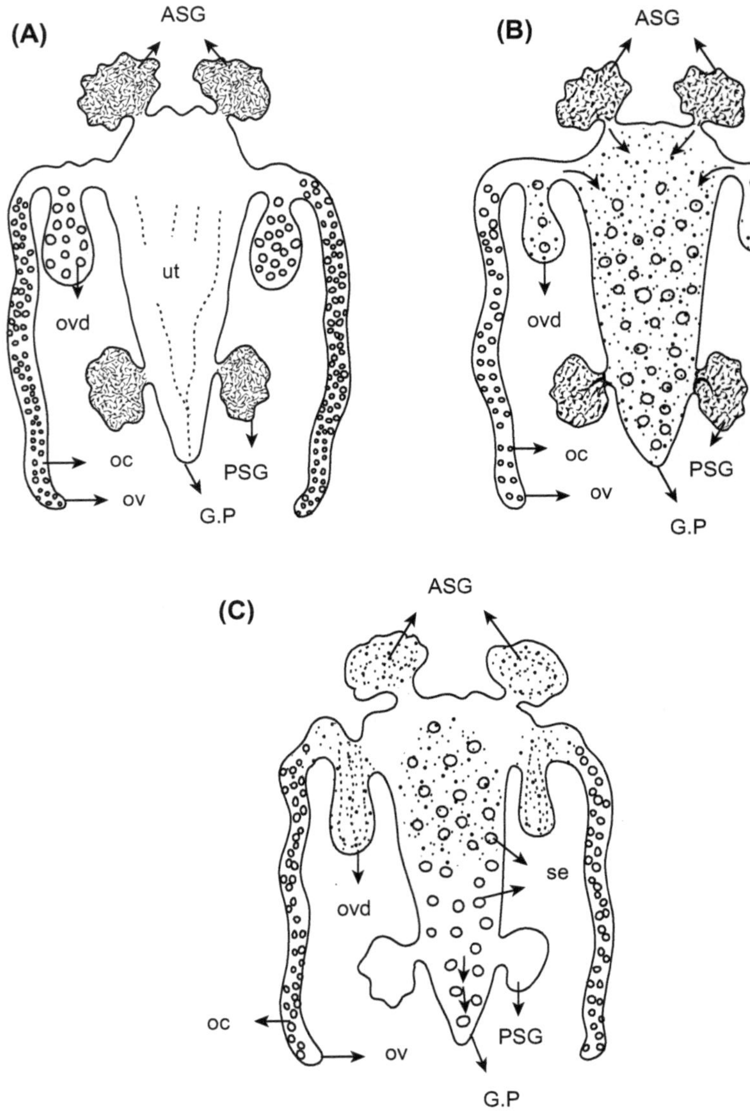

FIGURE 13.7 Schematic diagram showing the nature of shell glands, ovary, oviduct, and uterus before and after copulation in *Streptocephalus dichotomus.* (A) Before copulation the oocytes (OC), in the ovary (OV) are stocked in the oviducts (Ovd), the uterus (ut) is wide, and the anterior and posterior shell glands (ASG, PSG) are charged with secretory products; (B) Copulation determines the passage of the oocytes from the oviducts into the uterus and the shell gland secretions are poured (*arrows*) into the uterus as well; (C) After copulation the products of shell gland secretions are completely discharged to accomplish eggshell formation; shelled eggs extruded through the secretions are completely discharged to accomplish eggshell formation; shelled eggs are extruded through the genital opening (GP). *Modified from Munuswamy, N., Subramoniam, T., 1985. Influence of mating on ovarian and shell gland activity in a freshwater fairy shrimp Streptocephalus Dichotomus (Anostraca). Crustaceana 49, 225–232.*

are rich in protein containing aromatic groups. In addition, intense positivity to lipids (mainly phospholipids) was found in the shell gland and egg shell. The granules of the shell gland secretion showed positivity to diphenols and tyrosinase enzyme, the latter converting monophenolic tyrosine into diphenols. Further conversion of diphenols to quinone occurs in the uterus, where the egg shell formation is completed (Munuswamy and Subramoniam, 1984). Chromatographic analysis further revealed the presence of dityrosine in the egg shell, which is specifically absent in the shell gland. The detection of dityrosine in the egg shell implies a special mode of hardening by β-sclerotization. Further studies by nuclear magnetic resonance spectroscopy indicated the possible occurrence of β-sclerotization in the egg shell, as evidenced by the presence of aromatic substrates with aliphatic side chains (Munuswamy, 1986). Supporting evidence for quinone tanning in the egg shell came from the fact that the color of the egg shell membrane changed from white to brown during sclerotization. Dityrosine linkage is shown to precede the actual phenolic tanning involving dihydroxy phenol and quinone in insect cuticle (Brunet, 1980). The cross-linking of aliphatic side chains imparts strength and resistance against chemical and enzymatic digestion. Spontaneous oxidation of the tyrosine to quinone is also suggested in the egg shell of another freshwater anostracan *C. bundyi* (Linder, 1960). Taken together, egg shell hardening in the anostracans resembles those of insect ootheca in many respects (Brunet, 1980).

13.8 INTEGUMENTAL GLANDS

Tegumental glands of Crustacea have the primary function in releasing the enzyme, phenol oxidase, to bring about tanning and melanization of the new cuticle at ecdysis (Talbot and Demers, 1993). Tegumental glands, found in the pleopod of the American lobster, *Homarus americanus*, have been shown to exhibit a cyclical activity with molting (Talbot and Zao, 1991). Their role, if any, related to male reproduction was the subject of debate for the subsequent workers on the tegumental gland. However, several studies have indicated the occurrence of rosette-shaped tegumental glands inside the first gonopod of brachyuran crabs (Spalding, 1942; Ryan, 1967; Diesel, 1986a,b). In the snow crab, *C. opilio*, Beninger et al. (1991) studied the distribution pattern of the rosette-shaped tegumental gland within the first gonopod. By virtue of their restricted localization in the gonopods (the first pair of pleopods) in the adult males, these glands are named as gonopod tegumental glands. A complex duct network, emanating from these glands, leads to pores in the ejaculatory canals. This observation led to the supposition that these integumental glands produce seminal fluid, justifying their classification as ASG.

Further work on the distribution of these glands revealed their localized occurrence at the base of their endopodites of the first gonopod in the mature male crabs (Beninger et al., 1995). The gland cells secrete and release a mixture of acid mucopolysaccharides and neutral mucopolysaccharides into the cuticular ejaculatory duct. These secretions are believed to facilitate the easy transfer of the viscous seminal plasma, containing spermatophores into the oviduct during copulation. Having bacteriostatic properties, acid mucopolysaccharides present in the gonopod glands may have additional functions of protecting the transferred spermatophores within the spermatheca from opportunistic microbes (Beninger et al., 1995).

13.9 CEMENT GLANDS OF CRAYFISH

In certain crustaceans such as crayfishes and stomatopods, separate glandular structures of tegumental origin, with no connection to the oviduct, are found to release their contents through the cuticle in the ventral thoracic region. Their secretions are generally mucoid in nature, serving the function of agglutinating the amorphous egg mass, released during oviposition, and fastening them to the pleopods.

Stephens (1952) described the development of one such tegumental gland, the cement gland, appearing as a secondary sexual organ during the ovarian development of the crayfishes such as *Cambarus rusticus* and *Cambarus virilise*. The cement glands in these species are distributed ventrally beneath the abdominal sterna and in the abdominal appendages. In *C. rusticus*, the glands underlie the abdominal sterna (second to sixth sterna) and extend laterally into the anterior region of the associated pleura. They are also found to occur in the propodites, endopodites, and exopodites of the abdominal appendages. They discharge their secretions through the pores opening independently on the setae of the appendages.

During their early development, cement glands could be distinguished as milky white translucent, circular, or semicircular areas through the sterna cuticle. As many as 30–45 such areas could be seen in each ramus of the uropod. The glandular unit opens through the distal duct cells into tiny pores in the cuticle. With the onset of secretory phase, the glandular units become white lobate clusters, each surrounding the cement gland pores on the integument. Different duct cells from the glandular units open independently into one pore.

The extent of cement gland development was positively correlated to the ovarian development in *C. rusticus*. After egg laying, these glands disappear and reappear only at the onset of next ovarian cycle. In the mature female crayfish, ablation of eyestalk produced acceleration in the cement gland activity. A further increment in the cement gland activity was also noticed when the cerebral ganglia was implanted into the abdomen of the ablated female. Interestingly, cement gland activity synchronized with ovarian development, implying a common endocrine regulation for both the processes (Stephens, 1952).

13.9.1 Cement Gland of Stomatopods

In the stomatopod, *S. holoschista*, three patches of the cement glands, borne on the ventral side of the sixth, seventh, and eighth thoracic segments, are seen as dense white masses in the mature females (Deecaraman and Subramoniam, 1980). It is composed of two types of gland cells. The first type is elliptical in shape and possesses 15–20 rosette cells, arranged in a concentric manner. The second type is larger and composed of spherical glandular cells, arranged like tiles in a mosaic floor. Both types of gland cells possess minute collecting ducts, which join the main duct that opens on the sternum of the sixth to eighth thoracic segments. Whereas the first cell type secretes a fluid containing glycoprotein, the second type produces dense granular substance, rich in sulfated acid mucopolysaccharide. These two secretions are mixed together before they are released outside simultaneously to agglutinate the eggs laid onto the ventral sternum. The eggs are connected to one another by threads formed by cement gland secretions. The acid mucopolysaccharide nature of the secretions protects the externally developing eggs from infestation of various microbial epibionts.

The synchronous development of cement glands and the maturation of the ovary implied common endocrine regulation. In the crayfish, *Cambarus*, Stephens (1952) reported that a common endocrine principle could control both the cement gland and ovarian activity. Similarly, in the stomatopod, *S. holoschista*, Deecaraman and Subramoniam (1983c) provided histological evidence for the neurosecretory coordination by the brain and thoracic ganglia for cement gland and ovarian activity.

13.10 MALE ACCESSORY SEX GLANDS

Insects elaborate a variety of male ASG to assist semen production and spermatophore formation (see review by Chen, 1984). In crustaceans, only a few reports are available on the occurrence of male ASG. However, the proximal part of vas deferens in many species is lined internally with a highly secretory epithelium, producing copious quantities of seminal substances that play an important role in sperm transport and spermatophore formation (Subramoniam, 1993). In many crustaceans, spermatogenesis occurs in the testis, but the final maturation is completed only after their release into the vas deferens. For example, in the marine shrimp, *Sicyonia ingentis*, the spermatids become fully mature in the vas deferens by completing the morphogenesis of subacrosomal components, including the formation of the anterior spike from its primordium, underlying the apical acrosomal membrane (Shigekawa and Clark, 1986). Further morphological changes occur in the subacrosomal region subsequent to the transfer and storage of sperm in the female thelycum. The complex spermatophore formation is accomplished only in the vas deferens of many crustaceans (for details see chapter: Spermatophore and Sperm Transfer Mechanisms).

However, the presence of male ASG has been reported in many brachyuran crabs and stomatopods. They are either connected directly to the ejaculatory duct or open separately on the tip of the intromittent organ.

13.10.1 Male Accessory Glands in Brachyuran Crabs

Among the brachyuran crabs, the presence of male ASG has been reported in the land crab, *Cardisoma carnifex* (Kalyanaraman and Chandran, 1982), the ocypodid crab, *Ocypoda platytarsis* (Sukumaran, 1985), two estuarine grapsid crabs, *Metopograpsus messor* and *Sesarma quadratum* (Suganthi and Anilkumar, 1998), and the Chinese mitten crab, *Eriocheir sinensis* (Hou et al., 2010). In all these crabs, the ASG open directly into the ejaculatory duct. In the grapsid crabs, the accessory glands are found connected to the posterior region of the reproductive tract, as tubelike outgrowths. The accessory glands in both species consist of an outer connective tissue layer, a middle muscle layer, and an inner epithelial layer. In the adult males, the inner epithelium contains rough endoplasmic reticulum with bleblike formations and a profuse presence of Golgi complex, in addition to free ribosomes and mitochondria.

The lumen of the gland cells contains eosinophilic vesicular homogeneous materials, in which basophilic granules were embedded. At the ultrastructural level, the luminal contents showed vesicles with a spherical/oval dense homogeneous core and an electron-lucent peripheral portion. In *M. messor*, the accessory gland materials are emptied into the lumen by exocytosis, as determined by the presence of vesicular materials in the released contents. However, in *S. quadratum*, the secretion is merocrine mode as seen from the profuse presence of granular entities within the lumen in close proximity to the apical portion of the epithelium. The major component of the accessory gland secretion (AGS) of the two crabs is protein. The proteinaceous nature of the secretion enhances the dilution of the viscous seminal substances, reaching the ejaculatory duct, thereby aiding in the safe transmission of male gametes to the female. Although the chemical composition of the ASG secretions have not been investigated in the above crab species, Kalyanaraman and Chandran (1982) suggested a role for the dissolution of spermatophore wall by the secretions of male accessory glands in the land crab, *C. carnifex*.

In the Chinese mitten crab, *E. sinensis*, Hou et al. (2010) reported the occurrence of male ASG in the form of dendritic branches at the confluence of seminal vesicle with ejaculatory duct. During mating, the ASG secretions, along with spermatophores and seminal fluid from the seminal vesicle enter into the female spermatheca through the ejaculatory duct (Wang et al., 2013). The secretory materials of the accessory gland contain small and large vesicular materials containing a milky white secretion. The AGS has been shown to possess digestive ability of spermatophore wall within the spermatheca. The functional protein of the ASG involved in the spermatophore digestion is a 50 kDa protein with high cathepsin A activity. Cathepsin A is one of the important intracellular lysosomal hydrolases, existing in an inactive form (Liu et al., 2009) and in *E. sinensis*, it is probably activated by interaction with spermathecal fluid after mating. It is an acidic serine carboxypeptidase having additional deaminase and esterase activities. Cathepsin A of ASG secretion can release sperm within 30 min of treatment with spermatophores, with complete digestion of spermatophore wall taking place in 45 min time. In vitro incubation studies with cathepsin A inhibitor also revealed complete abolition of spermatophore digestion (Wang et al., 2013). Male AGS also interact with the proteins from spermatheca to increase acrosomal enzyme (acrosin) activity in the free sperm, in addition to enhancing sperm viability. Understandably, the male AGS possess sperm activating factors that could enhance the fertilizability of the spermathecal sperm. Wang et al. (2013) studied the seasonal activity of cathepsin A in the male accessory gland in different seasons and found the highest activity only during the mating season.

13.10.2 Male Accessory Gland of Stomatopods

A typical male accessory gland, as adjunct to the paired vas deferens, was described as early as 1920 by Komai in the stomatopod, *Squilla oratoria*. Deecaraman and Subramoniam (1980) studied the functional morphology of the ASG of another species, *S. holoschista*. The paired accessory glands are found on either side of the partly fused elongated testes, extending from the cephalothorax to the eighth thoracic segment. Each gland is a white, transparent, convoluted structure, with its anterior end enlarged into a club-shaped configuration (Fig. 13.8). The gland on the left side is longer than the right and lies curved over the right gland. The accessory gland duct run on either side of the midline to the eighth thoracic segment and opens separately at the tip of the intromittent organ (Fig. 13.9). The inner epithelium of the glands is ensheathed by a membranous layer. In the mature males, the glandular cells are filled with secretory granules which are released into the lumen by a process of macroapocrine, as reported in the accessory gland of the dipteran insect *Aedes triseriatus* (Ramalingam and Craig, 1978). The secretory granules are rich in protein and neutral mucopolysaccharides. In *S. holoschista*, the sperm is cemented together to form a sperm cord, in which form the spermatozoa are released. During copulation, when the sperm cord is transmitted to females through the intromittent organ, the AGS are also discharged simultaneously into the oviduct. As the sperm cord undergoes dissolution to liberate free sperm in the oviduct, it is suggested that the AGS might be involved in the dissolution of the cementing materials of the sperm cord. Deecaraman and Subramoniam (1983c) further investigated the postcopulatory role of ASG secretions in *S. holoschista*. Electrophoretic analysis of the hemolymph protein of the mated females revealed the presence of a new protein homologous to that of the AGS. Interestingly, the same protein fraction was visualized in the ovary, suggesting a possible translocation of the ASG protein into the ovary via sequestration of hemolymph proteins. Friedel and Gillot (1977) also demonstrated in the insect *Melanoplus sanguinipes* that

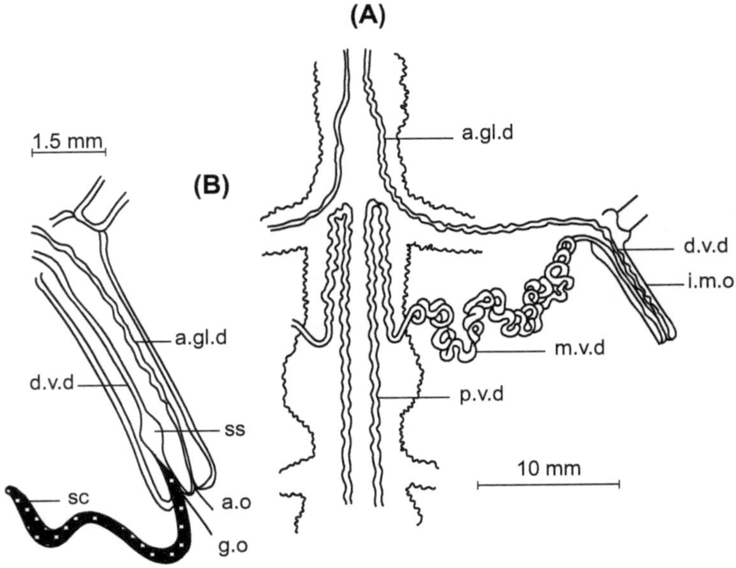

FIGURE 13.8 (A) Diagram showing the three regions of the vas deferens, accessory gland duct, and their entry into the intromittent organ of *Squilla holoschista.* (B) Intromittent organ. *a.gl.d.*, accessory gland duct; *a.o.*, accessory gland opening; *d.v.d.*, distal vas deferens; *g.o.*, genital orifice; *i.m.o.*, intromittent organ; *m.v.d.*, middle vas deferens; *p.v.d.*, proximal vas deferens; *s.c.*, sperm cord; *s.s.*, sperm sac. *Adapted from Deecaraman, M., Subramoniam, T., 1980. Cyclical cement glands activity in a stomatopod crustacean* Squilla holoschista. *Proc. First All India Sym. Inv. Repr. Uni. Mad. 68–76.*

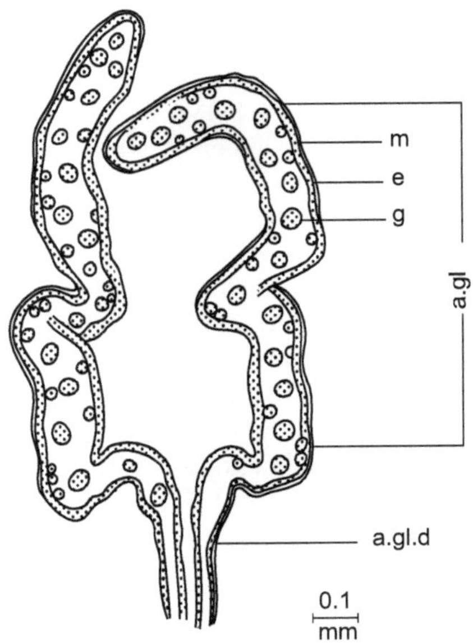

FIGURE 13.9 *Squilla holoschista*: The accessory glands and their ducts, showing the layers and granules. *a.gl.*, accessory gland; *a.gl.d.*, accessory gland duct; *e*, epithelial layer; *g*, granule; *m*, muscular layer. *Adapted from Deecaraman, M., Subramoniam, T., 1980. Cyclical cement glands activity in a stomatopod crustacean* Squilla holoschista. *Proc. First All India Sym. Inv. Repr. Uni. Mad. 68–76.*

the male accessory gland substance was absorbed into the female blood, before being incorporated into the ovary.

13.11 ACCESSORY GLAND SECRETIONS: A FUNCTIONAL EVALUATION

The foregoing account on the crustacean ASG has uncovered the vast diversity in the morphology and their relative location to gonadal ducts. However, chemical composition of their secretions in relation to the multifarious functions they play in the gamete transport and successful union of egg and sperm during fertilization has not been fully understood. In general, the female ASG produce mucoprotein substances that help in the formation of additional egg envelopes for protection and easy transport through the female reproductive tract. In addition, they are used to cement the egg mass to the pleopodal brood chamber for egg incubation. Extensive histochemical studies on the shell gland secretions revealed the presence of phenolic compounds and the enzymes necessary for phenolic tanning of egg shell of anostracans. On the contrary, understanding the chemical nature and functions of male ASG has much more promise to unravel the modulations of postmating female reproductive behavior. In the brachyuran crabs, the role of male ASG secretions in the spermatophore wall digestion is well established (see above). Digestion of spermatophore wall, before egg release from the ovary is crucial to the timing of fertilization in crabs. The role played by cathepsin A in the breaking down of spermatophore wall indicates the importance of ASG secretions in coordinating the reproductive processes of sperm release, activation, and their interaction with eggs. Although several earlier reports have suggested ASG in sperm maturational functions, it is only in the Chinese mitten crab that the impact of ASG secretions on sperm activation within the spermatheca has been investigated in detail (Hou et al., 2010). In insects, several molecules contained in the AGS are known to modulate many female behavioral and physiological processes, across a wide range of insect species (Avila et al., 2011). With recent technological advances in transcript and protein identification, it is now possible to identify molecular factors in the seminal secretions including the AGS from different crustacean species in controlling the behavioral physiology of mated females. Wang et al. (2012) identified ubiquitin genes (Es-UBS 27 and Es-UBL 40) in the testis, ovary, and male accessory glands of *E. sinensis*. Interestingly, their expression levels in the ASG peaked during the mating season, indicating their involvement in the regulation of cell cycle during gametogenesis and the sperm quality in *E. sinensis*. He et al. (2013) performed de novo transcriptome sequencing to produce a comprehensive transcript database for the ASG of the crab *E. sinensis*. Such transcriptome data analysis of ASG and testis in crabs is expected to provide fundamental information about the sperm maturation process during transport through ASG secretion and elucidate sperm modification mechanisms during acrosome reaction and sperm–egg interaction.

Chapter 14

Fertilization

14.1 INTRODUCTION

Fertilization is the process in which the mature male and female gametes fuse to form the zygote that initiates development. Although great strides have been made to understand the mechanisms underlying fertilization and early development, using sea urchin, amphibian, and mammalian models, investigations on the biology of fertilization in crustaceans are meager, probably due to the atypical nature of their gametes and the difficulties in making in situ observations on gamete interaction. However, events such as sperm capacitation, acrosome reaction, and cortical reaction draw parallel with those in mammalian and sea urchin models. As yet, studies using penaeiodean shrimps have shed much light on the mechanism and control of fertilization in crustacean species. Fertilization studies in the commercially significant aquaculture shrimp species are important in the context of developing in vitro fertilization protocols.

Crustaceans adopt many strategies to bring male and female gametes together to effect fertilization. In the majority of the malacostracans, the sperm is atypical and nonmotile, and hence free-spawning into the water is not practized in any of the crustaceans, despite being primarily aquatic. Therefore, crustaceans have evolved different mating systems in accordance with the environmental conditions as well as the reproductive morphology to accomplish fertilization (see chapter on mating systems). In crustaceans, fertilization is either external or internal. By far, fertilization of crustacean gametes involves initial gamete activation, gamete recognition, sperm–egg interaction followed by pronuclear fusion. The female gamete activation in crustaceans begins with the resumption of meiotic maturation that is held at metaphase I.

14.2 RESUMPTION OF MEIOTIC MATURATION

Meiosis produces progeny cells having haploid number of chromosomes, thus making sexual reproduction possible. In the gametic cells, there are two meiotic blocks that arrest the process of gametogenesis at specific stages. In the oocytes, the primary meiotic arrest occurs at prophase I with intact germinal vesicle, and the secondary meiotic arrest takes place at metaphase I. Secondary meiotic arrest also occurs at metaphase II, when the first polar body is extruded. In other cases, such as the eggs of sea urchin, complete maturation occurs before

Sexual Biology and Reproduction in Crustaceans. http://dx.doi.org/10.1016/B978-0-12-809337-5.00014-9

fertilization, but those of the starfish enter meiosis metaphase I immediately before their spawning but do not proceed with meiosis, unless they are fertilized (Wong and Wessel, 2006). Evidently, many marine invertebrates release their eggs at differently arrested meiotic stages.

In crustaceans, the oocyte development is arrested at the first meiotic prophase, followed by a protracted period of yolk accumulation. After completion of vitellogenesis, the oocytes resume meiosis to reach the first meiotic metaphase (metaphase I). Germinal vesicle breaks down at this stage and the meiotic apparatus migrates toward the peripheral cytoplasm from the center of the oocyte. At the time of spawning, the oocyte development is generally arrested at metaphase I with homologous chromosomes joined at chiasma, and the second resumption of meiosis takes place when the eggs are spawned into the water. In decapod crustaceans, fertilization takes place close to the timing of spawning. In all the aquatic crustaceans, fertilization occurs either internally within the seminal receptacle (spermatheca) or externally, when the eggs are spawned into the seawater. In either case, spawned eggs are exposed to the mineral ions dissolved in the seawater, which activate the eggs to initiate resumption of meiotic maturation. In the marine shrimp, *Sicyonia ingentis*, meiosis precedes fertilization, overruling the possible involvement of sperm activating meiotic maturation (Clark et al., 1984). Obviously, the mechanism underlying both reinitiation and the completion of meiotic maturation in crustaceans is different from that of well-studied marine invertebrate and vertebrate eggs, in which sperm does the entire egg activational role.

In the natantian decapod *Palaemon serratus*, oocyte development is interrupted by two meiotic arrests, one at prophase I and the other at metaphase I, when spawning occurs. Clédon (1985) observed that the meiotic resumption in this prawn was stimulated by fertilization, which caused an elevation in the internal Ca^{2+} level. Subsequent studies by Goudeau and Goudeau (1986), however, showed that in *P. serratus*, metaphase I meiotic resumption depends on the presence of external Mg^{2+} ions in the seawater. Incubation of ovulated oocytes in Mg^{2+}-free artificial seawater did not cause resumption of meiosis. Besides, external Ca^{2+} has not been shown to have any influence on meiotic resumption in *P. serratus* (Table 14.1; Fig. 14.1). Obviously, this meiotic resumption is independent of fertilization, as the eggs from a virgin, unmated female underwent meiotic resumption.

The assumption that external Mg^{2+} ions stimulate an increase of cytosolic free Ca^{2+} ions finds support from in vitro incubation of *P. serratus* eggs in Mg^{2+}-free artificial seawater and 5 μM calcium ionophore A23187 that stimulated resumption of meiosis in 78–100% oocytes (Table 14.2). Possibly, A23187 induced resumption of meiosis in the absence of external Mg^{2+}, but its action required external Ca^{2+}. Further inference from the experiment with A23187 is that entry of Mg^{2+} ions into the egg might increase the level of intracellular Ca^{2+}, which in turn induces meiotic resumption. Interestingly, requirement for external Mg^{2+} in meiotic maturation has also been reported

TABLE 14.1 Requirement for Extra Cellular Mg^{2+} but Not Ca^{2+} for Resumption of Meiotic Maturation of Ovulated Prawn Oocytes

Duration of Incubation (min) at 21°C	ASW (Artificial Seawater)	Percentage Resuming Meiosis	Standard Errors of the Mean	*N* = Number of Oocytes
25	Standard ASW	86	±2.5	228 (8♀)
	Ca^{2+}-free ASW	86	±4.9	151 (5♀)
	Mg^{2+}-free ASW	0	–	279 (7♀)
Confidence Limit				
25	Mg^{2+}-free ASW	0	0–24%	16 (1♀)
65	Mg^{2+}-free ASW	0	0–11%	28 (1♀)
90	Mg^{2+}-free ASW	0	0–11%	30 (1♀)
25	Mg^{2+}-free, Ca^{2+}-enriched ASW	0	0–10%	34 (1♀)

Reproduced from Goudeau, H., Goudeau, M., 1985. Fertilization in crabs: IV. The fertilization potential consists of a sustained egg membrane hyperpolarization. Gamete Res. 11, 1–17.

in the progesterone-induced maturation of *Xenopus* oocytes (Merriam, 1971) as well as the 1-methyladenine-induced maturation in starfish oocytes (Rosenberg and Lee, 1981).

Mg^{2+}-dependent meiotic resumption has been demonstrated in many penaeiodean shrimp species such as *S. ingentis*, *Penaeus aztecus*, and *Penaeus setiferus* (Clark et al., 1974, 1980; Clark and Lynn, 1977). In *S. ingentis*, resumption of meiosis, which is held at metaphase I, at the time of spawning, was triggered by exposure to seawater both in fertilized and unfertilized eggs. Similar to *P. serratus*, meiotic resumption was triggered by seawater Mg^{2+} for *S. ingentis* (Lindsay et al., 1992). However, if the gametes are not fertilized, both Mg^{2+} and Ca^{2+} are required for meiotic resumption. The requirement of Ca^{2+} in the unfertilized eggs to activate meiotic maturation as well as cortical reaction (see below) suggest that a calcium influx from the seawater medium is necessary for activation of unfertilized eggs (see Jaffe, 1985). In *S. ingentis*, meiotic maturation is completed soon after the formation of hatching envelope (HE). Thus, meiotic resumption and egg activation (cortical reaction and HE formation) occur simultaneously under the

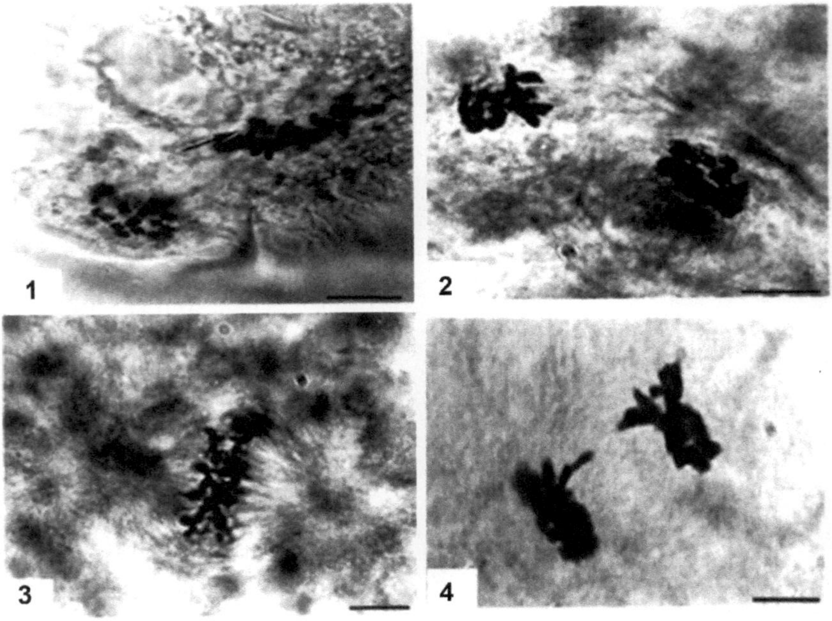

FIGURE 14.1 Resumption of meiosis. (1) Ovulated prawn oocyte incubated in standard ASW at 1040 Mosm, pH = 8.2 at 21°C for 25 min, first polar body and second meiotic metaphase (*arrow*) (×1300). (2) Ovulated prawn oocyte incubated in Ca^{2+}-free ASW at 1040 Mosm, pH = 8.2 at 21°C for 25 min, first polar body and second meiotic metaphase (×1500). (3) Ovulated prawn oocyte incubated in Mg^{2+} and Ca^{2+}-free ASW at 1040 Mosm, pH = 8.2 at 21°C for 25 min, first meiotic metaphase (×1300). (4) Ovulated prawn oocyte incubated in Mg^{2+}-free ASW + 5 μM ionophore A23187 at 1040 Mosm, pH = 8.2 at 21°C for 25 min, first meiotic telophase (×1500) (scale bar = 10 μm). *Reproduced from Goudeau, M., Goudeau, H., 1986. The resumption of meiotic maturation of oocyte of the prawn Palaemon serratus regulated by increase in extra cellular* Mg^{2+} *during spawning. Dev. Biol. 118, 361–370.*

influence of external Mg^{2+}. Incidentally, induction of these two egg activation processes does not require fertilization (Clark et al., 1984).

In the Pacific white shrimp, *Litopenaeus vannamei*, Hertzler (2005) studied the timing of meiotic maturation in relation to fertilization events. Similar to *S. ingentis*, meiotic maturation and cortical reaction occurred under the influence of Mg^{2+} ions in the seawater. By 5-min postspawning (ps), the oocyte reached anaphase I and the first polar body was visible in oocytes at 7–8 min ps. Incorporated sperm nuclei within the oocytes were observed at 12–15 min ps. Anaphase was reached at 15 min ps, and the formation of female pronuclei was completed at 17.5-min ps. Pronuclear migration occurred between 15 and 25 min ps and at 27.5-min ps the formation of the mitotic spindle at prophase was visualized. Such synchronization in the timing of meiosis and the fertilization events has also been observed in other penaeoidean species such as *S. ingentis* (Lynn et al., 1991).

TABLE 14.2 Stimulation of the Resumption of Meiosis by the Ionophore A23187 in Mg^{2+}-Free ASW

Duration of Incubation (min) at 21°C	ASW (Artificial Seawater)	Percentage of Resumption of Meiosis	Confidence Limit at $p=.05$	N = Number of Oocytes
25	Mg^{2+}-free ASW	0	0–30%	10 (1♀)
25	Mg^{2+}-free ASW + 5 μM A23187	93	65–98%	15 (1♀)
25	Standard ASW	86	70–96%	29 (1♀)
25	Mg^{2+}-free ASW	0	0–22%	16 (1♀)
25	Mg^{2+}-free ASW + 5 μM A23187	78	57–94%	18 (1♀)
65	Mg^{2+}-free ASW	0	0–13%	28 (1♀)
65	Mg^{2+}-free ASW + 5 μM A23187	100	90–100%	38 (1♀)
25	Mg^{2+} and Ca^{2+}-free ASW	0	0–10% (at $p=.1$)	37 (1♀)
25	Mg^{2+} and Ca^{2+}-free ASW + 5 μM A23187	17	9–31% (at $p=.1$)	52 (1♀)

Reproduced from Goudeau, H., Goudeau, M., 1985. Fertilization in crabs: IV. The fertilization potential consists of a sustained egg membrane hyperpolarization. Gamete Res. 11, 1–17.

14.3 MOLECULAR MECHANISMS OF MEIOTIC MATURATION

The cell division cycle regulatory protein, cyclin B, combines with catalysis subunit cyclin-dependent kinase 2 (Cdc2) to form the cyclin B–Cdc2 kinase complex, commonly referred to as maturation-promoting factor (MPF) (Nurse, 1990; Ledan et al., 2001). MPF plays a major role in meiotic resumption, making them possible marker genes of oocyte developmental competence (Han et al., 2012). Studies on the identification and expression patterns of factors controlling the cell cycle during meiotic division in the maturing oocytes of crustaceans are rather recent. Key factors such as cyclin B, Cdc2, as well as Cdc7 have been identified and characterized in the developing oocytes of penaeid shrimp, *Penaeus monodon* and the crab, *Eriocheir sinensis* (Phinyo et al., 2014; Qiu and Liu, 2009). Cyclin B, the chief regulatory protein that controls cell division cycle at M-phase, was first characterized in the kuruma prawn, *Marsupenaeus japonicus* by Qiu and Yamano (2005). Cyclin B mRNAs were detected both in the proliferating mitotic oogonia and the perinucleolus stage, when the nucleus transforms into germinal vesicle signifying the entry into meiotic cycle. Similarly, Han et al. (2012) reported the gene expression of cyclin B and Cdc2 in the cytoplasm of oogonia and previtellogenic and primary vitellogenic oocytes. Such cyclical fluctuations in the cyclin B and Cdc2 transcripts were also reported in *P. monodon* (Visudtiphole et al., 2009). In *P. monodon*, Cdc2 is characterized as a 299 amino acid polypeptide with the conserved Thr14 and Tyr15 and Thr161 residues. The expression level of pmCdc2 mRNA increased significantly in stage II (vitellogenic) and stage III (cortical rod formation) and peaked in stage IV (mature) ovaries. These results suggest that PmCdc2 protein would play an important role in the development and maturation of eggs in *P. monodon* (Phinyo et al., 2013).

In the Indian white shrimp, *Fenneropenaeus indicus*, highest levels of both Cdc2 and cyclin B mRNA expression were observed at stage IV in wild-caught females, correlated with the resumption of meiosis and germinal vesicle breakdown at the final stage of oocyte maturation (Tomy et al., 2016). Cdc2 kinase, a catalytic subunit of MPF and a central factor for inducing the meiotic maturation of oocytes has also been identified in the ovary of the Chinese mitten crab *E. sinensis* as two isoforms (Qiu and Liu, 2009). Similar to other subunits of MPF, Cdc2 kinase activity in the crab oocyte elicited a progressive increase from previtellogenic stage to germinal vesicle breakdown stage. Immunocytochemistry revealed that Cdc2 proteins are initially located in the ooplasm of the previtellogenic stage, translocated to the germinal vesicle at the vitellogenic stage, and then accumulate on meiotic spindle. Evidently, Cdc2 kinase plays a pivotal role in inducing germinal vesicle breakdown and generating meiotic spindle apparatus during crab oocyte maturation.

Another cell cycle controlling factor, Cyclin-dependent kinase 7 has been isolated and characterized in the tiger shrimp, *P. monodon* (pmCdk7). The expression of pmCdk7 steadily increased from stage I (previtellogenic) and stage II (vitellogenic) to stage IV (mature) ovaries and slightly reduced in the spent ovary (Phinyo et al., 2014). Interestingly, expression profiles of pmCdk7 are similar to that of pmCdc2, where peak levels were found in the mature ovary (Phinyo et al., 2013). Immunolocalization of pmCdk7 protein revealed its translocation from the ooplasm in the previtellogenic oocytes into the nucleus of the mature oocytes. After germinal vesicle breakdown, the Cdk protein was localized in the nucleocytoplasmic compartment, the cytoskeletal structures and cortical rods of the oocytes, indicating its role in the meiotic maturation of the shrimp oocytes.

14.4 HORMONAL CONTROL OF MEIOTIC MATURATION

It is well established that eyestalk removal results in the induction/acceleration of vitellogenesis in crustaceans. The lowering gonad-inhibiting hormone (GIH) levels by way of eyestalk ablation also have the same effect on promoting meiotic maturation in *F. indicus* (Tomy et al., 2016). Eyestalk ablation in the wild brood stock females induced expression level of a key maturation inducing factor, *pmCdc2* in stages I–IV ovaries of the shrimp, *P. monodon* (Phinyo et al., 2014). Apparently, factors controlling meiotic resumption after vitellogenesis are under the inhibitory control of eyestalk hormones such as GIH. Tomy et al. (2016) again found the increasing levels of gene expression related to meiotic maturation such as *cyclin B*, *Cdc2*, as well as thrombospondin (*tsp*) with advancement in maturation. These results, along with the data on eyestalk ablation resulting in induced vitellogenesis as well as ovarian maturation in several shrimp species clearly demonstrate the inhibitory effect of GIH on maintaining prophasic meiotic arrest that was set in before the onset of vitellogenesis.

On the contrary, influence on the resumption of meiotic maturation by stimulatory hormones in crustaceans is poorly understood. Arvy et al. (1954) found a rise in the hemolymph ecdysteroids coincident with the initial stages of oogonial and spermatogonial mitoses in the shore crab, *Carcinus maenas*. In the prawn *P. serratus*, Lanot and Clédon (1989) found that the first meiotic resumption at prophase I was initiated to undergo germinal vesicle breakdown, under a high titer of ecdysteroids in the eggs. These authors also induced meiotic maturation by incubating prawn oocytes with physiological doses of ecdysteorid. In *P. serratus*, however, the second meiotic resumption is triggered by spawning (Goudeau and Goudeau, 1986). A role for the vertebrate steroid, progesterone in inducing meiotic maturation was also indicated in the spiny lobster *Panulirus homarus* and the mud crab *Scylla serrata* (Kirubagaran et al., 2005; Warrier et al., 2001).

Continued investigation on the hormonal regulation of meiotic maturation has revealed that biogenic amines like serotonin/5-HT could stimulate resumption of meiotic maturation in penaeid shrimps and brachyuran crabs. Thus, injection of 5-HT promoted the expression level of Cdc2 and Cdk7, two key molecules involved in the MPF signaling pathway to control resumption of meiotic maturation in the penaeid shrimp, *P. monodon* (Phinyo et al., 2014). In the Indian white shrimp, *F. indicus,* Tomy et al. (2016) showed that injection of 5-HT resulted in an upregulated expression of ovarian genes involved in vitellogenesis (*vg* and *vgr*) and meiotic maturation (*cyclin B*, *Cdc2*, and *tsp*) especially at the maturation stage, demonstrating a possible dual regulatory role for 5-HT in vitellogenesis and meiotic maturation of the oocytes. However, these authors noticed significant enhancement in the acquisition of meiotic competence and oocyte maturation only when 5-HT treatment was given to eyestalk-ablated females, suggesting once again the overpowering inhibitory effect of VIH both on vitellogenesis and final meiotic maturation. These results, along with other studies (Wongprasert et al., 2006), in which 5-HT injection has induced ovarian maturation and spawning would bolster the view that serotonin possesses a general, but demonstrable, effect on vitellogenesis and oocyte maturation, at least in the marine shrimps. Furthermore, there seems to be a parallel pattern in the hormonal regulation of vitellogenesis and meiotic maturation in the decapod crustaceans. Both the maturational processes are controlled by an inhibitory and stimulatory hormone signaling systems, operating one against the other. The control by biogenic amines on both vitellogenesis and resumption of meiosis reveals a synergistic or one-after-another influencing effect on the two vital maturational processes.

14.5 EGG ACTIVATION

In the majority of the animals, the union of sperm and egg triggers an egg activation process to initiate development that leads to embryogenesis. Egg activation refers to a series of morphological and molecular changes that occur in the egg after fertilization. Egg activation in many marine invertebrates with broadcast fertilization is believed to prevent polyspermy and establish a microenvironment to support embryo development (Wong and Wessel, 2006). In sea urchin, the sperm-induced egg activation is accompanied by specialized events such as (1) a change in egg membrane potential, (2) a transient increase in intracellular calcium, and (3) fusion of male and female pronuclei.

Egg activation in crustaceans presents many peculiarities, inasmuch as the sperm is atypical and generally nonmotile; hence, sperm attachment is a passive process during fertilization. Nonetheless, the sequence of egg activation is the same as that of other gametes, although many of the early events are not sperm-induced; the mere exposure of egg to seawater is adequate to initiate activation. The presence of jelly layer surrounding the egg surface is a characteristic feature of many marine invertebrates such as echinoderms, where fertilization occurs externally in the seawater. Jelly layer enveloping the delicate vitelline envelope protects the spawned eggs from environmental hazards such as microbial infestations. Penaeiodean shrimps have evolved a surprisingly similar mechanism of egg activation, when spawned into the seawater. However, in the shrimps, the jelly precursor substance, which forms the jelly layer, is housed within the cortical crypts, released only upon exposure to seawater.

14.5.1 Cortical Reaction

Cortical reaction in crustacean eggs is a biphasic event in all penaeiodean shrimps, in which eruption of a cortical rod from the so-called cortical crypts occurs to release the jelly precursor substances to form the jelly layer. This is followed by a second phase of cortical reaction in which cortical granules or vesicles from the egg cortex are released into the perivitelline space to form the egg envelopes, as in sea urchin. Crustaceans other than penaeiodean shrimps undergo only the second phase of cortical reaction, as they lack cortical crypts.

14.5.1.1 Cortical Reaction in Penaeus aztecus

Clark et al. (1980) documented the morphological changes that occur in the cortex of the egg upon contact with seawater. In the unreacted eggs, the club-shaped cortical rods are located in the peripheral cytoplasm. The rods lie within the membranous cortical crypts and are separated from the exterior by a thin fibrous investment coat. The first reaction to the seawater exposure is the emergence of cortical rods from the surface of the egg, giving it a blebbed appearance. Between the emerging rods, the investment coat seems to stretch and undergo partial loss of integrity. Meiotic metaphase figures are seen in the peripheral cytoplasm of the eggs, indicating meiotic resumption. Sperm are seen closely associated with the egg surface. In the next stage, the integrity of investment coat is lost, so that the cortical rods are well out of their crypts. This is followed by a swelling in the basal and apical ends of the cortical rods, resulting in the release of the "bottlebrush" structures, which eventually disintegrate to form a flocculent material that surrounds the entire egg, like a corona. The second phase of cortical reaction follows the jelly layer formation. Fig. 14.2 shows the cortical rod emergence from the cortical crypts in *P. monodon* (Pongtippatee-Taweepreda et al., 2004).

14.5.1.2 Mg^{2+}-Dependent Egg Activation in Sicyonia ingentis

Using *S. ingentis*, as a model to investigate the biology of fertilization in marine shrimps, Clark and his associates have delineated a sequence of events occurring as a result of egg activation by external magnesium ions. They are (1) a dramatic release of jelly precursors from the extracellular cortical crypts, (2) transformation of the precursor material into a layer of jelly that forms around the eggs, (3) a rapid rise in intracellular Ca^{2+} concentration ($[Ca^{2+}]i$) in the form of a wave, (4) dramatic cortical contraction, (5) release from metaphase I arrest and completion of meiosis, and (6) the elevation and assembly of HE following exocytosis of vesicles from the egg cytoplasm (Clark et al., 1984; Pillai and Clark, 1987, 1988; Lindsay et al., 1992).

As spawning occurs, the spermathecal sperm are applied onto the oocytes; however, the entry of sperm into the eggs do not happen until after the oocyte activation has taken place in response to Mg^{2+} (Lindsay et al., 1992). Thus, in the absence of sperm, oocyte activation takes place normally, although mitotic cell divisions are abnormal and the development usually is arrested at cleavage (Pillai and Clark, 1987). The characteristic rise in the intracellular Ca^{2+} ($[Ca^{2+}]i$) in *S. ingentis* eggs is induced by seawater Mg^{2+} and not in response to sperm entry. The Mg^{2+}-induced ($[Ca^{2+}]i$) rise does not require extracellular Ca^{2+} either, and hence the source of it should be from internal storage sites, such as endoplasmic reticulum. In support of it, Lindsay et al. (1992) found no effect of Mg^{2+} on the rise of ($[Ca^{2+}]i$), when injected directly into the oocyte, suggesting that Mg^{2+} acts through a membrane associated receptor to activate a signal transduction pathway to induce calcium ion release from intracellular storage sites. Although a Mg^{2+} receptor or sensor has not been identified so far in penaeiodean shrimps, Linsday et al. (1992) showed that protease trypsin abolished the ability of Mg^{2+} to elevate ($[Ca^{2+}]i$) in the oocyte, suggesting that a surface protein is required for Mg^{2+} to exert its effects.

Injection of *S. ingentis* oocytes with the second messenger inositol 1,4,5-triphosphate (IP3), however, resulted in an immediate rise in the intracellular

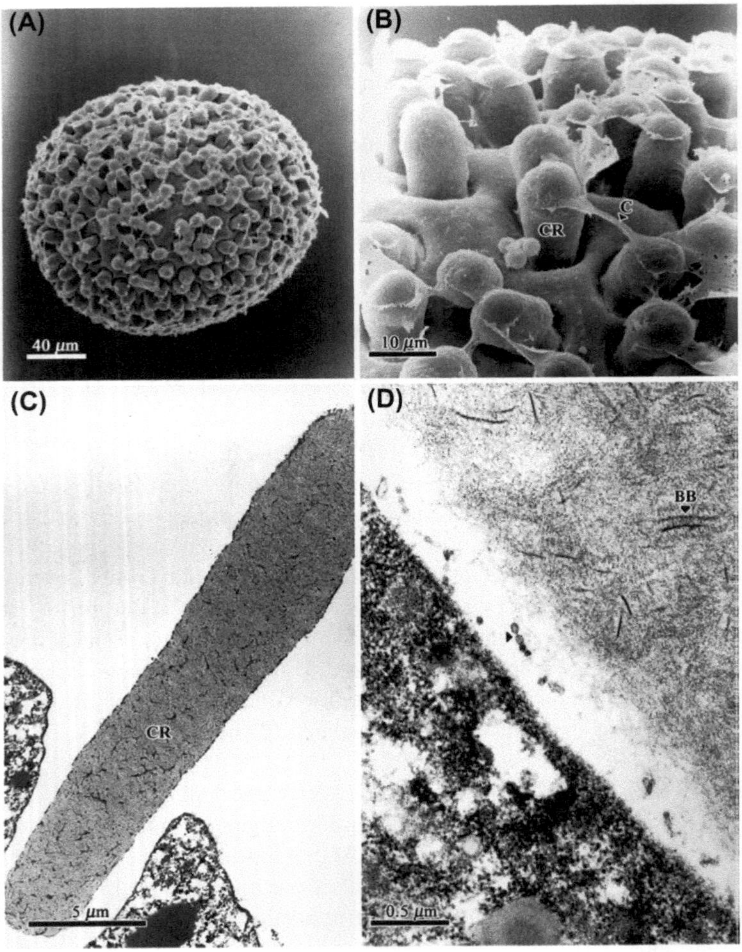

FIGURE 14.2 SEM and TEM micrographs of the egg at 15 s after spawning. (A) The cortical rods begin to protrude from the cortical crypts, with no apparent changes in the substructure of the rods. (B) Club-shaped cortical rods slightly expanded and protrudes out. (C) Extruded cortical rod enlarged in TEM. (D) Beads of granules with a size of 10–30 mm, were observed between the surface of the ooplasm and the cortical rods. *BB*, bottlebrush; *C*, investment coat; *CR*, corticol rod. *Reproduced from Pongtippatee-Taweepreda, P., Chavadej, J., Plodpai, P., Pratoomchart, B., Sobhon, P., Weerachatyanukul, W., et al., 2004. Egg activation in the black tiger shrimp Penaeus monodon. Aquaculture 234, 183–198.*

Ca^{2+}, followed by normal cortical reaction (Lindsay and Clark, 1994). Furthermore, these workers found evidence that the purported Mg^{2+} "receptor" activates the enzyme tyrosine kinase, which in turn increases production of IP3, finally to release intracellular Ca^{2+} stores. Evidently, entry of Mg^{2+} from seawater stimulates IP3 production, which has a known function in the release of free calcium, stored in the cytoplasmic organelles such as endoplasmic reticulum (Whitaker and Irvine, 1984). Intracellular calcium release within the egg seems to be a general feature of egg activation at fertilization in a wide variety of animal species (Wong and Wessel, 2006).

14.5.1.3 Cortical Reaction and Hatching Envelope Formation

On contact with the seawater, eggs of many penaeiodean shrimps undergo spontaneous activation in which two dramatic morphological changes occur in the cortex. The first is the eruption of cortical rods and the disruption and loss of the thin extracellular membrane that surrounds the cortical rods from the outside. The second is the elevation of the extracellular envelope that has been closely opposed to the oolemma, and its transformation into a substantial HE. After the release of cortical rods, the depth of the crypts is reduced and the crypts have disappeared completely to give the egg a smooth, round shape in *S. ingentis*, 10 min after spawning (Clark et al., 1990). The exposed extracellular material now assumes a fibrous coat, after detaching from the oolemma. During this time of egg activation, the subcortex of the egg undergoes dramatic change in producing materials for the formation of HE. In *S. ingentis*, this event of cortical reaction is independent of the preceding cortical crypt eruption and jelly layer formation.

It is important to note that cortical vesicles that will release the components of HE are not present until after formation of the jelly layer. By then, two types of cortical vesicles, viz., dense vesicles and ring vesicle appear in the periphery of the egg cytoplasm one after another. The dense vesicles migrate first to the cortex and line up beneath the plasmalemma and exocytozise the contents to give rise to the thin HE (Pillai and Clark, 1987). In the second exocytosis, the ring vesicles, contained in cisternal elements migrate to the cortical cytoplasm and undergo exocytosis. The coalescence of the material from the ring vesicles with the thin HE results in the formation of fully formed envelope. In all the shrimp species studied, the outer thin HE is electron-dense, whereas the inner thick layer is electron-translucent. The elaboration of HE is simultaneous around the entire egg and does not proceed in a wavelike fashion from a single point. The above-described events in the making of HE is similar in all shrimp species studied till date, although there might be differences in the time course of events in the elaboration of the envelope (Pongtippatee-Taweepreda et al., 2004)

The chemical nature of HE has not been fully elucidated in the shrimp eggs. Pillai and Clark (1990) reported the presence of chitin-like carbohydrates in the HE of *S. ingentis*. Lectin-binding assays demonstrated the presence of mannose and *N*-acetylglucosamine, the structural oligosaccharide found in the chitin. Although ovoperoxidase, the enzyme involved in the dityrosyl linkages between tyrosyl residues, has not been detected in the HE of shrimps, the occurrence of oxidase activities, associated with the elevation and "hardening" of the HE, has been reported in the penaeiodean shrimps, *S. ingentis* and *Trachypenaeus similis* eggs (Glas et al., 1995).

HE of marine shrimps is in many ways homologous to the sea urchin fertilization envelopes, taking origin from the exudates of cortical reaction. Both of them are lamellar in nature, with evidence of arthropod integument-like chemical composition and a phenolic tanning process imparting hardness. As early as Pryor et al. 1949, described a phenolically tanned protein with obvious absence of chitin in the ootheca (egg case) of the cockroach, *Periplaneta americana*. The presence of chitin–protein envelopes found in the spermatophores of the crab *S. serrata*, and the evidence for phenolic tanning in it lends further support to the existence of arthropod cuticle-like hardening in other reproductive structures, such as the HE in Crustacea (Uma and Subramoniam, 1979).

The general sequence of egg activation process described above for *S. ingentis* is similar in all other penaeiodean shrimps. Fig. 14.3 summarizes the sequence of events that occur as a result of egg activation in *P. monodon* (Pongtippatee-Taweepreda et al., 2004).

14.5.1.4 Cortical Reaction in Crabs

In all the penaeiodean shrimps and the prawn *P. serratus*, the cortical reaction takes place in two steps, the cortical crypt dissolution and jelly layer formation, and the cortical vesicle reaction to modify vitelline envelope into the thick fertilization envelope. However, in the crab *C. maenas*, there is no extracellular cortical crypts and hence there is no jelly layer formation. Furthermore, the fertilizing sperm induces specific surface changes in the egg. In the crab, cortical vesicle exocytosis to modify the vitelline into fertilization envelope is under the stimulus from the first contact between the sperm and the egg plasma membrane. The crab egg cytoplasm displayed two populations of structurally distinct cortical vesicles, releasing two different types of exudates (Goudeau and Becker, 1982). The first exudation lasts for 15 min. The exudates, consisting of fine granules, are deposited beneath the vitelline envelope to form envelope I. The second cortical reaction lasts for a longer period (7–8 h) and releases a mass of ring-shaped granules into the space beneath envelope I. These ring-shaped cortical granules coalesce together to form a new thick coating (envelope II) under envelope I. Goudeau (1984) found that the cortical vesicle containing ring-shaped granules take origin from the tubular endoplasmic reticulum and are directly released into perivitelline space, bypassing the Golgi complex. The elaboration of the tripartite capsule in the crab is similar to shrimp HE in protecting the developing embryo.

14.5.1.5 Cortical Reaction in Lobster and Crayfish

Similar to crabs, the oocytes of lobsters and crayfish lack cortical crypt and hence they do not form jelly layer. However, there is a similarity in cortical reaction and the formation of egg envelopes between lobsters and crayfishes, as well as other decapods described above.

In the noble crayfish, *Astacus astacus*, the mature ovarian oocytes are enclosed by an envelope that consists of two layers, all secreted by the ovary

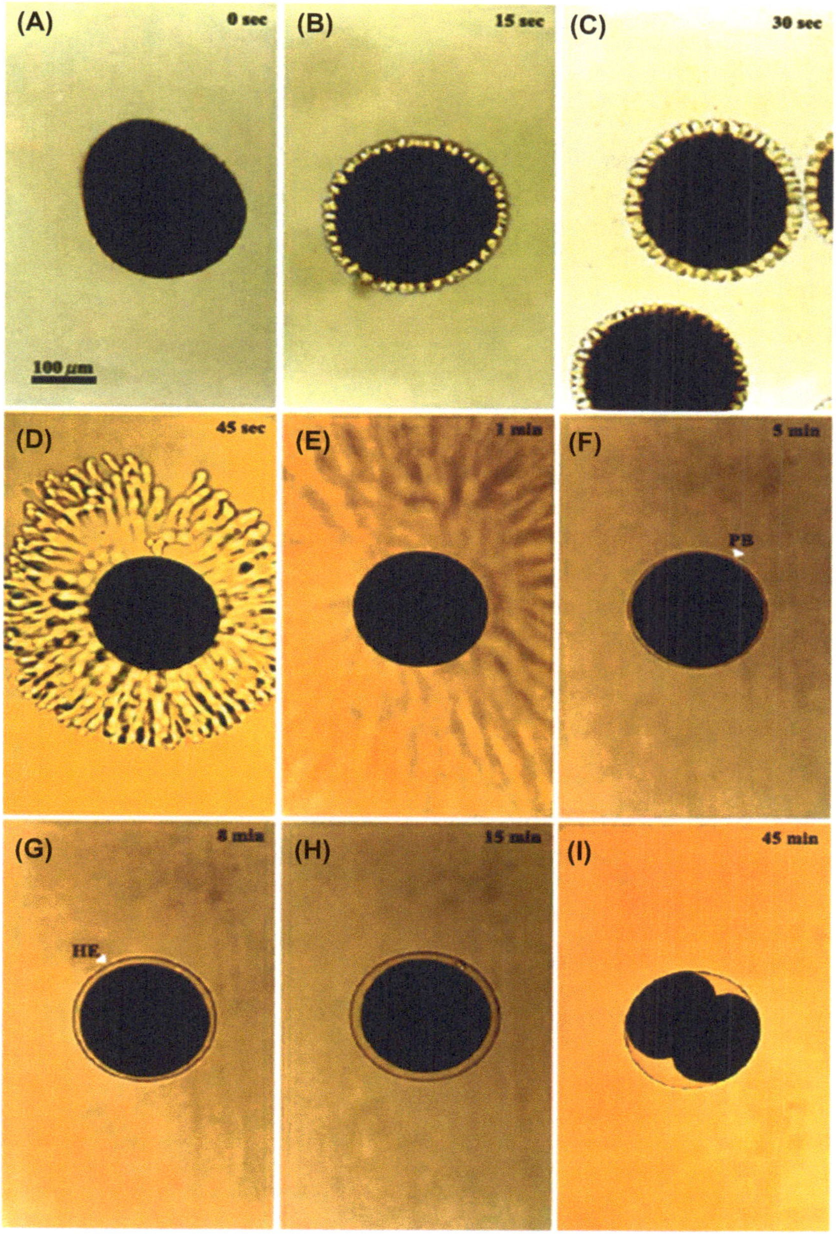

FIGURE 14.3 Light micrographs showing the cortical reaction of *Penaeus monodon* eggs. Mature eggs at the time of spawning (0 s) are not absolutely round and are approximately 270–280 nm in diameter (A). Within 15 s, the egg becomes rounded and the cortical rods begin to emerge from the egg (B). Once the cortical rods are expelled, they form a corona around the egg (C, D) which becomes dissipated within 45–60 s (D, E). Release of polar body and formation of a hatching envelope (F, G and H), to the first mitotic starts (I). *HE*, hatching envelope; *PB*, polar body. Bar = 100 μm. *Reproduced from Pongtippatee-Taweepreda, P., Chavadej, J., Plodpai, P., Pratoomchart, B., Sobhon, P., Weerachatyanukul, W., et al., 2004. Egg activation in the black tiger shrimp Penaeus monodon. Aquaculture 234, 183–198.*

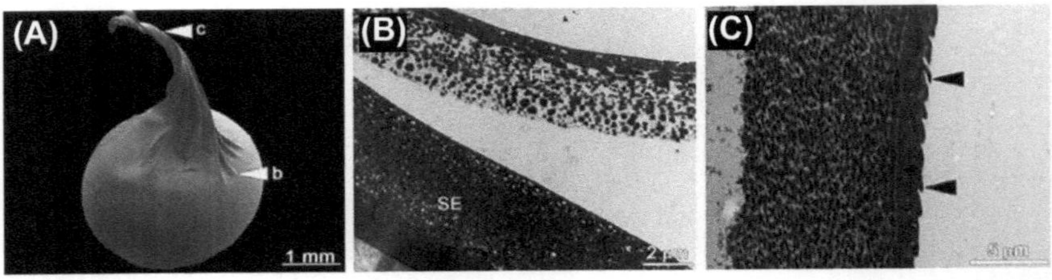

FIGURE 14.4 SEM and TEM micrographs of attachments stalk in the egg of the noble crayfish. (A) Fully developed attachment stalk, areas indicated by *arrowheads* as b and c are described in more detail in the next micrographs. (B) Site of attachment stalk derivation from first envelope. (C) Ultrastructure of upper part attachment stalk, *arrowheads* show serrated margin on the surface of the attachment stalk. *FE*, first envelope; *SE*, second envelope. *Reproduced from Niksirat, H., Kouba, A., Kozák, P., 2015. Ultrastructure of egg activation and cortical reaction in the noble crayfish Astacus astacus. Micron 68, 115–121.*

(Niksirat et al., 2015). Numerous pores are found on the surface of the outer layer of this envelope and the inner layer is made up of bottlebrush-shaped materials. On ovulation, the pores in the outer layer are closed, and the bottlebrush-like objects in the inner layer are replaced by a spongiform electron-lucent matrix, containing electron-dense vesicles. In the egg cortex, three types of vesicles are present. These cortical vesicles are either highly or moderately electron-dense or multilayered in nature. One hour after spawning, the inner layer of the first envelope is condensed. At the same time, the dense vesicles present in the egg cortex appear to have discharged their contents into the perivitelline space, where the vesicular-dense materials condense to form a second layer beneath the outer layer (Fig. 14.4). Twenty-four hours postovulation, the highly and moderately electron-dense vesicles continue to release their contents into the perivitelline space to stabilize the inner layer. The egg attachment stalk is derived from the outermost envelope (Fig. 14.4).

14.6 ELECTRICAL EVENTS AT EGG ACTIVATION

The earliest recognizable physiological change in the egg cell at fertilization is the recordable electrical events associated with egg activation. Electrophysiological studies carried out by Goudeau and Goudeau (1989) on crab, lobster, and the palaemonid prawn have provided the basic understanding of the electrical responses to fertilization and spontaneous activation (in seawater) for decapod crustacean eggs.

14.6.1 Electrical Response to Egg Activation in Crab

The electrical response of the crab oocyte membrane was examined during sperm-independent spontaneous activation by artificial seawater (ASW). Incubation of eggs in ASW for 100 min resulted in the hyperpolarization of the crab membrane. The membrane potential E_m slightly decreased from −50 to

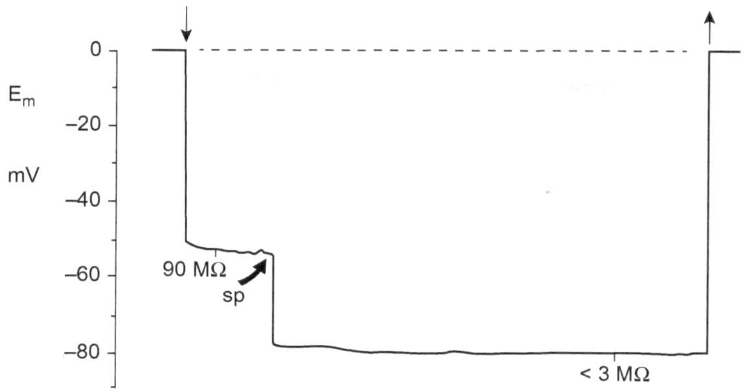

FIGURE 14.5 Fertilization potential of a crab egg. Recorded at slow speed. The *light and dark arrows* indicate when the microelectrode penetrated the egg and when the egg was inseminated (sp), respectively. Numbers on record indicate R_m. Not drawn to scale. *Modified from Goudeau, H., Goudeau, M., 1989. Electrical responses to fertilization and spontaneous activation in decapod crustacean eggs: characteristics and role. In: Nuccitelli, R., et al. (Eds.), Mechanism of Egg Activation. Plenum Press, New York. Permission from Springer.*

−37 mV during the first 50 min of incubation then became increasingly negative until it stabilized at −80 mV after 95 min of incubation. Membrane resistance R_m remained unchanged at 170 MΩ for the first 15 min of incubation, and then abruptly decreased to 36 MΩ after another 50 min. The large drop in R_m is correlated with the lifting of vitelline coat due to the first step of the cortical reaction (Goudeau and Goudeau, 1985). R_m value then decreased to reach a very low value in the range of 3–5 MΩ, after 80 min of incubation.

In the crab egg, the membrane potential change is biphasic; the hyperpolarization is elicited by ASW, with the concurrent, but slow increase in the membrane permeability to K^+. In the first phase, the membrane potential change occurs after 50 min of incubation with ASW. However, with the addition of sperm at the 50th min of incubation, the second phase of membrane potential change occurs very fast. Within 1 s of the application of a sperm pellet on to the vitelline envelope of the crab *Maia squinado* egg, an abrupt and long-lasting hyperpolarization of the egg plasma membrane takes place from −54 ± 2 to −80 ± 1 mV in less than 400 ms. Fig. 14.5 depicts the hyperpolarization response in the crab eggs after addition of sperm to the medium (Goudeau and Goudeau, 1989).

At resting potential, crab oocytes were more permeable to Cl^- than to K^+. Apparently, the acquisition of K^+ permeability is similar to the time course of membrane electrical response. Subsequently, there is a depolarization of the membrane to reach a positive value following the sperm-induced hyperpolarization. The membrane resistance of 90 MΩ, reached after the first step polarization, was reduced to 3 MΩ, after the addition of sperm. This kind of hyperpolarizing fertilization potential might be unique to crabs and other decapods investigated, as it has not been reported for other marine invertebrates such as sea urchin and echiuroid worms (Hagiwara and Jaffe, 1979). Nevertheless, in the sea urchin eggs, change in membrane potential produces a rapid and transient block to polyspermy (Jaffe, 1976). In the crabs, however, it is not clear whether such changes in the membrane potential resulted in the prevention of polyspermy. Furthermore, this purported electrically mediated polyspermy block in the crab is different from several other species in that it is fast, negative, and long-lasting.

14.6.2 Electrical Response in Lobster Eggs

In mature oocytes of the European lobster, *Homarus gammarus*, arrested at first meiotic metaphase, the plasma membrane was selectively permeable to Cl^-, and the resting potential (E_m) was −32 mV. Mature oocytes, when inseminated in vitro with sperm taken from the female thelycum, instantaneously triggered a sustained hyperpolarization of the plasma membrane that corresponded to the fertilization potential. The hyperpolarization was obviously caused by the increased membrane permeability to K^+. In the mature oocytes, before fertilization, the egg plasma membrane was permeable to chloride ions, as found in the brachyuran crabs, *M. squinado* and *C. maenas*. The sudden rise in the membrane permeability to K^+ may be due to an increase in ([Ca^{2+}]i), which might mediate the membrane hyperpolarization. In other species, fertilization triggers a dramatic rise in ([Ca^{2+}]i), promoting a Ca^{2+}-activated K^+ conductance (Meech, 1978). Concomitant to this electrical response, the oocyte resumed its meiotic maturation, which was arrested at first metaphase (Goudeau and Goudeau, 1986). The first polar body formation also occurred concurrent to membrane hyperpolarization.

14.6.3 Electrical Response in the Prawn *P. serratus*

In the crab and lobster, only fertilization (sperm attachment) can stimulate meiotic maturation, cortical reaction as well as the hyperpolarization of the oocyte membrane, due to an increase in the K^+ permeability (Goudeau et al., 1984). In contrast, in the palaemonid prawn *P. serratus*, such changes occur spontaneously by the interaction of eggs with Mg^{2+} ions in the seawater at spawning, thereby indicating a partial parthenogenetic activation of these externally fertilizing decapod crustaceans. When the ovulated oocytes are incubated in ASW or in natural seawater, their membrane hyperpolarizes and their resistance decreases, due to an increase in K^+ permeability. However, a threshold concentration of 10–15 mM Mg^{2+} in the seawater is required to trigger egg activation and hyperpolarization of the membrane (Goudeau and Goudeau, 1986). The effects of external Mg^{2+} on the K^+ permeability is specific, as the same effect was not observed when another divalent cation, Ca^{2+} was used in the place of Mg^{2+}. In addition, the increase in K^+ permeability occurring at spawning is independent of fertilization, as the unfertilized eggs also displayed a hyperpolarized membrane with electrophysiological characteristics identical to those observed in ovarian oocytes, incubated in ASW. The observed threshold effect (10–15 mM Mg^{2+} in the seawater) is necessary to prevent spontaneous oocyte activation within the ovary in which a low Mg^{2+} concentration of 10–12 mM is maintained (Goudeau and Goudeau, 1986). Furthermore, the lack of an electrical response to fertilization in prawn eggs indicates that the mechanism insuring monospermy is different from the electrically mediated block to polyspermy present in other species. Sperm entry into these eggs is rather slow, compared to sea urchins in which the process is faster, and hence a quick electrically mediated block is a physiological necessity.

The electrophysiological membrane characteristics of oocytes in the three decapod crustaceans described above are similar in all respect, except for the initiator of egg activation (Table 14.3). In the prawn *P. serratus*, the increased concentration of external Mg^{2+} ions is enough to bring about all the three important egg activation processes, namely, resumption of meiosis, changes in the membrane electrical properties, and cortical reaction. In the penaeid shrimp, *P. aztecus*, Clark et al. (1980) observed similar Mg^{2+}-dependent jelly extrusion, elaboration of jelly layer, and cortical reaction.

On the other hand, in crabs and lobsters the initiator of egg activation is fertilization. However, the pattern of electrical responses of egg membrane at egg activation is totally different from several other species, suggesting uniqueness in the prevention of polyspermy in the decapod crustaceans, investigated so far. Nevertheless, the possibility that the crustacean eggs are physiologically polyspermic could not be overruled. In fact, several fertilization cones and more than one sperm nucleus have been reported in the naturally fertilized crab eggs (Goudeau and Goudeau, 1985).

The increase in K^+ conductance causing the sustained hyperpolarization is a unique electrophysiological feature in lobster and crab eggs, but this gain of K^+ conductance during fertilization was reported for sea urchin, echiurians, amphibians, fish, and hamster eggs (see Hagiwara and Jaffe, 1979). Therefore, increased K^+ conductance seems to be a general ionic mechanism related to fertilization in a wide range of animal eggs.

14.7 SPERM ACTIVATION

14.7.1 Sperm Capacitation

Capacitation, the last phase of sperm maturation has been well studied in mammals. By this process, the sperm acquires the physiological ability to undergo acrosome reaction and fertilization competency after reaching the female genital tract (Bedford, 1970). In addition to morphological changes, the sperm undergoes ionic and biochemical changes, such as increase in the membrane fluidity, rise in the intracellular calcium and protein tyrosine phosphorylation during capacitation (Breitbart, 2002). In the crustacean sperm, physiological maturation

TABLE 14.3 Electrophysiological Characteristics of Oocytes in Decapods Crustacean Membrane

Species Name	Resting Potential of Egg Membrane in E_m	Fertilization Potential in E_m	Initiation	Concurrent Event	Permeability
Prawn—*Palaemon serratus*	E_m −42 mV	−70 mV	External Mg^{2+}	Resumption of meiosis	K^+
Crab—*Carcinus maenas*	E_m −41 mV	−62 mV	Insemination	Resumption of meiosis	K^+
Maia squamata	E_m −33.7 mV	−62.6 mV	Insemination	Resumption of meiosis	K^+
	E_m −53 ± mV				
Lobster—*Homarus gammarus*	E_m −32 mV	−68 mV	Insemination	Resumption of meiosis	K^+

by capacitation has not been thought about for a long time, although spermiogenesis is completed in the vas deferens in many decapods. Nevertheless, postmating spermatophore storage in several decapods, especially in the penaeoidean shrimps, is known to be accompanied by morphological and molecular changes that are presumed to be necessary for the spermatozoon to acquire fertilization ability. For example, in *S. ingentis*, Clark and Griffin (1988) reported morphological changes in the spike and cap region (subacrosome) during their storage in the female seminal receptacle. By virtue of these changes, the sperm obtained from the thelycum is able to undergo the complete biphasic acrosome reaction, whereas the sperm taken from vas deferens or terminal ampoule is incapable of completing the second phase (filament formation), thereby losing its ability to fertilize an egg (Griffin and Clark, 1990). Further studies by Wikramanayake et al. (1992) confirmed that sperm removed from males do not undergo acrosome reaction, although they are capable of recognizing the eggs and bind to it. These studies emphasize that subtle morphological rearrangements in the cap region occur during the capacitation of sperm inside the female seminal receptacle. Corresponding biochemical changes include an increase in the intracellular Ca^{2+} and a reduction in pH of the capacitated *S. ingentis* sperm (Lindsay et al., 1992).

In another closed thelycal shrimp species, *P. monodon*, the capacitated sperm recovered from thelycum has less condensed nucleus than the sperm taken from the male (Vanichviriyakit et al., 2004). Visconti et al. (1995) also found changes in the protein profiles of sperm membrane as well as an increased protein tyrosine phosphorylation, which is an index of capacitation in the mammalian sperm (Yanagimachi, 1994). It has been suggested that components of the thelycal fluid may play a significant role in inducing these biochemical changes within the stored sperm. In another closed thelycal shrimp, *Farfantepenaeus paulensis*, the nuclear chromatin became less condensed, and the acrosomal and subacrosomal regions became more electron-dense, while the acrosomal cap became less concave in the reacted sperm (Braga et al., 2014).

The sperm of open thelycal penaeiodeans also undergoes capacitation after transfer to the female. In *L. vannamei*, Alfaro et al. (2004) found evidence that final maturation of the sperm occurs after the deposition of spermatophore on to the external surface of the thelycum. Ultrastructural studies on three species of *Litopenaus* showed that the formation of a filamentous network between the nucleus and the hemispherical cap is completed only after the deposition of spermatophore on to the female sternum. In vitro induction of the acrosome reaction in sperm from males and females of *Litopenaus occidentalis* demonstrated that the acrosome reactivity to the egg water is significantly superior in sperm cells that have been attached to the open thelycum than those recovered from males. It seems likely that materials added to the filamentous network after mating are important for sperm capacitation and the subsequent acrosome reaction.

In spite of the evidence shown in favor of sperm capacitation in the open thelycal species, it is still unresolved whether the deposited sperm mass undergoes spontaneous maturation, or factors residing in the cuticular surface of the thelycal region could influence such capacitational changes. Aungsuchawan et al. (2011) suggest the possibility that the exposure to seawater may trigger both morphological and biochemical changes in the sperm. Recently, Niksirat et al. (2015) observed calcium deposits in the postmating sperm membrane of the crayfish *A. astacus*. In addition, numerous small calcium deposits found in the subacrosomal zone and nucleus of the stored sperm could suggest that calcium absorbed from the seawater may cause capacitational changes.

14.7.2 Acrosome Reaction

An important prerequisite for fertilization is the activation of sperm by undergoing acrosomal alterations, termed acrosomal reaction (Dan, 1952). Acrosomal reaction has been studied extensively in the marine shrimp, *S. ingentis* (Clark and Griffin, 1988). Among the multistellate spermatozoa, detailed ultrastructural investigation was carried out in the crab, *C. maenas* (Goudeau, 1982). Artificial induction of acrosome reaction in another brachyuran crab, *S. serrata* has been achieved in the context of a viability test for cryopreserved spermatozoa (Bhavanishankar and Subramoniam, 1977). Despite showing some common sequential steps in the acrosomal reaction with the well-studied models such as echinoderms, decapod sperm cells have evolved very different ways of sperm activation and interaction with the egg during fertilization.

14.7.2.1 Acrosome Reaction in the Marine Shrimp Sicyonia ingentis

By far, a comprehensive understanding of acrosome reaction and male gamete activation could be obtained from the work of Clark and his associates on *S. ingentis* (see Clark and Griffin, 1988). *S. ingentis* sperms are highly polarized cells with three visible anatomical regions: a posterior main body, a central cap, and an anterior spike (Clark et al., 1981). These parts can be further divided into three functional regions, viz., the nuclear region, the subacrosome, and the acrosomal vesicle. The main body is the posterior region of the sperm and contains the centrally located uncondensed nucleus. Nucleus lacks nuclear envelope, but it is surrounded posterolaterally by a cytoplasmic band, while the anterior margin of the nucleoplasm is bounded by an electron-dense plaque, termed nuclear plate. The central cap lies anterior to the main body and contains subacrosomal elements, and a portion of the acrosomal vesicle. The subacrosome is located between the posterior nuclear plate and the anterior acrosomal vesicle. The subacrosome consists of a posterior crystalline lattice and an extended dense saucer that originates at the crystalline lattice and terminates in 12–15 laterally extending petals. There is also a granular region which surrounds the extended saucer. The subacrosome is separated from the components of the acrosomal vesicle by the inner acrosomal membrane. At the crystalline lattice junction, the sperm plasma membrane forms a pentalaminar membrane that extends around the cap and spike, completing the acrosomal vesicle. Besides, the acrosomal vesicle contains membrane pouches that encircle the granular region of the subacrosome, anterior granule, and spike.

The induction of acrosome reaction in *S. ingentis* sperm was carried out using unreacted sperm, isolated from the female seminal receptacle, and incubated in 50 μg/mL (protein) of egg water. Egg water is made up of components released from the jelly layer during spawning. As in the sea urchin, the shrimp sperm are activated to undergo acrosome reaction by jelly coat substances. Under natural conditions, sperm binds to ova and undergo exocytosis and, some 10–20 min later, undergo acrosomal filament formation (Fig. 14.6).

The first phase of acrosome reaction consists of acrosomal exocytosis, due to retraction of the spike, swelling of the membrane pouches, and dehiscence of the pentalaminar membrane in the region of the spike. The spike retraction is facilitated by the depolymerization of microfilament contained in the spike of unreacted sperm (Clark et al., 1984; Brown et al., 1976). The acrosomal exocytosis is caused by the loss of pentalaminar membrane confluence in the region of the retracted spike. The remaining pentalaminar system (referred to as cap membranes) folds back facilitating the release of the acrosomal contents and exposing the anterior granule and subacrosomal components, which form a flocculent precipitate between the membrane of the cap and the oolemma (see Fig. 14.6). The anterior granule is now opposed to the egg surface. The petals, at the end of the subacrosome, undergo dramatic extension, subsequent to acrosomal exocytosis.

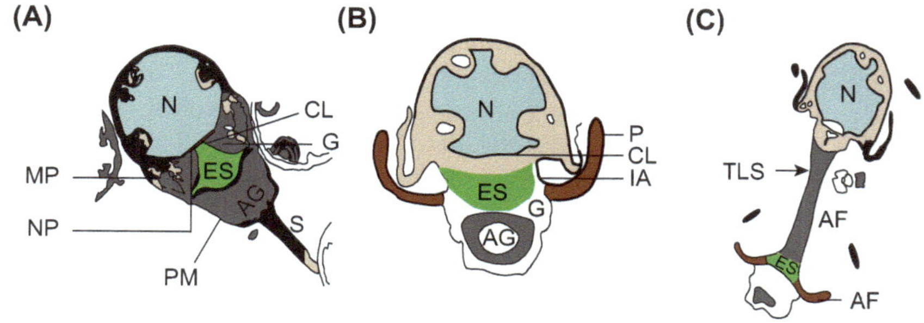

FIGURE 14.6 Acrosome reaction in *Sicyonia ingentis*. (A) Unreacted sperm. (B) Phase I. Depolymerized spike and acrosomal exocytosis. (C) Phase II. Formation of the acrosomal filament. *AF*, acrosomal filament; *AG*, anterior granule; *CL*, crystalline lattice; *ES*, extended saucer; *G*, granular region; *IA*, inner acrosomal membrane; *MP*, membrane pouches; *N*, nucleus; *NP*, nuclear plate; *P*, the petals; *PM*, pentalaminar membrane; *TLS*, tubular-like structures. *Adapted from Clark Jr., W.H., Griffin, F.J., 1988. The morphology and physiology of the acrosome reaction in the sperm of the decapod, Sicyonia ingentis. Dev. Growth Differ. 30 (5), 451–462; Braga, A., Nakayama, C. L., Poersch, L., Wasielesky Jr., W., 2013. Unistellate spermatozoa of decapods: comparative evaluation and evolution of the morphology. Zoomorphology 132 (3), 261–284.*

The second phase of acrosome reaction is the acrosomal filament formation that occurs within 10–20 min after the first phase of acrosomal granule exocytosis. The filament originates from the subacrosomal materials and is composed of a matrix of dense granular material that contains unique tubelike structures. Griffin et al. (1987) activated acrosome reaction using calcium ionophore A23187 to find out the temporal and sequential separation of the two phases of acrosome reaction under in vitro conditions. They found that the acrosomal reaction depended upon external Ca^{2+}, whereas the formation of the acrosomal filament was associated with a decrease in internal pH (pH_i) of the sperm. This decrease in pH_i also causes the outward movement of K^+ ions from the sperm cells. Ionic regulation of gamete activation has not been ascertained in the crustacean spermatozoa. However, it is speculated that the K^+ efflux, caused by calcium ionophore could alter the sperm membrane potential, similar to hyperpolarization of egg membrane during decapod egg activation (Griffin et al., 1987).

14.7.2.2 *Acrosome Reaction in the Penaeid Shrimp* Penaeus monodon

In *P. monodon*, the mature sperm is comprised of three major parts, the anterior spike, middle acrosome, and the posterior nucleus forming the main body. As in *S. ingentis*, the acrosome reaction could be initiated by egg water in *P. monodon* (Kruevaisayawan et al., 2008). Within 5 min of egg water addition, the anterior spike completely disappeared, followed by the rupture of the acrosome. However, there was no filament formation, even after prolonged treatment with the egg water. In this penaeid shrimp, a trypsin-like protease, present both in the egg water and the acrosomal membrane is held responsible for the accomplishment of acrosome reaction (Fig. 14.7). This hydrolytic enzyme, present in the egg water, binds to the receptors present on the sperm surface and triggers the fusion of sperm plasma membrane and the outer acrosome membrane. This membrane fusion creates pores on the membrane surrounding the acrosomal vesicle. Through these membrane pores, the trypsin-like enzyme present on the acrosomal vesicle are released and escapes into the egg water, initiating depolymerization of the anterior spike due to its hydrolytic activity. Following this event, rupturing of the entire acrosome and the exocytosis of cell content occurs. With the result, a dense spherical mass is exposed anterior to the nucleus. This spherical mass could be homologized with the acrosomal filament of *S. ingentis*. In the penaeid shrimps, further progression towards sperm–egg interaction has not been reported for any species. Possibly, the exposed spherical mass could draw the nuclear material together into the egg cytoplasm during sperm entry. This kind of monophasic acrosome reaction is characteristic of penaeid shrimps (Clark and Griffin, 1993).

14.7.2.3 *Acrosome Reaction in* Homarus americanus

In lobsters, a process called "sperm evagination" has been described by early workers (Pochon-Masson, 1965, 1968; Worley, 1939). This reaction is

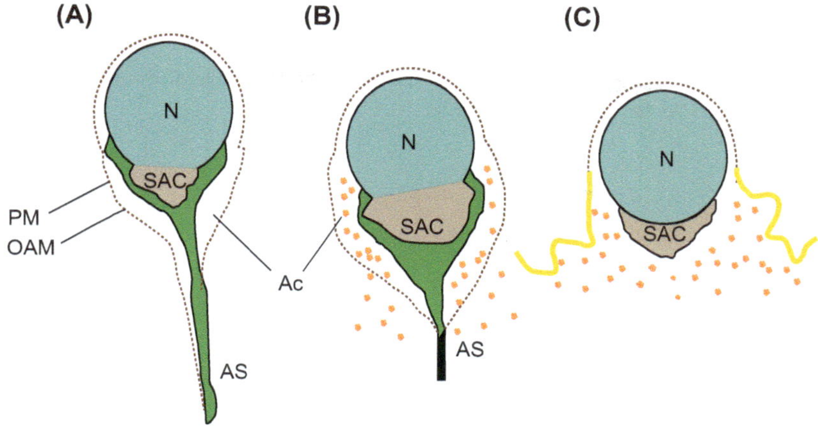

FIGURE 14.7 Acrosome reaction in *Penaeus monodon*. (A) Unreacted sperm. (B) Phase I. Depolymerized spike and acrosomal exocytosis. (C) Phase II. Extrusion of subacrosomal component. *AC*, membrane overlying the acrosome; *AS*, depolymerization of the anterior spike; *N*, Nucleus; *OAM*, outer acrosomal membrane; *PM*, plasma membrane; *SAC*, subacrosomal region. *Adapted from Kruevaisayawan, H., Vanichviriyakit, R., Weerachatyanukul, W., Iamsaard, S., Withyachumnarnkul, B., Basak, A., et al., 2008. Induction of the acrosome reaction in black tiger shrimp (Penaeus monodon) requires sperm trypsin-like enzyme activity. Biol. Reprod. 79 (1), 134–141; Braga, A., Nakayama, C. L., Poersch, L., Wasielesky Jr., W., 2013. Unistellate spermatozoa of decapods: comparative evaluation and evolution of the morphology. Zoomorphology 132 (3), 261–284.*

considered to be the acrosome reaction, during which the acrosome turns inside out, and the nucleus and spikes are drawn up into the space formerly occupied by the acrosome. During this process, a filament forms at the leading edge of the sperm. Later workers like Talbot and Chanmanon (1980b) added calcium ionophore A23187 in the seawater to induce acrosome reaction in *H. americanus*. The ionophore-induced acrosome reaction takes place in two stages: the first stage is the eversion of the acrosomal vesicle by the swelling of the apical cap, found at the tip of the acrosome. The second stage is ejection of the subacrosomal and nuclear material into the cavity created by acrosomal eversion. The everted acrosomal contents form a fuzzy coat on the outside of the sperm. At the same time, the subacrosomal materials rapidly condense to form a prominent filament at the leading edge of the sperm. The nuclear materials and part of the spikes are now pulled into the cavity surrounded by the everted acrosomal materials. The eversion of the acrosome as well as the extension of the filament has resulted in the forward movement of the reacted sperm toward the egg membrane by 18 μm. The spawned oocyte of *H. americanus* is surrounded by a thick chorion and hence acrosome-reacted sperm is supposed to release hydrolytic enzymes to digest the chorion for ejecting the nuclear components into the egg. The uncondensed nuclear material which is pliable, elastic, and hydrophobic makes the ejection through the chorion easy. In *H. americanus*, ionophore A23187 induced acrosome reaction, suggesting that the entry of external calcium ions is responsible for initiating acrosome reaction (Talbot and Chanmanon, 1980b; Gesteira and Halcrow, 1988).

14.7.2.4 Acrosome Reaction in Crabs

Among brachyuran decapods, eversion of acrosomal materials has been reported for many crabs such as *Callinectes sapidus* (Brown, 1966), *C. maenas* (Pochon-Masson, 1968), *Libinia emarginata* (Hinsch, 1972) and the anomuran crab, *Emerita talpoidea* (Barker and Austin, 1963). In these crabs, the acrosome reaction occurs, concomitant with sperm penetration into the oocyte membrane. Artificial induction of sperm eversion in brachyuran crabs as well as other decapods has been achieved by many investigators. In *C. maenas*, Goudeau (1982) made detailed investigations on the ultrastructural features of acrosome reaction, using in vitro experiments. During mating, spermatophores are introduced into the paired oviducts, where they are stored in the spermatheca (Spalding, 1942). Sperm and eggs are mixed just prior to the eggs exiting the female genital pores, effecting internal fertilization. In vitro incubation of ovulated eggs with sperm showed initial events of sperm attachment onto the vitelline envelope. During sperm penetration through egg envelopes, the acrosome reaction takes place, within a period of 5 min in vitro experiments.

In the second step, the reacted spermatozoa approached the egg plasma membrane. The acrosome reaction consists of the eversion of the acrosomal vesicle, which turns inside out by moving anteriorly through the apical cap. Concomitant with the acrosome eversion, an acrosomal cavity is formed. The acrosomal tubule, the complex membrane system, and the nucleus are now pulled into the acrosomal cavity one after another, with the acrosomal tubule or the percutor organ, forming the leading end. In *C. maenas*, penetration of the sperm nucleus through the vitelline envelope and its subsequent location near the oolemma, beneath the vitelline egg envelope has been shown to be the last step of acrosome reaction (Goudeau, 1982). Both light and electron microscopic studies on the acrosome reaction have reported on various brachyuran crab species, such as *Menippe mercenaria* (Binford, 1913), *L. emarginata* (Hinsch, 1972), *C. maenas* (Pochon-Masson, 1962), and *C. sapidus* (Brown, 1966).

In the mud crab, *S. serrata*, acrosome reaction was induced artificially by treatment with the divalent cation ionophore A23187 in the context of testing the viability of cryopreserved spermatozoa (Bhavanishankar and Subramoniam, 1977, 2008). The reacted spermatozoa were stained with the fluorochrome acridine orange and observed for fluorescence. The ionophore treatment resulted in a twofold swelling of nuclear mass, accompanied by retraction of nuclear arms into the nucleus. This reaction was followed by the eversion of the entire subacrosomal material through the apical acrosomal cap, culminating in the formation and extension of a slender acrosomal process. Furthermore, acridine

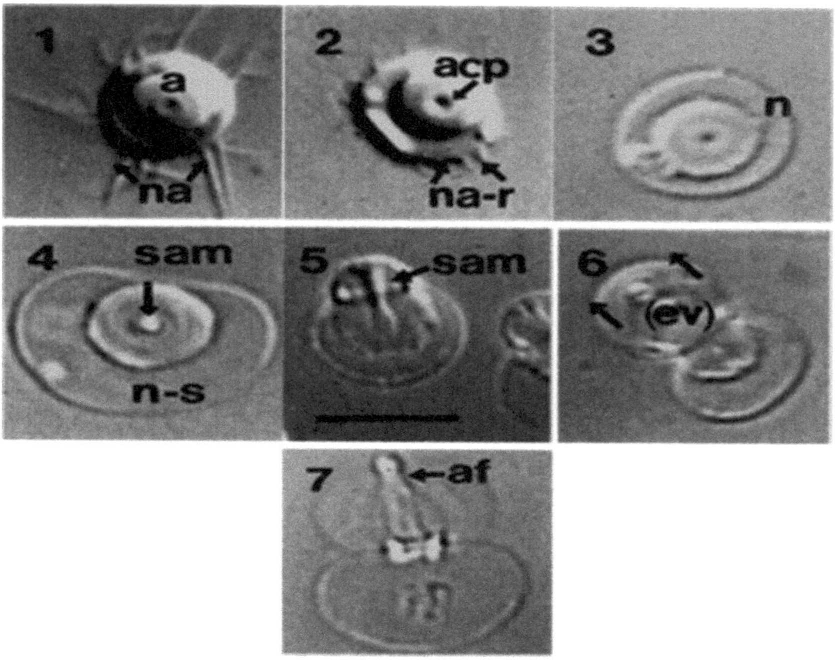

FIGURE 14.8 Acrosome reaction in the spermatozoa of *Scylla serrata*. (1) Normal spermatozoon. The reaction is initiated with retraction of nuclear arms (2–3) and swelling of nuclear mass (3–4), followed by eversion of the subacrosomal material through the apical cap (5–6) (*arrows* in (6) indicate the direction of progressive eversion) and culminates in the formation of acrosomal filament (7). Scale bar = 4 μm. *Reproduced from Bhavanishankar, S., Subramoniam, T., 1977. Cryopreservation of spermatozoa of the edible mud crab Scylla serrata (Forskal). J. Exp. Zool. 277, 326–336.*

orange staining showed the extension of nuclear material into the acrosomal filament. Different stages in the biphasic acrosome reaction in *S. serrata* sperm are depicted in Fig. 14.8.

14.8 SPERM–EGG INTERACTION AND PRONUCLEAR FUSION

Sperm–egg interaction during fertilization is well studied only in the free-spawning echinoderms. Among crustaceans, the dendrobranchiate penaeiodean shrimps are free-spawners but release only the fertilized eggs into the seawater. In the sea urchin, the acrosome-reacted sperm binds to the vitelline envelope through an interaction between a protein, called bindin on the vitelline envelope and a glycoproteinous-binding receptor found on the tip of the acrosomal process (Vacquier, 1979; Glabe and Vacquier, 1978). This brings about the membrane fusion between the acrosomal membrane and the egg vitelline/plasma membrane to facilitate sperm entry into the egg.

Work on sperm–egg interaction in crustaceans is limited to a small number of decapod species. However, as yet, penaeiodean shrimp, *S. ingentis*, has received considerable attention on various aspects of gamete activation. Clark et al. (1991) summarized various events in the sperm–egg activation and interaction chronologically in this shrimp (Fig. 14.9). The spawned eggs, upon contact with seawater, start resumption and completion of meiosis (Pillai and Clark, 1987). This is followed by a dramatic alteration and reorganization of extracellular jelly substances (Pillai and Clark, 1990). By this time, sperm are released from the seminal receptacle and brought into contact with the eggs. Once in contact with the eggs, the sperm bind to the vitelline envelope by the tip of their spikes and undergo acrosomal exocytosis (Clark et al., 1984). Upon entering into the perivitelline space, the sperm undergoes the second

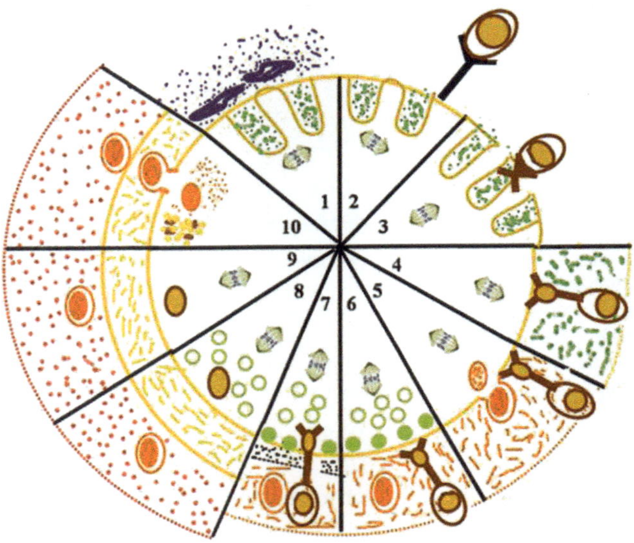

FIGURE 14.9 Chronological order of gamete activation and fertilization in *Sicyonia ingentis*. (1) Oocytes in the first meiotic metaphase is covered in acellular vitelline envelope containing jelly precursor (JP). (2) Binding of sperm using tip of their appendage to the egg's vitelline envelope. (3) Sperm bind to the egg surface glycocalyx, egg activation is evidenced by extrusion of JP. (4) Sperm begins generation of an acrosomal filament. (5) Egg extrudes its first polar body. (6) Two populations of cortical vesicles become evident in the egg cortex. (7) Cortical vesicles exocytase, coalescing with the lifting glycocalyx from the egg surface, marking the beginning of egg hatching envelope. (8) Male pronucleus becomes evidenced and remains until completion of meiotic maturation. (9) Second population of the cortical vesicles exocytose, their contents fuse to form hatching envelope. (10) Second polar body is released and both male and female pronuclei are observed. *Modified from Clark Jr., W.H., Chen, T.I., Pillai, M.C., Uhlinger, K., Shoffner-McGee, J., Griffin, F.J., 1991. The biology of gamete activation and fertilization in Sicyonia ingentis (Penaeoidea); present knowledge and future directions. Bull. Inst. Zool. Academia Sinica, Monograph 16, 553–571; Clark Jr., W.H., Yudin, A.I., Lynn, J.W., Griffin, F.J., Pillai, M.C., 1990. Jelly layer formation in Penaeoidean shrimp eggs. Biol. Bull. 178, 295–299. Permission from The Biological Bulletin.*

phase of acrosome reaction by generating an acrosomal filament, approximately 10–20 min postspawning. During the next 15 min postspawning, two distinct types of cortical vesicles appear in the egg cortex. Although the pronuclear fusion has not been observed in *S. ingentis*, the disintegration of the egg nuclear membrane occurs within 30–35 min. It is at this time, the formation of the HE is completed by the exocytosis of the cortical vesicles (Pillai and Clark, 1990).

Sperm–egg interaction has also been reported in a caridean prawn, *Rhynchocinetes typus* (Dupré and Barros, 2011). As in other caridean prawns, the *R. typus* sperm do not undergo acrosome reaction and hence are incorporated into egg as it is. Furthermore, there is no jelly coat formation in the egg. The sperm–egg interaction is established between the tip of the rigid spike and the outermost chorion of the egg (Barros et al., 1986). The first interaction occurs between protein receptor-type lectins present on the sperm spike and the sperm ligand localized on the oocyte chorion. This lectin-like molecule on the tip of the spike is capable of recognizing the oocyte ligands (sugars), which contain residues of α-Glu, Man(α1–3)Man, α- and β-GlcNAc and α-GalNAc terminal residues. In addition, a high concentration of *N*-acetyl glucosamine on the egg surface would also strengthen the bonding between sperm and egg. The protein molecule responsible for recognition of carbohydrates on the oocyte envelope of this prawn is similar to bindin of sea urchin spermatozoa (Glabe et al., 1982). It was found that α-Man and GlcNAc residues are the ligands that are directly related to the adhesion process and further penetration of sperm into the egg. After this first interaction, the spike enters the oocyte envelope by the action of a trypsin-like enzyme, rhyncocinecine (Ríos and Barros, 1997). Protease enzyme that breaks down the egg chorion has also been demonstrated in the penaeid shrimp, *P. monodon* (Pongtippatee-Taweepreda et al., 2007). Trypsin-like proteases was also identified cytochemically at the spike tip in another caridean prawn, *Macrobrachium rosenbergii*, suggesting that they are responsible for sperm incorporation into the egg (Lynn and Clark, 1983). After entering into the oocyte of *R. typus*, sperm and its arms penetrate the oocyte cytoplasm, where the chromatin contained in the arms begin to migrate into the central area of the sperm, condensing to a cup-shaped structure to become the male pronucleus. The appearance of first meiotic spindle and the extrusion of polar body were also observed at this time. Although final fusion of male and female pronuclei has not been observed, the first mitotic spindle was observable after 36 h of insemination (Dupré and Barros, 2011).

In the crab model, by the time the acrosome reaction was completed, the reacted sperm has already crossed the vitelline envelope, and the nucleus has been engulfed into the perivitelline space, near the oolemma (Goudeau, 1982). At the site of the sperm contact with the egg, the oolemma formed minute fertilization cone. In the interface between the acrosome tubule and the fertilization cone, certain bell-shaped corpuscles have been found. It is suggestive that these corpuscles could release a "bindin-like substance" to promote sperm–egg cell membrane coalescence and fusion, similar to a condition reported in the sea urchin (Vacquier and Moy, 1977).

Taken together, our understanding on the biology of fertilization in crustaceans is still inadequate to make generalizations. Crustaceans are known for their species-specific variability in sperm morphology, which is reflected on the complex acrosomal reaction that they undergo during fertilization. Besides, there is no uniformity in the egg envelope formation among crustaceans. Many malacostracans have internal fertilization and the fertilized eggs are immediately fastened to pleopodal hairs, making the observations on fertilization more cumbersome. Nevertheless, seawater-induced initial egg activation, together with jelly layer formation in the penaeiodean shrimps has parallels with other marine invertebrates such as sea urchin. Understanding the physiology and control of gamete interaction in the commercially important decapods is imperative to develop in vitro fertilization protocols.

Glossary

Acrosomal vesicle Membrane-bound organelle at anterior end of spermatid, derived from Golgi apparatus.

Acrosome Membrane-bound organelle closely applied to the surface of the nucleus of spermatid and spermatozoa. Derived from Golgi apparatus.

Adaptation Process of change by which an organism or species becomes better suited to its environment.

Allele Any of the alternative forms of a gene or other homologous DNA sequence located at a specific position on a specific chromosome.

Alloenzymes (or also called allozymes) The variant forms of an enzyme that are coded by different alleles at the same locus.

Amphigenous Crustacean female increasing in size by growth on all sides.

Anaphase Stage of mitosis where replicated chromosomes are split and the sister chromatids are moved to opposite poles of the cell.

Anaphase I Meiotic division that follows metaphase I of meiosis.

Androdioecious *see* **Androdioecy**.

Androdioecy Breeding system in which males and hermaphrodites coexist in a population.

Androgen Steroid hormone, such as testosterone or androsterone that controls the development and maintenance of masculine characteristics.

Androgenic gland Male sex hormone-producing gland in Crustacea.

Anecdysic Stop molting after sexual maturity.

Anisogamy Sexual reproduction by the fusion of dissimilar gametes.

Antagonistic Showing dislike or opposition.

Apomixis Asexual reproduction, without fertilization.

Apoptosis Form of cell death in which a programmed sequence of events leads to the elimination of cells.

Aquaculture Farming of aquatic organisms, such as fish, crustaceans, mollusks, and aquatic plants.

Archenteron Primitive gut that forms during gastrulation in the developing blastula. It develops into the digestive tract of an animal.

Astaxanthin Reddish pigment that belongs to carotenoids.

Atrophy Partial or complete wasting away of a part of the body.

Autosome Chromosome pairs not involved in sex determination.

Biodiversity Life's genetic heterogeneity, at any or all levels of biological organization.

Blastomere Type of cell produced by cleavage (cell division) of the zygote after fertilization.

Blastula Early embryonic form produced by cleavage of a fertilized ovum.

Brood Number of young ones produced or hatched at one time.

Capacitation Last phase of sperm maturation where the sperm acquires the physiological ability to undergo acrosome reaction and fertilization competency.

Carapace Shield extending from the head region and enveloping a smaller or larger part of the body and is a characteristic feature of many crustaceans.

Carotenoid Fat-soluble pigments.

Castration Any action, surgical, chemical, or otherwise, by which a male loses the functions of the testicles or a female loses the functions of the ovaries.

Centriole Small set of microtubules composed of protein called tubulin arranged in a specific way.

Chiasma Point at which paired chromosomes remain in contact during the first metaphase of meiosis, and at which crossing over and exchange of genetic material occur between the strands.

Chitin Carbohydrate polymer that makes up the crustacean exoskeleton.

Chorion Outermost membrane around the embryo.

Chromatin Complex of DNA and proteins that forms chromosomes within the nucleus of eukaryotic cells.

Cisterna Flattened membrane-enclosed sacs or tubelike structures.

Clade Monophyletic group of organisms.

Commensalism Association between two organisms in which one benefits and the other derives neither benefit nor harm.

Conspecific Member of the same species.

Copulation Sexual intercourse.

Cortical granules Regulatory secretory organelles found within oocytes and are associated with polyspermy prevention after fertilization.

Cuticle Layer of wax and chitin that covers the outermost surface of arthropods and is normally secreted by the epidermis.

Cytodifferentiation Process by which embryonic cells acquire biochemical and morphological properties, essential for specialization and diversification.

Cytokinesis Cytoplasmic division of a cell at the end of mitosis or meiosis, bringing about the separation into two daughter cells.

Cytoplasm Protoplasm enclosed by the plasma membrane of cell excluding the nucleus in eukaryotic cells and cellular DNA in prokaryotic cells.

Decapoda Order of crustaceans within the class Malacostraca, including many familiar groups, such as crayfish, crabs, lobsters, prawns, and shrimp.

Diapause Period of suspended development during unfavorable environmental conditions.

Dioecy Distinct male and female individual organisms.

Dosage compensation Hypertranscription of the single Z or X chromosome in the heterogametic sex to balance the ratio of sex-linked to autosomal gene products.

Ecdysteroids Molting hormones of arthropods.

Embryogenesis Process by which the embryo forms and develops.

Endochorion Innermost portion of the external oocyte envelope.

Endocytosis Process by which a living cell takes up molecules bound to its surface.

Endogenous Growing or originating from within an organism.

Endosomes Membrane-bound vesicles, formed through endocytosis.

Environmental sex determination (ESD) Sex determination caused by an environmental cue, such as temperature, light, etc.

Epibenthic Lives on the surface of sediments at the bottom of the sea.

Epigenetics Study of heritable changes in gene expression or cellular phenotype caused by mechanisms other than changes in the underlying DNA sequence.

Epithelium Thin tissue forming the outer layer of a body's surface.

Epizoic Living on or attached to the body of an animal.

Eusociality Highest level of organization of animal sociality and is defined by the cooperative brood care, overlapping generations within a colony of adults, and a division of labor into reproductive and nonreproductive groups.

Exocytosis Process by which the contents of a cell vacuole are released to the exterior through fusion of the vacuole membrane with the cell membrane.

Exogenous Growing or originating from outside an organism.

Exuviae Cast skins, shells, or other coverings of animals; in arthropods, the old cuticle.

Fecundity Actual reproductive rate of an organism or population, measured by the number of gametes (eggs), seed set, or asexual propagules.

Flagellum Slender threadlike structure, especially a microscopic whiplike appendage which enables many protozoa, bacteria, spermatozoa, etc., to swim.

Follicle Small secretory cavity, sac, or gland.

Folliculogenesis Follicle cells surrounding early vitellogenic oocytes.

Gamete Mature sexual reproductive cell, as a sperm or egg that unites with another cell to form a new organism.

Gametogenesis Process by which diploid or haploid precursor cells undergo cell division and differentiation to form mature haploid gametes.

Gene duplication Duplication of a region of DNA which can arise as products of several types of errors in DNA replication and repair machinery.

Genetic recombination Exchange of a segment of DNA between two homologous chromosomes during meiosis leading to a novel combination of genetic material in the offspring.

Genome Complete genetic constitution of an organism.

Germ cells Haploid cell containing half the number of chromosomes of a somatic cell, which forms the sperm or egg.

Germarium Part of an ovary and testis that produces eggs and sperm, respectively.

Germinal vesicle Enlarged nucleus of the egg before completion of meiosis.

Glycolipids Carbohydrate attached with lipids by a glycosidic bond.

Glycopeptides Peptides that contain carbohydrate moieties (glycans) covalently attached to the side chains of the amino acid residues that constitute the peptide.

Glycosylation Reaction in which a carbohydrate (ie, a glycosyl donor) is attached to a hydroxyl or other functional group of another molecule.

Gonad Sex gland or reproductive gland.

Gonadotropic hormones Hormone secreted by the anterior pituitary of vertebrates which stimulates and controls gonads reproductive activity.

Gonochorism Condition where the sex of an individual is sexually predetermined at birth or hatching and do not change sex. Determination and development of separate sexes, male and female, in a given species.

Gonocyte Germ cell responsible for spermatogenesis in males and oogenesis in females.

Gonopore Sometimes called a gonadopore is an opening through which the gametes are discharged.

Gynandromorphism Abnormal reproductive condition in which both female and male characteristics are displayed in one and the same individual.

Gynodioecy Breeding system with both hermaphrodites and females.

Harem Group of females sharing a single mate.

Hemocytes Free-floating cells within the hemolymph.

Hemolymph Fluid, analogous to the blood in vertebrates, that circulates in the interior of the body remaining in direct contact with the animal's tissues.

Hermaphroditic Organism that has both male and female reproductive organs.

Heterodimer Protein composed of two polypeptide chains differing in composition in the order, number, or kind of their amino acid residues.

Heterogametes Sex of a species in which the sex chromosomes are not the same.

Heterogamety *see* Heterogametes.

Heteromorphic Having different forms at different periods of the life cycle.

Heterosexual Sexual behavior between organisms of opposite sex.

Holoblastic Characterized by complete cleavage that divides the whole egg into distinct and separate blastomeres.

Homologous Having the same relation, relative position, or structure.

Homologue *see* Homologous.

Hormone Chemical substance produced by the body that produces a specific physiological response.

Hyperglycemic High blood sugar.

Hyperplasia Enlargement of an organ or tissue owing to an increase in the number of cells.

Hyperpolarization Change in a cell's membrane potential that makes the cell more negative.

Hypertrophy Increase in cell size causing an increase in the size of an organ or tissue.

Inclusive fitness Individual's own genetic fitness as well as his or her effects on the genetic fitness of close relatives.

Intersex Individual which is a mixture of male and female body parts.

Isoenzyme Variant forms of same enzyme that catalyze the same reaction but that may differ in activity, composition, or physical properties.

Isoform Several different forms of the same protein.

Isogamy Condition entailing the fusion of gametes of similar size.

Iteroparous Characterized by multiple reproductive cycles over the course of its lifetime.

Juvenile Refers to a young who is not yet old enough to be considered an adult.

Karyotype Pictorial representation of all the chromosomes from an individual cell.

Lecithotrophy Form of development in which the embryo receives no nutrition other than the yolk originally contained within its egg.

Lipogenesis Process by which acetyl-CoA is converted to fatty acids.

Lipovitellin Large crystallizable oocyte protein that contains 30% (w/w) lipid.

Malacostracan Largest of the six classes of Crustacea.

Marsupium Pouch that protects eggs, offspring, or reproductive structures.

Megalopa Larval form that appears after the zoea, which swims with its abdominal appendages.

Meiosis Cellular process whereby a diploid cell divides to form haploid gametes.

Meiotic sex-chromosome inactivation (MSCI): Inactivation of the X or Z chromosome in the latter stages of meiosis in the heterogametic sex. Sometimes called male germline X inactivation.

Meiotic spindle During meiosis, spindle-shaped structures attached to the kinetochore pull the chromosomes apart toward opposite poles.

Meiotic *see* Meiosis.

Meroistic Producing nutritive cells as well as eggs from oocytes.

Mesoderm One of the three primary germ layers in the very early embryo with the mesoderm as the middle layer between them.

Metamorphosis Process of transformation from an immature form to an adult form in two or more distinct stages.

Metaphase Second stage of mitosis, between prophase and anaphase, during which the chromosomes become attached to the spindle fibers.

Metaphase I Meiotic division that follows prophase I of meiosis I.

Metaphase II Second meiotic division of meiosis that follows prophase II.

Microfilaments Composed of contractile protein called actin found in the cytoplasm of eukaryotic cells.

Micropinocytosis Taking up into a cell of specific macromolecules by invagination of the plasma membrane which is then pinched off as small vesicles into the cytoplasm.

Microsporidia Group of spore-forming unicellular parasites.

Microtubules Composed of tubulin proteins and are the thinnest filaments of the cytoskeleton that are responsible for various kinds of movements in all eukaryotic cells.

Microvilli Microscopic cellular membrane protrusions that increase the surface area of cells.

Mitochondria Spherical or elongated double membrane organelle in the cytoplasm of nearly all eukaryotic cells.

Mitosis Part of the cell cycle in which chromosomes are separated into two identical sets of chromosomes with its own nucleus.

Mitotic recombination Type of genetic recombination resulting in the somatic crossing over between homologous sequences during mitosis.

Mitotic spindle Microtubule-based bipolar structure that segregates the chromosomes in mitosis.

Molt In invertebrates molting is shedding or sloughing of cuticle periodically to allow body growth.

Monogamous *see* Monogamy.

Monogamy Habit of having only one mate at a time.

Mucopolysaccharide Complex polysaccharides containing amino groups and are chief components of connective tissue.

Multiparous Producing many or more than one at a birth.

Nauplius Free-swimming first stage larva of certain crustaceans.

Neonate New-born larvae.

Neotenous Attainment of sexual maturity during the larval stage.

Neoteny *see* Neotenous.

Nuclear envelope Double lipid bilayer membrane surrounding the eukaryotic cell nucleus.

Nuclear membrane Also known as nuclear envelope.

Nuclear pore Protein-lined channel in the nuclear envelope that regulates the transportation of molecules between the nucleus and the cytoplasm.

Nucleoid Irregularly shaped region within the cell of a prokaryote that contains all or most of the genetic material.

Nucleolus Small, typically round granular body composed of protein and RNA in the nucleus of a cell.

Nucleus Membrane-enclosed organelle that contains hereditary information, or DNA, and controls cellular growth and reproduction.

Nurse cell Type of cell present in the ovary that supplies nourishment to the developing egg cell and is rudimentary in nature.

Ontogeny (also ontogenesis or ontogenetic) Origination and development of an organism, usually from the time of fertilization of the egg to the organism's mature form.

Oocytes Ovarian cell, which may undergo meiotic division to form an ovum.

Oogenesis Female form of gametogenesis and involves the development of various stages of immature ovum.

Oogonium Immature female reproductive cell that gives rise to primary oocytes by mitosis.

Oolemma Cell membrane of an oocyte.

Ooplasm Protoplasmic portion of the oocyte.

Oosorption Resorption of oocytes in the ovary.

Oostegites Modified thoracic legs which form a pouch for brooding embryos.

Ootheca Firm-walled egg case.

Ovariectomy Surgical removal of one or both ovaries.

Ovariole One of the tubes of which the ovaries are composed.

Ovary Female reproductive organ in which ova or eggs are produced.

Ovigerous Carrying or bearing eggs.

Oviparous Producing young ones by means of eggs which are hatched after they have been laid by the parent.

Ovum Female gamete.

Pachytene Third stage of the prophase of meiosis I, following zygotene, during which exchange of segments between chromatids occur.

Panoistic Ovaries, lacking nurse cells (in insects).

Papillae Small rounded protuberance on a part or organ of the body.

Paraphyletic Descended from a common evolutionary ancestor or ancestral group but not including all the descendant groups.

Parasite Organism that lives off or in another organism, obtaining nourishment and protection while offering no benefit in return.

Parthenogenesis Form of asexual reproduction in which growth and development of embryos occur without fertilization.

Parturition Process of giving birth to offspring.

Peduncle Elongated stalk of tissue.

Pelagic Living in open oceans or seas.

Perivitelline space Fluid-filled space between the fertilization membrane and the ovum after the entry of a sperm into the egg.

Phagocytosis Ingestion of a smaller cell or cell fragment, a microorganism, or foreign particles by means of local infolding of cell membrane.

Pheromones Chemicals secreted within an organism to impact the behavior of the receiving organism.

Photoperiod Period of time each day during which an organism receives illumination.

Pinocytosis Transport of fluid into a cell by means of local infoldings by the cell membrane so that a tiny vesicle or sac forms around each droplet which is then taken into the interior of the cytoplasm.

Planktonic Small or microscopic organisms that drift or swim weakly in a body of water, including bacteria, diatoms, jellyfish, and various larvae.

Polyandry Pattern of mating in which a female has more than one male mate.

Polyclonal antisera Antibodies that are secreted by different B cell lineages within the blood serum.

Polygenes Group of nonallelic genes that together control a quantitative characteristic in an organism.

Polygenic *see* Polygenes.

Polygynandry Reproductive strategy that occurs when two or more males have a sexual relationship with two or more females within their breeding group.

Polygynous *see* Polygyny.

Polygyny Pattern of mating in which a male has more than one female mate.

Polymorphism Two or more clearly different phenotypes existing in the same population of a species.

Polyspermy Egg fertilized by more than one sperm.

Predator Organism that lives by preying on other organisms.

Primiparous Giving or having given birth for the first time.

Prophase First stage of mitosis, before metaphase, during which the chromosomes become visible as paired chromatids and the nuclear envelope disappears.

Prophase I First stage in the meiotic division of meiosis I.

Protandrous *see* Protandry.

Protandry Type of hermaphroditism in which an individual is first male and then later in life switches to female.

Proteolysis Breakdown of proteins into smaller polypeptides or amino acids.

Pseudoautosomal region (PAR) Segments of sex chromosomes that recombine in the heterogametic sex.

Puberty Process of physical changes through which an organism matures into an adult capable of sexual reproduction.

Pyknosis Irreversible condensation of chromatin in the nucleus of a cell undergoing necrosis or apoptosis.

Receptor Protein molecule that receives and responds to a neurotransmitter, environmental stimuli, or other substance.

Regulatory gene Segment of DNA that exerts operational control over the expression of other genes.

Semelparous Characterized by a single reproductive episode before death.

Seminal fluid Fluid component of semen, excluding sperm.

Sequential hermaphrodites Individual born as one sex and changes sex at some point in their life which could be from a male to female (**protandry**) or from female to male (**protogyny**).

Sertoli cell Large stellate and columnar somatic cell present in the seminiferous epithelium of mammals.

Sexual dimorphism Differences in appearance between males and females of the same species caused by the inheritance of one or other sexual pattern in their genetic material.

Spawning Process of releasing the eggs and sperm.

Sperm Male gamete in animals.

Sperm ampulla *see* Spermatophore.

Sperm capacitation Physiological changes a spermatozoon undergoes in order to have the ability to penetrate and fertilize an egg.

Sperm plug Solidification of seminal secretions inside the spermatheca.

Spermatheca Receptacle in the female reproductive tract of invertebrates where sperm is stored after mating.

Spermatid Haploid male gametic cell that results from division of secondary spermatocytes.

Spermatocyte Cell produced at the second stage of spermatozoa formation from a spermatogonium and form spermatids after meiotic division.

Spermatogenesis Process in which spermatozoa are produced from male primordial germ cells by way of mitosis and meiosis.

Spermatogonia Undifferentiated male germ cell, originating in a seminiferous tubule and dividing into primary spermatocytes.

Spermatophore Capsule, packet, or mass enclosing spermatozoa that is extruded by the male and transferred to the reproductive tract of the female.

Spermatozoa Mature motile male sex cell of an animal, by which the ovum is fertilized.

Spermiation Process when mature spermatozoa are released from the testis into the lumen of seminiferous tubules.

Spermiogenesis Differentiation of spermatid to spermatozoon.

Spongocoel Central cavity of a sponge, which opens to the outside by way of the osculum.

Sternum Thickened ventral plate on each segment of the body of an arthropod.

Symbiotic Close association between two or more organisms of different species, often but not necessarily benefiting each member.

Syncytium Single cell or cytoplasmic mass containing several nuclei, formed by fusion of cells or by division of nuclei.

Synergism Interaction or cooperation of two or more organisms to produce a combined effect greater than the sum of their separate effects.

Testicles Male sex glands that produce and store sperm.

Thelycum External modification of the female's posterior cephalothoracic sternites and/or coxae that are used in sperm receipt and storage.

Thelygeny Phenomenon in which a pair produce mainly or exclusively females.

Transcription First step of gene expression, in which a particular segment of DNA is copied into complementary RNA (mRNA) by the enzyme RNA polymerase.

Transcripts Copies of mRNA obtained from transcription.

Transfection Process of artificial introduction of nucleic acids into cells of another organism.

Translation Process of translating messenger RNA (mRNA) sequence to a sequence of amino acids during protein synthesis.

Transposon Mobile genetic elements found in the genome of many organisms.

Trioecy Breeding system that represents an evolutionary unstable transitory phase.

Vas deferens Part of the male reproductive system of many vertebrates that transport sperm from the epididymis to the ejaculatory ducts.

Vesicle Membrane-bound structure within a cell in which molecules such as enzymes are transported or stored.

Vitellin Phosphoprotein in the yolk of eggs.

Vitelline membrane Extracellular coat surrounding the outer surface of the plasma membrane of an ovum.

Vitellogenesis (also known as yolk deposition) Process of yolk formation.

Vitellogenin Precursor protein of egg yolk.

Viviparity Growth of the fertilized egg within the maternal body as a larva.

Vulval External parts of the female genital organs.

Zoea Larval form which follows the naupliar stage that swims with its thoracic appendage.

Zygote Diploid cell resulting from the fusion of two haploid gametes; a fertilized ovum.

Zygotic *see* Zygote.

References

Abdu, U., Davis, C., Khalaila, I., Sagi, A., 2002. The vitellogenin cDNA of *Cherax quadricarinatus* encodes a lipoprotein with calcium binding ability, and its expression is induced following the removal of the androgenic gland in a sexually plastic system. Gen. Comp. Endocrinol. 127 (3), 263–272.

Abe, M., Fukuhara, H., 1996. Protogynous hermaphroditism in the brackish and freshwater isopod *Gnorimosphaeroma naktogense* (Crustacea: Isopoda, Sphaeromatidae). Zool. Sci. 13, 325–329.

Abele, L.G., Campanella, P.J., Salmon, M., 1986. Teritoriality and social organization of the semi-terrestial graspid crab *Pachygraspus transverses*. J. Exp. Mar. Biol. Ecol. 62, 1–24.

Achuthankutty, C.T., 1973. Sexual abnormalities in the genus *Acetus* (Sergetidae, Decapoda). Curr. Sci. 42, 827–828.

Acquistapace, P., Aquiloni, L., Hazlett, B.A., Gherardi, F., 2002. Multimodal communication in crayfish: sex recognition during mate search by male *Austropotamobius pallipes*. Can. J. Zool. 80, 2041–2045.

Adiyodi, K.G., Adiyodi, R.G., 1970. Endocrine control of reproduction in decapod Crustacea. Biol. Rev. 45, 121–165.

Adiyodi, K.G., Adiyodi, R.G., 1975. Morphology and cytology of the accessory sex glands in invertebrates. Int. Rev. Cytol. 43, 353–398.

Adiyodi, K.G., Anilkumar, G., 1988. Arthropoda-Crustacea. In: Adiyodi, K.G., Adiyodi, R.G. (Eds.), Reproductive Biology of Invertebrates: Accessory Sex Glands. John Wiley & Sons Ltd, New York, pp. 261–318.

Adiyodi, R.G., Subramoniam, T., 1983. Arthropoda-Crustacea. In: Adiyodi, K.G., Adiyodi, R.G. (Eds.), Reproductive Biology of Invertebrates. Oogenesis, Oviposition, and Oosorption, vol. 1. Wiley, Chischester, England, pp. 443–495.

Aflalo, E.D., Hoang, T.T.T., Nguyen, V.H., Lam, Q., Nguyen, D.M., Trinh, Q.S., Sagi, A., 2006. A novel two-step procedure for mass production of all-male populations of the giant freshwater prawn *Macrobrachium rosenbergii*. Aquaculture 256, 468–478.

Aflalo, E.D., Bakhrat, A., Raviv, S., Harari, D., Sagi, A., Abdu, U., 2007. Characterization of a vasa-like gene from the pacific white shrimp *Litopenaeus vannamei* and its expression during oogenesis. Mol. Reprod. Dev. 74, 172–177.

Afzelius, B., 1971. Spermatozoon of nemertine *Malacobdella-grossa*. J. Submicrosc. Cytol. 3 (2), 181–198.

Afzelius, B.A., 1972. Sperm morphology and fertilization biology. In: Beatty, R.A., Gluecksohn-Waelsch, S. (Eds.), The Genetics of the Spermatozoon, pp. 131–143. Edinburgh, August 16–20, 1971: Proc. Int. Symp.

Aggio, J., Derby, C.D., 2011. Chemical communication in lobsters. In: Breithaupt, T., Thiel, M. (Eds.), Chemical Communication in Crustaceans. Springer, New York, pp. 239–256.

Aguilar, M.B., Quackenbush, L.B., Hunt, D.T., Shabanowitz, J., Huberman, A., 1992. Identification, purification and initial characterization of the vitellogenesis-inhibiting hormone from the Mexican crayfish *Procambarus bouvieri* (Ortmann). Comp. Biochem. Physiol. B. 102, 491–498.

Aiken, D.E., Waddy, S.L., 1980. Reproductive biology. In: Cobb, J.C., Phillips, B.F. (Eds.), The Biology and Management of Lobsters, vol. 1. Academic Press, New York, pp. 215–276.

Aktaş, M., Kumlu, M., 2005. Gonadal maturation and spawning in *Penaeus semisulcatus* de Hann, 1844 by hormone injection. Turkish J. Zool. 29, 193–199.

Aktaş, M., Kumlu, M., Eroldogan, O.T., 2003. Off-season maturation and spawning of *Penaeus semisulcatus* by eyestalk ablation and/or temperature–photoperiod regimes. Aquaculture 228 (1), 361–370.

Alcock, J., 1980. Natural selection and mating systems of solitary bees. Am. Sci. 68, 146–153.

Alexander, R.D., Noonan, K.M., Crespi, B.J., 1991. The evolution of eusociality. In: Sherman, P.W., Jarvis, J.U.M., Alexander, R.D. (Eds.), The Biology of the Naked Mole-rat. Princeton University Press, Princeton, NJ, pp. 3–44.

Alexander, R.D., 1964. The evolution of mating behaviour in Arthropods. In: Highnam, K.C. (Ed.), Insect Reproduction. Royal Entomological Society of London, pp. 78–94.

Alexander, R.D., 1991. Some unanswered questions about naked mole-rats. In: Sherman, P.W., Jarvis, J.U.M., Alexander, R.D. (Eds.), The Biology of the Naked Mole-rat. Princeton University Press, Princeton, NJ, pp. 446–465.

Alfaro, J., Zúñiga, G., Komen, J., 2004. Induction of ovarian maturation and spawning by combined treatment of serotonin and a dopamine antagonist, spiperone in *Litopenaeus stylirostris* and *Litopenaeus vannamei*. Aquaculture 236 (1), 511–522.

Alfaro, J., Ulate, K., Vargas, M., 2007. Sperm maturation and capacitation in the open thelycum shrimp *Litopenaeus* (Crustacea: Decapoda: Penaeoidea). Aquaculture 270, 436–442.

Alfaro-Montoya, J., Lozano, X., 1993. Development and deterioration of spermatophores in pond-reared *Penaeus vannamei*. J. World Aquacult. Soc. 24 (4), 522–529.

Alfaro-Montoya, J., 2010. The reproductive conditions of male shrimps, genus *Penaeus*, sub-genus *Litopenaeus* (open thelyca penaeoid shrimps): a review. Aquaculture 300 (1), 1–9.

Allen, J.A., 1959. On the biology of *Pandalus borealis* Kroyer, with reference to a population of Northumberland coast. J. Mar. Biol. Assoc. U.K. 38, 189–220.

Allsop, D.J., West, S.A., 2003a. Changing sex at the same relative body size. Nature 425, 783–784.

Allsop, D.J., West, S.A., 2003b. Constant relative age and size at sex change in sequentially hermaphroditic fish. J. Evol. Biol. 16, 921–929.

Allsop, D.J., West, S.A., 2004. Sex-ratio evolution in sex changing animals. Evolution 58, 1019–1027.

Almeida, A.O., Buckup, L., 2000. Occurrence of protandric hermphfroditism in a population of the Neotropical freshwater crayfish *Parastacus brasiliensis* (Parastacidae). J. Crust. Biol. 20, 224–230.

Amma, K., 1911. Über die Differenzierung der Keimbahnzellen bei den Copepoden. Arch. Zelloforsch 6, 497–576.

Anders, F., 1957. Über die Geschlechtsbeeinflüssende Wirkung von Farbfaktoren bei *Gammarus pulex* ssp. *Subteraaneus* (Schheider). Zeitschrift für inductive Abstammung-und- Vererbungslehre 88, 291–332.

Anderson, D.T., 2013. Embryology and Phylogeny in Annelids and Arthropods: International Series of Monographs in Pure and Applied Biology. Zoology, vol. 50, Elsevier.

Anderson, E., Lochhead, J.H., Lochhead, M.S., Huebner, E., 1970. The origin and structure of the tertiary envelope in thick-shelled eggs of the brine shrimp, *Artemia*. J. Ultra Struct. Res. 32, 497–525.

Ando, H., Makioka, T., 1998. Structure of the ovary and mode of oogenesis in a freshwater crayfish, *Procambarus clarkii* (Girard). Zool. Sci. 15 (6), 893–901.

Anger, K., 1996. Physiological and biochemical changes during lecithotrophic larval development and early juvenile growth in the northern stone crab, *Lithodes maja* (Decapoda: Anomura). Mar. Biol. 126 (2), 283–296.

Anilkumar, G., Adiyodi, K.G., 1980. Ovarian growth, induced by eyestalk ablation during the pre-breeding season, is not normal in the crab, *Paratelphusa hydrodromous* (Herbst). Int. J. Invertebr. Reprod. 2 (2), 95–105.

Anilkumar, G., Sudha, K., Anitha, E., Subramoniam, T., 1996. Aspects of sperm metabolism in the spermatheca of the brachyuran crab *Metopograpsus messor* (Forskål). J. Crust. Biol. 16 (2), 310–314.

Anilkumar, G., Sudha, K., Subramoniam, T., 1999. Spermatophore transfer and sperm structure in the brachyuran crab *Metapograpsus messor* (Decapoda: Grapsidae). J. Crust. Biol. 19, 361–370.

Antony Raja, B.T., 1963. Observations on the rate of growth, sexual maturity and breeding of four sedentary organisms from Madras Harbour. J. Mar. Biol. Assoc. India 5, 113–132.

Aoto, T., 1952. Sexual phases in the prawn, *Pandalus kessleri* Czer niavski, with special reference to the reversal of sex. J. Fac. Sci. Hokkaido Univ. Ser. VI Zool. 11, 1–20.

Aquacop, 1979. Penaeid reared brood stock: closing the cycle of *P. monodon, P. stylirostris* and *P. vannamei*. Proc. World Maricult. Soc. 10, 445–452.

Araki, A., Matsuura, S., 1995. Gynandromorph of a freshwater crab, *Geothelphusa dehaani* (White). Sci. Bull. Fac. Agric. Kyushu Univ. 49, 133–137.

Arcangeli, A., 1925. Ermafroditismo e partenogenesi negli ispodi terrestn. Monogr. Zool. 36, 105–122.

Arcos, G.F., Ibarra, A.M., Vazquez-Boucard, C., Palacios, E., Racotta, I.S., 2003. Haemolymph metabolic variables in relation to eyestalk ablation and gonad development of Pacific white shrimp *Litopenaeus vannamei* Boone. Aquacult. Res. 34 (9), 749–755.

Arnstein, D., Beard, T.W., 1975. Induced maturation of the prawn *Penaeus orientalis* Kishinouye in the laboratory by means of eyestalk removal. Aquaculture 5 (4), 411–412.

Arvy, C., Eclacier, G., Gabe, M., 1954. Modifications de la gonade de *Carcinides maenas* L. apres abaltion bilaterale de lorgane. Y. C. R. Acad. Sci. Paris 239, 1852–1855.

Asai, N., Fusetani, N., Matsunaga, S., Sasaki, J., 2000. Sex pheromones of the hair crab *Erimacrus isenbeckii*. Part 1: isolation and structures of novel ceramides. Tetrahedron 56, 9895–9899.

Asakura, A.K.I.R.A., 2009. The evolution of mating systems in decapod crustaceans. Crustaceana 18, 121–182.

Atella, G.C., Silva-Neto, M.A.C., Golodne, D.M., Arefin, S., Shahabuddin, M., 2006. *Anopheles gambiae* lipophorin: characterization and role in lipid transport to developing oocyte. Insect Biochem. Mol. Biol. 36 (5), 375–386.

Atema, J., Steinbach, M., 2007. Chemical communication and social behaviour of the lobster *Homarus americanus* and other decapod Crustacea. In: Duffy, J.E., Thiel, M. (Eds.), Evolutionary Ecology of Social and Sexual Systems. Crustaceans as Model Organisms. Oxford University Press, pp. 115–144.

Aungsuchawan, S., Browdy, C.L., Withyachumnarnkul, B., 2011. Sperm capacitation of the shrimp *Litopenaeus vannamei*. Aquacult. Res. 42 (2), 188–195.

Avarre, J.-C., Michelis, R., Tietz, A., Lubzens, E., 2003. Relationship between vitellogenin and vitellin in a marine shrimp (*Penaeus semisulcatus*) and molecular characterization of vitellogenin complementary DNAs. Biol. Reprod. 69, 35–364.

Avila, F.W., Sirot, L.K., LaFlamme, B.A., Rubinstein, C.D., Wolfner, M.F., 2011. Insect seminal fluid proteins: identification and function. Ann. Rev. Entomol. 56, 21.

Avise, J.C., 2007. Conservation genetics of marine turtles—10 years later. Wildl. Sci. 295.

Avise, J., 2011. Hermaphroditism: A Primer on the Biology, Ecology, and Evolution of Dual Sexuality. Columbia University Press.

Ayasse, M., Paxton, R.J., Tengö, J., 2001. Mating behavior and chemical communication in the order Hymenoptera. Annu. Rev. Entomol. 46, 31–78.

Azzauna, A., Greve, P., Martin, G., 2004. Sexual differentiation traits in functional males with female genital apertures (♂fga) in the woodlice *Armadillidium vulgare* Latr. (Isopoda, Crustacea). Gen. Comp. Endocrinol. 138, 42–49.

Babin, P.J., Bogerd, J., Kooiman, F.P., Van Marrewijk, W.J., Van der Horst, D.J., 1999. Apolipophorin II/I, apolipoprotein B, vitellogenin, and microsomal triglyceride transfer protein genes are derived from a common ancestor. J. Mol. Evol. 49 (1), 150–160.

Babin, P.J., Carnevali, O., Lubzens, E., Schneider, W.J., 2007. Molecular Aspects of Oocyte Vitellogenesis in Fish. Springer, Netherlands, pp. 39–76.

Babu, D.T., Antony, S.P., Joseph, S.P., Bright, A.R., Philip, R., 2013. Marine yeast *Candida aquaetextoris* S527 as a potential immunostimulant in black tiger shrimp *Penaeus monodon*. J. Invertebr. Pathol. 112 (3), 243–252.

Baccetti, B., 1970. The sperm cell as index of arthropod phylogenesis. Accad. Naz. Lincei 137, 169–182.

Baeza, J.A., Anker, A., 2008. *Lysmata hochi* n.sp, a new hermaphroditic shrimp from the southwestern Caribbean sea (Caridea: Hippolytidae). J. Crust. Biol. 28, 148–155.

Baeza, J.A., Piantoni, C., 2010. Sexual system, sex ratio and group living in the shrimp *Thor amboinensis* (De Man): relevance to resource-monopolization and sex-allocation theories. Biol. Bull. 219, 151–165.

Baeza, J.A., Thiel, M., 2000. Host use pattern and life history of *Liopetrolisthes mitra*, a crab associate of the black sea urchin *Tetrapygus niger*. J. Mar. Biol. Assoc. U.K. 80, 639–645.

Baeza, J.A., Thiel, M., 2007. The mating system of symbiotic crustaceans. A conceptual model based on optimality and ecological constraints. In: Duffy, J.E., Thiel, M. (Eds.), Evolutionary Ecology of Social and Sexual Systems: Crustaceans as Model Organism. Oxford Univ. Press, pp. 249–267.

Baeza, J.A., 2006. Testing three models on the adaptive significance of protandric simultaneous hermaphroditism in a marine shrimp. Evolution 60, 1840–1850.

Baeza, J.A., 2008. Protandric simultaneous hermaphroditism in the shrimps *Lysmata bahia* and *Lysmata intermedia*. Invertebr. Biol. 127, 181–188.

Baeza, J.A., 2008. Social monogamy in the shrimp *Pontonia margarita*, a symbiont of *Pinctada mazatlantica*, in the tropical eastern Pacific coast. Mar. Biol. 153, 387–395.

Baffoni, G.M., 1962. Sul reassorbimento degli ovociti in un Crostaceo decapode di acqua dolce. Acta Embryol. Morphol. Exp. 5, 124–125.

Baggio, R.A., Pil, M.W., Boeger, W.A., Patella, L.A., Ostrensky, A., Pie, M.R., 2011. Genetic evidence for multiple paternity in the mangrove land crab *Ucides cordatus* (Decapoda: Ocypodidae). Mar. Biol. Res. 7, 520–524.

Bagøien, E., Kiørboe, T., 2005. Blind dating—mate finding in planktonic copepods. I. Tracking the pheromone trail of *Centropages typicus*. Mar. Ecol. Prog. Ser. 300, 105–115.

Baker, R.C., Rosof, J.A., 1928. Spermatogenesis in *Branchipus Vernalis*: Part III. Secondary spermatocyte, spermatid and spermatozoon. The Ohio J. Sci. 28 (6), 315–332.

Bandi, C., Dunn, A.M., Hurst, G.D.D., Rigaud, T., 2001. Inherited microorganisms, sex specific virulence and reproductive parasitism. Trends Parasitol. 17, 88–94.

Banner, D.M., Banner, A.H., 1975. The alpheid shrimp of Australia. Part 2: the genus *Synalpheus*. Rec. Aust. Mus. 29, 267–389.

Banner, A.H., Banner, D.M., 1983. An Annotated Checklist of the Alpheid Shrimp from the Western Indian Ocean (No. 645). IRD Editions.

Banta, M., Wood, T.R., Brown, L.A., Ingle, L., 1939. Studies on the physiology, genetics, and evolution of some Cladocera. Carnegie Inst. Wash. Publ. 513.

Barata, C., Hontoria, F., Amat, F., Browne, R., 1996. Demographic parameters of sexual and parthenogenetic Artemia: temperature and strain effects. J. Exp. Mar. Biol. Ecol. 196 (1), 329–340.

Barker, K.R., Austin, C.R., 1963. Sperm morphology of *Emerita talpoda*. Biol. Bull. 125, 361–362.

Barki, A., Karplus, I., Manor, R., Sagi, A., 2006. Intersexuality and behavior in crayfish: the demasculinization effects of androgenic gland ablation. Horm. Behav. 50, 322–331.

Barnes, H., Blackstock, J., 1977a. Studies on the reproduction of cirripedes. III. Ovisacase; properties of the enzyme released by the late stage embryos. J. Exp. Mar. Biol. Ecol. 27, 233–249.

Barnes, H., Blackstock, J., 1977b. Studies on the reproduction of cirripedes. IV. The protein amino-acid composition and gel electrophoresis of the oviducal sac and egg case. J. Exp. Mar. Biol. Ecol. 27, 251–259.

Barnes, H., Crisp, D.J., 1956. Evidence of self-fertilization in certain species of barnacles. J. Mar. Biol. Assoc. U.K. 35, 631–639.

Barnes, N.B., Wenner, A.M., 1968. Seasonal variation in the sand crab *Emerita analoga* (Decapoda, Hippidae) in the Santa Barbara area of California. Limnol. Oceanogr. 13, 465–476.

Barros, C., Dupré, E., Viveros, L., 1986. Sperm-egg interactions in the shrimp *Rynchocinetes typus*. Gamete Res. 14, 171–180.

Batham, E.J., 1945. Description of female, male and larval forms of a tiny stalked barnacle, *Ibla idiotica*, n. sp. Trans. R. Soc. N.Z. 75, 347–356.

Bauchau, A.G., Mangeot, J.C., 1966. Serotonin et glycenic chez le Crustacea. Experientia 92, 238–259.

Bauer, R.T., Abdalla, J.H., 2001. Male mating tactics in the shrimp *Palaemonetes pugio* (Decapoda, Caridea): precopulatory mate guarding vs. pure searching. Ethology 107, 185–199.

Bauer, R.T., Cash, C.E., 1991. Spermatophore structure and anatomy of the ejaculatory duct in *Penaeus setiferus, P. duorarum*, and *P. aztecus* (Crustacea: Decapoda): homologies and functional significance. Trans. Am. Microsc. Soc. 144–162.

Bauer, R.T., Holt, G.J., 1998. Simultaneous hermaphroditism in the marine shrimp *Lysmata wurdemanni* (Caridea: Hippolytidae): an undescribed sexual system in the decapod Crustacea. Mar. Biol. 132, 223–235.

Bauer, R.T., Min, L.J., 1993. Spermatophores and plug substance of the marine shrimp *Trachypenaeus similis* (Crustacea: Decapoda: Penaeidae): formation in the male reproductive tract and disposition in the inseminated female. Biol. Bull. 185 (2), 174–185.

Bauer, R.T., Newman, W.A., 2004. Protandric simultaneous hermaphroditism in the marine shrimp *Lysmata californica* (Caridea: Hippolytidae). J. Crust. Biol. 24, 131–139.

Bauer, R.T., Thiel, M., 2011. First description of a pure-search mating system and protandry in the shrimp *Rhynchocinetes uritai* (Decapoda: Caridea). J. Crust. Biol. 31, 286–295.

Bauer, R.T., VanHoy, R., 1996. Variation in sexual systems (protandry, gonochorism) and reproductive biology among three species of the shrimp genus *Thor* (Decapoda: Caridea). Bull. Mar. Sci. 59, 53–73.

Bauer, H., 1939. Über die Chromosomen der bisexuellen und der parthenogenetischen Rasse des Ostracoden *Heterocypris incongruens* Ramd. Chromosoma 1 (1), 620–637.

Bauer, R.T., 1976. Mating behaviour and spermatophore transfer in the shrimp *Heptacarpus pictus* (Stimpson) (Decapoda: Caridea: Hippolytidae). J. Nat. Hist. 10, 315–440.

Bauer, R.T., 1978. Antifouling adaptations of caridean shrimp: cleaning of the antennal flagellum and general body grooming. Mar. Biol. 49, 69–82.

Bauer, R.T., 1979. Sex attraction and recognition in the caridean shrimp *Heptacarpus paludicola* Holmes (Decapoda: Hippolytidae). Mar. Freshwater Behav. Physiol. 6, 157–174.

Bauer, R.T., 1986a. Sex change and life history pattern in the shrimp *Thor manningi* (Decapoda: Caridea): a novel case of partial protandric hermaphroditism. Biol. Bull. 170, 11–31.

Bauer, R.T., 1986b. Phylogenetic trends in sperm transfer and storage complexity in decapod crustaceans. J. Crust. Biol. 6, 313–325.

Bauer, R.T., 1991. Sperm transfer and storage structures in penaeoid shrimps: a functional and phylogenetic perspective. In: Bauer, R.T., Martin, J.W. (Eds.), Crustacean Sexual Biology, Columbia University Press, New York, p. 18S207.

Bauer, R.T., 1996. A test of hypotheses on male mating systems and female molting in decapod shrimp, using *Sicyonia dorsalis* (Decapoda: Penaeoidea). J. Crust. Biol. 16, 429–436.

Bauer, R.T., 2002. Reproductive ecology of a protandric simultaneous hermaphrodite, the shrimp *Lysmata Wurdermanni* (Decapoda: Caridea: Hippolytidae). J. Crust. Biol. 22, 742–749.

Bauer, R.T., 2004. Remarkable Shrimp: Adaptations and Natural History of Carideans. University of Oklahoma Press, Norman.

Bauer, R.T., 2007. Hermaphroditism in Caridean Shrimps: Mating Systems, Sociobiology, and Evolution, with Special Reference to *Lysmata*. Evolutionary Ecology of Social and Sexual Systems: Crustaceans as Model Organism. Oxford Univ. Press, pp. 232–248.

Beach, D., Talbot, P., 1987. Ultrastructural comparison of sperm from the crayfishes *Cherax tenuimanus* and *Cherax albidus*. J. Crust. Biol. 205–218.

Beams, H.W., Kessel, R., 1962. Intracisternal granules of the endoplasmic reticulum in the crayfish oöcyte. J. Cell Biol. 13 (1), 158–162.

Beard, T.W., Wickins, J.F., 1980. Breeding of *Penaeus monodon* Fabricius in laboratory recirculating systems. Aquaculture 20, 79–89.

Becker, C., Brandis, D., Storch, V., 2011. Morphology of the female reproductive system of European Pea Crabs (Crustacea, Decapoda, Brachyura, Pinnotheridae). J. Morphol. 272, 12–26.

Bedford, J.M., 1970. Sperm capacitation and fertilization in Mammals. Biol. Reprod. Suppl. 2, 128–158.

Beer, C.G., 1959. Notes on the behavior of two estuarine crab species. Trans. Roy. Soc. N.Z. 86, 197–203.

Beermann, W., 1954. Weib1iche Heterogametie bei Copepoden. Chromosoma 6, 381–396.

Belk, D., 1978. A gynandromorphy of the fairy shrimp *Branchinecta lindalhi* Packard, 1883 (Anostraca). Crustaceana 35, 319–321.

Belk, D., 1991. Anostracan mating behavior: a case of scramble-competition polygyny. In: Bauer, R.T., Martin, J.W. (Eds.), Crustacean Sexual Biology. Columbia University Press, New York, pp. 111–125.

Benesh, D.P., Tellervo Valtonen, E., Jormalainen, V., 2007. Reduced survival associated with precopulatory mate guarding in male *Asellus aquaticus* (Isopoda). Ann. Zool. Fennici 44, 425–434.

Benhalima, K., Moriyasu, M., 2000. Structure and function of the posterior vas deferens of the snow crab, *Chionoecetes opilio* (Brachyura, Majidae). Inv. Reprod. Dev. 37 (1), 11–23.

Benhalima, K., Moriyasu, M., 2001. Prevalence of bacteria in the spermathecae of female snow crab, *Chionoecetes opilio* (Brachyura: Majidae). Hydrobiologia 449, 261–266.

Benijts, F., Vanvoorden, E., Sorgeloos, P., 1976. Changes in the biochemical composition of the early larval stages of the brine shrimp, *Artemia salina* L. In: Proceedings of the 10th European Symposium on Marine Biology, vol. 1, pp. 1–9.

Beninger, P.G., Elner, R.W., Foyle, T.P., Odense, P.H., 1988. Functional anatomy of the male reproductive system and the female spermatheca in the snow crab *Chionoecetes opilio* (O. Fabricius) (Decapoda: Majidae) and a hypothesis for fertilization. J. Crust. Biol. 322–332.

Beninger, P.G., Elner, R.W., Poussart, Y., 1991. Gonopods of the majid crab *Chionoecetes opilio* (O. Fabricius). J. Crust. Biol. 11, 217–228.

Beninger, P.G., Lanteigne, C., Elner, R.W., 1993. Reproductive processes revealed by spermatophore dehiscence experiments and by histology, ultrastructure, and histochemistry of the female reproductive system in the snow crab *Chionoecetes opilio* (O. Fabricius). J. Crust. Biol. 13, 1–16.

Beninger, P.G., Ferguson, A., Lanteigne, C., 1995. The gonopod tegumental glands of snow crab *Chionoectes opilio* (Fabricius, 1788) are accessory reproductive glands. J. Shellfish Res. 14, 365–370.

Benvenuto, C., Knott, B., Weeks, S.C., 2009. Mate guarding approach in clam shrimp: a field approach. Behav. Ecol. 20, 1125–1132.

Berglund, A., 1986. Sex change by a polychaete: effects of social and reproductive costs. Ecology 67, 837–845.

Bergreen, P.W., Meenakshi, V.R., Scheer, B.T., 1961. The oxidation of glucose by crustaceans. Comp. Biochem. Physiol. 2 (3), 218–220.

Bergstrom, B.I., 1997. Do protandric pandalid shrimp have environmental sex determination. Mar. Biol. 128, 397–407.

Bergström, B.I., 2000. The biology of *Pandalus*. Adv. Mar. Biol. 38, 55–245.

Berkeley, A., 1929. Sex reversal in *Pandalus danae*. Am. Nat. 63, 571–573.

Berkeley, A.A., 1930. The post-embryonic development of the common pandalids of British Columbia. Contr. Can. Biol. N. 5 (6), 79–163.

Berman, H., Lasky, L., 1985. Engineering glycoproteins for use as pharmaceuticals. Trends Biotech. 3, 51–53.

Berreur-Bonnenfant, J., Inagaki, H., 1973. Développement post-embryonnaire de la gonade et de l'appendix masculina chez *Ligia italica* Fabr. Crustacé Isopode. Extrait du Bulletin de la Société Zoologique de France 4, 533–540.

Berrill, M., Arsenault, M., 1982. Mating behavior of the green shore crab *Carcinus maenas*. Bull. Mar. Sci. 32, 632–638.

Berry, P.F., Heydorn, A.E.F., 1970. A Comparison of the Spermatophoric Masses and Mechanisms of Fertilization in Southern African Spiny Lobsters (Palinuridae) (No. 25), vol. 25. Oceanographic Research Institute, Durban, pp. 1–18 (Investigative report).

Berry, P.F., 1970. Mating Behaviour, Oviposition and Fertilization in the Spiny Lobster: *Panulirus* Gilchristi, vol. 24. Oceanographic Research Institute, Durban, pp. 1–16 (Investigative report).

Berticat, O., Negre_Sadargues, G., Castillo, R., 2000. The metabolism of astaxanthin during the embryonic development of the crayfish *Astacus leptodactylus* eschscholtz (Crustacea, Astacidea). Comp. Biochem. Physiol. 127B, 309–318.

Bhavanishankar, S., Subramoniam, T., 1977. Cryopreservation of spermatozoa of the edible mud crabn *Scylla serrata* (Forskal). J. Exp. Zool. 277, 326–336.

Bhavanishankar, S., Subramoniam, T., 2008. Protocol development for the cryopreservation of the spermatozoa and spermatophores of the mud crab *Scylla serrata*. In: Cabrita, et al. (Ed.), Methods in Reproductive Aquaculture: Marine and Freshwater Species. Taylor and Francis Group LLC, Florida, USA, pp. 513–535.

Billinghurst, Z., Clare, A.S., Matsumura, K., Depledge, M.H., 2000. Induction of cypris major protein in barnacle larvae by exposure to 4-n-nonylphenol and 17β-oestradiol. Aquat. Toxicol. 47, 203–212.

Bilodeau, A.L., Felder, D.L., Neigel, J.E., 2005. Multiple paternity in the thalassinidean ghost shrimp, *Callichirus islagrande* (Crustacea: Decapoda: Callianassidae). Mar. Biol. 146, 381–385.

Binford, R., 1913. The germ cells and the process of fertilization in the crab, *Manippe mercenaria*. J. Mophol. 24, 147–200.

Bingru, Y., Hong, S., 1990. Appropriate contents of protein, fat and caloric-protein ratio in the diet for freshwater shrimp *(Macrobrachium nipponense)*. J. Fisheries China (China). 14 (4), 321–327.

Blades-Eckelbarger, P.I., 1991. Functional morphology of spermatophores and sperm transfer in calanoid copepods. In: Bauer, R.T., Martin, W.J. (Eds.), Crustacean Sexual Biology. Columbia University Press, New York, pp. 246–270.

Blanchet-Tournier, M.F., 1982. Qudgues aspects des interactions hormonales entre la mue et la viteelogenese chez le crustace Amphipode *Orchestia gammarellus* (Pallas). Reprod. Nutr. Dev. 22, 325–344.

Bliss, D.E., van Montifrans, J., van Montifrans, M., Boyer, J.R., 1978. Behaviour and growth of the land crab *Gecarcinus lateralis* (Freminville) in southern Florida. Am. Mus. Nat. Hist. Bull. 160, 111–152.

Bloch, F., 1935. Contribution a l'etude des gametes et de la fecondation chez les crustaces decapodes. Travaux de la Station Zoologique de Wimereux 12, 185–279.

Böcking, D., Dircksen, H., Keller, R., 2002. The crustacean neuropeptides of the CHH/MIH/GIH family. In: Structures and Biological Activities, the Crustacean Nervous System. Springer, Berlin Heidelberg, pp. 84–97.

Boddeke, R., Bosschieter, J.R., Goudswaard, P.C., 1991. In: Bauer, R.T., Martin, J.W. (Eds.), Sex Change, Mating, and Sperm Transfer in *Crangon crangon* (L.) Crustacean Sexual Biology. Columbia University Press, New York, pp. 164–182.

Bollache, L., Cézilly, F., 2004. Sexual selection on male body size and assortative pairing in *Gammarus pulex* (Crustacea: Amphipoda): field surveys and laboratory experiments. J. Zool. 264, 135–141.

Bomirski, A., Klek-Kawinska, 1976. Estimation of oogenesis in the sand shrimp *Crangon crangon*. Gen. Comp. Endocrinol. 30, 234–242.

Bomirsky, A., Arendarczyk, M., Kawinska, E., Kleinholz, L.H., 1981. Partial characterization of crustacean gonad inhibiting hormone. Int. J. Invertebr. Reprod. 3, 213–219.

Boolootian, R.A., Giese, A.C., Farmanfarmaian, A., Tucker, J., 1959. Reproductive cycles of five west coast crabs. Physiol. Zool. 32, 213–220.

Borowsky, B., 1986. Laboratory observations of the pattern of reproduction of *Elasmopus levis* (Crustacea: Amphipoda). Mar. Behav. Physiol. 12, 245–256.

Borowsky, B., 1991. Patterns of reproduction in some amphipod crustaceans and insights into the nature of their stimuli. In: Bauer, R.T., Martin, J.W. (Eds.), Crustacean Sexual Biology. Columbia University Press, New York, pp. 33–49.

Borst, D.W., Laufer, H., Landau, M., Chang, E.S., Hertz, W.A., Baker, F.C., Schooley, D.A., 1987. Methyl farnesoate (MF) and its role in crustacean reproduction and development. Insect Biochem. 17, 1123–1127.

Bouchard, S., Sainte-Marie, B., McNeil, J.N., 1996. Indirect evidence indicates female semiochemicals release male precopulatory behaviour in the snow crab, *Chionoecetes opilio* (Brachyura: Majidae). Chemoecology 7, 39–44.

Bouchon, D., Cordaux, R., Grève, P., 2008. Feminizing Wolbachia and the evolution of sex determination in isopods. Insect Symbiosis 3, 273–294.

Bovbjerg, R.V., 1960. Courtship behavior of the lined shore crab, *Pachygraspus crassipes* Randall. Pac. Sci. 14, 421–422.

Bowen, S.T., Hanson, J., 1962. A gynandromorphy of the brine shrimp *Artemia salina*. Genetics 47, 277–280.

Bowen, S.T., Sterling, G., 1978. Esterase and malate dehydrogenase isozyme polymorphisms in 15 *Artemia* populations. Comp. Biochem. Physiol. Part B 61, 593–595.

Bowen, S.T., Hanson, J., Dowling, P., Poon, M.C., 1966. The genetics of *Artemia salina*. VI. Summary of Mutations. Biol. Bull. 131, 230–250.

Bowen, J.A., Bazer, F.W., Burghardt, R.C., 1996. Spatial and temporal analyses of integrin and Muc-1 expression in porcine uterine epithelium and trophectoderm in vivo. Biol. Reprod. 55, 1098–1106.

Bowen, S.T., 1963. The genetics of *Artemia salina*. II. White eye, a sex-linked mutation. Biol. Bull. 124, 17–23.

Bowman, T.E., Abele, L.G., 1982. Classification of the recent Crustacea. Biol. Crust. 1, 1–27.

Bowman, T.E., 1973. *Cymodetta gamhosa*, a new sphaeromatid isopod (Crustacea) from Australia, with notes on its mating behaviour. Rec. Aust. Mus. 29, 235–242.

Boxshall, G.A., 1992. Copepoda. Microscop. Anat. Invertebr. 9, 347–384.

Braga, A.A., Laura, S., Santos, C.D., Fransozo, A., 2009. Morphological evidence for protandric simultaneous hermaphroditism in the caridean *Exhippolysmata oplophoroides*. J. Crust. Biol. 29, 34–41.

Braga, A., Nakayama, C.L., Poersch, L., Wasielesky Jr., W., 2013. Unistellate spermatozoa of decapods: comparative evaluation and evolution of the morphology. Zoomorphology 132 (3), 261–284.

Braga, A., de Castro, L.A.S., Poersch, L.H., Wasielesky, W., 2014. Spermatozoal capacitation of pink shrimp *Farfantepenaeus paulensis*. Aquaculture 430, 207–210.

Breitbart, H., 2002. Intracellular calcium regulation in sperm capacitation and acrosomal reaction. Mol. Cell. Endocrinol. 187, 139–144.

Breithaupt, T., Thiel, M., 2011. Chemical Communication in Crustaceans. Springer, New York, NY.

Brian, J.V., 2005. Inter-population variability in the reproductive morphology of the shore crab (*Carcinus maenas*): evidence of endocrine disruption in a marine crustacean? Mar. Poll. Bull. 50, 410–416.

Brockerhoff, A.M., McLay, C.L., 2005a. Mating behaviour, female receptivity and male-male competition in the intertidal crab *Hemigrapsus sexdentatus* (Brachyura: Grapsidae). Mar. Ecol. Progr. Ser. 290, 179–191.

Brockerhoff, A.M., McLay, C.L., 2005b. Comparative analysis of the mating strategies in grapsid crabs with special references to the intertidal crabs *Cyclograpsus lavauxi* and *Helice crassa* (Decapoda: Grapsidae) from New Zealand. J. Crust. Biol. 25, 507–520.

Brockerhoff, A.M., McLay, C.L., 2005c. Factors influencing the onset and duration of receptivity of female purple rock crabs, *Hemigrapsus sexdentatus* (H. Milne Edwards, 1837) (Brachyura: Grapsidae). J. Exp. Mar. Biol. Ecol. 314, 123–135.

Brody, S., 1968. Bioenergetics and Growth. Hafner Publishing Corporation, New York, p. 1023.

Brook, H.J., Rawlings, T.A., Davies, R.W., 1994. Protogynous sex change in the intertidal isopod *Gnorimosphaeroma oregonense* (Crustacea: Isopoda). Biol. Bull. 187, 99–111.

Brook, M. de L., 1981. Size as a factor influencing the ownership of copulation burrows by the ghost crab (*Ocypode ceraphtothalamus*). Z. Tierpsychol. 53, 63–78.

Browdy, C.L., Samocha, T.M., 1985b. Maturation and spawning of ablated and non-ablated *Penaeus semisulcattrs* de Haan, 1844. In: Proc. 16th Annu. Meet. World Maricult. Soc.vol. 16. , pp. 236–249.

Browdy, C.L., Samocha, T.M., 1985a. The effect of eyestalk ablation on spawning, molting and mating of *Penaeus semisulcatus* de Haan. Aquaculture 49 (1), 19–29.

Browdy, C.L., Fainzilber, M., Tom, M., Loya, Y., Lubzens, E., 1990. Vitellin synthesis in relation to oogenesis in in vitro-incubated ovaries of *Penaeus semisulcatus* (Crustacea, Decapoda, Penaeidae). J. Exp. Zool. 255 (2), 205–215.

Brown, G.G., Metz, C.B., 1967. Ultrastructural studies on the spermatozoa of two primitive crustaceans *Hutchinsoniella macracantha* and *Derocheilocaris typicus*. Z. Zellforschung 80, 78–92.

Brown Jr., A., Kleve, M.G., Clark Jr., W.H., 1976. Evidence for the presence of actin in Natantian sperm. Am. Zool. 16, 180.

Brown, A., McVey, J., Middleditch, B.S., Lawrence, A.L., 1979. Maturation of white shrimp (*Penaeus setiferus*) in captivity. In: Proceedings of the World Mariculture Society, vol. 10. Blackwell Publishing Ltd, pp. 435–444. no. 1–4.

Brown, G.G., 1966. Ultrastructural studies of sperm morphology and sperm-egg interaction in the decapod *Callinectes sapidus*. J. Ultrastruct. Res. 14, 425–440.

Brown, G.G., 1970. Some ultrastructural aspects of spermatogenesis and sperm morphology in the brine shrimp *Artemia salina* Leach (Crustacea, Branchiopoda). Proc. Iowa Acad. Sci. 76, 473–486.

Browne, W.E., Price, A.L., Gerberding, M., Patel, N.H., 2005. Stages of embryonic development in the amphipod crustacean, *Parhyale hawaiensis*. Genesis 42 (3), 124–149.

Brunet, P.C.J., 1980. The metabolism of the aromatic amino acids concerned in the cross-linking of insect cuticle. Insect Biochem. 10, 467–500.

Brusca, R.C., 1978. Studies on the cymothoid fish symbionts of the eastern Pacific (Isopoda, Cymothoidae). I. Biology of *Nerocila californica*. Crustaceana 34, 141–154.

Bublitz, R., Sainte-Marie, B., Newcomb-Hodgetts, C., Fletcher, N., Smith, M., Hardege, J., 2008. Interspecific activity of the sex pheromone of the European shore crab (*Carcinus maenas*). Behaviour 145, 1465–1478.

Buchanan, J.B., 1963. The biology of *Calocaris macandreae* (Crustacea: Thalassinidea). J. Mar. Biol. Assoc. U.K. 43, 729–748.

Bückle-Ramirez, L.F., 1965. Untershuchungen uber die Biologie von *Heterotanais oerstedi Krôyer* (Crustacea, Tanaidacea). Z. Morphol. Oekol. Tiere 55, 714–782.

Buckmann, D., 1989. The significance of ecdysone in comparative physiology. In: Koolman, J. (Ed.), Ecdysone–From Chemistry to Mode of Action. Thieme, Stuttgart, pp. 20–26.

Bull, J.J., 1983. Evolution of Sex Determining Mechanisms. The Benjamin/Cummings Publishing Company, Inc.

Bulmer, M.G., Bull, J.J., 1982. Models of polygenic sex determination and sex ratio control. Evolution 13–26.

Bulnheim, H.P., Vavra, J., 1968. Infection by the microsporidian *Octosporea effeminans* sp. n., and its sex determining influence in the amphipod *Gammarus duebeni*. J. Parasitol. 241–248.

Bulnheim, H.P., 1965. Untersuchungen u¨ber intersexualita¨t bei *Gammarus duebeni* (Crustacea, Amphipoda). Helgolander Wissenschaftliche Meeresuntersuchungen 12, 349–394.

Bulnheim, H.P., 1967. On the influence of the photoperiod on the sex realization in *Gammarus duebeni*. Helgolander Wiss. Meeresunt. 16, 69–83.

Bulnheim, H.P., 1969. Zur Analyse geschlechtsbestimmender Faktoren bei *Gammarus duebeni* (Crustacea, Amphipoda). Zool. Anz. Suppl. Dtsch. 32, 244–260.

Bulnheim, H.P., 1977. Sex transformation in *Gammarus duebeni* under the influence of hormonal and parasitic factors. Biol. Zentralbl. 96, 61–78.

Bulnheim, H.P., 1978a. Variability of the modes of sex determination in littoral Amphipods. In: B. Battaglia, Beardmore, J.A. (Eds.), Marine Organisms. Plenum Press, New York and London, pp. 529–548.

Bulnheim, H.P., 1978b. Interaction between genetic, external and parasitic factors in the crustacean amphipod *Gammarus deubeni*. Helgolander Wissenscaftliche Meerusuntersuchengen 31, 1–33.

Büning, J., 1994. The Insect Ovary: Ultrastructure, Previtellogenic Growth and Evolution. Springer Science & Business Media.

Burbanck, M.P., Burbanck, W.D., 1974. Sex reversal of female *Cyanthura polita* (Stimpson, 1855) *(Isopoda, Anthuridae)*. Crustaceana 26, 110–112.

Buřič, M., Hulák, M., Kouba, A., Petrusek, A., Kozák, P., 2011. A successful crayfish invader is capable of facultative parthenogenesis: a novel reproductive mode in decapod crustaceans. PLoS ONE 6, e20281.

Burkenroad, M.D., 1934. The Penaeidea of Louisiana, with a discussion of their world relationships. Bull. Am. Mus. Nat. Hist. 68, 61–143.

Burkenroad, M.D., 1936. The Aristaeinae, Solenocerinae, and pelagic Penaeinae of the Binghan Oceanographic Collection. Materials for a revision of the oceanic Penaeidae. Bull. Bingham Oceanogr. Collect. Yale Univ. 5, 1–151.

Burkenroad, M.D., 1947. Reproductive activities of decapod Crustacea. Am. Nat. 81, 392–398.

Burkenroad, M.D., 1981. The higher taxonomy and evolution of Decapoda (Crustacea). Trans. San. Diego. Soc. Nat. Hist. 19, 251–268.

Burns, B.G., Sangalang, G.B., Freeman, H.C., McMenemy, M., 1984. Isolation and identification of testosterone from the serum and testes of the American lobster (*Homarus americanus*). Gen. Comp. Endocrinol. 54 (3), 429–432.

Burtis, K.C., Baker, B.S., 1989. Drosophila doublesex gene controls somatic sexual differentiation by producing alternatively spliced mRNAs encoding related sex-specific polypeptides. Cell 56, 997–1010.

Bushmann, P.J., Atema, J., 2000. Chemically mediated mate location and evaluation in the lobster, *Homarus americanus*. J. Chem. Ecol. 26, 883–899.

Buss, L.W., Iverson, E.W., 1981. A new genus and species of Sphaeromatidae (Crustacea; Isopoda) with experiments and observations on its reproductive biology, interspecific interactions and color polymorphisms. Postilla 184, 1–3.

Butler, T.H., 1964. Growth, reproduction and distribution of Pandalid shrimps in British Columbia. J. Fish. Res. Board Canada 21, 1403–1452.

Butler, T.H., 1980. Shrimps of the pacific coast of Canada. Can. Bull. Fish. Aquat. Sci. 202, 280.

Bykova, L.P., Markevich, G.I., 1979. Interesting case of gynandromorphism in *Daphnia longispina* (Cladocera, Daphniidae). Zoologichesky Zhurnal. 58, 1407–1409.

Byrne, B.M., Gruber, M.A.B.G., Ab, G., 1989. The evolution of egg yolk proteins. Prog. Biophys. Mol. Biol. 53 (1), 33–69.

Caldwell, R.L., 1988. Interspecific interactions among selected intertidal stomatopods. In: Behavioral Adaptation to Intertidal Life. Springer, USA, pp. 371–385.

Caldwell, R.L., 1991. Variation in reproductive behavior in stomatopod. In: Bauer, R.T., Martin, J.W. (Eds.), Crustacean Sexual Biology. Columbia University Press, New York, pp. 67–90.

Campos-Ramos, R., Maeda-Martínez, A.M., Obregón-Barboza, H., Murugan, G., Guerrero-Tortolero, D.A., 2006. First report of gynandromorphy in parthenogenetic *Artemia* (Branchiopoda: Anostraca). J. Crust. Biol. 26, 107–112.

Campos-Ramos, R., Maeda-Martínez, A.M., Obregón-Barboza, H., Murugan, G., Guerrero-tortolero, D.A., 2005. First report of gynandromorphy in parthenogenetic *Artemia* (Branchiopoda: Anostraca). J.Crust. Biol. 26, 107–112.

Carayon, J., 1941. Morphologie et structure de lappareil genital femelle chez quelques *Pagures*. Bull. Soc. Zool. Fr. 66, 95–122.

Cardoso, A.M., Barros, C.M., Correia, A.J.F., Cardoso, J.M., Cortez, A., Carvalho, F., Baldaia, L., 1997. Identification of vertebrate type steroid hormones in the shrimp *Penaeus japonicus* by tandem mass spectrometry and sequential product ion scanning. J. Am. Soc. Mass Spectrom. 8 (4), 365–370.

Carlisle, D.B., 1959. On the sexual biology of *Pandalus borealis* (Crustacea Decapoda). The termination of the male phase. J. Mar. Biol. Assoc. U.K. 38, 481–491.

Carpenter, A., 1978. Protandry in the freshwater shrimp, *Paratya curvirostris* (Heller, 1862) (Decapoda: Atyidae), with a review of the phenomenon and its significance in the Decapoda. J. R. Soc. N.Z. 8, 343–358.

Carvalho, H.A., Pereira, M.C.G., 1981. Descrição dos estádios ovarianos de *Macrobrachium acanthurus* (Wiegmann, 1836) (Crustacea, Palaemonidae) durante o ciclo reprodutivo. Ciência e Cultura 33 (10), 1353–1359.

Caskey, J.L., Bauer, R.T., 2005. Behavioral tests for a possible contact sex pheromone in the caridean shrimp *Palaemonetes pugio*. J. Crust. Biol. 25, 571–576.

Caskey, J.L., Hasenstein, K.H., Bauer, R.T., 2009. Studies on contact sex pheromones of the caridean shrimp *Palaemonetes pugio*: I. Cuticular hydrocarbons associated with mate recognition. Invertebr. Reprod. Dev. 53, 93–103.

Cave, M.D., 1982. Morphological manifestation of ribosomal DNA amplification during insect oogenesis. In: King, R.C., Akai, H. (Eds.), Insect Ultrastructure, vol. 1. Plenum Press, New York; London, pp. 86–117.

Ceccaldi, H.J., 1967. Transport des pigments caroténoides dans l'hémolymphe de *Carcinus maenas* Linné. C. R. Séanc Soc. Biol. 161, 1105–1110.

Chace Jr., F.A., Moore, G.M., 1959. A bicolored gynandromorph of the lobster, *Homarus americanus*. Biol. Bull. 116, 226–231.

Chace, F.A., 1972. The shrimps of the Smithsonian–Bredin Caribbean expeditions with a summary of the west Indian shallow-water species (Crustacea: Decapoda: Natantia). Smith. Contr. Zool. 98, 1–179.

Chaigneau, J., 1972. Particularités ultrastructurales de l'organe de Bellonci d'*Anilocra frontalis* Milne Edwards, Crustacé Isopode Flabellifère Cymothoide. C. R. Acad. Sci. 274, 1194–1197.

Chaix, J.C., De Reggi, M., 1982. Ecdysteroid levels during ovarian development and embryogenesis in the spider crab *Acanthonyx lunulatus*. Gen. Comp. Endocrinol. 47, 7–14.

Chamberlain, G.W., Gervais, N.F., 1984. Comparison of unilateral eyestalk ablation with environmental control for ovarian maturation of *Penaeus stylirostris*. J. World Maric. Soc. 15, 29–30.

Chamberlain, G.W., Lawrence, A.L., 1981. Effect of light intensity and male and female eyestalk ablation on reproduction of *Penaeus stylirostris* and *P. vannamei*. J. World Maric. Soc. 12 (2), 357–372.

Chan, S.M., Mak, A.S., Choi, C.L., Ma, T.H., Hui, J.H., Tiu, S.H., 2005. Vitellogenesis in the red crab, *Charybdis feriatus*: contributions from small vitellogenin transcripts (CfVg) and farnesoic acid stimulation of *CfVg* expression. Ann. New York Acad. Sci. 1040 (1), 74–79.

Chang, E.S., Hertz, W.A., Prestwich, G., 1992. NOAA Technical Report. Reproductive Endocrinology of the Shrimp *Sicyonia Ingentis*: Steroids, Peptides and Terpenoid Hormones, vol. 106. NMFS, pp. 1–6.

Charlesworth, B., Charlesworth, D., 1978. A model for the evolution of dioecy and gynodioecy. Am. Nat. 112, 988.

Charniaux-Cotton, H., Cazes, M., 1979. Premiére masculinisation des femelles d'un Crustacé Decapodgeonochorique, *Palaemonetes varians* (Leach), par greffesde glandes androgens. C. R. Hebd. Seanc. Acad. Sd. Ser. D 288, 1707–1709.

Charniaux-Cotton, H., Payen, G., 1985. Sexual differentiation. In: Bliss, D.E., Mantel, L.H. (Eds.), The Biology of Crustacea, vol. 9. Academic Press, New York, pp. 217–299.

Charniaux-Cotton, H., Payen, G., 1988. Reproduction in malacostraca Crustacea. Endocrinology of selected invertebrate types. Invertebr. Endocrinol. 2, 279–303.

Charniaux-Cotton, H., Touir, A., 1973. Controle de la previtellogenese et de la vitellogenese chez la crevette hermaphrodite *Lysmata seticaudata* Risso. C. R. Hebd Seanc. Acad. Sci. 276, 2717–2720.

Charniaux-Cotton, H., Zerbib, C., Meusy, J.J., 1966. Monographie de la glande androgene des Crustaces Superieurs. Crustaceana (Leiden) 10, 113–136.

Charniaux-Cotton, H., 1952. Castration chirurgicale chez un Crustacea Amphipode: *Orchestia gammarella* Pallas et determinisme des caract6res sexuels secondaires. Premiers résultats. C. R. Acad. Sci. Paris 234, 2570–2572.

Charniaux-Cotton, H., 1955. Le determinisme hormonal des characteres sexuals d'*Orchestia gammarella* (Crustace Amphipode). Comptes rendu hebdomadaires des sOances de I'Acad~mie des Sci. 240, 1487–1489.

Charniaux-Cotton, H., 1956. Déterminisme hormonal de la différenciation sexuelle chez les Crustacés. Ann. Biol. 32, 371–399.

Charniaux-Cotton, H., 1957. Croissance, régénération et déterminisme endocrinien des caractères sexuels secondaires d'*Orchestia gammarella* (Pallas) crustacé amphipode, vol. 11, (Masson).

Charniaux-Cotton, H., 1959. Etude compare du developpement post embryonnaire de l'appreil genital et de la glande androgene chez *Orchestia mediterranea* (Crustaces: Amphipodes). Autodifferenciation ovarienne. Bull. Soc. Zool. Fr. 84, 105–115.

Charniaux-Cotton, H., 1960. Sexual determination. In: Waterman, T.H. (Ed.), The Physiology of Crustacea. Metabolism and Growth, vol. 1. Academic Press, London, pp. 411–447.

Charniaux-Cotton, H., 1962. Androgenic gland of crustaceans. Gen. Comp. Endocrinol. 1, 241–247.

Charniaux-Cotton, H., 1965. Hormonal control of sex differentiation in invertebrates. Organogenesis 701–740.

Charniaux-Cotton, H., 1975. Hermaphroditism and gynandromorphism in malacostracan Crustacea. In: Reinboth, R. (Ed.), Intersexuality in the Animal Kingdom. Springer-Verlag, Berlin/Heidelberg, pp. 91–105.

Charniaux-Cotton, H., 1980. Experimental studies of reproduction in Malacostraca crustaceans. In: Clark, W.H., Adams, T.S. (Eds.), Description of Vitellogenesis and of Its Endocrine Control in Adv. Inv. Reprod. Elsevier, North Holland, pp. 177–185.

Charniaux-Cotton, H., 1985. Vitellogenesis and its control in malacostracan Crustacea. Am. Zool. 25, 197–206.

Charnov, E.L., Bull, J.J., 1989. Non-Fisherian sex ratios with sex change and environmental sex determination. Nature 338, 148–150.

Charnov, E.L., Hannah, R.W., 2002. Shrimp adjust their sex ratio to fluctuating age. Evol. Ecol. Res. 4, 239–246.

Charnov, E.L., Skúladóttir, U., 2000. Dimensionless invariants for the optimal size (age) of sex change. Evol. Ecol. Res. 2, 1067–1071.

Charnov, E.L., Gotshall, D.W., Robinson, J.G., 1978. Sex ratio: an adaptive response to population fluctuations in pandalid shrimps. Science (NY) 200, 204–206.

Charnov, E.L., 1979a. Natural selection in Pandalid shrimp: a test of life history theory. Am. Nat. 134, 824–827.

Charnov, E.L., 1979b. Simultaneous hermaphroditism and sexual selection. Proc. Nat. Acad. Sci. U.S.A. 76, 2480–2484.

Charnov, E.L., 1981. Sex reversal in *Pandalus borealis*: effect of a shrimp fishery. Mar. Biol. Lett. 2, 53–57.

Charnov, E.L., 1982. The Theory of Sex Allocation. Princeton University Press, New Jersey, p. 355.

Charnov, E.L., 1993. Life History Invariants. Oxford University Press.

Cheesman, D.F., Lee, W.L., Zagalsky, P.F., 1967. Carotenoproteins in invertebrates. Biol. Rev. 42 (1), 131–160.

Chen, M.H., Yang, Y.W., Soong, K., 1998. Preliminary observations on change of sex by the coral-inhabiting snails *Coralliophila violacea* (Lamarck) (Gastropoda: Coralliophilidae). J. Exp. Mar. Biol. Ecol. 264, 207–212.

Chen, Y.N., Tseng, D.Y., Ho, P.Y., Kuo, C.M., 1999. Site of vitellogenin synthesis determined from a cDNA encoding a vitellogenin fragment in the freshwater giant prawn, *Macrobrachium rosenbergii*. Mol. Reprod. Dev. 54 (3), 215–222.

Chen, Y.N., Fanm, H.F., Hsiehm, S.L., Kuom, C.M., 2003. Physiological involvement of DA in ovarian development of the freshwater giant prawn, *Macrobrachium rosenbergii*. Aquaculture 228, 383–395.

Chen, T., Zhang, L.P., Wong, N.-K., Zhong, M., Ren, C.H., Hu, C.Q., 2014. Pacific white shrimp (*Litopenaeus vannamei*) vitellogenesis-inhibiting hormone (VIH) is predominantly expressed in the brain and negatively regulates hepatopancreatic vitellogenin (VTG) gene expression. Biol. Reprod. 47, 1–10.

Chen, P.S., 1984. The functional morphology and biochemistry of insect male accessory glands and their secretions. Ann. Rev. Entomol. 29, 233–255.

Cheung, T.S., 1968. Trans-molt retention of sperm in the female stone crab, *Menippe Mercenaria.* Crustaceana 15, 117–120.

Chevaillier, P., 1967. Miss en evidence et etude cytochimique d'une proteine basique extranucleaire dans les spermatozoides des Crustaces Decapodes. J. Cell Biol. 32, 547.

Chiba, S., Goshima, S., Mizushima, T., 2000. Factors affecting the occurrence of early maturing males in the protandrous pandalid shrimp *Pandalus latirostris*. Mar. Ecol. Prog. Ser. 203, 215–224.

Chiba, S., Goshima, S., Shinomiya, Y., 2003. Male-male competition selects for delayed sex change in the protandrous shrimp *Pandalus latirostris*. Mar. Biol. 142, 1153–1157.

Chiba, S., Aoki, L., Ogata, T., 2004. Response of the pandalid shrimp *Pandalus latirostris* to dissolved oxygen, salinity and turbidity. Fisheries Sci. 70, 1174–1176.

Chow, S., Ogasawara, Y., Taki, Y., 1982. Male reproductive system and fertilization of the palaemonid shrimp *Macrobrachium rosenbergii*. Bull. Jpn. Soc. Sci. Fish 48, 177–183.

Chow, S., Dougherty, M.M., Dougherty, W.J., Sandifer, P.A., 1991. Spermatophore formation in the white shrimps *Penaeus setiferus* and *P. vannamei*. J. Crust. Biol. 201–216.

Christofferson, J.P., 1978. Evidence for the controlled release of a crustacean sex pheromone. J. Chem. Ecol. 4, 633–639.

Christy, J.H., 1980. The Mating System of the Sand Fiddler Crab, *Uca pugilator* (Ph.D. thesis). Cornell University, Ithaca, New York, p. 322.

Christy, J.H., 1982. Burrow structure and use in the sand fiddler crab, *Uca pugilator*. Anim. Behav. 30, 487–494.

Christy, J.H., 1983. Female choice in the resource-defense mating system of the sand fiddler crab, *Uca pugilator*. Behav. Ecol. Sociobiol. 12, 169–180.

Christy, J.H., 1987. Competitive mating, mate choice and mating associations of brachyuran crabs. Bull. Mar. Sci. 41, 177–191.

Christy, J.H., 2007. Predation and the reproductive behavior in the fiddler crabs (genus *Uca*). In: Duffy, J.E., Thiel, M. (Eds.), Evolutionary Ecology of Social and Sexual Systems. Oxford University Press, Oxford, pp. 211–231.

Chung, A.C.K., Durica, D.S., Hopkins, P.M., 1998. Tissue-specific patterns and steady-state concentrations of ecdysteroid receptor and retinoid-X-receptor mRNA during the molt cycle of the fiddler crab, *Uca pugilator*. Gen. Comp. Endocrinol. 109, 375–389.

Chung, J.S., Katayama, H., Dircksen, H., 2012. New functions of Arthropod Bursicon: inducing deposition and thickening of new cuticle and hemocyte granulation in the Blue Crab, *Callinectes sapidus*. PLoS ONE 7 (9), e46299.

Cipriani, R., Collin, R., 2005. Life-history invariants with bounded variables cannot be distinguished from data generated by random processes using standard analyses. J. Evol. Biol. 18, 1613–1618.

Clapper, D.L., Brown, G.G., 1980. Sperm motility in the horseshoe crab, *Limulus polyphemus* L: I. Sperm behavior near eggs and motility initiation by egg extracts. Dev. Biol. 76 (2), 341–349.

Clark, L.S., Bowen, S.T., 1976. The genetics of *Artemia salina*. VII. Reproductive isolation. J. Hered. 67, 385–388.

Clark Jr., W.H., Griffin, F.J., 1988. The morphology and physiology of the acrosome reaction in the sperm of the decapod, *Sicyonia ingentis*. Dev. Growth Differ. 30 (5), 451–462.

Clark, W.H., Griffin, F.J., 1993. Acquisition and manipulation of penaeoidean gametes. In: McVey, J.P. (Ed.), Crc-hand Book of Mariculture, vol. 1. second ed. CRC Press, London, pp. 133–151.

Clark, W.H., Lynn, J.W., 1977. A Mg^{++}-dependent cortical reaction in the eggs of penaeid shrimp. J. Exp. Zool. 200, 177–183.

Clark, W.H., Persyn, H., Yudin, A.I., 1974. A microscopic analysis of the cortical specialization and their activation in the egg of the prawn, *Penaeus* sp. Am. Zool. 14, 1251.

Clark, W.H., Lynn, J.W., Yudin, A.I., Persyn, H.O., 1980. Morphology of the cortical reaction in the eggs of *Penaeus aztecus*. Biol. Bull. 158, 175–186.

Clark Jr., W.H., Kleve, M.G., Yudin, A.T., 1981. An acrosome reaction in Natantian sperm. J. Exp. Zool. 218–279.

Clark Jr., W.H., Yudin, A.I., Griffin, F.J., Shigekawa, K., 1984. The control of gamete activation and fertilization in the marine penaeidae, *Sicyonia ingentis*. In: Engels, W., et al. (Ed.), Advances in Invertebrate Reproduction. vol. 3. Elsevier, Amsterdam, pp. 459–472.

Clark Jr., W.H., Yudin, A.I., Lynn, J.W., Griffin, F.J., Pillai, M.C., 1990. Jelly layer formation in Penaeoidean shrimp eggs. Biol. Bull. 178, 295–299.

Clark Jr., W.H., Chen, T.I., Pillai, M.C., Uhlinger, K., Shoffner-McGee, J., Griffin, F.J., 1991. The biology of gamete activation and fertilization in *Sicyonia ingentis* (Penaeoidea); present knowledge and future directions. Bull. Inst. Zool. Acad. Sinica Monogr. 16, 553–571.

Clark, A., 1977. A preliminary investigation of the lipids of *Chorismus antarcticus* (Pfeffer) (Crustacea, Decapoda) at South Georgia. Adaptations within Antarctic Ecosystems. In: Llano, G.A. (Ed.), Proceedings of the 3rd SCAR.

Clark Jr., W.H., Yudin, A.H., Shigekawa, K.J., Griffin, F.J., 1986. Cellular events of fertilization in the Penaeoid *Sicyonia ingentis*; in Indian ocean-Biology of benthic marine organisms. In: Thompson, M.F., Sarojini, R., Nagabhushanam, R. (Eds.). Oxford and IBH, New Delhi, pp. 44–149.

Clarke, A., Skadsheim, A., Holmes, L.J., 1985. Lipid biochemistry and reproductive biology in two species of Gammaridae (Crustacea: Amphipoda). Mar. Biol. 88, 247–263.

Clédon, P., 1985. Cytological and experimental analysis of oocyte maturation and activation in the prawn *Palaemon serratus* (Crustaceae, Decapoda Natantia). C. R. Acad. Sci. Paris 301, 317–322.

Clutton-Brock, T.H., Parker, G.A., 1992. Potential reproductive rates and the operation of sexual selection. Quart. Rev. Biol. 437–456.

Coccia, E., De Lisa, E., Di Cristo, C., Paolucci, M., 2010. Effects of estradiol and progesterone on the reproduction of the freshwater crayfish *Cherax albidus*. Biol. Bull. 218 (1), 36–47.

Cohen, A.C., Morin, J.G., 1990. Patterns of reproduction in ostracodes: a review. J. Crust. Biol. 10, 184–211.

Collin, R., McLellan, M., Gruber, K., Bailey-Jourdain, C., 2005. Effects of conspecific associations on size at sex change in three species of calyptraeid gastropods. Mar. Ecol. Prog. Ser. 293, 89–97.

Conan, G.Y., Moriyasu, M., Wade, E., 1988. Assessment and Spatial Distribution Surveys of Snow Crab Stocks by Geostatistics. International Council for the Exploration of the Sea (ICES).

Conan, G.Y., 1978. Life History, Growth, Production and Biomass Modelling of *Emerita analoga, Nephrops norvegicus* and *Homarus Vulgaris* (Crustacea, Decapoda) (Ph.D. thesis). University of California, Sandiago.

Conides, A., Glamuzina, B., Jug-Dujakovic, J., Papaconstantinou, C., Kapiris, K., 2006. Age, growth, and mortality of the karamote shrimp, *Melicertus kerathurus* (Forskal, 1775), in the east Ionian Sea (western Greece). Crustaceana 79, 33–52.

Cordaux, R., Bouchon, D., Greve, P., 2011. The impact of endosymbionts on the evolution of host sex-determination mechanisms. Trends Genet. 27, 332–341.

Correa, C., Thiel, M., 2003. Mating systems in caridean shrimp (Decapoda: Caridea) and their evolutionary consequences for sexual dimorphism and reproductive biology. Rev. Chi. Hist. Nat. 76, 187–203.

Correa, C., Baeza, J.A., Dupré, E., Hinojosa, I.A., Thiel, M., 2000. Mating behavior and fertilization success of three ontogenetic stages of male rock shrimp *Rhynchocinetes typus* (Decapoda: Caridea). J. Crust. Biol. 20, 628–640.

Correa, C., Baeza, J.A., Hinojosa, I.A., Thiel, M., 2003. Male dominance hierarchy and mating tactics in the rock shrimp *Rhynchocinetes typus* (Decapoda: Caridea). J. Crust. Biol. 23, 33–45.

Costa, R.S., 1979. Bioecologia do caranguejo-uçá, *Ucides cordatus* (Linnaeus, 1763)-crustáceo, decápode-no nordeste brasileiro. Boletim Cearense de Agronomia 20, 1–74.

Costlow, J.D., Bookhout, C.G., 1970. Marine larvae-the neglected link in biological resources. Proc. Ser. Symp. Hydrobiol. 8, 228–239.

Cotelli, F., Ferraguti, F., Lanzavecchia, G., Lamiadonin, C., 1976. The spermatozoon of Peracarida. 1. The spermatozoon of terrestrial isopods. J. Ultrastruct. Res. 55, 378–390.

Cothran, R., 2003. Precopulatory mate guarding affects predation risk in two freshwater amphipod species. Anim. Behav. 68, 1133–1138.

Cothran, R.D., 2004. Precopulatory mate guarding affects predation risk in two freshwater amphipod species. Anim. Behav. 68, 1133–1138.

Couch, E.F., Hagino, N., Lee, J.W., 1987. Changes in 17ß- estradiol and progesterone immunoreactivity in tissues of the lobster, *Homarus americanus*, with developing and immature ovaries. Comp. Biochem. Physiol. 85A, 765–770.

Courrier, R., 1921. Sur l'existence d'une glande interstitielle dans le testicule des poissons. Compt. Rend. Soc. Biol. 85, 939.

Cowan, D.F., Atema, J., 1990. Molt staggering and serial monogamy in American lobsters *Homarus americanus*. Anim. Behav. 39, 1199–1206.

Cox, G.W., Dudley, G.H., 1968. Seasonal pattern of reproduction of the sand crab, *Emerita analoga*, in southern California. Ecology 49, 746–751.

Crane, J., 1941. On the growth and ecology of brachyuran crabs of the genus Ocypode. Zoologica 26, 297–310.

Crane, J., 1975. Fiddler Crabs of the World. Princeton Univ. Press, p. 736.

Crespi, B.J., 1994. Three conditions for the evolution of eusociality: are they sufficient. Ins. Soc. 41, 395–400.

Criel, G.R., Macrae, T.H., 2002. Reproductive Biology of *Artemia*. In *Artemia*: Basic and Applied Biology. Springer, Netherlands, pp. 39–128.

Criel, G., 1980. Morphology of the female genital apparatus of *Artemia*: a review. Brine Shrimp Artemia 1, 75–86.

Criel, G.R., 1991. Morphology of *Artemia*. Artemia Biol. 155–185.

Crisp, D.J., Davies, P.A., 1955. Observations in vivo on the breeding of *Elminius modestus* grown on glass slides. J. Mar. Biol. Assoc. U.K. 34, 357–380.

Crisp, D.J., 1983. *Chelonobia patula* (Ranzani), a pointer to the evolution of the complemental male. Mar. Biol. Lett. 4, 281–294.

Croisille, Y., Junera, H., Meusy, J.J., Charniaux-Cotton, H., 1974. The female-specific protein (vitellogenic protein) in Crustacea with particular reference to *Orchestic gammarella* (Amphipoda). Am. Zool. 14 (4), 1212–1228.

Cronin, L.E., 1947. Anatomy and histology of the male reproductive system of *Callinectes sapidus* Rathbun. J. Morphol. 81, 209–239.

d'Udekem d'Acoz, C., 2003. *Lysmata seticaudata* (Risso, 1816) and L. nilita Dohrn & Holthuis, 1950 are protandrous simultaneous hermaphrodites (Decapoda, Caridea, Hippolytidae). Crustaceana 75, 1149–1152.

Dall, W., Smith, D.M., Moore, L.E., 1995. Carotenoids in the tiger prawn *Penaeus esculentus* during ovarian maturation. Mar. Biol. 123 (3), 435–441.

Dan, J.C., 1952. Studies on the acrosome reaction. I. Reaction to egg-water and other stimuli. Biol. Bull. 107, 54–66.

Das, R., Krishna, G., Priyadarshi, H., Gireesh-Babu, P., Pavan-Kumar, A., Rajendran, K.V., Chaudhari, A., 2015. Captive maturation studies in *Penaeus monodon* by GIH silencing using constitutively expressed long hairpin RNA. Aquaculture 448, 512–520.

Davey, K.G., Sevala, V.L., Gordon, B.R.B., 1993. The action of juvenile hormone and antigonadotropin on the follicle cells of *Locusta migratoria*. Inv. Reprod. Dev. 24, 39–45.

De Kleijn, D.P.V., Herp, F.V., 1995. Molecular biology of neurohormone precursors in the eyestalk of Crustacea. Comp. Biochem. Physiol. B 112, 573–579.

de Kleijn, D.P., de Leeuw, E.P., van den Berg, M.C., Martens, G.J., van Herp, F., 1995. Cloning and expression of two mRNAs encoding structurally different crustacean hyperglycemic hormone precursors in the lobster *Homarus americanus*. Biochim. Biophys. Acta, Gene Struct. Expression 1260 (1), 62–66.

De Kleijn, D.P.V., Coenen, T., Laverdure, A.M., Tensen, C.P., Van Herp, F., 1992. Localisation of messenger RNAs encoding crustacean hyperglycemic hormone and gonad inhibiting hormone in the X-organ sinus gland complex of the lobster, *Homarus americanus*. Neuroscience 51, 121–128.

De La Vaulx, R., 1919. L'interscxualith chcz un crustace Cladoc&re: Daphne Atkinsoni Baird. Compt. Rend. Acad. Sci. Paris 169, 97–99.

De Rivera, C.E., 2005. Long searches for male-defended breeding burrows allow female fiddler crabs, *Uca crenulata*, to release larvae on time. Anim. Behav. 70, 289–297.

De Souza, P.L., Feitosa Silva, J.R., 2009. Morphology of the female reproductive system of the red-clawed mangrove tree crab (*Goniopsis cruentata* Latreille, 1803). Scientia Marina 73 (3), 527–539.

Deecaraman, M., Subramoniam, T., 1980a. Cyclical cement glands activity in a stomatopod crustacean *Squilla holoschista*. In: Proc. First All India Sym. Inv. Repr. Uni. Mad, pp. 68–76.

Deecaraman, M., Subramoniam, T., 1980b. Male reproductive tract and accessory glands of a stomatopod, *Squilla holoschista*. Int. J. Invertebr. Reprod. 2, 175–188.

Deecaraman, M., Subramoniam, T., 1983a. Mating and its effect on female reproductive physiology with special reference to the fate of male accessory sex gland secretion in the stomatopod *Squilla holoschista*. Mar. Biol. 77, 161–170.

Deecaraman, M., Subramoniam, T., 1983b. Synchronous development of the ovary and the female accessory sex glands of a crustacean, *Squilla holoschista*. Proc. Indian Acad. Sci. Anim. Sci. 92, 179–184.

Deecaraman, M., Subramoniam, T., 1983c. Histochemistry of the neurosecretory systems in a stomatopod crustacean *Squilla holoschista*. Proc. Indian Acad. Sci. (Anim. Sci.) 95, 387–398.

Dellinger, T., Davenport, J., Wirtz, P., 1997. Comparisons of social structure of Columbus crabs living on loggerhead sea turtles and inanimate flotsam. J. Mar. Biol. Assoc. U.K. 77, 185–194.

Demeusy, N., 1962. Role de la glande de mue dans l'evolution ovarienne du crabe *Carcinus maenas* Linne. Cah. Biol. Mar. 3, 37–56.

Dhadiall, T.S., Raikhel, A.S., 1990. Biosynthesis of mosquito vitellogeneis. J. Biol. Chem. 265, 9924–9933.

Dharani, G., Altaff, K., 2002. Facultative sex reversal in the freshwater plankton *Sinodiaptomus* (*Rhinediaptomus*) *indicus* Calanoida: Copepoda. Curr. Sci. 82, 7–10.

Dhas, M.A., Subramoniam, T., Varadarajan, S., Govindharajulu, P., 1980. Germinal zone activity and oocyte differentiation in the marine crab *Portunus pelagicus*. Proc. Indian Natl. Sci. Acad. B Biol. Sci. 46, 287–292.

Di Cosmo, A., Di Cristo, C., 1998. Neuropeptidergic control of the optic gland of *Octopus vulgaris*: FMRF-amide and GnRH immunoreactivity. J. Comp. Neurol. 398, 1–12.

Di Fiore, M.M., Rastogi, R.K., Ceciliani, F., Messi, E., Botte, V., Botte, L., Pinelli, C., D'Aniello, B., D'Aniello, A., 2000. Mammalian and chicken I forms of gonadotropin-releasing hormone in the gonads of a protochordate, *Ciona intestinalis*. Proc. Natl. Acad. Sci. U.S.A. 97, 2343–2348.

Díaz, E.R., Thiel, M., 2004. Chemical and visual communication during mate searching in rock shrimp. Biol. Bull. 206, 134–143.

Diaz, H., 1980. The mole crab *Emerita talpoida* (Say): a case of changing life history pattern. Ecol. Monogr. 437–456.

Didderen, K., Fransen, C.H., De Voogd, N.J., 2006. Observations on sponge-dwelling colonies of *Synalpheus* (Decapoda, Alpheidae) of Sulawesi, Indonesia. Crustaceana 79, 961–975.

Diesel, R., Schubart, C.D., Duffy, J., Thiel, M., 2007. The social breeding system of the Jamaican bromeliad crab *Metopaulias depressus*. In: Duffy, J.E., Thiel, M. (Eds.), Evolutionary Ecology of Social and Sexual Systems: Crustaceans as Model Organisms. Oxford press, pp. 365–386.

Diesel, R., 1986a. Population dynamics of the symbolic spider crab *Inachus phalangium* (Decapoda, Majidae). Mar. Biol. 91, 481–489.

Diesel, R., 1986b. Optimal mate searching strategy in the symbolic spider crab *Inachus phalangium* (Decapoda). Ethology 72, 311–328.

Diesel, R., 1989. Structure and function of the reproductive system of the symbiotic spider crab *Inachus phalangium*: observations on sperm transfer, sperm storage, and spawning. J. Crust. Biol. 9, 266–277.

Diesel, R., 1991. Sperm Competition and the Evolution of Mating Behavior in Brachyura, with Special Reference to Spider Crabs (Decapoda, Majidae). Crustacean Sexual Biology. Columbia University Press, New York, pp. 145–163.

Dietz, R., 1954. Multiple geschlechtschromosomen bei dem ostracoden *Notodromas monacha*. Chromosoma 6, 397–418.

Dietz, R., 1958. Multiple geschlechtschromosomen bei den cypriden ostracoden, ihre evolution und ihr teilungsverhalten. Chromosoma 9, 359–440.

Ding, X., Nagaraju, G.P.C., Novotney, D., Lovett, D.L., Borst, D.W., 2010. Yolk protein expression in the green crab, *Carcinus maenas*. Aquaculture 298 (3), 325–331.

Dohrn, P.F.R., 1950. Studi Sulla *Lysmata seticaudata* Risso (Hippolytidae). Publ. Stn. Zool. Napoli 22, 257–272.

Donahue, J.K., 1940. Occurrence of estrogens in the ovaries of certain marine invertebrates. Endocrinology 27, 149–152.

Donahue, J.K., 1948. Fluorimetric and biological determination of estrogens in the eggs of the American lobster (*Homarus americanus*). Exp. Biol. Med. 69 (1), 179–181.

Donahue, J.K., 1952. Studies on Ecdysis in the American Lobster (*Homarus americanus*). I. The Lobster Egg as a Source of Estrogenic Hormone. State of Maine Dept. of Sea and Shore Fisheries, Res. Bull. No. 8.

Donahue, J.K., 1955. Studies on Ecdysis in the American Lobster (*Homarus americanus*). IV. Estrogenic Hormone as a Possible Moult-inhibitor in the Egg-bearing Female. State of Maine Dept. of Sea and Shore Fisheries Res. Bull. No. 24.

Donahue, J.K., 1957. Chromatographic Identification of Lobster Estrogen. State Mar. Dept. Sea Shore Fish Res. Bull. 28.

Dougherty, W.J., Dougherty, M.M., Harris, S.G., 1986. Ultrastructural and histochemical observations on electroejaculated spermatophores of the palaemonid shrimp, *Macrobrachium rosenbergii*. Tissue Cell 18 (5), 709–724.

Dougherty, W.J., 1987. Oriented spermatozoa in the spermatophore of the palaemonid shrimp. *Macrobrachium rosenbergii*. Tissue Cell 19 (1), 145–152.

Dudenhausen, E.E., Talbot, P., 1982. An ultrastructural analysis of mature sperm from the crayfish *Pacifastacus leniusculus* Dana. Int. J. Invertebr. Reprod. 5, 149–159.

Dudenhausen, E.E., Talbot, P., 1983. An ultrastructural comparison of soft and hard spermatophores from the crayfish *Pacifastacus leniuseula* (Dana). Can. J. Zool. 61, 182–194.

Duffy, J.E., Macdonald, K.S., 1999. Colony structure of the social snapping shrimp *Synalpheus filidigitus* in Belize. J. Crust. Biol. 19, 283–292.

Duffy, J.E., Thiel, M., 2007. Evolutionary Ecology of Social and Sexual Systems. Oxford University Press.

Duffy, J.E., Morrison, C.L., Rios, R., 2000. Multiple origins of eusociality among sponge-dwelling shrimps (*Synalpheus*). Evolution 54, 503–516.

Duffy, J.E., Morrison, C.L., Macdonald, K.S., 2002. Colony defense and behavioral differentiation in the eusocial shrimp *Synalpheus regalis*. Behav. Ecol. Sociobiol. 51, 488–495.

Duffy, J.E., 1992. Host use patterns and demography in a guild of tropical sponge-dwelling shrimps. Mar. Ecol. Prog. Ser. 90, 127–138.

Duffy, J.E., 1996. *Synalpheus regalis*, new species, a sponge-dwelling shrimp from the Belize Barrier Reef, with comments on host specificity in *Synalpheus*. J. Crust. Biol. 16, 564–573.

Duffy, E.J., 1998. On the frequency of eusociality in snapping shrimps (Decapoda: Alpheidae), with description of a second eusocial species. Bull. Mar. Sci. 63, 387–400.

Dunham, E.J., Hurshman, A.M., 1991. Precopulatory mate guarding in aquatic Crustacea: *Gammarus lawrencianus* as a model system. In: Bauer, R.T., Martin, J.W. (Eds.), Crustacean Sexual Biology. Columbia University Press, New York, USA, pp. 50–66.

Dunham, T., Alexander, A., Hurshman, 1986. Precopulatory mate guarding in an amphipod, *Gammarus lawrencianus* Bousfield. Anim. Behav. 34, 1680–1686.

Dunham, P.J., 1978. Sex pheromones in Crustacea. Biol. Rev. 53, 555–583.

Dunn, A.M., Adams, J., Smith, J.E., 1993. Is intersexuality a cost of environmental sex determination in *Gammarus duebeni*. J. Zool. 231, 383–389.

Dupont, S., Ortega-Martı´nez, O., Thorndyke, M., 2010. Impact of near-future ocean acidification on echinoderms. Ecotoxicology 19, 449–462.

Dupré, E.M., Barros, C., 2011. In vitro fertilization of the rock shrimp, *Rhynchocinetes typus* (Decapoda, Caridea): a review. Biol. Res. 44, 125–133.

Durica, D.S., Wu, X., Anilkumar, G., Hopkins, P.M., Chung, A.C.-K., 2002. Characterization of crab EcR and RXR homologs and expression during limb regeneration and oocyte maturation. Mol. Cell. Endocrinol. 189, 59–76.

Dutrieu, J., 1960. Observations biochimiques et physiologiques sur le développement d' *Artemia salina* Leach. Centre national de la recherche scientifique. Arch. Zool. Exp. Gen. 82, 110–118.

Eberhard, W.G., 1996. Female Control: Sexual Selection by Cryptic Female Choice. Princeton University Press.

Edomi, P., Azzon, E., Mettulio, R., Pandolfelli, N., Ferrero, E.A., Giulianini, P.G., 2002. Gonad-inhibiting hormone of the Norway lobster (*Nephrops norvegicus*): cDNA cloning, expression, recombinant protein production and immunolocalisation. Gene 284, 93–102.

Edwards, E., 1966. Mating behavior in the European edible crab (*Cancer pagarus* L.). Crustaceana 10, 23–30.

Efford, I.E., 1967. Neoteny in sand crabs of the genus *Emerita* (Anomura, Hippidae). Crustaceana 13, 81–93.

Efford, I.E., 1970. Recruitment to sedentary marine populations as exemplified by the sand crab, *Emerita analoga* (Decapoda, Hippidae). Crustaceana 18, 293–308.

Eickstaedt, L.L., 1969. The Reproductive Biology of the Sand Crab *Emerita analoga* (Stimpson) Doctoral dissertation. Stanford University.

Elner, R., Gass, C.A., Campbell, A., 1985. Mating behavior of the Jonah crab, *Cancer borealis* Stimpsonn (Decapoda, Brachyura). Crustaceana 48, 34–39.

Elner, R., Elner, J.K., 1980. Observations on a simultaneous mating embrace between a male and two female rock crabs *Cancer irrotatus*. (Decapoda, Brachyura). Crustaceana 38, 96–98.

Emlen, S.T., Oring, L.W., 1977. Ecology, sexual selection, and the evolution of mating systems. Science 197, 215–223.

Emmerson, W.D., 1980. Induced maturation of prawn *Penaeus indicus*. Mar. Ecol. Prog. Ser. 2 (2), 121–131.

Emmerson, W.D., 1983. Maturation and growth of ablated and unablated *Penaeus monodon* Fabricius. Aquaculture 32 (3), 235–241.

Engelmann, F., 1970. The physiology of insect reproduction. Int. Ser. Monogr. Pure Appl. Biol. 44 (Pergamen Press).

Engelstädter, J., Hurst, G.D., 2009. The ecology and evolution of microbes that manipulate host reproduction. *Annual Review of* Ecol. Evol. Syst. 40, 127–149.

Engstrom, N.A., Lucenti, J.M., 1984. Time-lapse movies and the diurnal time budget and activity patterns of *Cataleptodius floridanus*, a tropical intertidal xanthid crab. J. Crust. Biol. 4, 266–276.

Erdman, R.B., Blake, N.J., 1987. Population dynamics of the sponge-dwelling alpheid *Synalpheus longicarpus*, with observations on *S. brooksi* and *S. pectiniger*, in shallow-water assemblages of the eastern Gulf of Mexico. J. Crust. Biol. 328–337.

Erickson, G.F., Hsueh, A.J., 1978. Stimulation of aromatase activity by follicle stimulating hormone in rat granulosa cells in vivo and in vitro. Endocrinology 102, 1275–1282.

Erkan, M., Tunalı, Y., Sancar-Bas, S., 2009. Male reproductive system morphology and spermatophore formation in *Astacus leptodactylus* (Eschscholtz, 1823) (Decapoda: Astacidae). J. Crust. Biol. 29 (1), 42–50.

Eurenius, L., 1973. An electron microscope study on the developing oocytes of the crab *Cancer pagurus* L. with special reference to yolk formation (Crustacea). Zoomorphology 75 (3), 243–254.

Extavour, C.G., Akam, M., 2003. Mechanisms of germ cell specification across the metazoans: epigenesis and preformation. Development 130, 5869–5884.

Extavour, C.G., 2005. The fate of isolated blastomeres and formation of germ cells in the amphipod crustacean *Parhyale hawaiensis*. Dev. Biol. 277, 387–402.

Ezhilarasi, S., Subramoniam, T., 1982. Spermathecal activity and ovarian development in *Scylla serrata* (Forskal) (Decapoda: Portunidae). Prog. Invertebr. Reprod. Aquacult. 77–78.

Ezhilarasi, S., Subramoniam, T., 1986. Serological studies on the egg maturation in the edible crab, *Scylla serrata*. J. Singap. Natl. Acad. Sci. 15, 21–25.

Ezquieta, B., Vallejo, C.G., 1985. The tripsin-like proteinase of *Artemia*: yolk localization and developmental activation. Comp. Biochem. Physiol. 82, 731–736.

Fahrenbach, W.H., 1962. The biology of a harpacticoid copepod. Cellule 62, 303–376.

Fairs, N.J., Evershed, R.P., Quinlan, P.T., Goad, L.J., 1989. Detection of unconjugated and conjugated steroids in the ovary, eggs, and haemolymph of the decapod crustacean *Nephrops norvegicus*. Gen. Comp. Endocrinol. 74 (2), 199–208.

Fairs, N.J., Quinlan, P.T., Goad, L.J., 1990. Changes in ovarian unconjugated and conjugated steroid titres during vitellogenesis in *Penaeus monodon*. Aquaculture 89, 83–99.

Farazmand, A., Inanloo, K., Agh, N., 2010. Expression of Dmrt family genes during gonadal differentiation in two species of *Artemia* (Branchiopoda, Anostraca) from Urmia Lake (Iran). Crustaceana 83, 1153–1165.

Farfante, I.P., 1978. Intersex anomalies in shrimp of the genus *Penaeopsis* (Crustacea: penaeidae). US Nat. Mar. Fish Ser. Fish Bull. 76, 687–691.

Farmer, A.S., 1972. A bilateral gynandromorph of *Nephrops norvegicus* (Decapoda: Nephropidae). Mar. Biol. 15, 344–349.

Farmer, A.S.D., 1974. Reproduction in *Nephrops norvegicus* (Decapoda; Nephropidae). J. Zool. Lond. 174, 161–183.

Fawcett, D.W., 1961. Cilia and flagella. In: Brachet, J., Mirsky, A.E. (Eds.), The Cell. Academic Press, New York, pp. 217–297.

Felgenhauer, B.E., Abele, L.G., 1991. Morphological Diversity of Decapod Spermatozoa. Crustacean Sexual Biology. Columbia University Press, New York, pp. 322–341.

Feng, Z.F., Zhang, Z.F., Shao, M.Y., Zhu, W., 2011. Developmental expression pattern of the Fc-vasa-like gene, gonadogenesis and development of germ cell in Chinese shrimp, *Fenneropenaeus chinensis*. Aquaculture 314, 202–209.

Férézou, J.P., Barbier, M., Berreur-Bonnenfant, J., 1978. Biosynthèse de la farnésylacétone- (E, E) par les glandes androgènes du crabe *Carcinus maenas*. Helvetica Chim. Acta 61, 669–674.

Ferguson, E., 1967. New ostracods from the playa lakes in eastern New Mexico and Western Texas. Trans. Am. Microsc. Soc. 86 (3), 244–250.

Fiedler, G.C., Rhyne, A., Segawa, R., Aotsuka, T., Schizas, N., 2010. The evolution of euhermaphroditism in caridean shrimps: a molecular perspective of sexual systems and systematics. BMC Evol. Biol. 10, 297.

Fiedler, G.C., 1998. Functional, simultaneous hermaphroditism in female-phase *Lysmata amboinensis* (Decapoda: Caridea: Hippolytidae). Pac. Sci. 52, 161–169.

Fielder, D.R., Eales, A.J., 1972. Observations on courtship, mating and sexual maturity in *Portunus pelagicus* (L, 1766) (Crustacea, Portunidae). J. Nat. Hist. 6, 273–277.

Fielder, D.R., 1965. The spiny lobster, *Jasus lalandei* (H. Milne-Edwards), in South Australia. III. Food, feeding, and locomotor activity. Mar. Fres. Res. 16 (3), 351–368.

Fingerman, M., Nagabhushanam, R., Sarojini, R., Reddy, P.S., 1994. Biogenic amines in crustaceans: identification, localization, and roles. J. Crust. Biol. 14, 413–437.

Fingerman, M., 1997. Roles of neurotransmitters in regulating reproductive hormone release and gonadal maturation in decapod crustaceans. Invertebr. Reprod. Dev. 31 (1–3), 47–54.

Fittschen, U.E.A., 2001. Identifizierung von Naturstoffprofilen aus der Nordseekrabbe Crangon crangon (Ph.D. thesis). University of Hamburg.

Flaherty, C.M., Dodson, S.I., 2005. Effects of pharmaceuticals on *Daphnia* survival, growth, and reproduction. Chemosphere 61, 200–207.

Fleminger, A., 1985. Dimorphism and possible sex change in copepods of the family Calanidae. Mar. Biol. 88, 273–294.

Ford, A.T., Fernandes, T.F., Read, P.A., Robinson, C.D., Davies, I.M., 2004. The costs of intersexuality: a crustacean perspective. Mar. Biol. 145, 951–957.

Ford, A.T., Rodgers-Gray, T.P., Davies, I.M., Dunn, A.M., Read, P.A., Robinson, C.D., Smith, J.E., Fernandes, T.F., 2005. Abnormal gonadal morphology in intersex, *Echinogammarus marinus* (Amphipoda): a possible cause of reduced fecundity? Mar. Biol. 147, 913–918.

Ford, A.T., Fernandes, T.F., Robinson, C.D., Davies, I.M., Read, P.A., 2006. Can industrial pollution cause intersexuality in the amphipod, *Echinogammarus marinus*. Mar. Pollut. Bull. 53, 100–106.

Foster, B.A., 1983. Complemental males in the barnacle *Bathylasma alearum* (Cirripedia: Pachylasmidae). In: Lowry, J.K. (Ed.), Papers from the Conference on the Biology and Evolution of Crustacea. Australian Museum Memoir. The Australian Museum, Sydney, New South Wales, pp. 133–139.

Fraenkel, G., Hsiao, C., 1965. Bursicon, a hormone which mediates tanning of the cuticle in the adult fly and other insects. J. Insect Physiol. 11, 513–556.

Franke, H.D., 1993. Mating system of the commensal marine isopod *Jaera hopeana* (Crustacea). Mar. Biol. 115, 65–73.

Fréchette, J.G., Corrivault, G.W., Couture, R., 1970. Hermaphrodisme protérandrique chez une crevette de la famille des crangonidés, *Argis dentata* Rathbun. Naturalistecan 97, 805–822.

Frentz, R., 1960. Contribution a l'étude biochimique du milieu intrieur de *Carcinus maenas* Linn. Bull. Soc. Sci. Nancy Nile Série 19, 1–176.

Frey, D.G., 1965. Differentiation of *Alona costata* Sars from two related species (Cladocera, Chydoridae). Crustaceana 8, 159–173.

Fricke, H., Fricke, S., 1977. Monogamy and sex change by aggressive dominance in coral reef fish. Nature (London) 266, 830–832.

Friedel, T., Gillott, C., 1977. Contribution of male-produced proteins to vitellogenesis in *Melanoplus sanguinipes*. J. Insect Physiol. 23, 145–151.

Fryer, G., 1960. The spermatophores of *Dolops ranarum* (Crustacea, Branchiura): their structure, formation, and transfer. Quar. J. Microscop. Sci. 3, 407–432.

Fukuhara, H., 1999. Sex change in Crustacea. Aquabiology 125, 487–494 (in Japanese with English abstract).

Fyffe, W.E., O'Connor, J.D., 1974. Characterization and quantification of a crustacean lipovitellin. Comp. Biochem. Physiol. Part B 47 (4), 851–867.

Fyhn, U.E., Costlow, J.D., 1977. Histology and histochemistry of the ovary and oogenesis in *Balanus amphitrite* L. and *B. eburneus* Gould (Cirripedia, Crustacea). Biol. Bull. 152 (3), 351–359.

Galil, B., Tom, M., 1990. Histology and external morphology of a gynandromorphy of *Quadrella coronate* data (Decapoda, Trapizillidae). J. Crust. Biol. 10, 112–113.

Ganion, L.R., Kessel, R.G., 1972. Intracellular synthesis, transport, and packaging of proteinaceous yolk in oocytes of *Orconectes immunis*. J. Cell Biol. 52 (2), 420–437.

García-Guerrero, M., Racotta, I.S., Villarreal, H., 2003. Variation in lipid, protein and carbohydrate content during the embryonic development of the crayfish *Cherax quadricarinatus* (Decapoda: Parastacidae). J. Crust. Biol. 23 (1), 1–6.

García-Velazco, H., Obregón-Barboza, H., Rodríguez-Jaramillo, C., Maeda-Martínez, A.M., 2009. Reproduction of the tadpole shrimp *Triops* (Notostraca) in Mexican waters. Curr. Sci. 96, 91–97.

Garza-Torres, R., Campos-Ramos, R., Maeda-Martínez, A.M., 2009. Organogenesis and subsequent development of the genital organs in female and male Pacific white shrimp *Penaeus (Litopenaeus) vannamei*. Aquaculture 296, 136–142.

Gavio, M.A., Orensanz, J.M., Armstrong, D., 2006. Evaluation of alternative life history hypotheses for the sand shrimp *Crangon franciscorum* (Decapoda: Caridea). J. Crust. Biol. 26, 295–307.

George, M.J., 1962. On the breeding of penaeids and the recruitment of their postlarvae into the backwaters of Cochin. Ind. J. Fisher. 9, 110–116.

George, M.J., 1963. Note on an abnormality in the penaeid prawn *Metapenaeus monoceros* Fabricius. J. Mar. Biol. Assoc. India 5, 145–146.

Gesteira, T.C., Halcrow, K., 1988. Influence of some external factors on the acrosome reaction in the spermatozoa of *Homarus americanus* Milne Edwards, 1837. J. Crust. Biol. 8, 317–321.

Gharagozlou-van-Ginneken, I.D., Pochon-Masson, J., 1979. Infrastructure, nature et formation des grains de secretion dans le canal deferent de *Tisbe holothuriae* (Copepode: harpacticoide). Arch. Biol. 90, 59–81.

Gherardi, F., Calloni, C., 1993. Protandrous hermaphroditism in the tropical shrimp *Athanus indicus* (Decapoda: Caridea), a symbiont of sea urchins. J. Crust. Biol. 13, 675–689.

Ghiselin, M.T., 1969. The evolution of hermaphroditism among animals. Q. Rev. Biol. 44, 189–208.

Ghosh, D., Ray, A.K., 1993. Subcellular action of estradiol-17β in a freshwater prawn, *Macrobrachium rosenbergii*. Gen. Comp. Endocrinol. 90 (3), 274–281.

Giese, A.C., Pearse, J.S., 1974. Introduction: general principles. In: Giese, A.C., Pearse, J.S. (Eds.), Reproduction of Marine Invertebrates, vol. 1. Academic Press, New York, pp. 1–49.

Gilchrist, B., Lee, W.L., 1967. Carotenoids and carotenoid metabolism in *Carcinus maenas* (Crustacea: Decapoda). J. Zool. 151, 171–180.

Gilchrist, B.M., Lee, W.L., 1972. Carotenoid pigments and their possible role in reproduction in the sand crab *Emerita analoga* (Stimpson 1857). Comp. Biochem. Physiol. 42B, 263–294.

Gilchrist, B.M., 1978. Scanning electron microscope studies of the egg shell in some Anostraca (Crustacea: Branchiopoda). Cell Tissue Res. 193, 337–351.

Ginsburger-Vogel, T., Charniaux-Cotton, H., 1982. Sex determination. Biol. Crustacea 2, 257–281.

Ginsburger-Vogel, T., 1972. Démonstration expérimentale de l'homogamétie femelle chez *Orchestia cavimana* (Heller) (Crustacés Amphipodes Talitridae). Acad. Sci. Paris C. R. Ser. D 275, 3606–3609.

Ginsburger-Vogel, T., 1975. Temperature-sensitive intersexuality and its determination in *Orchestia gammarella* Pallas. In: Reinboth, E. (Ed.), Intersexuality in the Animal Kingdom. Springer-Verlag, Berlin, West Germany, pp. 106–120.

Ginsburger-Vogel, T., 1991. Intersexuality in *Orchestia mediterranea Costa*, 1853, and *Orchestia aestuarensis Wildish*, 1987 (Amphipoda) – a consequence of hybridization or parasitic infestation. J. Crust. Biol. 11, 530–539.

Girish, B.P., Swetha, C.H., Reddy, P.S., 2015. Induction of ecdysteroidogenesis, methyl farnesoate synthesis and expression of ecdysteroid receptor and retinoid X receptor in the hepatopancreas and ovary of the giant mud crab, *Scylla serrata* by melatonin. Gen. Comp. Endocrinol. 217–218.

Glabe, C.G., Vacquier, V.D., 1978. Egg surface glycoprotein receptor for sea urchin sperm bindin. Proc. Natl. Acad. Sci. U.S.A. 75, 881–885.

Glabe, C.G., Grabel, L.B., Vacquier, V.D., Rosen, S.D., 1982. Carbohydrate specificity of sea urchin binding: a cell surface lectin mediated sperm egg adhesion. J. Cell Biol. 94, 123–128.

Glas, P.S., Green, J.D., Lynn, J.W., 1995. Oxidase activity associated with the elevation of the Penaeoid shrimp hatching envelope. Biol. Bull. 189, 13–21.

Gleeson, R.A., Adams, M.A., Smith III, A.B., 1984. Characterization of a sex pheromone in the blue crab, *Callinectes sapidus*: crustecdysone studies. J. Chem. Ecol. 10, 913–921.

Gleeson, R.A., Adams, M.A., Smith, A.B., 1987. Hormonal modulation of pheromone-mediated behavior in a crustacean. Biol. Bull. 172, 1–9.

Gleeson, R.A., 1980. Pheromone communication in the reproductive behavior of the blue crab, *Callinectes sapidus*. Mar. Behav. Physiol. 7, 119–134.

Gleeson, R.A., 1982. Morphological and behavioural identification of the sensory structures mediating pheromone reception in the blue crab, *Callinectes sapidus*. Biol. Bull. 163, 162–171.

Gleeson, R.A., 1991. Intrinsic factors mediating pheromone communication in the blue crab, *Callinectes sapidus*. In: Bauer, R.T., Martin, J.W. (Eds.), Crustacean Sexual Biology. Columbia University Press, New York, NY, pp. 17–32.

Goldberg, E., Hawtrey, C., 1967. The ontogeny of sperm specific lactate dehydrogenase in mice. J. Exp. Zool. 164 (3), 309–316.

Gomez, R., 1965. Acceleration of development of gonads by implantations of brain in the crab *Paratelphusa hydrodromous*. Naturwiss 52, 216.

Gomez, E.D., 1975. Sex determination in *Balanus (conopea) galeatus* (L.) (Cirripedia Thoracica). Crustaceana 28, 105–107.

González-Baró, M.R., Heras, H., Pollero, R.J., 2000. Enzyme activities involved in lipid metabolism during embryonic development of *Macrobrachium borellii*. J. Exp. Zool. 286 (3), 231–237.

González-Pisani, X., Baron, P., Lopez Greco, L.S., 2012. Functional anatomy of the female reproductive systems of two spider crabs (Decapoda, Majoidea). Invertebr. Biol. 131, 61–74.

Goodbody, I., 1965. Continuous breeding in populations of two tropical crustaceans, *Mysidium columbiae* (Zimmer) and *Emerita portoricensis* Schmidt. Ecology 195–197.

Goodwin, T.W., Horn, D.H.S., Karlson, P., Koolman, J., Nakanishi, K., Robbuns, W.E., Siddall, J.B., Takemoto, T., 1978. Ecdysteroids: a new generic term. Nature 272, 122.

Goodwin, T.W., 1984. Carotenoid—protein complexes. In: The Biochemistry of the Carotenoids. Springer, Netherlands, pp. 1–21.

Gosselin, T., Sainte-Marie, B., Bernatchez, L., 2003. Patterns of sexual cohabitation and female ejaculate storage in the American lobster (*Homarus americanus*). Behav. Ecol. Sociobiol. 55, 151–160.

Goswami, U., Goswami, S.C., 1974. Cytotaxonomical studies on some Calanoids copepods. Nucleus 17, 242–243.

Goudeau, M., Becker, J., 1982. Fertilization in a crab. II. Cytological aspects of the cortical reaction and fertilization envelope elaboration. Tissue Cell 14, 273–282.

Goudeau, H., Goudeau, M., 1985. Fertilization in crabs: IV. The fertilization potential consists of a sustained egg membrane hyperpolarization. Gamete Res. 11, 1–17.

Goudeau, M., Goudeau, H., 1986. The resumption of meiotic maturation of oocyte of the prawn *Palaemon serratus* regulated by increase in extra cellular Mg^{2+} during spawning. Dev. Biol. 118, 361–370.

Goudeau, H., Goudeau, M., 1989. Electrical responses to fertilization and spontaneous activation in Decapod crustacean eggs: characteristics and role. In: Nuccitelli, R., et al. (Ed.), Mechanism of Egg Activation. Plenum Press, New York.

Goudeau, M., Lachaise, F., 1980. Fine structure and secretion of the capsule enclosing the embryo in a crab (*Carcinus maenas*(L)). Tissue Cell 12, 287–336.

Goudeau, H., Kubisz, P., Goudeau, M., 1984. Mise en évidence du potentiel de fécondation chez les Crustacés Décapodes Brachyoures *Carcinus maenas* et Maia squinado. C. R. Acad. Sci. Série 3 Sciences de la vie 299 (6), 167–172.

Goudeau, M., Lachaise, F., Carpenter, G., Goxe, B., 1990. High titres of ecdysteroids are associated with the secretory process of embryonic envelopes in the European lobster. Tissue Cell 22, 269–281.

Goudeau, M., 1982. Fertilization in a crab (*Carcinus maenas*). 1. Early events in the ovary, and cytological aspects of the acrosome reaction and gamete contacts. Tissue Cell 14, 97–112.

Goudeau, M., 1984. Fertilization in a crab: III. Cytodifferentiation of vesicles enclosing ring-shaped elements involved in the cortical reaction. Gamete Res. 9, 409–424.

Graeve, M., Wehrtmann, I., 2003. Lipid and fatty acid composition of Antarctic shrimp eggs (Decapoda: Caridea). Polar Biol. 26 (1), 55–61.

Grafen, A., Ridley, M., 1983. A model of mate guarding. J. Theor. Biol. 102, 549–567.

Greenspan, B.N., 1980. Male size and reproductive success in the communal courtship system of the fiddler crab *Uca rapax*. Anim. Behav. 28, 387–392.

Greenwood, J.G., 1972. The male reproductive system and spermatophore formation in *Pagurus novae-zealandiae* (Dana) (Anomura: Paguridea). J. Nat. Hist. 6 (5), 561–574.

Greve, P., Sorokine, O., Berges, T., Lacombe, C., Dorsselaer, A.V., Martin, G., 1999. Isolation and amino acid sequence of a peptide with vitellogenesis- inhibiting activity from the terrestrial isopod *Armadillidium vulgare* (Crustacea). Gen. Comp. Endocrinol. 115, 406–411.

Griffin, F.J., Clark Jr., W.H., 1990. Induction of acrosomal filament formation in the sperm of *Sicyonia ingentis*. J. Exp. Zool. 254, 296.

Griffin, F.J., Clark Jr., W.H., Crowe, J.H., Crowe, L.M., 1987. Intracellular pH decreases during the in-vitro induction of the acrosome reaction in the sperm of *Sicyonia ingentis*. Biol. Bull. 173, 311.

Griffin, D.J.G., 1968. Social and maintenance behaviour in two Australian ocypodid crabs (Crustacea: Brachyura). J. Zool. 156, 291–305.

Grygier, M.J., 1981. *Goronolaureus muzikae* sp. nov. (Crustacea: Ascothoracida) parasitic on a Hawaiian gorgonian, with special reference to its protandric hermaphroditism. Nat. Hist. 15, 1019–1045.

Gu, P., Tobe, S., Chow, B., Chu, K., He, J., Chan, S., 2002. Characterization of an additional molt inhibiting hormone-like neuropeptide from the shrimp *Metapenaeus ensis*. Peptides 23, 1875–1883.

Guan, Z.B., Shui, Y., Zhou, X., Xu, Z.H., Zhao, C.Y., Song, C.M., Liao, X.R., 2013. Participation of calmodulin in ovarian maturation induced by eyestalk ablation in red swamp crayfish *Procambarus clarkii*. Aquacult. Res. 44, 1625–1631. http://dx.doi.org/10.1111/are.12090.

Guewuch, W.T., Croker, R.A., 1973. Microfauna of northern new England marine sand: I. The biology of *Mancocuma stellifera* Zimmer (Crustacea: Cumacea). Can. J. Zool. 51, 1011–1020.

Guiasu, R.C., Dunham, D.W., 1998. Inter-form agonistic contests in male crayfishes, *Cambarus robustus* (Decapoda, Cambaridae). Invertebr. Biol. 144–154.

Gunamalai, V., Subramoniam, T., 2002. Synchronisation of molting and oogenic cycles in a continuously breeding population of the sand crab *Emerita asiatica* on the Madras Coast, South India. J. Crust. Biol. 22, 398–410.

Gunamalai, V., Kirubagaran, R., Subramoniam, T., 2004. Hormonal coordination of molting and female reproduction by ecdysteroids in the mole crab *Emerita asiatica*. Gen. Comp. Endocrinol. 138, 128–138.

Gunamalai, V., Kirubagaran, R., Subramoniam, T., 2006. Vertebrate steroids and the control of female reproduction in two decapod crustaceans, *Emerita asiatica* and *Macrobrachium rosenbergii*. Curr. Sci. 90 (1), 119–123.

Gunamalai, V., 1993. Enzyme Studies in the Developing Egg of *Emerita Asiatica* (Milne Edwards) with Special Reference to Carbohydrates (M. Phil. Dissertation). University of Madras, India, p. 43.

Gunamalai, V., 2002. Studies on the Moulting and Female Reproductive Biology of the Intertidal Mole Crab *Emerita Asiatica* (Milne Edwards) (Ph.D. thesis). University of Madras, India.

Gusev, O., Zabotin, Y., 2007. Observation of intersexuality in land hermit crabs (Anomura: Coenobitidae). J. Mar. Biol. Assoc. U.K. 87, 533–536.

Gusmão, L.F.M., McKinnon, A.D., 2009. The effect of food type and quantity on egg production and nucleic acid content of *Acartia sinjiensis*. Aquaculture 296 (1), 71–80.

Hagiwara, S., Jaffe, L.A., 1979. Electrical properties of egg membranes. Annu. Rev. Biopys. Bioeng. 8, 385–416.

Haj, J.E.A., Tamone, S.L., Peake, M., Chang, E.S., 1997. An Ecdysteroid-responsive gene in a lobster - a potential crustacean member of the steroid hormone receptor superfamily. Gene 201 (1–2), 127–135.

Halder, D.D., 1978. Induced maturation and breeding of *Penaeus monodon* (Fabricius) under brackish water pond conditions by eyestalk ablation. Aquaculture 15, 171–174.

Haley, S.R., 1984. Spermatogenesis and spermatophore production in the Hawaiian red lobster *Enoplometopus occidentalis* (Randall) (Crustacea, Nephropidae). J. Morphol. 180 (3), 181–193.

Hallberg, E., Skog, M., 2011. Chemosensory sensilla in crustaceans. In: Breithaupt, T., Thiel, M. (Eds.), Chemical Communication in Crustaceans. Springer, New York, pp. 103–121.

Hamel, J.F., Ng, P.K., Mercier, A., 1999. Life cycle of the pea crab *Pinnotheres halingi* sp. nov., an obligate symbiont of the sea cucumber *Holothuria scabra* Jaeger. Ophelia 50, 149–175.

Hamilton, W.D., 1964. The genetical evolution of social behaviour. J. Theor. Biol. 7, 1–16.

Hamon, M., 1937. Les mécanismes produisant la déhiscence des spermatophores d'Eupagurus prideauxi Leach. C. R. Acad. Sci. Paris 204, 1504–1506.

Han, C.H., Okumura, T., Suzuki, Y., Aida, K., Hanyu, I., 1994. Immunocytochemical identification of the site of vitellogenin synthesis in the freshwater prawn *Macrobrachium nipponense*. Fisheries Sci. 60 (2), 149–154.

Han, K., Dai, Y., Zou, Z., Fu, M., Wang, Y., Zhang, Z., 2012. Molecular characterization and expression profiles of cdc2 and cyclin B during oogenesis and spermatogenesis in green mud crab (*Scylla paramamosain*). Comp. Biochem. Physiol. Part B 163 (3), 292–302.

Hanamura, Y., Ohtsuka, S., 2003. Occurrence of intersex individuals in the sergestid shrimp, *Acetes sibogae*, in Darwin harbour, northern territory, Australia. Crustaceana 76, 749–754.

Hanström, B., 1937. Die sinusdrüse und der hormonal bedingte fabwechesl der crustaceen. Kungl. Svenska Vetenskap. Handl. 3, 1–99.

Hardege, J.D., Terschak, J.A., 2011. Identification of crustacean sex pheromone. In: Breithaupt, T., Thiel, M. (Eds.), Chemical Communication in Crustaceans. Springer, New York, pp. 103–121.

Hardege, J.D., Jennings, A., Hayden, D., Muller, C.T., Pascoe, D., Bentley, M.G., Clare, A.S., 2002. Novel behavioural assay and partial purification of a female-derived sex pheromone in *Carcinus maenas*. Mar. Ecol. Prog. Ser. 244, 179–189.

Hartnoll, R.G., 1960. Swimming in spider crabs of the genus *Macropodia*. Nature 185, 181.

Hartnoll, R.G., 1968. Reproduction in the burrowing crab, *Corystes cassivelaunus* (Pennant, 1777) (Decapoda, Brachyura). Crustaceana 15, 165–170.

Hartnoll, R.G., 1969. Mating in the Brachyura. Crustaceana 16, 161–181.

Hartnoll, R.G., 1975. The Grapsidae and Ocypodidae (Decapoda: Brachyura) of Tanzania. J. Zool. 177 (3), 305–328.

Hartnoll, R.G., 1979. The phyletic implications of spermathecal structure in the Raninidae (Decapoda: Brachyura). J. Zool. 187, 75–83.

Hartnoll, R.G., 2000. Evolution of brachyuran mating behavior: relation to the female molting pattern. Crust Issue. 12, 519–525.

Hartnoll, R.G., 2006. Reproductive investment in Brachyura. Hydrobiologia 557, 31–40.

Hasegawa, Y., Katakura, Y., 1983. Masculinization of WW females in the isopod crustacean, *Armadillidium vulgare*. Annot. Zool. Jpn. 56, 163–166.

Hasegawa, Y., Haino-Fukushima, K., Katakura, Y., 1987. Isolation and properties of androgenic gland hormone from the terrestrial isopod, *Armadillidium vulgare*. Gen. Comp. Endocrinol. 67, 101–110.

Hayakawa, H., Andoh, T., Watanabe, T., 2006. Precursor structure of egg proteins in the coral *Galaxea fascicularis*. Biochem. Biophys. Res. Commun. 344 (1), 173–180.

Hayashi, K.I., Ogawa, Y., 1987. *Spongicola levigata* sp. nov., a new shrimp associated with a hexactinellid sponge from the East China Sea (Decapoda, Stenopodidae). Zool. Sci. 4, p367–373.

Hayden, D., Jennings, A., Mu¨ller, C., Pascoe, D., Bublitz, R., Webb, H., Breithaupt, T., Watkins, L., Hardege, J., 2007. Sex-specific mediation of foraging in the shore crab, *Carcinus maenas*. Horm. Behav. 52, 162–168.

Hazel, C.M., 1986. Steroidogenesis in the Female Crab, *Carcinus maenas* (Ph.D. thesis). University of Liverpool, p. 81121.

He, L., Jiang, H., Cao, D., Liu, L., Hu, S., Wang, Q., 2013. Comparative transcriptome analysis of the accessory sex gland and testis from the Chinese mitten crab (*Eriocheir sinensis*). PLoS ONE 8, e53915.

Heberer, G., 1932. Untersuchungen uber Bau und Funktion der Copepoden. I. Der mannliche Genitalapparat der Calanoiden Copepoden. Z. fiir Mikroscopische Anatornie 31, 25M24.

Heegaard, P., 1971. *Penaetts keriilhitnts* Fbrskal, a protandric hermaphrodite. Bulletin Institut National Scientifique et Technique d'Oceanographic et de Peche de Salammbo 2, 257–266.

Heergaard, P., 1967. On behavior, sex-ratio and growth of *Solenocera membranacea* (Risso) (Decapoda, Penaeidae). Crustaceana 13, 227–237.

Helfman, G.S., 1977. Copulatory behavior of the coconut or robber crab *Birgus latro* (L.) (Decapoda, Anomura, Paguridea, Coenobitidae) Crustaceana 33 (2), 198–202.

Henning, H.G., 1975. Kampf-, Fortpflanzung-und Haeuctungsverhalten- Wachstum and Geschlechts-reife von *Cardisoma guanhumi*. Forma et Functio 8, 463–510.

Heras, H., Gonzalez-Baro, M.R., Pollero, R.J., 2000. Lipid and fatty acid composition and energy partitioning during embryo development in the shrimp *Macrobrachium borelli*. Lipids 35, 645–651.

Herberholz, J., Schmitz, B., 2001. Signaling via water currents in behavioural interactions of snapping shrimp *(Alpheus heterochaelis)*. Biol. Bull. 201, 6–16.

Herborg, L.M., Bentley, M.G., Clare, A.S., Last, K.S., 2006. Mating behaviour and chemical communication in the invasive Chinese mitten crab *Eriocheir sinensis*. J. Exp. Mar. Biol. Ecol. 329, 1–10.

Hertzler, P.L., 2005. Timing of meiotic maturation and first mitosis in the Pacific white shrimp *Litopenaeus vannamei*. J. World Aqua. Soc. 36, 570–576.

Hess, G.S., Bauer, R.T., 2002. Spermatophore transfer in the hermit crab *Clibanarius vittatus* (Crustacea, Anomura, Diogenidae). J. Morphol. 253 (2), 166–175.

Hessler, R.R., Elofsson, R., Hessler, A.Y., 1995. Reproductive system of *Hutchinsoniella macracantha* (Cephalocarida). J. Crust. Biol. 15 (3), 493–522.

Heydorn, A.E.F., 1969. The rock lobster of the South African West Coast *Jasus lalandii* (H. Milne-Edwards): population studies, behaviour, reproduction, moulting, growth and migration. S. Afr. Div. Sea Fish. Invest. Rep. 71, 1–32.

Hicks, J.W., 1985. The breeding behavior and migrations of the terrestrial crab *Gecarcoidea natalis* (Decapoda: Brachyura). Aust. J. Zool. 33, 127–142.

Highsmith, R.C., 1983. Sex reversal and fighting behavior: coevolved phenomena in a tanaid crustacean. Ecology 64, 719–726.

Hines, A.H., Johnson, E.G., Darnell, M.Z., Rittschof, D., Miller, T.J., Bauer, L.J., Rodgers, P., Aguilar, R., 2010. Predicting effects of climate change on blue crabs in Chesapeake Bay. In: Kruse, G.H., Eckert, G.L., Foy, R.J., Lipcius, R.N., Sainte-Marie, B., Stram, D.L., Woodby, D. (Eds.), Biology and Management of Exploited Crab Populations under Climate Change. Alaska Sea Grant, University of Alaska Fairbanks. http://dx.doi.org/10.4027/bmecpcc.2010.22.

Hinojosa, I., 2001. Habilidad de apareamientos múltiples de los machos y elección críptica de la hembra en el camarón de roca *Rhynchocinetes typus* Milne-Edwards, 1837 (Crustacea: Decapoda: Rhynchocinetidae) (Unpublished thesis). Universidad Católica del Norte, Coquimbo.

Hinsch, G.W., Bennett, D.C., 1979. Vitellogenesis stimulated by thoracic ganglion implants into destalked immature spider crabs, *Libinia emarginata*. Tissue Cell 11, 345–351.

Hinsch, G.W., Clark, W.H., 1973. Comparative fine structure of Cnidaria spermatozoa. Biol. Reprod. 8 (1), 62–73.

Hinsch, G.W., Cone, M.V., 1969. Ultrastructural observations of vitellogenesis in the spider crab, *Libinia emarginata* L. J. Cell Biol. 40 (2), 336–342.

Hinsch, G.W., Walker, M.H., 1974. The vas deferens of the spider crab, *Libinia emarginata*. J. Morphol. 143 (1), 1–19.

Hinsch, G.W., 1970. Possible role of intranuclear membranes in nuclear-cytoplasmic exchange in spider crab oocytes. J. Cell Biol. 47 (2), 531–535.

Hinsch, G.W., 1972. Some factors controlling reproduction in the spider crab, *Libinia emarginata*. Biol. Bull. 143, 358–366.

Hinsch, G.W., 1973. Sperm structure of *Oxyrhyncha*. Can. J. Zool. 51, 421–426.

Hinsch, G.W., 1980. Effect of mandibular organ implants upon the spider crab ovary. Trans. Am. Microsc. Soc. 99, 317–322.

Hinsch, G.W., 1986. A comparison of sperm morphologies, transfer, and sperm mass storage between two species of crab, *Ovalipes ocellatus* and *Libinia emarginata*. Int. J. Invertebr. Reprod. Dev. 10, 79–87.

Hinsch, G.W., 1988. Morphology of the reproductive tract and seasonality of reproduction in the golden crab *Geryon fennari* from the eastern Gulf of Maxico. J. Crust. Biol. 8, 254–261.

Hinsch, G.W., 1993. Ultrastructure of spermatogonia, spermatocytes, and Sertoli cells in the testis of the crayfish, *Procambarus paeninsulanus*. Tissue Cell 25 (5), 737–742.

Hinton, C.W., 1955. The behavior of an unstable ring chromosome of *Drosophila melanogaster*. Genetics 40, 951.

Hinton, H.E., 1977. Function of the shell structures of pig louse and how egg maintains a low equilibrium temperature in direct sunlight. J. Insect Physiol. 23, 785–800.

Hiransuchalert, R., Thamniemdee, N., Khamnamtong, B., Yamano, K., Klinbunga, S., 2013. Expression profiles and localization of vitellogenin mRNA and protein during ovarian development of the giant tiger shrimp *Penaeus monodon*. Aquaculture 412, 193–201.

Hobbs Jr., H.H., Harvey, M.C., Hobbs III, H.H., 2007. A comparative study of functional morphology of the male reproductive systems in the Astacidea with emphasis on the freshwater crayfishes (Crustacea: Decapoda). Smithson. Contrib. Zool. 624, 1–69.

Hoffman, D.L., 1972. The development of the ovotestis and copulatory organs in a population of protandric shrimp. *Pandalus plalyceros* from Lopez Sound. Wash. Biol. Bull. 142, 251–270.

Holland, D.L., East, J., Gibson, K.H., Clayton, E., Oldfield, A., 1985. Identification of the hatching factor of the barnacle *Balanus balanoides* as the novel eicosanoid 10, 11, 12-trihydroxy-5, 8, 14, 17-cicosatctraenoic acid. Prostaglandins 29 (6), 1021–1029.

Holland, D.L., 1978. Lipid reserves and energy metabolism in the larvae of benthic marine invertebrates. Biochem. Biophys. Perspect. Mar. Biol. 4, 85–123.

Holldobler, B., Wilson, E.O., 1990. The Ants. Springer Verlag, Berlin.

Hollenbeck, V.C., Weeks, S.C., Gould, W.R., Zucker, N., 2002. Maintenance of androdioecy in the freshwater shrimp *Eulimnadia texana*: sexual encounter rates and out crossing success. Behav. Ecol. 13, 561–570.

Hopkins, C.C.E., 1978. The male genital system, and spermatophore production and function in *Euchaeta norvegica* Boeck (Copepoda: Calanoida). J. Exp. Mar. Biol. Ecol. 35 (3), 197–231.

Hort-Legrand, C., Berreur-Bonnenfant, J., Ginsurger-Vogel, T., 1974. Étude anatomique et histologique comparee de la différenciation des gonades chez les mâles et les femelles chez *Orchestia gammarella* Pallas (Crustacé Amphipode) pendant le période post-embryonnaire. Bull. Soc. Zool. Fr. 99, 521–524.

Hou, X.L., Mao, Q., He, L., Gong, Y.N., Qu, D., d Wang, Q., 2010. Accessory sex gland proteins affect spermatophore digestion rate and spermatozoa acrosin activity in *Eriocheir sinensis*. J. Crust. Biol. 30, 435–440.

Hough, A.R., Bannister, N.J., Naylor, E., 1992. Intersexuality in the mysid *Neomysis integer*. J. Zool. Lond. 226, 585–588.

Howard, R.W., Blomquist, G.J., 2005. Ecological, behavioural, and biochemical aspects of insect hydrocarbons. Annu. Rev. Entomol. 50, 371–393.

Huang, J., Zhang, Y., Li, M., Wang, S., Liu, W., Couble, P., Huang, Y., 2007. RNA interference-mediated silencing of the bursicon gene induces defects in wing expansion of silkworm. FEBS Lett. 581 (4), 697–701.

Huang, H., Ye, H., Li, S., Wang, G., 2008. Immunocytological evidence for the presence of vertebrate FSH- and LH-like substances in the brain and thoracic ganglion of the swimming crab, *Portunus trituberculatus*. Prog. Nat. Sci. 18 (11), 1453–1457.

Huber, M.E., 1985. Nonrandom mating with respect to mate size in the crab *Trapezia* (Brachyura, Xanthidae). Mar. Behav. Physiol. 12, 19–32.

Hufnagl, M., Temming, A., Danhardt, A., 2010. Hermaphroditism in brown shrimp: lessons from field data and modeling. Mar. Biol. 157, 2097–2108.

Huner, J.V., Lindqvist, O.V., 1984. Effects of temperature and diet on reproductively active male noble crawfish (*Astacus astacus*) subjected to bilateral eyestalk ablation. J. World Maric. Soc. 15 (1–4), 138–141.

Ignace, D.D., Dodson, S.I., Kashian, D.R., 2011. Identification of the critical timing of sex determination in *Daphnia magna* (Crustacea, Branchiopoda) for use in toxicological studies. Hydrobiologia 668, 117–123.

Ikuta, K., Makioka, T., 1997. Structure of the adult ovary and oogenesis in *Argulus japonicus* Thiele (Crustacea: Branchiura). J. Morphol. 231 (1), 29–39.

Ikuta, K., Makioka, T., 1999. Ovarian structure and oogenesis in the myodocopid ostracod *Vargula hilgendorfii*. J. Crust. Biol. 19 (4), 730–737.

Ikuta, K., Makioka, T., 2004. Structure of ovary and oogenesis in the halocyprid ostracod *Conchoecia imbricata*. J. Crust. Biol. 24, 72–80 (BioOne).

Ikuta, K., Makioka, T., Amikura, R., 1997. Eggshell ultrastructure in *Argulus japonicus* (Branchiura). J. Crust. Biol. 17, 45–51.

Ikuta, K., Maruo, F., Tsutsumi, T., Makioka, T., 2007. Structure of the ovary and "nurse cells" in a freshwater ostracod, *Cyprinotus uenoi* Brehm (Podocopida: Cypridoidea). Zool. Sci. 24 (9), 906–912.

Ishimaru, S., 1984. Sexual dimorphism and protogynous hermaphroditism in *Leptochelia neapolitana* (Tanaidacea, Paratanaidae). Zool. Sci. 1, 997.

Isler, O., 1971. In: Isler, O. (Ed.), Carotenoids. Birkhauser Verlag, Basel, p. 931.

Itaya, P.W., 1979. Electron microscopic investigation of the formation of spermatophores of *Armadillidium vulgare*. Cell Tissue Res. 196 (1), 95–102.

Ivanov, B.G., Sokolov, V.I., 1997. Abnormalities in development of secondary sexual characters in the deepwater prawn *Pandalus borealis* (Crustacea, Decapoda, Pandalidae). Zool. Zh. 76, 133–141.

Iwaki, T., 1975. Breeding and settlement of *Chthamalus challengeri* Hoek on the southern coast of Hokkaido. Bull. Facul. Fish. 26 (1), 1–10 Hokkaido University.

Iwasaki, Y., 1970. Consistency condition for propagators. Phys. Rev. D 2 (10), 2255.

Jackson, H.C., 1913. 'Eupagurus', L.M.B.C. Memoirs. , p. 21.

Jackson, H.G., 1928. Hermaphroditism in *Rhyscotus*, a terrestrial isopod. Quart. Microsc. Sci. 71, 527–539.

Jaffe, L.A., 1976. Fast block to polyspermy in sea urchin eggs is electrically mediated. Nature (London) 26, 68–71.

Jaffe, L.F., 1985. The role of calcium explosion. waves and pulses in activating eggs. In: Metz, C.B., Monroy, A. (Eds.), Biology of fertlliiatlon. The Fertilization Response of the Egg, vol. 3. Academic press, Orlando, pp. 128–157.

Jahnke, M., Smith, J.E., Dubuffet, A., Dunn, A.M., 2013. Effects of feminizing microsporidia on the masculinizing function of the androgenic gland in *Gammarus duebeni*. J. Invertebr. Pathol. 112, 146–151.

Jamieson, B., 1989. Ultrastructural comparison of the spermatozoa of *Ranina ranina* (Oxystomata) and of other crabs exemplified by *Portunus pelagicus* (Brachygnatha) (Crustacea, Brachyura). Zoomorphology 109, 103–111.

Jamieson, B.G.M., 1991. Ultrastructure and phylogeny of crustacean spermatozoa. Mem. Queensland Mus. 31, 109–142.

Jamieson, B.G.M., 1993. Ultrastructure of the spermatozoon of *Potamonautes perlatus sidneyii* (Heterotremata, Brachyura, Crustacea). S. Afr. J. Zool. 28 (1), 40–45.

Jamieson, B.G.M., 1994. Phylogeny of the Brachyura with particular reference to the Podotremata: evidence from a review of spermatozoal ultrastructure (Crustacea, Decapoda). Philos. Trans. R. Soc. Lond. B 345, 373–393.

Jarvis, J.U.M., O'Riain, M.J., Bennett, N.C., Sherman, P.W., 1994. Mammalian eusociality: a family affair. Trends Ecol. Evol. 9, 47–51.

Jayasankar, V., Subramoniam, T., 1997. Proteolytic activity in the seminal plasma of the mud crab *Scylla serrata* (Forskal). Comp. Biochem. Physiol. 116B, 347–352.

Jayasankar, V., Tsutsui, N., Jasmani, S., Saido-Sakanaka, H., Yang, W.J., Okuno, A., Wilder, M.N., 2002. Dynamics of vitellogenin mRNA expression and changes in hemolymph vitellogenin levels during ovarian maturation in the giant freshwater prawn *Macrobrachium rosenbergii*. J. Exper. Zool. 293 (7), 675–682.

Jayasankar, V., 1996. Biochemical Studies on the Seminal Secretions of the Mud Crab *Scylla serrata* (Forskal) (Decapoda: Portunidae), with Special Reference to Protease and Glycosidase Activities (Ph.D. thesis). submitted to the University of Madras, India.

Jeng, S.S., Wan, W.C., Chang, C.F., 1978. Existence of an estrogen-like compound in the ovary of the shrimp *Parapenaeus fissurus*. Gen. Comp. Endocrinol. 36, 211–214.

Jeyalectumie, C., Subramoniam, T., 1987. Biochemical composition of seminal secretions with special reference to LDH activity in the reproductive tissues of the field crab, *Paratelphusa hydrodromous* (Herhst). Exp. Biol. 46, 231–236.

Jeyalectumie, C., Subramoniam, T., 1991. Biochemistry of seminal secretions of the crab *Scylla serrata* with reference to sperm metabolism and storage in the female. Mol. Reprod. Dev. 30 (1), 44–55.

Jivoff, P., Hines, A.H., 1998. Female behaviour, sexual competition and mate guarding in the blue crab, *Callinectes sapidus*. Anim. Behav. 55, 589–603.

Jivoff, P., Hines, A.H., Quackenbush, L.S., 2007. Reproductive biology and embryonic development. In: Kennedy, V.S., Cronin, L.E. (Eds.), The Blue Crab, *Callinectes sapidus*. Maryland, Maryland Sea Grant, College Park, pp. 255–298.

Jivoff, P., 1997. Sexual competition among male blue crab, *Callinectes sapidus*. Biol. Bull. 193, 368–380.

Johnson, M.E., Atema, J., 2005. The olfactory pathway for individual recognition in the American lobster *Homarus americanus*. J. Exp. Biol. 208, 2865–2872.

Johnson, P.T., Otto, S.V., 1981. Histology of a bilateral gynandromorphy of the blue crab, *Callinectes sapidus* Rathbun (Decapoda, Portunidae). Biol. Bull. 161, 236–245.

Johnson, W.S., Stevens, M., Watling, L., 2001. Reproduction and development of marine peracaridans. Adv. Mar. Biol. 39, 105–260.

Johnson, G., 1961. Contribution a/l'etude de la determination du sexe chez les Oniscoides: phenomenes d'hermaphrodisme et de monogenie. Bulletin Biologique de la France et de la Belgique 95, 177–271.

Johnson Jr., V.R., 1969. Behavior associated with pair formation in the banded shrimp *Stenopus hispidus* (Olivier). Pac. Sci. 23, 40–50.

Johnson, P.T., 1980. Histology of the Blue Crab, *Callinectes sapidus*: A Model for the Decapoda. Praeger Publishers.

Johnson, C., 1985. Mating behavior of the terrestrial isopod, *Venezillo evergladensis* (Oniscoidea, Armadillidae). Am. Midl. Nat. 114, 216–224.

Jones, D.R., Hartnoll, R.G., 1997. Mate selection and mating behaviour in spider crabs. Est. Coast. Shelf Sci. 44, 185–193.

Jones, D.A., 1972. Aspects of the ecology and behaviour of *Ocypode ceratophthalmus* (Pallas) and *O. kuhlii* de Haan (Crustacea: Ocypodidae). J. Exp. Mar. Biol. Ecol. 8, 31–43.

Jormalainen, V., Merilaita, S., 1995. Female resistance and duration of mate-guarding in three aquatic peracarids (Crustacea). Behav. Ecol. Sociobiol. 36, 43–48.

Jormalainen, V., Merilaita, S., Tuomi, J., 1994. Male Choice and male-male competition in *Idotea baitica* (Crustacea, Isopoda). Ethology 96, 46–57.

Jormalainen, V., Merilaita, S., Tuomi, J., 1995. Differential predation on sexes affects colour polymorphism of the isopod *Idotea baltica* (Pallas). Biol. J. Linn. Soc. 55, 45–68.

Jormalainen, V., Shuster, S.M., Wildey, H.C., 1999. Reproductive anatomy, precopulatory mate guarding, and paternity in the Socorro isopod, *Thermosphaeroma thermophilum*. Mar. Fres. Behav. Physiol. 32, 39–56.

Jormalainen, V., 1998. Precopulatory mate guarding in crustaceans: male competitive strategy and intersexual conflict. Quart. Rev. Biol. 275–304.

Jormalainen, V., 2007. Mating Strategies in Isopods: From Mate Monopolization to Conflicts. Evolutionary Ecology of Social and Sexual Systems: Crustaceans as Model Organisms. Oxford University Press Inc., New York, pp. 167–190.

Jormaleinen, V., Shuster, S.M., 1999. Female Reproductive cycle and sexual conflict over precopulatory mate-guarding in *Thermosphaeroma* (Crustacea: Isopoda). Ethology 233–346.

Jossart, Q., Wattier, R.A., Kastally, C., Aron, S., David, B., De Ridder, C., Rigaud, T., 2014. Genetic evidence confirms polygamous mating system in a crustacean parasite with multiple hosts. PLoS ONE e9 680.

Josso, N., di Clemente, N., Gouédard, L., 2001. Anti-Müllerian hormone and its receptors. Mol. Cell. Endocrinol. 179, 25–32.

Juchault, P., Legrand, J.J., 1964. Biologie experimentale-transformation de femelles genetiques en males physiologiques chez les oniscoides porcellio dilatatus et helleria brevicornis. Comptes Rendus Hebdomadaires Des Seances De L Academie Des Sciences 258, 2197.

Juchault, P., Legrand, J.J., 1979. Analyse génétique et physiologique de la détermination du sexe dans une population du Crustacé Isopode Oniscoïde *Armadillidium nasatum* Budde-Lund. Arch. Zool. Exp. Gén. 120, 25–43.

Juchault, P., Legrand, J.J., 1981. Contribution a l'etude qualitative et quantitative des facteurs controlant le sexe dans les populations du crustace isopode terrestre *Armadillidium vulgare* Latreille. II. Populations hebergeant le facteur feminisant F (Bacteroide intracytoplasmique). Archives de Zoologie Experimentale et Generale 122, 65–74.

Juchault, P., Rigaud, T., 1995. Evidence for female heterogamety in two terrestrial crustaceans and the problem of sex chromosome evolution in isopods. Heredity 75, 466–471.

Juchault, P., Martin, G., Legrand, J.J., 1980. Induction par la temperature d'une physiologie male-chez les neo-femelles et les intersexues du Crustace Oniscoide *Armadillidium vulgare* Latr. hebergeant un bacteroide a action feminisante. Int. J. Invertebr. Reprod. 2, 223–235.

Juchault, P., Louis, C., Martin, G., Noulin, G., 1991. Masculinization of female isopods (Crustacea) correlated with non-Mendelian inheritance of cytoplasmic viruses. Proc. Nat. Acad. Sci. U.S.A 88, 10460–10464.

Juchault, p., 1967. Contribution a letude de la differenciation sexuelle male chez les crustaces isopodes. Ann. Biol. 6, 191–212.

Jugan, P., Soyez, D., 1985. Demonstration in vitro de l'inhibition de l'endocytose ovocytaire par un extrait de sinus chez la crevette *Macrobrachium rosenbergii*. C. R. Acad. Sci. Paris 300, 705–709.

Jugan, P., Van Herp, F., 1989. Introductory study of an oocyte membrane protein that specifically binds vitellogenin in the crayfish *Orconectes limosus*. Invertebr. Reprod. Dev. 16 (1–3), 149–154.

Junera, H., Croisille, Y., 1980. Recherche du lieu de synthese de la vitellogenine chez le Crustace Amphipode *Orchestia gammarella* (Pallas). Mise en evidence d'une activation de is synthese proteique dans le tissue adipeux sous-epidermique en liaison avec la production de vitellogenine. C. R. Acad. Sci. Paris 290, 703–706.

Junéra, H., Zerbib, C., Martin, M., Meusy, J.J., 1977. Evidence for control of vitellogenin synthesis by an ovarian hormone in *Orchestia gammarella* (Pallas), Crustacea; Amphipoda. Gen. Comp. Endocrinol. 31 (4), 457–462.

Kaestner, A., 1970. Invertebrate Zoology. Crustacea, vol. III. Intersceicne, New York.

Kalavathy, Y., Mamatha, P., Reddy, P.S., 1999. Methyl farnesoate stimulates testicular growth in the freshwater crab *Oziotelphusa senex senex* Fabricius. Naturwissenschaften 86 (8), 394–395.

Kalyanaram, A.J., Chandran, M.R., 1982. On the occurrence of male accessory sex gland in the land crab, *Cardisoma carnifex* (Herbst). In: Subramonim, T., Varadharajan, S. (Eds.), Progress in Invertebrate Reproduction and Aquaculture. Indian Society of Invertebrate Reproduction, Madras, pp. 61–67.

Kamalaveni, S., 1947. Biology of *Cilbanarius olivaceous*. M.Sc dissertation. Madras University.

Kamiguchi, Y., 1972. Mating behavior in the freshwater prawn, *Palaemon paucidens*.: a study of the sex pheromone and its effect on males. J. Fac. Sci. Hokkaido Univ. Ser. VI Zool. 18 (3), 347–355.

Kamio, M., Derby, C.D., 2011. Approaches to a molecular identification of sex pheromones in blue crabs. In: Breithaupt, T., Thiel, M. (Eds.), Chemical Communication in Crustaceans. Springer, New York, pp. 393–412.

Kamio, M., Matsunaga, S., Fusetani, N., 2002. Copulation pheromone in the crab *Telmessus cheiragonus* (Brachyura: Decapoda). Mar. Ecol. Prog. Ser. 234, 183–190.

Kang, B.J., Nanri, T., Lee, J.M., Saito, H., Han, C.H., Hatakeyama, M., Saigusa, M., 2008. Vitellogenesis in both sexes of gonochoristic mud shrimp, *Upogebia major* (Crustacea): analyses of vitellogenin gene expression and vitellogenin processing. Comp. Biochem. Physiol. Part B 149 (4), 589–598.

Kannupandi, T., Krishnan, T., Soundarapandian, P., Shanmugam, A., 1999. Yolk utilization in an estuarine edible crab *Thalamita crenata* (Latreille). Ind. J. Fish. 46 (3), 289–294.

Karlson, P., Lüscher, M., 1959. 'Pheromones': a new term for a class of biologically active substances. Nature 183, 55–56.

Karplus, I., Harpaz, S., 1990. Preliminary observations on behavioral interactions and distribution patterns of freshwater prawns *Macrobrachium rosenbergii* under semi-natural conditions (Decapoda, Caridea). Crustaceana 59 (2), 193–203.

Karplus, I., Malecha, S.R., Sagi, A., 2000. The biology and management of size variation. In: New, M.B., Valenti, W.C. (Eds.), Freshwater Prawn Culture: The Farming of *Macrobrachium rosenbergii*. Blackwell Science, Malden, pp. 259–289.

Karplus, I., Sagi, A., Khalaila, I., Barki, A., 2003. The influence of androgenic gland implantation on the agonistic behavior of female crayfish (*Cherax quadricarinatus*) in interactions with males. Behaviour 140, 649–663.

Katakura, Y., Fujimaki, Y., Unno, K., 1975. Partial purification and characterization of androgenie gland hormone from the isopod crustacean, *Armadillidium vulgare*. Annot. Zool. Jpn. 48, 203–209 (Katakura, Y).

Katakura, Y., 1959. Masculinization through implanting testes into the female *Armadillidium vulgare*, an isopod crustacean. Proc. Jpn. Acad. 35, 95–98.

Katakura, Y., 1961. Hormonal Control of development of sexual characters in the isopod Crustacean, *Armadillidium vulgare*. Annot. Zool. Jpn. 34, 60–71.

Katakura, Y., 1982. Sex differentiation in the pill bug, hormonal control of sex differentiation in the terrestrial isopod, *Armadillidium vulgare*. Cunma Symp. Endocrinol. 4, 49–64.

Katakura, Y., 1984. Sex differentiation and androgenic gland hormone in the terrestrial isopod *Armadillidium vulgare*. In: Sutton, S.L., Holdich, D.M. (Eds.), The Biology of Terrestrial Isopods. Oxford Univ. Press, Oxford, pp. 127–142.

Kato, Y., Kobayashi, K., Oda, S., Colbourn, J.K., Tatarazako, N., Watanabe, H., Iguchi, T., 2008. Molecular cloning and sexually dimorphic expression of DM-domain genes in *Daphnia magna*. Genomics 91, 94–101.

Kato, Y., Kobayashi, K., Oda, S., Tatarazako, N., Watanabe, H., Iguchi, T., 2010. Sequence divergence and expression of a transformer gene in the branchiopod crustacean, *Daphnia magna*. Genomics 95, 160–165.

Kato, Y., Kobayashi, K., Watanabe, H., Iguchi, T., 2011. Environmental sex determination in the branchiopod crustacean *Daphnia magna*: deep conservation of a Doublesex gene in the sex-determining pathway. PLoS Genet. 7, e1001345.

Kelemec, J.A., Smith, I.R., 1984. Effects of low temperature storage and eyestalk enucleation of gravid eastern king prawns, *Penaeus plebejus*, on spawning, egg fertilisation and hatching. Aquaculture 40, 67–76.

Keller, R., 1992. Crustacean neuropeptides: structures, functions and comparative aspects. Experientia 48, 439–448.

Kelly, A., Dunn, A.M., Hatcher, M.J., 2002. Incomplete feminisation by the microsporidian sex ratio distorter, *Nosema granulosis*, and reduced transmission and feminisation efficiency at low temperatures. Int. J. Paras 32 (7), 825–831.

Kelly, A., Hatcher, M.J., Dunn, A.M., 2004. Intersexuality in the amphipod *Gammarus duebeni* results from incomplete feminisation by the vertically transmitted parasitic sex ratio distorter *Nosema granulosis*. Evol. Ecol. 18, 121–132.

Kensley, B., 1982. Deep-water Atlantic Anthuridea (Crustacea: Isopoda). Smithson. Contrib. Zool. 346, 1–51.

Kerr, M.S., 1969. The hemolymph proteins of the blue crab, *Callinectes sapidus*: II. A lipoprotein serologically identical to oocyte lipovitellin. Dev. Biol. 20 (1), 1–17.

Kessel, R.G., Beams, H.W., 1968. Intranucleolar membranes and nuclear-cytoplasmic exchange in young crayfish oocytes. J. Cell Biol. 39 (3), 735.

Kessel, R.G., 1968. Mechanisms of protein yolk synthesis and deposition in crustacean oocytes. Z. Zellforsch. Mikroskop. Anat. 89, 17.

Kessel, R.G., 1983. Intranuclear membranes (vesicles, lamellae, annulate lamellae) in oocytes of the ascidian, *Styela partita*. J. Submicrosc. Cytol. 15 (3), 773–785.

Kettlewell, J.R., Raymond, C.S., Zarkower, D., 2000. Temperature-dependent expression of turtle Dmrt 1 prior to sexual differentiation. Genesis 26, 174–178.

Khalaila, I., Katz, T., Abdu, U., Yehezkel, G., Sagi, A., 2001. Effects of implantation of hypertrophied androgenic glands on sexual characters and physiology of the reproductive system in the female red claw crayfish, *Cherax quadricarinatus*. Gen. Comp. Endocrinol. 121, 242–249.

Khalaila, I., Manor, R., Weil, S., Granot, Y., Keller, R., Sagi, A., 2002. The eyestalk–androgenic gland–testis endocrine axis in the crayfish *Cherax quadricarinatus*. Gen. Comp. Endocrinol. 127 (2), 147–156.

Khalaila, I., Peter-Katalinic, J., Tsang, C., Radcliffe, C.M., Aflalo, E.D., Harvey, D.J., Dwek, R.A., Rudd, P.M., Sagi, A., 2004. Structural characterization of the N-glycan moiety and site of glycosylation in vitellogenin from the decapod crustacean *Cherax quadricarinatus*. Glycobiology 14 (9), 767–774.

Khayat, M., Shenker, O., Funkenstein, B., Tom, M., Lubzens, E., Tietz, A., 1994. Fat transport in the Penaeid shrimp *Penaeus-semisulcatus* (dehaan). Israel J. Aquacult. Bamidgeh 46 (1), 22–32.

Khayat, M., Yang, W.-J., Aida, K., Nagasawa, H., Tietz, A., Funkenstein, B., Lubzens, E., 1998. Hyperglycaemic hormones inhibit protein and mRNA synthesis in in vitro-incubated ovarian fragments of the marine shrimp *Penaeus semisulcatus*. Gen. Comp. Endocrinol. 110, 307–318.

Khayat, M., Babin, P.J., Funkenstein, B., Sammar, M., Nagasawa, H., Tietz, A., Lubzens, E., 2001. Molecular characterization and high expression during oocyte development of a shrimp ovarian cortical rod protein homologous to insect intestinal peritrophins. Biol. Reprod. 64 (4), 1090–1099.

Kim, D.H., Jo, Q., Kim, B.K., Han, C.H., 2002. Eyestalk ablation-induced androgenic gland activity and gonad development in the freshwater prawns *Macrobrachium nipponense* (De Haan, 1849). Invertebr. Reprod. Dev. 42, 35–42.

Kim, Y.K., Kawazoe, I., Tsutsui, N., Jasmani, S., Wilder, M.N., Aida, K., 2004. Isolation and cDNA cloning of ovarian cortical rod protein in kuruma prawn *Marsupenaeus japonicus* (Crustacea: Decapoda: Penaeidae). Zool. Sci. 21 (11), 1109–1119.

Kim, Y.K., Tsutsui, N., Kawazoe, I., Okumura, T., Kaneko, T., Aida, K., 2005. Localization and developmental expression of mRNA for cortical rod protein in kuruma prawn *Marsupenaeus japonicus*. Zool. Sci. 22 (6), 675–680.

Kim, K., Kotov, A.A., Taylor, D.J., 2006. Hormonal induction of undescribed males resolves cryptic species of cladocerans. Proc. R. Soc. Lond. B 273, 141–147.

Kim, K.Y., Lee, S.Y., Cho, Y.S., Bang, I.C., Kim, K.H., Kim, D.S., Nam, Y.K., 2007. Molecular characterization and mRNA expression during metal exposure and thermal stress of copper/zinc-and manganese-superoxide dismutases in disk abalone, *Haliotis discus*. Fish Shellfish Immunol. 23 (5), 1043–1059.

King, J.E., 1948. A study of the reproductive organs of the common marine shrimp, *Penaeus setiferus* (Linnaeus). Biol. Bull. 94 (3), 244–262.

King, D.S., 1964. Fine structure of the androgenic gland of the crab, *Pachygrapsus crassipes*. Gen. Comp. Endocrinol. 4, 533–544.

Kirubagaran, R., Peter, D.M., Dharani, G., Vinithkumar, N.V., Sreeraj, G., Ravindran, M., 2005. Changes in vertebrate-type steroids and 5-hydroxytryptamine during ovarian recrudescence in the Indian spiny lobster, *Panulirus homarus*. N. Z. J. Mar. Fres. Res. 39, 527–537.

Kittredge, J.S., Terry, M., Takahashi, F.T., 1971. Sex pheromone activity of the molting hormone, crustecdysone, on male crabs (*Pachygrapsus crassipes*, *Cancer antennarius* and *C. anthonyi*). Fish. Bull. 69, 337–342.

Klann, M., Scholtz, G., 2014. Early embryonic development of the freshwater shrimp *Caridina multidentata* (Crustacea, Decapoda, Atyidae). Zoomorphology 133 (3), 295–306.

Kleiman, D.G., 1977. Monogamy in mammals. Quart. Rev. Biol. 39–69.

Klepal, W., Barnes, H., 1977. Studies on the reproduction of cirripedes. V. *Pollicipes cornucopia* Leach and *Balanus balanus* (L.); an electron microscope investigation of the structure of the oviducal sacs. J. Exp. Mar. Biol. Ecol. 27 (3), 261–287.

Kleve, M.G., Yudin, A.I., Clark Jr., W.H., 1980. Fine structure of the unistellate sperm of the shrimp, *Sicyonia ingentis* (Natantia). Tissue Cell 12, 29–45.

Knowlton, N., 1980. Sexual selection and dimorphism in two demes of a symbiotic, pair-bonding snapping shrimp. Evolution 34, 161–173.

Knudsen, J., 1964. Observations of the reproductive cycles and ecology of the common Brachyura and crablike Anomura of Puget Sound. Wash. Pacif. Sci. 18, 3–33.

Kokoza, V.A., Martin, D., Mienaltowski, M.J., Ahmed, A., Morton, C.M., Raikhel, A.S., 2001. Transcriptional regulation of the mosquito vitellogenin gene via a blood meal-triggered cascade. Gene 274 (1–2), 47–65.

Komai, T., 1920. Spermatogenesis of *Squilla oratoria* de Haan. J. Morphol. 34, 306–333.

Komatsu, M., Ando, S., 1992. A novel low-density lipoprotein with large amounts of phospholipid found in the egg yolk of sand crayfish *Ibacus ciliatus*: its function as vitellogenin degrading proteinase. Biochem. Biophys. Res. Commun. 186 (1), 498–502.

Kooda-Cisco, M.J., Talbot, P., 1982. A structural analysis of the freshly extruded spermatophore from the lobster, *Homarus americanus*. J. Morphol. 172 (2), 193–207.

Koolman, J., 1990. Ecdysteroids. Zool. Sci. 7, 563–580.

Korczynski, R.E., 1988. Additional records on intersexuality in the isopod *Mesidotea sibirica* (Birula, 1896). Can. J. Zool. 66, 1896–1898.

Kour, D.V.R., Subramoniam, T., 1992. Carotenoid metabolism during embryonic development of a marine crab, *Emerita asiatica* (Milne-Edwards). Invertebr. Reprod. Dev. 21 (2), 99–106.

Kozloff, E.N., 1987. Marine Invertebrates of the Pacific Northwest. University of Washington Press, Seattle.

Kramer, von P., 1967. Beobachtungen zur Biologie and zum Verhalten der Klippenkrabbe *Grapsus grapsus* L. (Brachyura: Graspidae) auf Galapagos und am ekuadorianischen Festland. Z. Tierpshychol. 24, 385–402.

Krång, A.S., Baden, S.P., 2004. The ability of the amphipod *Corophium volutator* (Pallas) to follow chemical signals from con-specifics. J. Exp. Mar. Biol. Ecol. 310, 195–206.

Krinsky, N.I., 1964. Canthaxanthin, the major carotenoid of the crustacean, *Artemia salina*. In: Abstracts, VI Int. Cong. Biochem., I.U.B, vol. 32. , p. 582.

Krishnakumar, R., 1985. Studies on Spermatheca of Some Decapod Crustaceans (Ph.D. thesis). University of Calicut, Thenhipalam, Kerala, India.

Kronenberger, K., Brandis, D., Türkay, M., Storch, V., 2004. Functional morphology of the reproductive system of *Galathea intermedia* (Decapoda: Anomura). J. Morphol. 262 (1), 500–516.

Kruangkum, T., Chotwiwatthanakun, C., Vanichviriyakit, R., Tinikul, Y., Anuracpreeda, P., Wanichanon, C., Hanna, P.J., Sobhon, P., 2013. Structure of the olfactory receptor organs, their GABAergic neural pathways, and modulation of mating behavior, in the giant freshwater prawn, *Macrobrachium rosenbergii*. Microsc. Res. Tech. 76 (6), 572–587.

Kruevaisayawan, H., Vanichviriyakit, R., Weerachatyanukul, W., Iamsaard, S., Withyachumnarnkul, B., Basak, A., Sobhon, P., 2008. Induction of the acrosome reaction in black tiger shrimp (*Penaeus monodon*) requires sperm trypsin-like enzyme activity. Biol. Reprod. 79 (1), 134–141.

Kubrakiewicz, J., Jaglarz, M.K., Iliffe, T.M., Bilinski, S.M., Koenemann, S., 2012. Ovary structure and early oogenesis in the remipede, *Godzilliognomus frondosus* (Crustacea, Remipedia): phylogenetic implications. Zoology 115 (4), 261–269.

Kühn, A., 1913. Die Sonderung der Keimesbezirke in der Entwicklung der Sommereier von *Polyphemus pediculus* De Geer. Zool. Jb Anat. 35, 243–340.

Kumari, S.S., Pandian, T.J., 1991. Interaction of ration and unilateral eyestalk ablation on energetics of female *Macrobrachium nobilii*. Asian Fisher. Sci. 4, 227–244.

Kung, S.Y., Chan, S.M., Hui, J.H.L., Tsang, W.S., Mak, A., He, J.G., 2004. Vitellogenesis in the sand shrimp, *Metapenaeus ensis*: the contribution from the hepatopancreas-specific vitellogenin gene (MeVg2). Biol. Reprod. 71 (3), 863–870.

Kunigelis, S.C., Saleuddin, S.M., 1986. Reproduction in the fresh water gastropod, *Helisoma*: involvement of prostaglandin in egg production. Int. J. Invertebr. Reprod. Dev. 10, 159–167.

Kuo, C.M., Chen, Y.N., Fan, H.F., Chuang, H.S., Hsieh, S.L., 2009. Enhancement of vitellogenin synthesis by serotonin in the giant fresh water prawn *Macrobrachium rosenbergii* (de Man). Zool. Stud. 48, 597–606.

Kurihara, H., Shimode, S., Shirayama, Y., 2004. Effects of raised CO_2 concentration on the egg production rate and early development of two marine copepods (*Acartia steueri* and *Acartia erythraea*). Mar. Pollut. Bull. 49, 721–727.

Kurihara, H., Matsui, M., Furukawa, H., Hayashi, M., Ishimatsu, A., 2008. Long-term effects of predicted future seawater CO 2 conditions on the survival and growth of the marine shrimp *Palaemon pacificus*. J. Exp. Mar. Biol. Ecol. 367 (1), 41–46.

Kuttner, 1909. Untcrsuchungen ubcr Fortpflanzungsverhaltnisse und Vcrcrbung bci Cladoceren. Int. Rev. Ges. Hydrobiol. 2, 633–667.

Lachaise, F., Hoffmann, J.A., 1982. Ecdysteroids and embryonic development in the shore crab, *Carcinus maenas*. Hoppe Seyler's Z. Physiol. Chem. 363, 1059–1067.

Lachaise, F., Lafont, R., 1984. Ecdysteroid metabolism in a crab, *Carcinus maenas*. Steroids 43, 243–260.

Lachaise, F., Gordeau, M., Heru, C., Kappler, C., Hoffman, J.A., 1981. Ecdysteroids and ovarian development in the shore crab *Carcinus maenas*. Hoppe Seyler's Z. Physiol. Chem. 362, 521–529.

Lack, D., 1968. Ecological Adaptations for Breeding in Birds. (Methuen London).

Lacombe, C., Gréve, P., Martin, G., 1999. Overview on the subgrouping of the crustacean hyperglycemic hormone family. Neuropeptides 33, 71–80.

Lagueux, M., Sall, C., Hoffmann, J.A., 1981. Ecdysteroids during embryogenesis in *Locusta migratoria*. Am. Zool. 21 (3), 715–726.

Lagueux, M., Hoffmann, J.A., Goltzene, F., Kappler, C., Tsoupras, G., Hetru, C., Luu, B., 1984. Ecdysteroids in ovaries and embryos of *Locusta migratoria*. In: Biosynthesis, Metabolism and Mode of Action of Invertebrate Hormones. Springer, Berlin Heidelberg, pp. 168–180.

Langreth, S.G., 1969. Spermiogenesis in Cancer crabs. J. Cell Biol. 43 (3), 575–603.

Lanot, R., Clédon, P., 1989. Ecdysteroids and meiotic reinitiation in *Palaemon serratus* (Crustacea, Decapoda, Natantia) and in *Locusta migratoria* (Insecta,Orthoptera) — A comparative study. Invertebr. Reprod. Dev. 16, 169–175.

Lattin, De Gustaf, 1951. Variations in the expression of sexuality in the normally protandric gastropod *Crepidula plana* Say. J. Exp. Zool. 108, 155–170.

Laubier, A., Chim, L., Payen, G.G., 1983. Morphogenèse sexuelle et régulation hormonale de l'activité génitale chez la crevette *Penaeus japonicus* en élevage. Actes de Colloques IFREMER 1, 195–206.

Laufer, H., Biggers, W.J., 2001. Unifying concepts learned from methyl farnesoate for invertebrate reproduction and post-embryonic development. Am. Zool. 41 (3), 442–457.

Laufer, H., Landau, M., 1991. Endocrine control of reproduction in shrimp and other Crustacea. In: Deloach, P.F., Dougherty, W.J., Davidson, M.A. (Eds.), Frontiers of Shrimp Research. Developments in Aquaculture and Fisheries Science, vol. 22. Elsevier, Amsterdam Oxford, New York, Tokyo, p. 294.

Laufer, H., Landau, M., Borst, D., Homola, E., 1986. The synthesis and regulation of methyl farnesoate, a new juvenile hormone for crustacean reproduction. In: Porchet, M., Andries, J.C., Dhainaut, A. (Eds.), Advances in Invertebrate Reproduction, vol. 4. Elsevier, Amsterdam, pp. 135–143.

Laufer, H., Demir, N., Pan, X., Stuart, J.D., Ahl, J.S., 2005. Methyl farnesoate controls adult male morphogenesis in the crayfish, *Procambarus clarkii*. J. Insect Physiol. 51 (4), 379–384.

Laulier, M., Demeusy, N., 1974. Histological study of an ovarian cycle by crab *Carcinus maenas* (Crustacea-Decapoda). Cahiers de Biologie Marine 15 (3), 343.

Laverdure, A.M., Soyez, D., 1988. Vitellogenin receptors from lobster oocyte membrane: solubilization and characterization by a solid phase binding assay. Int. J. Invertebr. Reprod. Dev. 13 (3), 251–266.

Lawrence, A.J., Soame, J.M., 2004. The effects of climate change on the reproduction of coastal invertebrates. Int. J. Avi. Sci. 146, 29–39.

Lawrence, A.L., Akamine, Y., Middleditch, B.S., Chamberlain, G., Hutchins, D., 1980. Maturation and reproduction of *Penaeus setiferus* in captivity. WMS 11 (1–4), 479–487 (Blackwell Publishing Ltd).

Le Roux, A., 1968. Description d'organes mandibulaires nouveaux chez les crustacés Decapodés. C. R. Hebd Séances Acad. Sci. Ser. D Sci. Nat. 266, 1414–1417.

LeBlanc, G.A., Mu, X., Rider, C.V., 2000. Embryotoxicity of the alkylphenol degradation product 4-nonylphenol to the crustacean *Daphnia magna*. Environ. Healt. Perspec. 108 (12), 1133.

LeBlanc, G.A., 2007. Crustacean endocrine toxicology: a review. Ecotoxicology 16, 61–81.

Lecher, P., 1963. Contribution á l'étude chromosomique de *Jaera marina* syie Bocquet (Isopode Asellote). C. R. Hebd Seances Acad. Sci. 257, 521–523.

Ledan, E., Polanski, Z., Terret, M.E., Maro, B., 2001. Meiotic maturation of the mouse oocyte requires an equilibrium between cyclin B synthesis and degradation. Dev. Biol. 232, 400–413.

Lee, C.L., Fielder, D.R., 1982. Maintenance and reproductive behaviour in the freshwater prawn *Macrobrachium australiense* Holthuis (Crustacea: Decapoda: Palaemonidae). Aust. J. Mar. Freshw. Res. 33, 629–646.

Lee, R.f., Puppione, D.L., 1998. Lipoproteins I and II from the hemolymph of the blue crab *Callinectes sapidus*: lipoprotein II associated with vitellogenesis. J. Exp. Zool. 248, 278–279.

Lee, F.Y., Shih, T.W., Chang, C.F., 1997. Isolation and characterization of the female-specific protein (vitellogenin) in mature female hemolymph of the freshwater prawn, *Macrobrachium rosenbergii*: comparison with ovarian vitellin. Gen. Comp. Endocrinol. 108 (3), 406–415.

Lee, S.O., Jeon, J.M., Oh, C.W., Kim, Y.M., Kang, C.K., Lee, D.S., Kim, H.W., 2011. Two juvenile hormone esterase-like carboxylesterase cDNAs from a Pandalus shrimp (*Pandalopsis japonica*): cloning, tissue expression, and effects of eyestalk ablation. Comp. Biochem. Physiol. B 159 (3), 148–156.

Lee, C.M., 1972. Structure and function of the spermatophore and its coupling device in the Centropagidae (Copepoda: Calanoida). Bull. Mar. Ecol. 8, 1–20.

Leelatanawit, R., Sittikankeaw, K., Yocawibun, P., Klinbunga, S., Roytrakul, S., Aoki, T., Menasveta, P., 2009. Identification, characterization and expression of sex-related genes in testes of the giant tiger shrimp *Penaeus monodon*. Comp. Biochem. Physiol. Part A 152, 66–76.

Legrand, J.J., Juchault, P., 1960. Structure and origin of the androgenic gland in *Helleria brevicornis* Ebner (Oniscoidea, Tylidae). Comptes Rendus des séances de la Société de Biologie et de Ses Filiales 154, 676.

Legrand, J.J., Juchault, P., 1963. Mise en évidence d'un hennaphrodisme protogynique fonctionnel chez l'Isopode Anthuride *Cyanthura carinata (Kroyer) et* étude du mecanisme de l'inversion sexuelle. C. R. Acad. Sci. 256, 2931–2933.

Legrand, J.J., Juchault, P., 1970. Controle de la sexualité chez les crustacés isopodes gonochoriques et hermaphrodites. Bull. Soc.Zool. Fr 95, 551–563.

Legrand, J.J., Juchault, P., 1972. Le contrôle humoral de la sexualité chez les Crustacés Isopodes gonochoriques. Hormones et differenciation sexuelle chez les invertebres. Gordon and Breach, Paris, France, pp. 179–218.

Legrand, J.J., Juchault, P., 1984. Nouvelles données sur le déterminisme génétique et épigénétique de la monogénie chez le crustacé isopode terrestre *Armadillidium vulgare*. Latr. Gén Sél Evol. 16, 57–84.

Legrand, J.J., Vandel, A., 1948. Le développement post-embryonnaire de la gonade chez les Isopodes terrestres normaux et intersexués. Evolution morphologique de la gonade. Bull. Biol. Fr. Belg 82, 79–94.

Legrand, J.J., Legrand-Hamelin, E., Juchault, P., 1987. Sex determination in Crustacea. Biol. Rev. 62, 439–470.

Legrand-Hamelin, E., Legrand, J.J., 1982. Genetic determinism of the female sex inversion accompanying the crossing of two subspecies of *Idotea balthica* (Pallas). II.–Results on the bilineata-lineata phenotype and its association with the albafusca phenotype. Reprod. Nutr. Dev. 22, 889.

Leopold, R.A., 1976. Role of male accessory glands in insect reproduction. Ann. Rev. Entomol. 21,199–221.

Leung-Trujillo, J.K., Lawrence, A.L., 1985. The effect of eyestalk ablation on spermatophore and sperm quality in *Penaeus vannamei*. J. World Maric. Soc. 16 (1–4), 258–266.

Li, K., Chen, L., Zhou, Z., Li, E., Zhao, X., Guo, H., 2006. The site of vitellogenin synthesis in Chinese mitten-handed crab *Eriocheir sinensis*. Comp. Biochem. Physiol. Part B 143 (4), 453–458.

Li, C.J., Liu, L., Chen, X.H., Zhang, T., Gan, F., Cheng, B.L., 2010. Identification of a vasa homologue gene in grass carp and its expression pattern in tissues and during embryogenesis. Comp. Biochem. Physiol. Part A 157, 159–166.

Li, S., Li, F., Sun, Z., Xiang, J., 2012. Two spliced variants of insulin-like androgenic gland hormone gene in the Chinese shrimp, *Fenneropenaeus chinensis*. Gen. Comp. Endocrinol. 177, 246–255.

Lindberg, W.J., 1980. Behaviour of the Oregon mud crab, *Hemigraspus oregonensis* (Dana) Brachyura, Grapsidae). Crustaceana 39, 263–281.

Linder, H.J., 1959. Studies on the fresh water fairy shrimp *Chirocephalopsis bundyi* (Forbes). I. Structure and histochemistry of the ovary and accessory reproductive tissues. J. Morphol. 104, 1–59.

Linder, J.H., 1960. Studies on the freshwater fairy shrimp *Chirocephalopsis bundyi* (Forbes).2. Histochemistry of the egg-shell formation. J. Morphol. 107, 259–282.

Lindsay, L.L., Clark, W.H., 1994. Signal transduction during shrimp oocyte activation by extracellular Mg2+: roles of inositol 1, 4, 5-trisphosphate, tyrosine kinases and G-proteins. Development 120 (12), 3463–3472.

Lindsay, L.L., Hertzler, P.L., Clark Jr., W.H., 1992. Extracellular Mg^{2+} induces an intra-cellular $Ca2^{+}$ wave during oocyte activation in the marine shrimp, *Sicyonia ingentis*. Dev. Biol. 152, 94–102.

Linsenmair, K.E., 1967. Konstruktion and Signalfunktion der Sandpyramide der Reiterkrabbe *Ocypode saratan* Frosk. (Decapoda Brachura Ocypodidae). Z. Tierpsychol. 24, 403–456.

Linsenmair, K.E., 1989. Sex-specific reproductive patterns in some terrestrial isopods. In: Rasa, A.E., Vogel, C., Voland, E. (Eds.), The Sociobiology of Sexual and Reproductive Strategies. Chapman and Hall, London, pp. 19–47.

Linsenmair, K.E., 2007. Sociobiology of terrestrial isopods. In: Duffy, J.E., Thiel, M. (Eds.), Evolutionary Ecology of Social and Sexual Systems – Crustaceans as Model Organisms. Oxford University Press, Oxford.

Lisk, D.R., 1961. Estradiol-17β in the eggs of the American lobster, *Homarus americanus*. Can. J. Biochem. Physiol. 39, 659–662.

Litulo, C., Tudge, C., 2005. Population structure and breeding season of the hermit crab *Diogenes brevirostris* Stimpson, 1858 (Decapoda, Anomura, Diogenidae) from southern Mozambique. J. Nat. Hist. 39, 2887–2899.

Liu, L., Laufer, H., 1996. Isolation and characterization of sinus gland neuropeptides with both mandibular organ inhibiting and hyperglycemic effects from the spider crab *Libinia emarginata*. Arch. Insect Biochem. Physiol. 32, 375–385.

Liu, X., Mccarron, R.C., Nordin, J.H., 1996. A cystine protease that processes insect vitellin. J. Biol. Chem. 271, 3344–3351.

Liu, Q., Yang, X., Cheng, Y., Wu, X., Fan, L., Huang, Z., 2009. Comparison of sperm morphology acrosin activity and immune indices between precocious and normal *Eriocheir sinensis*. J. Fish Sci. China 16, 183–189.

Locke, M., 1961. Pore canals and related structures in insect cuticle. J. Biophys. Biochem. Cytol. 10, 589–618.

Longhurst, A.R., 1955. The reproduction and cytology of the Notostraca (Crustacea, Phyllopoda). Proc. Zool. Soc. (Lond.) 125, 671–680.

Lonsdale, D.J., Frey, M.A., Snell, T.W., 1998. The role of chemical signals in copepod reproduction. J. Mar. Syst. 15, 1–12.

López Greco, L.S., Lo Nostro, F.L., 2008. Structural changes in the spermatophore of the freshwater 'red claw'crayfish *Cherax quadricarinatus* (Von Martens, 1898) (Decapoda, Parastacidae). Acta Zool. 89 (2), 149–155.

López Greco, L.S., Vazquez, F., Rodríguez, E.M., 2007. Morphology of the male reproductive system and spermatophore formation in the freshwater 'red claw'crayfish *Cherax quadricarinatus* (Von Martens, 1898) (Decapoda, Parastacidae). Acta Zool. 88 (3), 223–229.

Lopez-Greco, L.S., 2013. Functional anatomy of the reproductive system. In: Walting, L., Thiel, M. (Eds.), Functional Morphology and Diversity. Oxford University Press, pp. 413–450.

Lowry, J.K., Stoddart, H.E., 1986. Protandrous hermaphrodites among the Lysianassoid Amphipoda. J. Crust. Biol. 6, 742–748.

Lu, J.F., Chang, G.L., Wu, X.G., Yang, X.Z., Zhao, W.X., Cheng, Y.X., 2010. Hormonal regulations of ovarian development and vitellogenesis in Chinese mitten crab *Eriocheir sinensis* fed on two different diets. Oceanol. Limnol. Sin 41, 505–512.

Lubzens, E., Ravid, T., Khayat, M., Daube, N., Tietz, A., 1997. Isolation and characterization of the high-density lipoproteins from the hemolymph and ovary of the penaeid shrimp *Penaeus semisulcatus* (de Haan): apoproteins and lipids. J. Exp. Zool. 278 (6), 339–348.

Lui, C.W., O'Connor, J.D., 1977. Biosynthesis of crustacean lipovitellin. III. The incorporation of labeled amino acids into the purified lipovitellin of the crab *Pachygrapsus crassipes*. J. Exp. Zool. 199 (1), 105–108.

Lynn, J.W., Clark Jr., W.H., 1983. A morphological examination of sperm-egg interaction in the freshwater prawn, *Macrobrachium rosenbergii*. Biol. Bull. 164, 446–458.

Lynn, J.W., Clark Jr., W.H., 1983a. The fine structure of the mature sperm of the freshwater prawn, *Macrobrachium rosenbergii*. Biol. Bull. 164, 459–470.

Lynn, J.W., Pillai, M.C., Glas, P.S., Green, D.J., 1991. Comparative morphology and physiology of egg activation in selected penaeoidea. In: Davidson, M.A., Dougherty, W.J. (Eds.), Fronriers in Shrimp Research. Elsevier, Amsterdam, pp. 47–64.

Ma, W.M., Qian, Y.Q., Wang, M.R., Yang, F., Yang, W.J., 2010. A novel terminal ampulla peptide is involved in the proteolytic activity of sperm in the prawn, *Macrobrachium rosenbergii*. Reproduction 140, 235–245.

MacGinitie, G.E., 1938. Movements and mating habits of the sand crab, *Emerita analoga*. Am. Mid. Nat. 471–481.

Mahalingam, A., Kirubagaran, R., Dharani, G., Subramoniam, T., 2013. Probing the architecture of testis and morphology of male germinal cells in the mud crab *Scylla serrata* with the atomic force microscopy. Turk. J. Biol. 37 (5), 507–513.

Majer-Baranyi, K., Adányi, N., Nagy, A., Bukovskaya, O., Szendrő, I., Székács, A., 2015. Label-free immunosensor for monitoring vitellogenin as a biomarker for exogenous oestrogen compounds in amphibian species. Int. J. Environ. Anal. Chem. 95, 481–493.

Mak, A.S.C., Choi, C.L., Tiu, S.H.K., Hui, J.H.L., He, J.G., Tobe, S.S., Chan, S.M., 2005. Vitellogenesis in the red crab *Charybdis feriatus*: hepatopancreas-specific expression and farnesoic acid stimulation of vitellogenin gene expression. Mol. Reprod. Dev. 70 (3), 288–300.

Makinouchi, S., Honculada-Primavera, J., 1987. Maturation and spawning of *Penaeus indicus* using different ablation methods. Aquaculture 62 (1), 73–81.

Makioka, T., 1988. Ovarian structure and oogenesis in chelicerates and other arthropods. Proc. Arthro. Embryol. Soc. Jpn. 23 (1), 11.

Malati, E.F., Heidari, B., Zamani, M., 2013. The variations of vertebrate-type steroid hormones in the freshwater narrow-clawed crayfish *Astacus leptodactylus* (Eschscholtz, 1823) (Decapoda, Astacidae) during oocyte development. Crustaceana 86 (2), 129–138.

Malecha, S.R., Nevin, P.A., Ha, P., Barck, L.E., Lamadrid-Rose, Y., Masuno, S., Hedgecock, D., 1992. Sex-ratios and sex-determination in progeny from crosses of surgically sex-reversed freshwater prawns, *Macrobrachium rosenbergii*. Aquaculture 105, 201–218.

Manar, R., Bessi, H., Vasseur, P., 2009. Reproductive effects and bioaccumulation of chlordane in *Daphnia magna*. Environ. Toxicol. Chem. 28, 2150–2159.

Mann, R., 1984. On the selection of aquaculture species: a case study of marine molluscs. Aquaculture 39, 345–353.

Manor, R., Aflalo, E.D., Segall, C., Weil, S., Azulay, D., Ventura, T., Sagi, A., 2004. Androgenic gland implantation promotes growth and inhibits vitellogenesis in *Cherax quadricarinatus* females held in individual compartments. Invertebr. Reprod. Dev. 45, 151–159.

Manor, R., Weil, S., Oren, S., Glazer, L., Aflalo, E.D., Ventura, T., Sagi, A., 2007. Insulin and gender: an insulin-like gene expressed exclusively in the androgenic gland of the male crayfish. Gen. Comp. Endocrinol. 150, 326–336.

Mansecal, R., 1999. Chitin. In: Mark, J.E. (Ed.), Polymer Data Handbook. Oxford University Press, New York, NY, pp. 67–69.

Mantelatto, F.L., Dominciano, L.C.C., 2002. Pattern of shell utilization by the hermit crab *Paguristes tortugae* (Diogenidae) from Anchieta island, southern Brazil. Sci. Mar. 66, 265–272.

Mantelatto, F.L., Sousa, L.M., 2000. Population biology of the hermit crab *Paguristes tortugae* Schmitt, 1933 (Anomura, Diogenidae) from Anchieta island, Ubatuba, Brazil. Nauplius 8 (2), 185–193.

Marco, H.G., Avarre, J.C., Lubzens, E., Gäde, G., 2002. In search of a vitellogenesis-inhibiting hormone from the eyestalks of the South African spiny lobster, *Jasus lalandii*. Invertebr. Reprod. Dev. 41, 143–150.

Markham, J.C., 1975. Biological results of the University of Miami deep sea expeditions. Part 113. A review ofthe biopyrid isopod genus *Munidion*, parasitic on galatheid crabs in the Atlantic and Pacific. Bull. Mar. Sci. 25, 422–441.

Martin, G.G., Herzig, C., Narimatsu, G., 1987. Fine structure and histochemistry of the freshly extruded and hardened spermatophore of the spiny lobster *Panulirus interruptus*. J. Morphol. 192, 237–246.

Martin, G., Sorokine, O., Moniatte, M., Bulet, P., Hetru, C., Van Dorsselaer, A., 1999. The structure of a glycosylated protein hormone responsible for sex determination in the isopod, *Armadillidium vulgare*. Eur. J. Biochem. 262, 727–736.

Martin, P., Kohlmann, K., Scholtz, G., 2007. The parthenogenetic Marmorkrebs (marbled crayfish) produces genetically uniform offspring. Naturwissenschaften 94, 843–846.

Martin, J.W., 1992. Branchiopoda. In: Harrison, F.W., Humes, A.G. (Eds.), Microscopic Anatomy of Invertebrates. Crustacea, vol. 9. Wiley-Liss, New York, pp. 25–244.

Martins, J., Ribeiro, K., Rangel-Figueiredo, T., Coimbra, J., 2007. Reproductive cycle, ovarian development, and vertebrate-type steroids profile in the freshwater prawn *Macrobrachium rosenbergii*. J. Crust. Biol. 27 (2), 220–228.

Masuda, Y., oshida, M., ori, K., 2002. Synthesis of (2S, 2'R, 3S, 4R)-2-(2'-hydroxy-21'-methyldocosanoylamino)-1,3,4-pentadecanetriol, the ceramide sex pheromone of the female hair crab *Erimacrus isenbeckii*. Biosci. Biotechnol. Biochem. 66, 151–1537.

Mathews, L.M., 2003. Tests of the mate-guarding hypothesis for social monogamy: male snapping shrimp prefer to associate with high-value females. Behav. Ecol. 14, 63–67.

Matos, E., Matos, P., Oliveira, E., Azevedo, C., 1993. Aspectosmorfolo´gicos e ultraestruturais do espermatozo´ide de ermita˜o *Clibanarius vittatus* Bosc, 1802 (Crustacea, Decapoda) do litoral norte do Brasil. Revista Brasileira de Ciências Morfológicas 10, 126–131.

Matthews, D.C., 1951. The origin, development and nature of the spermatophoric mass of the spiny lobster *Panulirus penicillatus* (Oliver). Pac. Sci. 8, 115–120.
Matthews, D.C., 1953. The development of the pedunculate spermatophore of a hermit crab, *Dardanus asper* (De Haan). Pac. Sci. 8, 28–33.
Matthews, D.C., 1954. The development of sperrnatophoric masses of the rock lobster *Purribacus antarcticus* (Lund). Pac. Sci. 8, 28–34.
Matthews, D.C., 1956. The origin of the spermatophoric mass of the sand crab *Hippa Pacifica*. Q. J. Microsc. Sci. 97, 257–268.
Matthews, H.L., 1962. In: Ovary, T., Zuckermann, S. (Eds.), The Structure of the Ovary –Invertebrates. Academic Press, New York, pp. 89–120.
Mattson, M.P., Spaziani, E., 1985. Characterization of molt-inhibiting hormone MIH action on crustacean Y-organ segments and dispersed cells in culture and a bioassay for MIH activity. J. Exp. Zool. 236, 93–101.
Mauchline, J., 1980. Biology of mysids and euphausids. Adv. Mar. Biol. 18, 1–369.
Mauchline, J., Fisher, L., 1969. The biology of euphausiids. Adv. Mar. Biol. 7, 1–454.
Mazurová, E., Hilscherová, K., Triebskorn, R., Köhler, H.R., Maršálek, B., Bláha, L., 2008. Endocrine regulation of the reproduction in crustaceans: identification of potential targets for toxicants and environmental contaminants. Biologia 63, 139–150.
McCarthy, J.F., Skinner, D.M., 1979. Changes in ecdysteroids during embryogenesis of the blue crab, *Callinectes sapidus* Rathbun. Dev. Biol. 69, 627–633.
McCarthy, J.F., 1982. Ecdysone metabolism in premolt and land crabs (*Gecarcinus lateralis*). Gen. Comp. Endocrinol. 47 (3), 323–332.
McLay, C.L., Greco, L.S.L., 2011. A hypothesis about the origin of sperm storage in the Eubrachyura, the effects of seminal receptacle structure on mating strategies and the evolution of crab diversity: how did a race to be first become a race to be last. Zool. Anz. J. Comp. Zool. 250, 378–406.
McWhinnie, M.A., Saller, R.N., 1960. Analysis of blood sugars in the crayfish *Orconectes virilis*. Comp. Biochem. Physiol. 1 (2), 110–122.
Medesani, D.A., Greco, L.S.L., Rodríguez, E.M., 2004. Interference of cadmium and copper with the endocrine control of ovarian growth in the estuarine crab *Chasmagnathus granulata*. Aquat. Toxicol. 69, 165–174.
Medina, A., Rodríguez, A., 1992. Structural changes in sperm from the fiddler crab *Uca tangeri* (Crustacea, Brachyura), during the acrosome reaction. Mol. Reprod. Dev. 33, 195–201.
Medina, A., 1994. Spermiogenesis and sperm structure in the shrimp *Parapenaeus longirostris* (Crustacea Dendrobranchiata): comparative aspects among decapods. Mar. Biol. 119, 449–460.
Medina, A., 1995. Spermatozoal ultrastructure in Dendrobranchiata (Crustacea, Decapoda): taxonomic and phylogeny considerations. In: Jamieson, B.G.M., Ausio, J., Justine, J.L. (Eds.), Advances in Spermatozoa Phylogeny and Taxonomy. Me´moires du Muse´um National d'Histoire Naturalle, Paris, pp. 231–242.
Meech, R.W., 1978. Calcium-dependent potassium activation in nervous tissues. Annu. Rev. Biophys. Bioeng. 7, 1–18.
Meeratana, P., Withyachamnarnkul, B., Damrongphol, P., Wongprasert, K., Suseangtham, A., Sobhon, P., 2006. Serotonin induces ovarian maturation in giant freshwater prawn broodstock, *Macrobrachium rosenbergii* (de Man). Aquaculture 260, 315–325.
Mekuchi, M., Ohira, T., Kawazoe, I., Jasmani, S., Suitoh, K., Kim, Y.K., Jayasankar, V., Nagasawa, H., Wilder, M.N., 2008. Characterization and expression of the putative ovarian lipoprotein receptor in the Kuruma prawn, *Marsupenaeus japonicus*. Zool. Sci. 25, 428–437.
Menon, M.K., 1933. The life-histories of decapod Crustacea from Madras. Bull. Madras Govt. Mus. New Ser. 3, 1–45.

Menon, M.K., 1934. The oogenesis of *Emerita asiatica*. J. Madras Univ. 6, 1–15.

Merlin, J., Mohanlal, D.L., Balasubramanian, C.P., Tomy, S., Subramoniam, T., Syamadayal, J., Ravichandran, P., Ponniah, A.G., Gopal, C., Vijayan, K.K., 2015. Induction of vitellogenesis and reproductive maturation in tiger shrimp, *Penaeus monodon* by 17ß-estradiol and 17α-hydroxyprogesterone: in vivo and in vitro studies. Invertebr. Reprod. Develop 59 (3), 166–175.

Merriam, R.W., 1971. Progesterone-induced maturational events in oocytes of *Xenopus laevis*: I. Continuous necessity for diffusible calcium and magnesium. Exp. Cell Res. 68 (1), 75–80.

Meusy, J.J., Junera, H., Croisille, Y., 1971. Recherche de la « fraction protéique femelle chez les Crustacés Amphipodes *Orchestia gammarella* ayant subi une inversion expérimentale du sexe. C. R. Acad. Sci. Paris 273, 592–594 Ser. D.

Meusy, J.J., Junera, H., Cledon, P., Martin, M., 1983. La Vitellogenine chez Crustace Natantia Palaemon serratus Pennant. Mise en evidence comparison immunologique avee. Les viellines. Site de synthese et role des pedoncules oculaires. Reprod. Nutr. Dev. 23, 625–640.

Meusy, J.J., Martin, G., Soyez, D., Van Deijnen, J.E., Gallo, J.M., 1987. Immunochemical and immunocytochemical studies of the crustacean vitellogenesis inhibiting hormone (VIH). Gen. Comp. Endocrinol. 67, 333–341.

Meusy, J.J., 1963. La gametogenbse d'orchestia gammarellus p, Crustace Amphipode. Bull. Sot. Zool. Fr. 88, 197–220.

Meusy, J.J., 1968. Ultrastructure de la zone germinative et des gonies du testicule et de l'ovaire *d'Orchestia gammarella* P., (Crustacé, Amphipode). Ann. Sci. Nat. Zool. Biol. Anim. 10, 101–116.

Mich, J.K., Blaser, H., Thomas, N.A., Firestone, A.J., Yelon, D., Raz, E., Chen, J.K., 2009. Germ cell migration in zebrafish is cyclopamine-sensitivebutSmoothenedindependent. Dev. Biol. 328, 342–354.

Michaud, M., de Meeûs, T., Renaud, F., 2004. Environmental sex determination in a parasitic copepod: checking heterogeneity and unpredictability of the environment. Mar. Ecol. Prog. Ser. 269, 163–171.

Micheli, F., 1991. Bilateral gynandromorph of the fresh-water crab *Potamon fluviatile* Herbst (Decapoda, Brachyura). J. Crust. Biol. 11, 561–568.

Michener, C.D., Brothers, D.J., 1974. Were workers of eusocial Hymenoptera initially altruistic or oppressed? Proc. Nat. Acad. Sci. U.S.A. 71, 671–674.

Miki, W., Yamaguchi, K., Konosu, S., 1982. Comparison of carotenoids in the ovaries of marine fish and shellfish. Comp. Biochem. Physiol. B 71 (1), 7–11.

Millar, R.P., 2005. GnRHs and GnRH receptors. Anim. Reprod. Sci. 88, 5–28.

Miller, C.B., Crain, J.A., Marcus, N.H., 2005. Seasonal variation of male-type antennular setation in female *Calanus finmarchicus*. Mar. Ecol. Prog. Ser. 301, 217–229.

Mills, B.J., McCloud, P.I., 1983. Effects of stocking and feeding rates on experimental pond production of the crayfish *Cherax destructor* Clark (Decapoda: Parastacidae). Aquaculture 34, 51–72.

Mirheydari, S.M., Paolucci, M., Matinfar, A., Soltani, M., Kamali, A., Asadpour-Ousalou, Y., 2014. Occurrence of vertebrate-like steroids in the male narrow-clawed crayfish *Astacus leptodactylus* (Eschscholtz, 1823) from Iran during the Annual reproductive cycle. Glob. Veterinaria 13 (2), 247–254.

Mitchell, S.E., 2001. Intersex and male development in *Daphnia magna*. Hydrobiologia 442, 145–156.

Mittal, 0. P., Pahwa, S., 1980. Cytogenetical studies on two terrestrial isopods (Crustacea), *Philoscia lodnensis* and *Porcellio rathkei* Brandt. Cytobios 29, 17–24.

Mittal, 0. P., Pahwa, S., 1981. Germ cell chromosomes in two species of terrestrial isopods from India. Experientia 37, 343–345.

Mittwoch, U., 1996. Sex-determining mechanisms in animals. Trends Ecol. Evol. 11, 63–67.

Mocquard, J.P., Legrand-Hamelin, E., Soulty, C., 1978. Infl uence de la salinite et de la temperature sur la fecundite des femelles et l'intersexualite *d'Idotea balthica* (Isopoda, Valvifera). Bul. de la Soci. Zoolo. de France 103, 45–57.

Modlin, R.F., Hams, P.A., 1989. Observations on the natural history and experiments on the reproductive strategy of *Hageria rapax* (Tanaidacea). J. Crust. Biol. 9, 57–58.

Mohamed, K.S., Diwan, A.D., 1994. Vitellogenesis in the Indian white prawn *Penaeus indicus* (Crustacea: Decapoda: Penaeidae). J. Aquacult. Tropics 9, 157–172.

Montalenti, G., 1941. Studi sull' ermafroditismo dci cimotoidi-I. *Emetha audouinii (M. Edw.) e Anilocra physodes (L.).* Pubbl Stn. Zoo. Napoli. 18, 337–394.

Moore, C.G., Stevenson, J.M., 1991. The occurrence of intersexuality in harpacticoid copepods and its relationship with pollution. Mar. Pollut. Bull. 22, 72–74.

Moore, C.G., Stevenson, J.M., 1994. Intersexuality in benthic harpacticoid copepods in the Firth of Forth. Scotland. J. Nat. Hist. 28, 1213–1230.

Moraes-Riodades, P.M., Valenti, W.C., 2004. Morphotypes in male Amazon river prawns, *Macrobrachium amazonicum*. Aquaculture 236, 297–307.

Morais, S., Narciso, L., Calado, R., Nunes, M.L., Rosa, R., 2002. Lipid dynamics during the embryonic development of *Plesionika martia martia* (Decapoda; Pandalidae), *Palaemon serratus* and *P. elegans* (Decapoda; Palaemonidae): relation to metabolic consumption. Mar. Ecol. Prog. Ser. 242, 195–204.

Moreno-Mendoza, N., Harley, V.R., Merchant-Larios, H., 1999. Differential expression of SOX9 in gonads of the sea turtle *Lepidochelys olivacea* at male-or female-promoting temperatures. J. Exp. Zool. 284, 705–710.

Moriyasu, M., Mallet, P., Comeau, M., Benhalima, K., Ghica, O., 1988. Occurrence of Pseudohermaphroditism in the rock crab, *Cancer irrotatus*. Say, 1817, in the southern gulf of St. Lawrence, Canada (Decapoda, Brachyura). Crustaceana 71, 655–662.

Moriyasu, M., Benhalima, K., Duggan, D., Lawton, P., Robichaud, D., 2002. Reproductive biology of male Jonah crab, *Cancer borealis* Stimpson, 1859 (Decapoda, Cancridae) on the Scotian shelf, northwestern Atlantic. Crustaceana 75 (7), 891–913.

Morrell, L.J., Backwell, P.R., Metcalfe, N.B., 2005. Fighting in fiddler crabs *Uca mjoebergi*: what determines duration. Anim. Behav. 70, 653–662.

Motoh, H., 1981. Studies on the fisheries biology of the giant tiger prawn, *Penaeus monodon*. Tech. Rep. Philippines Aquacult. Department Southeast Asian Fish. Dev. Centre 7, 1–10.

Mouchet, S., 1931. Spermatophores des Crustacés Décapodes Anomoures et Brachyoures et castration parasitaire chez quelques *Pagures*. Annales de la Station Oceanographique de Salammbo 6, 1–203 (No. 6). E. Blondel La Rougery.

Müller, Y., Ammar, D., Nazari, E., 2004. Embryonic development of four species of palaemonid prawns (Crustacea, Decapoda): pre-naupliar, naupliar and post-naupliar periods. Revista Brasileira de Zoologia 21 (1), 27–32.

Müller, G.W., 1894. Die Ostracoden des Golfes von Neapel und der angrenzenden Meeresabschnitte, vol. 21. R. Friedlander.

Muller, von K., 1983. Population biology activity rhythms and distribution of *Uca tangeri* (Decapoda, Brachyura). Zool. Jb. Syst. 221–226.

Munday, P.L., Buston, P.M., Warner, R.R., 2006. Diversity and flexibility of sex-change strategies in animals. Trends Ecol. Evol. 21, 89–95.

Munn, E.A., Barnes, H., 1970. The structure of the axial filament complex of the spermatozoa of *Balanus balanus*. Exp. Cell Res. 60 (2), 277–284.

Munuswamy, N., Subramoniam, T., 1984. Egg envelopes of *Streptocephalus dichotomus* Baird: a structural and histochemical study. Hydrobiologia 114, 17–28.

Munuswamy, N., Subramoniam, T., 1985. Influence of mating on ovarian and shell gland activity in a freshwater fairy shrimp *Streptocephalus dichotomus* (Anostraca). Crustaceana 49, 225–232.

Munuswamy, N., 1986. Sclerotization in the egg shell of *Streptocephalus dichotomus* (Crustacea: Anostraca). Microbios Lett. 31, 103–105.

Murai, M., Goshima, S., Nakasone, Y., 1983. Adaptive droving behavior observed in the fiddler crab *Uca vocans*. Mar. Biol. 76, 159–164.

Murai, M., Goshima, S., Henmi, Y., 1987. Analysis of the mating system of the fiddler crab, *Uca lactea*. Anim. Behav. 35, 1334–1342.

Murai, M., Koga, T., Yong, H.S., 2002. The assessment of female reproductive state during courtship and scramble competition in the fiddler crab, *Uca paradussumieri*. Behav. Ecol. Sociobiol. 52, 137–142.

Murugadass, S., Mathavan, S., Marian, M.P., Pandian, T.J., 1988. Influence of Eyestalk Ablation on Growth and Egg Production in *Macrobrachium Malcomsoni*, vol. 1. , pp. 111–114.

Murugan, T., 1985. Studies on the Biology of the Sand Crabs of Kerala (Ph.D. thesis). University of Kerala, India. 393 pp..

Murugesan, P., Rajagopal, S., Khan, A.S., 1998. Utility of eyestalk ablation in farming and broodstock development of the shrimp *Penaeus monodon* Fabricus. Indian J. Exp. Biol. 36, 699–702.

Muthu, M.S., Laxminarayana, A., 1977. Induced maturation and spawning of Indian penaeid prawns. Indian J. Fish. 24, 172–180.

Muthuraman, A., Adiyodi, K.G., 1998. Effect of ecdysial cycle on the Semen and Spermatophore production of the Indian white prawn, *Penaeus indicus* (HM Edwards). In: Current and Emerging Trends in Acquaculture: Proceedings of the National Seminar on Current and Emerging Trends in Aquaculture and Its Impact on Rural Development, 14–16 February 1995. Daya Books, p. 195.

Muthuvelu, S., Sundhari, S., Purusothaman, S., Sivaraj, S., 2013. Biochemical and enzymatic fluctuation during ovarian development of edible marine crab *Portunus pelagicus* [Linnaeus, 1758]. Int. J. Pharm. Pharma. Sci. 5 (4), 638–642.

Nagabhushanam, R., Kulkarni, K.M., 1977. Seasonal changes in biochemical composition of the sand crab *Emerita holthuisi*. J. Mar.Biol. Assoc. India 19, 52–57.

Nagamine, C., Knight, A.W., Maggenti, A., Paxman, G., 1980. Effects of androgenic gland ablation on male primary and secondary sexual characteristics in the Malaysian prawn, Ma*crobrachium rosenbergii* (de Man) (Decapoda, Palaemonidae), with first evidence of induced feminization in a non hermaphroditic decapod. Gen. Comp. Endocrinol. 41, 423–441.

Nagaraju, G.C.P., Rajitha, B., Borst, D.W., 2011. Molecular cloning and sequence of retinoid X receptor in the green crab *Carcinus maenas*: a possible role in female reproduction. J. Endocrinol. 210, 379–390.

Nakamura, K., Matsuzaki, N., Yonekura, K.I., 1992. Organogenesis of genital organs and androgenic gland in the kuruma prawn. Nippon Suisan Gakk 58, 2261–2267.

Nakashima, Y., 1987. Reproductive strategies in a partially protandrous shrimp, *Athanas kominatoensis* (Decapoda: Alpheidae): sex change as the best of a bad situation for subordinates. J. Ethol. 5, 145–159.

Nakkrasae, L.I., Damrongphol, P., 2007. A vasa-like gene in the giant freshwater prawn, *Macrobrachium rosenbergii*. Mol. Reprod. Dev. 74, 835–842.

Narita, S., Pereira, R.A., Kjellberg, F., Kageyama, D., 2010. Gynandromorphs and intersexes: potential to understand the mechanism of sex determination in arthropods. Terr. Arthropod Rev. 3, 63–96.

Nascimento, I.A., Bray, W.A., Leung-Trujillo, J.R., Lawrence, A., 1991. Reproduction of ablated and unablated *Penaeus schmitti* in captivity using diets consisting of fresh-frozen natural and dried formulated feeds. Aquaculture 99, 387–398.

Nates, S.F., McKenney, C.L., 2000. Ontogenetic changes in biochemical composition during larval and early postlarval development of *Lepidophthalmus louisianensis*, a ghost shrimp with abbreviated development. Comp. Biochem. Physiol. Part B 127 (4), 459–468.

Nath, V., 1965. Animal Gametes (Male). A Morphological and Cytochemical Account of Spermatogenesis. Asia Publishing House, Bombay, p. 78.

Needham, J., 1950. Biochemistry and Morphogenesis. University Press, Cambridge, p. 785.

Nelis, H.J., Deleenheer, A.P., Criel, G.R., Vasteenberge, M.M., Sorgeloos, P., Lavens, P., 1988. Qualitative and quantitative changes in the carotenoids during development of the brine shrimp *Artemia*. J. Lipid Res. 29, 491–499.

Newman, W.A., 1974. Two new deep-sea Ciripedia (Ascothoracica and Acrothoracica) from the Atlantic. J. Mar. Biol. Assoc. U.K. 54, 437–456.

Ngernsoungnern, A., Ngernsoungnern, P., Weerachatyanuku, W., Chavadej, J., Sretarugsa, S.P., 2008a. The existence of gonadotropin-releasing hormone (GnRH) immunoreactivity in the ovary and the effects of GnRHs on the ovarian maturation in the black tiger shrimp *Penaeus monodon*. Aquaculture 279, 197–220.

Ngernsoungnern, P., Ngernsoungnern, A., Kavanaugh, S., Sobhon, P., Sower, S.A., Sretarugsa, P., 2008b. The presence and distribution of gonadotropin - releasing hormone like factor in the central nervous system of the black tiger shrimp *Penaeus monodon*. Gen. Comp. Endocrinol. 155, 613–622.

Nicholls, A.G., 1931. Studies on *Ligia oceanica*. IA Habitat and effect of change of environment on respiration. B. Observations on moulting and breeding. J. Mar. Biol. Asso. U.K. (New Ser.) 17, 655–673.

Niiyama, H., 1941. The XO type of sex-chromosome found in *Ovalipes punctatus* (de Haan) (Crustacea: Decapoda). Jpn. J. Genet. 17, 41–45.

NIiyama, H., 1950. The XY mechanism of the sex-chromosome in the male of *Anisogammarus annandalei* (Tattersall) (Crustacea: Amphipoda). Annot. Zool. Jpn. 23, 58–62.

Niiyama, H., 1956. Cytological demonstration of an XO sex-mechanism in males of *Tecticeps japonicus*, an isopod Crustacea. Cytologia 21, 38–43.

Nikitima, S.M., Savehenko, O.N., Kogan, M.E., Ezhkova, N.S., 1977. Preparative isolation of progesterone, testosterone and estrogens from the tissues of marine invertebrates. Zh. Evol. Biokhum. Fiziol. 13, 443–447.

Niksirat, H., Kouba, A., Kozák, P., 2015. Ultrastructure of egg activation and cortical reaction in the noble crayfish *Astacus astacus*. Micron 68, 115–121.

Nöel, P., 1973. Cycle biologique et inversion sexuelle du crustacé décapode Natantia *Processa edulis*. Cah. Biol. Mar. 14 (2), 17–227.

Noël, P., 1976. L'evolution des caracteres sexuels chez *Processa edulis* Risso (Decapode, Natantia). Vie Milieu 26, 65–104.

Nørrevang, A., 1983. Pentastomida. In: Adiyodi, K.G., Adiyodi, R.G. (Eds.), Reproductive Biology of Invertebrates, vol. 1. , pp. 521–534.

Nouvel, H., 1937. Observations de l'accouplement chez une espece de Mysis, *Pruunus pexuosus*. Comptes Rendus Hebdomudaire des seunces de I'Acudemie des Sciences 205, 1184–1186.

Nurse, P., 1990. Universal control mechanism regulating onset of M-phase. Nature 344, 503–508.

Nye, P., 1977. Reproduction, growth and distribution of the graspid crab *Helice crassa* (Dana, 1851) in the southern part of New Zealand. Crustaceana 33, 75–89.

Oberdörster, E., Brouwer, M., Hoexum-Brouwer, T., Manning, S., McLachlan, J.A., 2000. Long-term pyrene exposure of grass shrimp, *Palaemonetes pugio*, affects molting and reproduction of exposed males and offspring of exposed females. Environ. Health Perspect. 108, 641–651.

Oda, S., Tatarazako, N., Watanabe, H., Morita, M., Iguchi, T., 2005. Production of male neonates in *Daphnia magna* (Cladocera, Crustacea) exposed to juvenile hormones and their analogs. Chemosphere 61, 1168–1174.

Ohira, T., Hasegawa, Y., Tominaga, S., Okuno, A., Nagasawa, H., 2003. Molecular cloning and expression analysis of cDNAs encoding androgenic gland hormone precursors from two porcellionidae species, *Porcellio Scaber* and *P. dilatatus*. Zool. Sci. 20, 75–81.

Ohtsuka, S., Horiguchi, T., Hanamura, Y., Nagasawa, K., Suzaki, T., 2003. Intersex in the mysid *Siriella japonica izuensis* Ii: the possibility it is caused by infestation with parasites. Plankton Biol. Ecol. 50, 65–70.

Okazaki, R.K., Chang, E.S., 1991. Ecdysteroids in the embryos and sera of the crabs, *Cancer magister* and *C. antonyi*. Gen. Comp. Endocrinol. 81, 174–186.

Okumura, T., Aida, K., 2000. Hemolymph vitellogenin levels and ovarian development during the reproductive and non-reproductive molt cycles in the giant freshwater prawn *Macrobrachium rosenbergii*. Fisheries Sci. 66 (4), 678–685.

Okumura, T., Aida, K., 2001. Effects of bilateral eyestalk ablation on molting and ovarian development in the giant freshwater prawn, *Macrobrachium rosenbergii*. Fisheries Sci. 67 (6), 1125–1135.

Okumura, T., Sakiyama, K., 2004. Hemolymph levels of vertebrate-type steroid hormones in female kuruma prawn *Marsupenaeus japonicus* (Crustacea: Decapoda: Penaeidae) during natural reproductive cycle and induced ovarian development by eyestalk ablation. Fisheries Sci. 70 (3), 372–380.

Okumura, T., Han, C.H., Suzuki, Y., Aida, K., Hanyu, I., 1992. Changes in hemolymph vitellogenin and ecdysteroid levels during the reproductive and non-reproductive molt cycles in the freshwater prawn *Macrobrachium nipponense*. Zool. Sci. 9, 37–45.

Okumura, T., Yoshida, K., Nikaido, H., 2004. Ovarian development and hemolymph vitellogenin levels in laboratory-maintained protandric shrimp, *Pandalus hypsinotus*: measurement by a newly developed time-resolved fluoroimmunoassay (TR-FIA). Zool. Sci. 21 (10), 1037–1047.

Okumura, T., Nikaido, H., Yoshida, M., Kotaniguchi, M., Tsuno, Y., Seto, Y., Watanbe, T., 2005. Changes in gonadal development, androgenic gland cell structure, and hemolymph vitellogenin levels during male phase and sex change in laboratory –maintained protandric shrimp, *Pandalus hypsinotus* (Crustacea: Caridea: Pandalidae). Mar. Biol. 148, 347–361.

Okumura, T., Kim, Y.K., Kawazoe, I., Yamano, K., Tsutsui, N., Aida, K., 2006. Expression of vitellogenin and cortical rod proteins during induced ovarian development by eyestalk ablation in the kuruma prawn, *Marsupenaeus japonicus*. Comp. Biochem. Physiol. A 143, 246–253.

Okumura, T., 2006. Effects of cyclic nucleotides, calcium ionophore, and phorbol ester on vitellogenin mRNA levels in incubated ovarian fragments of the kuruma prawn *Marsupenaeus japonicus*. Gen. Comp. Endocrinol. 148 (2), 245–251.

Okumura, T., 2007. Effects of lipopolysaccharide on gene expression of antimicrobial peptides (penaeidins and crustin), serine proteinase and prophenoloxidase in haemocytes of the Pacific white shrimp, *Litopenaeus vannamei* Shell. Fish. Immunol. 22, 68–76.

Okuno, A., Hasegawa, Y., Nagasawa, H., 1997. Purification and properties of androgenic gland hormone from the terrestrial isopod *Armadillidium vulgare*. Zool. Sci. 14, 837–842.

Okuno, A., Hasegawa, Y., Ohira, T., Katakura, Y., Nagasawa, H., 1999. Characterization and cDNA cloning of androgenic gland hormone of the terrestrial isopod *Armadillidium vulgare*. Biochem. Biophys. Res. Commun. 264, 419–423.

Okuno, A., Katayama, H., Nagasawa, H., 2000. Partial characterization of vitellin and localization of vitellogenin production in the terrestrial isopod, *Armadillidium vulgare*. Comp. Biochem. Physiol. Part B 126 (3), 397–407.

Okuno, A., Yang, W.J., Jayasankar, V., Saido-Sakanaka, H., Huong, D.T.T., Jasmani, S., Atmomarsono, M., Subramoniam, T., Tsutsui, N., Ohira, T., Kawazoe, I., 2002. Deduced primary structure of vitellogenin in the giant freshwater prawn, *Macrobrachium rosenbergii*, and yolk processing during ovarian maturation. J. Exp. Zool. 292 (5), 417–429.

Olive, P.J.W., 1983. Annelida: *Polychaeta oogenesis*, oviposition and oosorption. In: Adiyodi, K.G., Adiyodi, R.G. (Eds.), Reproductive Biology of Invertebrates. John Wiley & sons, New York, pp. 357–422.

Ollevier, F., De Clerck, D., Diederik, H., De Loof, A., 1986. Identification of nonecdysteroid steroids in hemolymph of both male and female *Astacus leptodactylus* (Crustacea) by gas chromatography—mass spectrometry. Gen. Comp. Endocrinol. 61 (2), 214–228.

Ollivaux, C., Gallois, D., Amich, M., Boscaméric, M., Soyez, D., 2009. Molecular and cellular specificity of post-translational aminoacyl isomerization in the crustacean hyperglycaemic hormone family. FEBS J. 276, 4790–4802.

Olmstead, A.W., Leblanc, G.A., 2002. Juvenoid hormone methyl farnesoate is a sex determinant in the crustacean *Daphnia magna*. J. Exp. Zool. 293, 736–739.

Olmstead, A.W., LeBlanc, G.A., 2007. The environmental-endocrine basis of gynandromorphism (intersex) in a crustacean. Int. J. Biol. Sci. 3, 77–84.

Ombretta, M., Cesari, M., Eder, E., Scanabissi, F., Mantovani, B., 2005. Chromosomes in sexual populations of notostracan and conchostracan taxa (Crustacea, Branchiopoda). Caryologia 58, 164–170.

Omori, M., 1974. The biology of pelagic shrimps in the ocean. Adv. Mar. Biol. 12, 233–324.

Ongvarrasopone, C., Roshorm, Y., Somyong, S., Pothiratana, C., Petchdee, S., Tangkhabuanbutra, J., Sophasan, S., Panyim, S., 2006. Molecular cloning and functional expression of the *Penaeus monodon* 5-HT receptor. Osmoregulation, moulting, and reproduction. Gen. Comp. Endocrinol. 175, 217–233.

Orensanz, J.M., Parma, A.M., Armstrong, D.A., Armstrong, J., Wardrup, P., 1995. The breeding ecology of *Cancer gracilis* (Crustacea: Decapoda: Cancridae) and the mating systems of cancrid crabs. J. Zool. 235, 411–437.

Orlando, E., 1973. Heterochromosomes in *Mytilicola intestinalis* Steuer (Copepoda). Genetica 44 (2), 244–248.

Osorio, C., Bahamonde, N., López, M.T., 1967. El limache *Emerita analoga*, Stimpson*)* en Chile. Museo Nat. Hist. Nat. 29, 61–116.

Otsu, T., 1963. Bihormonal control of sexual cycle in the freshwater crab, *Potamon dehaani*. Embryologia 8 (1), 1–20.

Özhan-Kizil, G., Havemann, J., Gerberding, M., 2009. Germ cells in the crustacean *Parhyale hawaiensis* depend on Vasa protein for their maintenance but not for their formation. Dev. Biol. 327, 230–239.

Pai, S., 1949. Sexuelle Stufen and Chromosomen Grösse der keimzellen von *Chirocephalus nankinensis*. Exp. Cell Res. (Suppl. 1), 143–145.

Palacios, E., Carrefio, D., Rodriguez-Jaramillo, M.C., Racotta, I.S., 1999a. Effect of eyestalk ablation on maturation, larval performance, and biochemistry of white pacific shrimp, *Penaeus vannamei*. Broodstock. J. Appl. Aquacult. 9 (3), 1–23.

Palacios, E., Rodríguez-Jaramillo, C., Racotta, I.S., 1999b. Comparison of ovary histology between different-sized wild and pond-reared shrimp *Litopenaeus vannamei (=Penaeus vannamei)*. Invertebr. Reprod. Dev. 35 (3), 251–259.

Pan, M.L., Bell, W.J., Telfer, W.H., 1969. Vitellogenic blood protein synthesis by insect fat body. Science 165 (3891), 393–394.

Pandian, T.J., 1970. Ecophysiological studies on the developing eggs and embryos of the European lobster *Homarus gammarus*. Mar. Biol. 7, 249–254.

Panikkar, N.K., Aiyar, R.G., 1939. Observations on breeding in brackish-water animals of Madras. Proc. Indian Acad. Sci. B 9, 343–364.

Panouse, J.B., 1943. Influence du peduoncule oculare sur la eroissance de louaire chez la crevette *Leander serratus*. Ann. Inst. Oceranogr. 23, 65–147.

Paolucci, M., Di Cristo, C., Di Cosmo, A., 2002. Immunological evidence for progesterone and estradiol receptors in the freshwater crayfish *Austropotamobius pallipes*. Mol. Reprod. Dev. 63, 55–62.

Papathanassiou, E., King, P.E., 1984. Ultrastructural studies on gametogenesis of the prawn *Palaemon serratus* (Pennant). II. Spermiogenesis. Acta Zool. Stockh. 65, 33–40.

Parker, G., 1970. Sperm competition and its evolutionary consequences in the insects. Biol. Rev. 45, 525–567.

Parker, G.A., 1974. Assessment strategy and the evolution of fighting behaviour. J. Theor. Biol. 47, 223–243.

Parnes, S., Khalaila, I., Hulata, G., Sagi, A., 2003. Are intersex crayfish *Cherax quadricarinatus* (Von Martens) genetically females? Genet. Res. 82, 107–116.

Pateraki, L.E., Stratakis, E., 1997. Characterization of vitellogenin and vitellin from land crab *Potamon potamios*: identification of a precursor polypeptide in the molecule. J. Exp. Zool. 279 (6), 597–608.

Paul, M.D., 1942. Studies on growth and breeding of certain sedentary organisms in the Madras harbour. Proc. Indian Acad. Sci. B 15, 1–42.

Payen, G., 1973. Etude descriptive des principales etapes de la morphogenese sexuelle chez un Crustace Decapodes & developpement condense, I'Ecrevisse *Pontastacus leptodactylus* Ieptodactylrrs (Eschscholtz, 1823). Ann. Embryo. Morphogen 6, 179–206.

Payen, G., 1974. "Morphogenese sexuelle de quelques Brachyoures (Cyclometopes) au cours du development embryonnaire, larvaire et postlarvaire. Bull. Mus. Hist. Nat. Zool. 209 (139), 201–262.

Payen, G.G., 1981. Aspects foundamentaux de l'endocrinologie de la reproduction chez les crustaces marins. Oceanis 6, 309–339.

Peiró, D.F., 2012. Status taxonômico de *Callichirus major* (Say, 1818) sensu lato (Crustacea, Decapoda, Axiidea, Callianassidae) da costa brasileira: taxonomia, sistemática molecular, biologia populacional e reprodutiva. (Doctoral dissertation, Universidade de São Paulo).

Perez Farfante, I., 1975. Spermatophores and thelyca of the America white shrimps, genus *Penaeus*, subgenus *Litopenaeus*. U.S. Nat. Mar. Fish Ser. Fish. Bull. 73, 463–486.

Perkins, H.C., 1972. Developmental rates at various temperatures of embryos of the northern lobster (*Homarus americanus* Milne-Edwards). Fish. Bull. 70, 95–99.

Perry, D.M., 1980. Factors influencing aggregation patterns in the sand crab *Emerita analoga* (Crustacea: Hippidae). Oecologia 45, 379–384.

Petit, H., Sance, S., Negre-Sadargues, G., Castillo, R., Trilles, J.P., 1991. Ontogeny of carotenoid metabolism in the prawn *Penaeus japonicus* Bate (1888) (Crustacea Penaeidea). A qualitative approach. Comp. Biochem. Physiol. Part B 99 (3), 667–671.

Pfaller, J.B., Alfaro-Shigueto, J., Giffoni, B., Ishihara, T., Mangel, J.C., Peckham, S.H., Baeza, J.A., 2014. Social monogamy in the crab *Planes major*, a facultative symbiont of loggerhead sea turtles. J. Exp. Mar. Biol. Ecol. 461, 124–132.

Philippen, M.K., Webster, S.G., Chung, J.S., Dircksen, H., 2000. Ecdysis of decapod crustaceans is associated with a dramatic release of crustacean cardioactive peptide into the haemolymph. J. Exp. Biol. 203, 521–536.

Phinyo, M., Visudtiphole, V., Roytrakul, S., Phaonakrop, N., Jarayabhand, P., Klinbunga, S., 2013. Characterization and expression of cell division cycle 2 (Cdc2) mRNA and protein during ovarian development of the giant tiger shrimp *Penaeus monodon*. Gen. Comp. Endocrinol. 193, 103–111.

Phinyo, M., Nounurai, P., Hiransuchalert, R., Jarayabhand, P., Klinbunga, S., 2014. Characterization and expression analysis of Cyclin-dependent kinase 7 gene and protein in ovaries of the giant tiger shrimp *Penaeus monodon*. Aquaculture 432, 286–294.

Phiriyangkul, P., Puengyam, P., Jakobsen, I.B., Utarabhand, P., 2007. Dynamics of vitellogenin mRNA expression during vitellogenesis in the banana shrimp *Penaeus (Fenneropenaeus) merguiensis* using real-time PCR. Mol. Reprod. Dev. 74 (9), 1198–1207.

Picaud, J.L., 1980. Vitellogenin synthesis by the fat body of *Porcellio dilatatus* Brandt (Crustacea, Isopoda). Int. J. Invertebr. Reprod. 2 (6), 341–349.

Pillai, M.C., Clark Jr., W.H., 1987. Oocyte activation in the marine shrimp, *Sicyonia ingentis*. J. Exp. Zool. 244, 325–329.

Pillai, M.C., Clark Jr., W.H., 1988. Hatching envelope formation in shrimp (*Sicyonia ingenlis*) ova: origin and sequential exocytosis of cortical vesicles. Tissue Cell 20, 941–952.

Pillai, M.C., Clark Jr., W.H., 1990. Development of cortical vesicles in *Sicyonia ingentis* ova: their heterogeneity and role in elaboration of the hatching envelope. Mol. Reprod. Dev. 26, 78–89.

Pillai, C.K., Subramoniam, T., 1984. Monsoon-dependent breeding in the field crab *Paratelphusa hydrodromous* (Herbst). Hydrobiologia 119, 7–14.

Pillai, C.K., Subramoniam, T., 1985. Yolk utilization as an adaptive strategy of terrestrialization in the freshwater crab, *Paratelphusa hydrodromous* (Herbst). Physiol. Zool. 58, 445–457.

Pillai, R.S., 1960. Studies on the shrimp *Caridina laevis* (Heller). II. The reproductive system. J. Mar. Biol. Assoc. India 2, 226–236.

Pillay, K.K., Nair, N.B., 1971. The annual reproductive cycles of *Uca annulipes, Portunus pelagicus* and *Metapenaeus affinis* (Decapoda: Crustacea) from the south-west coast of India. Mar. Biol. 11, 152–166.

Pinheiro, M.A.A., Fransozo, A., 2002. Reproduction of the speckled swimming crab *Arenaeus cribrarius* (Brachyura: Portunidae) on the Brazilian coast near 23 30′ S. J. Crust. Biol. 22, 416–428.

Pinn, E.H., Atkinson, R.J.A., Rogerson, A., 2001. Sexual dimorphism and intersexuality in *Upogebia stellata* (Crustacea: Decapoda: Thalassinidea). J. Mar. Biol. Assoc. U.K. 81, 1061–1062.

Plaistow, J.S., Ballache, L., Cézilly, F., 2003. Energetically costly precopulatory mate guarding in the amphipod *Gammarus pulex*: causes and consequences. Anim. Behav. 65, 683–691.

Pochon-Masson, J., Gharagozlou-van-Ginneken, I.D., 1977. Peculiar features of the sperm of a harpacticoid copepod, *Tisbe holothuriae*. In: Adiyodi, K.G., Adiyodi, R.G. (Eds.), Advances in Invertebrate Reproduction, vol. 1. Peralam-Kennoth, Kerala, India, pp. 61–75.

Pochon-Masson, J., Gharagozlou-van-Ginneken, I.D., 1978. Contribution a l'etude du cell-coat a partir du spermatozoide d'un arthropode: *Tisbe holothruriae* (Crustace Copepode). Ann. Sci. Nat. Zool. 12, 389–399.

Pochon-Masson, J., 1962. Chondriofusomes of the male gametes of the decapod crustacean, Carcinus maenas. C.R. Hebdomadaires des Seances de l'Academie des Sciences 254, 4076–4078.

Pochon-Masson, J., 1963. Origine et formation de la vesicule du spermatozoide d'*Eupagurus bernhardus* (Decapode Anomoure). C. R. Acad. Sci. Paris 256, 2226–2228.

Pochon-Masson, J., 1965. Schema general du spermatozoide vesiculaire des decapods. C. R. Acad. Sci. Paris 260, 5093–5095.

Pochon-Masson, J., 1968. L'ultrastructure des Spermatozoides Chez les crustaces decapodes avant et anu cours de leur devagination expermentale. II. Macroures. Discussion et conclusions. Ann. Sci. Nat. (Zool.) 10, 367–454.

Pochon-Masson, J., 1983. Arthropoda Crustacea. In: Adiyodi, K.G., Adiyodi, R.G. (Eds.), Reproductive Biology of Invertebrates. Spermatogenesis and Sperm Function, vol. II. Wiley, Chichester, pp. 407–449.

Polijaroen, J., 2010. Spermatogenesis and Spermiogenesis in the Giant Freshwater Prawn, *Macrobrachium rosenbergii* (Ph.D. thesis). Mahidol University, Bangkok.

Poljaroen, J., Vanichviriyakit, R., Tinikul, Y., Phoungpetchara, I., Linthong, V., Weerachatyanukul, W., Sobhon, P., 2010. Spermatogenesis and distinctive mature sperm in the giant freshwater prawn, *Macrobrachium rosenbergii* (De Man, 1879). J. Comp. Zool. 249 (2), 81–94.

Pomiankowski, A., Nöthiger, R., Wilkins, A., 2004. The evolution of the *Drosophila* sex-determination pathway. Genetics 166, 1761–1773.

Pongtippatee-Taweepreda, P., Chavadej, J., Plodpai, P., Pratoomchart, B., Sobhon, P., Weerachatyanukul, W., Withyachumnarnkul, B., 2004. Egg activation in the black tiger shrimp *Penaeus monodon*. Aquaculture 234, 183–198.

Pongtippatee-Taweepreda, P., Vanichviriyakit, R., Chavadej, J., Plodpai, P., Pratoomchart, B., Sobhon, P., Withyachumnarnkul, B., 2007. Acrosome reaction in the sperm of the black tiger shrimp *Penaeus monodon* (Decapoda, Penaeidae). Aqua. Res. 38, 1635–1644.

Ponnuchamy, R., Reddy, S.R., Shakuntala, K., 1981. Effects of eyestalk ablation on growth and food conversion efficiency of the freshwater prawn *Macrobrachium lanchesteri* (De Man). Hydrobiologia 77 (1), 77–80.

Prakash, S., Ajithkumar, T.T., Bauer, R., Thiel, M., Subramoniam, T., 2015. Reproductive morphology and mating behaviour in the hingebeak shrimp *Rhynchocinetes durbanensis* Gordon, 1936 (Decapoda: Caridea: Rhynchocinetidae) in India. J. Mar. Biol. Assoc. U.K. 1–10. http://dx.doi.org/10.1017/S0025315415001083.

Preechaphol, R., Klinbunga, S., Yamano, K., Menasveta, P., 2010. Molecular cloning and expression of progestin membrane receptor component 1 (Pgmrc1) of the giant tigershrimp *Penaeus monodon*. Gen. Comp. Endocrinol. 168, 440–449.

Primavera, J.H., Borlongan, E., 1978. Ovarian rematuration of ablated sugpo prawn *Penaeus monodon* Fabricius. In: Annales de Biologie Animale Biochimie Biophysique 18 (4), 1067–1072.

Primavera, J.H., Caballero, R.M.V., 1992. Light color and ovarian maturation in unablated and ablated giant tiger prawn *Penaeus monodon* (Fabricius). Aquaculture 108, 247–256.

Primavera, J.H., 1978. Induced maturation and spawning in five-month old *Penaeus monodon* Fab, by eyestalk ablation. Aquaculture 13, 355–359.

Primavera, J.H., 1985. A review of maturation and reproduction in closed thelycum penaeids. In: The Proceedings of the First International Conference on the Culture of Penaeid Prawns/shrimps, pp. 47–64 (Illoilo City, Philippines).

Pryor, M.G.M., Russell, P.B., Todd, A.R., 1946. Protocatechuic acid, the substance responsible for the hardening of the cockroach ootheca. Biochem. J. 40, 627–628.

Przeslawski, R., Ahyong, S., Byrne, M., Woerheide, G., Hutchings, P.A.T., 2008. Beyond corals and fish: the effects of climate change on noncoral benthic invertebrates of tropical reefs. Glob. Change Biol. 14, 2773–2795.

Pudadera, R.A., Primavera, J.H., 1981. Effect of light quality and eyestalk ablation on ovarian maturation in *Penaeus monodon*. Kalikasan Philipp. J. Biol. 10, 231–241.

Qiu, G.F., Liu, P., 2009. On the role of Cdc2 kinase during meiotic maturation of oocyte in the Chinese mitten crab, *Eriocheir sinensis*. Comp. Biochem. Physiol. B. Biochem. Mol. Biol. 152 (3), 243–248.

Qiu, G.-F., Yamano, K., 2005. Three forms of cyclin B transcripts in the ovary of the kuruma prawn *Marsupenaeus japonicus*: their molecular characterizations and expression profiles during oogenesis. Comp. Biochem. Physiol. B 141, 186–195.

Quackenbush, L.S., Herrnkind, W.F., 1981. Regulation of molt and gonadal development in the spiny lobster, *Panulirus argus* (Crustacea: Palinuridae): effect of eyestalk ablation. Comp. Biochem. Physiol. Part A 69 (3), 523–527.

Quackenbush, L.S., Keeley, L.L., 1988. Regulation of vitellogenesis in the fiddler crab, *Uca pugilator*. Biol. Bull. 175, 321–331.

Quackenbush, L.S., 1989. Vitellogenesis in the shrimp *Penaeus vannamei*: in vitro studies of the isolated hepatopancreas and ovary. Comp. Biochem. Physiol. 94 (Part B), 253–261.

Quesnel, V.C., 1975. Breeding season and breeding size of female *Emerita portoricensis* Schmitt (Crustacea: Anomura) in Trinidad, West Indies. J. Trinidad Tobago Field Nat. Club 1, 54–59.

Quinitio, E.T., Yamauchi, K., Hara, A., Fuji, A., 1991. Profiles of progesterone and estradiol-like substances in the hemolymph of female *Pandalus kessleri* during an annual reproductive cycle. Gen. Comp. Endocrinol. 81 (3), 343–348.

Quinitio, E.T., Hara, A., Yamauchi, K., Nakos, S., 1994. Changes in the steroid hormone and vitellogenin levels during the gametogenic cycle of the giant tiger shrimp, *Penaeus monodon*. Comp. Biochem. Physiol. B 109, 21–26.

Ra'Anan, Z., Sagi, A., 1985. Alternative mating strategies in male morphotypes of the freshwater prawn *Macrobrachium rosenbergii* (De Man). Biol. Bull. 169, 592–601.

Radha, T., Subramoniam, T., 1985. Origin and nature of spermatophoric mass of the spiny lobster *Panulirus homarus*. Mar. Biol. 86 (1), 13–19.

Radu, D.V.G., Cracium, C., 1976. The ultrastructure of the androgenic gland in *Porcellio scaber* Latr. (terrestrial isopods). Cell Tissue Res. 175, 245–263.

Rahaman, A.A., 1967. Reproductive and nutritional cycles of the crab *Portunus pelagicus* (Linn) Decapoda, Brachyura of Madras coast. Proc. Indian Acad. Sci. B 65, 76–82.

Rahman, N., Dunham, D.W., Govind, C.K., 2003. Social monogamy in the big-clawed snapping shrimp, *Alpheus heterochelis*. Ethology 109, 457–473.

Raikhel, A.S., Dhadialla, T.S., 1992. Accumulation of yolk proteins in insect oocytes. Ann. Rev. Entomol. 37 (1), 217–251.

Ramachandran, N., 1992. Phospholipase Activity in the Developing Eggs of *Emerita asiatica* (Milne Edwards). M. Phil. Dissertation. University of Madras, India, p. 57.

Ramalingam, S., Craig, G.B., 1978. Fine structure of the male accessory glands in *Aedes triseriatus*. J. Insect Physiol. 24, 251257–255259.

Ramos-Trujillo, L., Flores, G.A., 1983. Induction artificial a la maduracion gonadal en hembras de *Penaeus notialis* Perez Farfante, 1967 por oculotomia. Rev. Inv. Mar. Univ. Hab. 4, 33–61 (in Spanish with English abstract).

Rani, K., Subramoniam, T., 1997. Vitellogenesis in the mud crab *Scylla serrata*-an in vivo isotope study. J. Crust. Biol. 17, 659–665.

Rankin, S.M., Davis, R.W., 1990. Ultrastructure of oocytes of the shrimp, *Penaeus vannamei*: cortical specialization formation. Tissue Cell 22 (6), 879–893.

Rateau, J.G., Zerbib, C., 1978. Ultrastructure of the oocyte follicles in the crustacean amphipod *Orchestia gammarellus* (Pallas). Comptes rendus hebdomadaires des seances de l'Academie des sciences. Serie D: Sciences Naturelles 286 (1), 65–68.

Rawson, P.D., Macnamee, R., Frick, M.G., Williams, K.L., 2003. Phylogeography of the coronulid barnacle, *Chelonibia testudinaria*, from loggerhead sea turtles, *Caretta caretta*. Mol. Ecol. 12, 2697–2706.

Raymont, J.E.G., Krishnaswamy, S., Woodhouse, M.A., Griffin, R.L., 1974. Studies on the fine structure of Copepoda. Observations on *Calanus finmarchicus* (Gunnerus). Proc. R. Soc. Lond. B 185 (1081), 409–424.

Reddy, P.S., Ramamurthy, R., 1998. Methyl farnesoate stimulates ovarian maturation in the freshwater crab *Oziotelphusa senex senex*. Curr. Sci. 74, 68–70.

Reddy, P.R.C., Nagaraju, G.P.C., Reddy, P.S., 2004. Involvement of methyl farnesoate in regulation of molting and reproduction in freshwater crab, *Oziotelphusa senex senex*. J. Crust. Biol. 24, 511–515.

Reddy, P.R., Kiranmayi, P., Kumari, K.T., Reddy, P.S., 2006. 17α-Hydroxyprogesterone induced ovarian growth and vitellogenesis in the freshwater rice field crab *Oziotelphusa senex senex*. Aquaculture 254 (1), 768–775.

Reese, E.S., 1968. Annual breeding seasons of three sympatric species of tropical intertidal hermit crabs, with a discussion of factors controlling breeding. J. Exp. Mar. Biol. Ecol. 2, 308–318.

Reger, J.F., Fain-Maurel, M.A., 1973. A comparative study on the origin, distribution, and fine structure of extracellular tubules in the male reproductive system of species of isopods, amphipods, schizopods, copepods, and cumacea. J. Ultrastruct. Res. 44 (3), 235–252.

Reger, J.F., Florendo, N.T., 1969. Studies on motile, nontubule-containing, filiform spermatozoa of the ostracod cypridopsis: I. Spermiogenesis. J. Ultrastruct. Res. 28 (3), 235–249.

Reger, J.F., 1970. Some aspects of the fine structure of filiform spermatozoa (Ostracoda, Cypridopsis) lacking tubule substructure. In: Baccetti, B. (Ed.), Comparative Spermatology. Academic Press, New York, pp. 237–245.

Reppond, K., Rugolo, L., de Oliveira, A.C.M., 2008. Change in biochemical composition during development of snow crab, *Chionoecetes opilio* embryos. J. Crust. Biol. 28 (3), 519–527.

Reverberi, G., Pittoti, M., 1942. Il ciclo biologico e la determinazione fenotipica del sesso di Ione thoracica Montagu, Bopyride parassita si *Callianassa laticauda* Otto. Pubblicazioni della Stazione Zoologica di Napoli 19, 111–184.

Ribeiro, E.A., Genofre, G.C., McNamara, J.C., 2001. Identification and quantification of carotenoid pigments during the embryonic development of the freshwater shrimp *Macrobrachium olfersii* (Crustacea, Decapoda). Mar. Freshwater Behav. Physiol. 34 (2), 105–116.

Rice, S.A., 1978. Spermatophores and sperm transfer in spionid polychaetes. Trans. Am. Microscop. Soc. 97, 160–170.

Richards, A.G., 1951. The Integument of Arthropods. University of Minnesota Press, Minneapolis, p. 411.

Richardson, H.G.D., Decaraman, M., Fingerman, M., 1991. The effect of biogenic amines on ovarian development in fiddler crab, *Uca pugilator*. Comp. Biochem. Physiol. 99, 53–56.

Rider, C.V., Gorr, T.A., Olmstead, A.W., Wasilak, B.A., LeBlanc, G.A., 2005. Stress signaling: coregulation of hemoglobin and male sex determination through a terpenoid signaling pathway in a crustacean. J. Exp. Boil. 208, 15–23.

Ridley, M., 1983. The Explanation of Organic Diversity: The Comparative Method and Adaptations for Mating. Clarendon Press, Oxford, pp. 1–272.

Rigaud, T., Moreau, J., 2004. A cost of *Wolbachia*-induced sex reversal and female-biased sex ratios: decrease in female fertility after sperm depletion in a terrestrial isopod. Proc. R. Soc. Lond. B 271, 1941–1946.

Rigaud, T., Juchault, P., Mocquard, J., 1997. The evolution of sex determination in isopod crustaceans. Bioessays 19, 409–416.

Ríos, M., Barros, C., 1997. Trypsin-like enzymes during fertilization in the shrimp *Rhynchocinetes typus*. Mol. Reprod. Dev. 46, 581–586.

Robinett, C.C., Vaughan, A.G., Knapp, J.M., Baker, B.S., 2010. Sex and the single cell. II. There is a time and place for sex. PLoS Biol. 8 (5), e1000365.

Robinson, B.W., Doyle, R.W., 1985. Trade-off between male reproduction (amplexus) and growth in the amphipod *Gammarus lawrencianus*. Biol. Bull. 168, 482–488.

Rodgers-Gray, T.P., Smith, J.E., Ashcroft, A.E., Isaac, R.E., Dunn, A.M., 2004. Mechanisms of parasite-induced sex reversal in *Gammarus duebeni*. Int. J. Parasitol. 34, 747–753.

Rodriguez, A.M., Roca, M.P., Palou, A., 2002. Opposite actions of testosterone and progesterone on UCP1 mRNA expression in cultured brown adipocytes. Cell. Mol. Life Sci. 59, 1714–1723.

Rodríguez, E.M., Medesani, D.A., Fingerman, M., 2007. Endocrine disruption in crustaceans due to pollutants: a review. Comp. Biochem. Physiol. Part A 146, 661–671.

Rojas, R., Morales, M.C., Rivadeneira, M.M., Thiel, M., 2012. Male morphotypes in the Andean river shrimp *Cryphiops caementarius* (Decapoda: Caridea): morphology, coloration and injuries. J. Zool. 288, 21–32.

Rosen, O., Manor, R., Weil, S., Gafni, O., Linial, A., Aflalo, E.D., Ventura, T., Sagi, A., 2010. A sexual shift induced by silencing of a single insulin-like gene in crayfish: ovarian upregulation and testicular degeneration. PLoS ONE 5, e15281.

Rosenberg, M.P., Lee, H.H., 1981. The role of Ca and Mg in star fish oocyte maturation induced by 1-methyladenine. J. Exp. Zool. 217, 389–397.

Ross, P.M., 2001. Larval supply, settlement and survival of barnacles in a temperate mangrove forest. Mar. Ecol. Prog. Ser. 215, 237–249.

Roth, Z., Khalaila, I., 2012. Identification and characterization of the vitellogenin receptor in *Macrobrachium rosenbergii* and its expression during vitellogenesis. Mol. Reprod. Dev. 79 (7), 478–487.

Ruangpanit, N., Maneevong, S., Pechmanee, T., Tanon, T., 1981. Induced ovarian maturation and rematuration of *Penaeus monodon* Fabricius by eyestalk ablation. Ann. Rep. Natl. Inst. Coastal Aquacult. Fish. Dep. (Thailand) 82–106.

Rudolph, E., Verdi, A., Tapia, J., 2001. Intersexuality in the burrowing crayfish *Parastacus varicosus* Faxon, 1898 (Decapoda, Parastacidae). Crustaceana 74, 27–37.

Rudolph, E.H., 1995. Partial protandric hermaphroditism in the burrowing crayfish *Parastacus nicoleti* (Philippi, 1882) (Decapoda, Parastacidae). J. Crust. Biol. 15, 720–732.

Rudolph, E.H., 2002. New records of intersexuality in the freshwater crayfish *Samastacus spinifrons* (Decapoda, parastacidae). J. Crust. Biol. 22, 377–389.

Rusch, M.E., 1958. Weibliche heterogametie bei *Cyclops viridus* Jurine. Naturwissenschaften 45, 69.

Rusch, M.E., 1960. Untersuchungen uber Geschlechtsbestirnmungsmechanism bei Copepoden. Chromosoma 11, 419–432.

Ryan, E.P., 1966. Pheromone: evidence in a decapod crustacean. Science 151 (3708), 340–341.

Ryan, E.P., 1967. Structure and function of the reproductive system of the crab *Portunus sanguinolentus* (Herbst) (Brachyura: Portunidae). II. The female system. In: Proc. Symp. Crustacea, vol. 2. , pp. 522–544.

Ryan, E.P., 1984. Spermatogenesis and male reproductive system in the Hawaiian crab *Ranina ranina* (L.). In: Clark Jr., W.H., Engels, W. (Eds.), Advances in Invertebrate Reproduction. Elsevier, New York, p. 629.

Sunil, I., Senthil Murugan, T., Venugopalan, V.P., Subramoniam, T., Munuswamy, N., Van der Velde, G., 2006. Larval development in sand crab *Emerita emeritus* (L., 1767) (Anomura, Hippoidae) reared in the laboratory. Crustaceana 70 (4), 441–458.

Saetan, J., Senarai, T., Tamtin, M., Weerachatyanukul, W., Chavadej, Hanna, P.J., Parhar, I., Sobhon, P., Sretarugsa, P., 2013. Histological organization of the central nervous system and distribution of a gonadotropin-releasing hormone-like peptide in the blue crab, *Portunus pelagicus*. Cell Tissue Res. 353, 493–510.

Sagawa, K., Yamagata, H., Shiga, Y., 2005. Exploring embryonic germ line development in the water flea, *Daphnia magna*, by zinc-finger-containing VASA as a marker. Gene Exp. Patterns 5, 669–678.

Sagi, A., Cohen, D., 1990. Growth, maturation and progeny of sex-reversed. World Aquacult. 21, 87.

Sagi, A., Homola, E., Laufer, H., 1993. Distinct reproductive types of male spider crabs *Libinia emarginata* differ in circulating and synthesizing methyl farnesoate. Biol. Bull. 185, 168–173.

Sagi, A., Rise, M., Isam, K., Arad, S.M., 1995a. Carotenoids and their derivatives in organs of the maturing female crayfish *Cherax quadricarinatus*. Comp. Biochem. Physiol. Part B 112 (2), 309–313.

Sagi, A., Silkovsy, J., Fleisher-Berkovich, S., Danon, A., Chayoth, R., 1995b. Prostaglandin E2 in previtellogeic ovaries of the prawn *Macrobrachium rosenbergii*: synthesis and effect on the level of cAMP. Gen. Comp. Endocrinol. 100 (3), 308–313.

Sagi, A., Khalaila, I., Barki, A., Hulata, G., Karplus, I., 1996. Intersex red claw crayfish, *Cherax quadricarinatus* (von Martens): functional males with pre-vitellogenic ovaries. Biol. Bull. 190, 16–23.

Sagi, A., Manor, R., Segall, C., Davis, C., Khalaila, I., 2002. On intersexuality in the crayfish *Cherax quadricarinatus*: an inducible sexual plasticity model. Invertebr. Reprod. Dev. 41, 27–33.

Sainath, S.B., Reddy, P.S., 2011. Effect of selected biogenic amines on reproduction in the fresh water edible crab, *Oziotelphusa senex senex*. Aquaculture 313 (1), 144–148.

Sainte-Maire, G., Sainte-Marie, B., 1998. Morphology of the spermatheca, oviduct, intermediate chamber, and vagina of the adult snow crab (*Chionoecetes opilio*). Can. J. Zool. 76, 1589–1604.

Sainte-Marie, G., Sainte-Marie, B., 1999. Reproductive processes in the adult snow crab (*Chionoecetes opilio*). I. Observations on spermiogenesis and spermatophore formation in the vas deferens. Can. J. Zool. 77 (3), 440–450.

Salmon, M., 1984. The courtship, aggression and mating system of a "primitive" fiddler crab (*Uca vocans*: Ocypodidae). Trans. Zool. Soc. Lond. 37, 1–50.

Salmon, M., 1987. On the reproductive behavior of the fiddler crab *Uca thayeri*, with comparisons to *U. pugilator* and *U. vocans*: evidence for behavioural convergence. J. Crust. Biol. 7, 25–44.

Salvador, J.A., Jaramillo, B., Romero, Z., Lucien-Brun, H., Lee, D.O.'.C., 1988. A comparative analysis of sperm quality, between wild caught males and unilaterally eyestalk ablated regenerated males in a commercial *Penaeus vannamei* facility. J. World Aquact. Soc. 19 (1), 61A.

Sandeman, D., Sandeman, R., Derby, C.D., Schmidt, M., 1992. Morphology of the brain of crayfish, crabs and spiny lobsters: a common nomenclature for homologous structures. Biol. Bull. 183, 304–326.

Sanders, M.B., Billinghurst, Z., Depledge, M.H., Clare, A.S., 2005. Larval development and vitellin-like protein expression in *Palaemon elegans* larvae following xeno-oestrogen exposure. Integr. Comp. Biol. 45 (1), 51–60.

Santhanakumar, J., Mohanraju, R., Kirubagaran, R., Subramoniam, T., 2014. Mate guarding behaviour in the supralittoral isopod, *Ligia dentipes* (Oniscidea) from the Andaman and Nicobar islands. Invertebr. Repr. Dev. 58, 128–137.

Santiago, A.C., 1977. Successful spawning of cultured *Penaeus monodon* Fab. after eyestalk ablation. Aquaculture 11, 185–196.

Sappingotn, T.W., Oishi, K., Raikhel, A.S., 2002. Structural characteristics of insect vitellogenin. In: Raikhel, A.S., Sappington, T.W. (Eds.), Reproductive Biology of Invertebrates, vol. XII (Part A), Science Publishers Inc., Enfield, pp. 69–101.

Sarika, S.N., Anilkumar, G., 2014. DNA binding domain of retinoid receptor gene (RXR) from a field crab inhabiting the Indian Peninsula. Indian J. Biotechnol. 13, 52–56.

Sarojini, R., Jayalakshmi, K., Sambashivarao, S., 1986. Effect of external steroids on ovarian development in freshwater prawn, *Macrobrachium rosenbergii*. J. Adv. Zool. 7 (1), 50.

Sarojini, R., Sambasiva Rao, S., Jayalakshmi, K., Nagabhushanam, R., 1989. Effect of PG F1a on ovarian maturation in a penaeid prawn *Metapenaeus affinis*. J. Zool. Res. 2, 7–11.

Sarojini, R., Nagabhushanam, R., Fingerman, M., 1995. *In vivo* inhibition by dopamine of 5-hydroxytryptamine-stimulated ovarian maturation in the red swamp crayfish, *Procamburus clarkii*. Experientia 51, 156–158.

Sarojini, R., Nagabhushanam, R., Fingerman, M., 1996. In vitro inhibition by dopamine of 5-hydroxytryptamine stimulated ovarian maturation in the red swamp crayfish, *Procambarus clarkii*. Cell. Mol. Life Sci. 52, 707–709.

Sarojini, R., Nagabhushanam, R., Fingerman, M., 1997. An in vitro study of the inhibitory action of methionine enkephalin on ovarian maturation in the red swamp crayfish *Procambarus clarkii*. Comp. Biochem. Physiol. 117C, 207–210.

Sasikala, S.L., Subramoniam, T., 1987. On the occurrence of acid mucopolysaccharides in the spermatophores of two marine prawns, *Penaeus indicus* (Milne-Edwards) and *Metapenaeus monoceros* (Fabricius) (Crustacea: Macrura). J. Exp. Mar. Biol. Ecol. 113 (2), 145–153.

Sasikala, S.L., Subramoniam, T., 1991. Influence of juvenile hormone JH III on ovarian activity of adult paddy field crab *Paratelphusa hydrodromous* Herbst. Indian J. Exp. Biol. 29 (5), 426–429.

Sassaman, C., Fugate, M., 1997. Gynandromorphism in Anostraca: multiple mechanisms of origin. Hydrobiologia 359, 163–169.

Sassaman, C., Weeks, S.C., 1993. The genetic mechanism of sex determination in the conchostracan shrimp *Eulimnadia texana*. Am. Nat. 314–328.

Sassaman, C., Simovich, M.A., Fugate, M., 1997. Reproductive isolation and genetic differentiation in North American species of *Triops* (Crustacea: Branchiopoda: Notostraca). Hydrobiologia 359, 125–147.

Sassaman, C., 1991. Sex ratio variation in female-biased populations of Notostracans. In: Studies on Large Branchiopod Biology and Aquaculture. Springer, Netherlands, pp. 169–179.

Sastry, A.N., 1983. Ecological aspects of reproduction. The biology of Crustacea. Environ. Adapt. 8, 179–270.

Sathapondecha, P., Panyim, S., Udomkit, A., 2015. A novel function of bursicon in stimulation of vitellogenin expression in black tiger shrimp, *Penaeus monodon*. Aquaculture 446, 80–87.

Scelzo, M.A., Mantelatto, F.L., Tudge, C.C., 2004. Spermatophore morphology of the endemic hermit crab *Loxopagurus loxochelis* (Anomura, Diogenidae) from the southwestern Atlantic—Brazil and Argentina. Invertebr. Rep. Dev. 46 (1), 1–9.

Schaller, F., 1971. Indirect sperm transfer by soil arthropods. Ann. Rev. Entomol. 16 (1), 407–446.

Schaller, F., 1980. Significance of sperm transfer and formation of spermatophores in arthropod Phylogeny. In: Gupta, A.P. (Ed.), Arthropod Physiology. Van Nostrand-Reinhold, New York, pp. 587–607.

Schmidt, M., Derby, C.D., 2005. Non-olfactory chemoreceptors in asymmetric setae activate antennular grooming behavior in the Caribbean spiny lobster *Panulirus argus*. J. Exp. Biol. 208, 233–248.

Schneider, W.J., 1996. Vitellogenin receptors: oocye-specific members of the low density lipoprotein receptor supergene family. Int. Rev. Cytol. 166, 103–137.

Scholtz, G., 2000. Evolution of the nauplius stage in malacostracan crustaceans. J. Zool. Syst. Evol. Res. 38 (3), 175–187.

Schram, F.R., Jenner, R.A., 2001. The origin of Hexapoda: a crustacean perspective. Ann. Soc. Entomol. Fr. 37 (1–2), 243–264. Société entomologique de France.

Schurko, A.M., Logsdon, J.M., Eads, B.D., 2009. Meiosis genes in *Daphnia pulex* and the role of parthenogenesis in genome evolution. BMC Evol. Biol. 9, 78.

Seeger, E., 1996. Christmas crabs. Wildl. Conserv. 99, 28–35.

Seibt, U., Wickler, W., 1979. The biological significance of the pair-bond in the shrimp *Hymenocera picta*. Z. für Tierpsychologie 50, 166–179.

Seibt, U., 1973. Sense of smell and pair-bond in *Hymenocera picta* Dana. Micronesica 9, 231–236.

Seifert, P., 1982. Studies on the sex pheromone of the shore crab, *Carcinus maenas*, with special regard to ecdysone excretion. Ophelia 21 (2), 147–158.

Seiple, W., Salmon, M., 1982. Cooperative social behavior of two graspid crabs, *Sesarma reticulatum* (Say) and *S. cinereum* (Bosc). J. Exp. Mar. Biol. Ecol. 62, 1–24.

Sellars, M.J., Lyons, R.E., Grewe, P.M., Vuocolo, T., Leeton, L., Coman, G.J., Preston, N.P., 2007. A PL10 vasa-like gene in the kuruma shrimp, *Marsupenaeus japonicus*, expressed during development and in adult gonad. Mar. Biotechnol. 9, 377–387.

Sexton, E.W., 1924. The moulting and growth stages of *Gammarus* with descriptions of the normals and intersexes of *G. chevreuxi*. J. Mar. Biol. Assoc. U.K. 13, 340–401.

Sharabi, O., Manor, R., Weil, S., Aflalo, E.D., Lezer, Y., Levy, T., Sagi, A., 2015. Identification and characterization of an insulin-like receptor involved in crustacean reproduction. Endocrinology 157 (2), 928–941 en-2015.

Shigekawa, K., Clark Jr., W.H., 1986. Spermiogenesis in the marine shrimp, *Sicyonia ingentis*. Dev. Growth Differ. 28, 95–112.

Shin, H.S., An, K.W., Park, M.S., Jeong, M.H., Choi, C.Y., 2009. Quantitative mRNA expression of sox3 and DMRT1 during sex reversal, and expression profiles after GnRHa administration in black porgy, *Acanthopagrus schlegeli*. Comp. Biochem. Physiol. B Biochem. Mol. Biol. 154, 150–156.

Shuster, S.M., Caldwell, R.L., 1989. Male defense of the breeding cavity and factors affecting the persistence of breeding pairs in the stomatopod, *Gonodactylus bredini* (Manning) (Crustacea: Hoplocarida). Ethology 82, 192–207.

Shuster, S.M., Wade, M.J., 1991. Female copying and sexual selection in a marine isopod crustacean, *Paracerceis sculpta*. Anim. Behav. 41, 1071–1078.

Shuster, S.M., Wade, M.J., 2003. Mating Systems and Strategies. Princeton University Press.

Shuster, S.M., 1987. Alternative reproductive behaviors: three discrete male morphs in *Paracerceis sculpta*, an intertidal isopod from the northern Gulf of California. J. Crust. Biol. 318–327.

Sigvardt, Z.M., Olesen, J., 2014. Mating behaviour in laevicaudatan clam shrimp (Crustacea, Branchiopoda) and functional morphology of male claspers in a phylogenetic context: a video-based analysis. PLoS ONE e84021.

Silas, E.G., Sankarankutty, C., 1967. Field investigations on the shore crabs of the Gulf of Mannar and Palk Bay, with special reference to the ecology and behavior of the pellet crab *Scopimera proxima* Kemp. Mar. Biol. Assoc. India Symp. Ser. 2, 1008–1025.

Silkovsky, J., Chayoth, R., Sagi, A., 1998. Comparative study of effects of Prostaglandin E 2 on cAMP levels in gonads of the prawn, *Macrobrachium rosenbergii* and the Crayfish *Cherax quadricarinatus*. J. Crust. Biol. 18 (4), 643.

Silva, J.R.F., Cruz-Landim, D.C., 2006. Macroscopic aspects and scanning electron microscopy of the ovaries of the spiny lobsters *Panulirus* (Crustacea: Decapoda). Braz. J. Morphol. Sci. 23 (3/4), 479–486.

Simeó, C.G., Ribes, E., Rotllant, G., 2009. Internal anatomy and ultrastructure of the male reproductive system of the spider crab *Maja brachydactyla* (Decapoda: Brachyura). Tissue Cell 41 (5), 345–361.

Sinclair, M.E., 1977. Agonistic behavior of the stone crab, *Menippe mercenaria* (Say). Anim. Behav. 25, 193–207.
Smith, D.G., 1967. An occurrence of a female intersex of *Asellus communis* (Isopoda, Asellidae). Crustaceana 32, 89–90.
Snedden, W.A., 1990. Determinants of male mating success in the temperate crayfish *Orconectes rusticus*: chela size and sperm competition. Behaviour 115, 101–113.
Sneddon, L.U., Huntingford, F.A., Taylor, A.C., Clare, A.S., 2003. Female sex pheromone-mediated effects on behavior and consequences of male competition in the shore crab (*Carcinus maenas*). J. Chem. Ecol. 29, 55–70.
Snow, C.D., Nielson, J.R., 1966. Premating and mating behavior of the Dungeness crab (*Cancer magister Dana*). J. Fish. Res. Bd. Canada 23, 1319–1323.
Sokolowicz, C.C., López-Greco, L.S., Gonçalves, R., Bond-Buckup, G., 2007. The gonads of *Aegla platensis* Schmitt (Decapoda, Anomura, Aeglidae): a macroscopic and histological perspective. Acta Zool. 88 (1), 71–79.
Soroka, Y., Milner, Y., Sagi, A., 2000a. The hepatopancreas as a site of yolk protein synthesis in the prawn *Macrobrachium rosenbergii*. Invertebr. Reprod. Dev. 37 (1), 61–68.
Soroka, Y., Sagi, A., Khalaila, I., Abdu, U., Milner, Y., 2000b. Changes in protein kinase C during vitellogenesis in the crayfish *Cherax quadricarinatus*—possible activation by methyl farnesoate. Gen. Comp. Endocrinol. 118, 200–208.
Soundarapandian, P., Dinakaran, G.K., Varadharajan, D., 2013. Biochemical Composition during the embryonic development and freshly hatched zoea of *Macrobrachium idella* (Hilgendorf, 1898). Adv. Tech. Biol. Med. 1, 111.
Souty, C., Picaud, J.L., 1981. Vitellogenin synthesis in the fat body of the marine crustacean Isopoda, *Idotea balthica* during vitellogenesis. Reprod. Nutr. Dév. 21 (1), 95–101.
Soyez, D., Van Deijnen, J.E., Martin, M., 1987. Isolation and characterization of a vitellogenesis – inhibiting factor from sinus glands of the lobster *Homarus americanus*. J. Exp. Zool. 244, 479–484.
Soyez, D., LeCaer, J.P., Noel, P.Y., Rossier, J., 1991. Primary structure of two isoforms of the vitellogenesis inhibiting hormone from the lobster *Homarus americanus*. Neuropeptides 20, 25–32.
Spalding, J.F., 1942. The nature and formation of the spermatophore and sperm plug in *Carcinus maenas*. Quart. J. Microsco. Sci. 83, 399–422.
Sparkes, T.C., Keogh, D.P., Pary, R.A., 1996. Energetic costs of mate guarding behavior in male stream-dwelling isopods. Oecologia 106, 166–171.
Sparkes, T.C., Keogh, D.P., Haskins, K.E., 2000. Female resistance and male preference in a stream-dwelling isopod: effects of female molt characteristics. Behav. Ecol. Sociobiol. 47, 145–155.
Sparkes, T.C., Keogh, D.P., Orsburn, T.H., 2002. Female resistance and mating outcomes in a stream-dwelling isopod: effects of male energy reserves and mating history. Behaviour 139, 875–895.
Spaziani, E.P., Hinsch, G.W., Edwards, S.C., 1993. Changes in prostaglandin E(2) and F2-alpha during vitellogenesis in the Florida crayfish. J. Comp. Physiol. B 163 (7), 541–545.
Spindler, K.D., Van Wormhoudt, A., Sellos, D., Spindler-Barth, M., 1987. Ecdysteroids levels during embryogenesis in the shrimp, *Palaemon serratus* (Crustacea: Decapoda)- Quantitative and qualitative changes. Gen. Comp. Endocrinol. 66, 116–122.
Spitschakoff, T., 1912. *Lysmata seticaudata* Risso, als Beispiel einesechten Hermaphroditismus beiden Decapoden. Z. Wiss. Zool. 100, 190–209.
Sroyraya, M., Hanna, P.J., Changklungmoa, N., Senarai, T., Siangcham, T., Tinikul, Y., Sobhon, P., 2013. Expression of the male reproduction-related gene in spermatic ducts of the blue swimming crab, *Portunus pelagicus*, and transfer of modified protein to the sperm acrosome. Micro. Res. Tech. 76 (1), 102–112.

Staiger, H., Bocquet, C., 1954. Cytological demonstration of female heterogamety in isopods. Experientia 10, 64–66.

Stebbing, P.D., Bentley, M.G., Watson, G.J., 2003. Mating behaviour and evidence for a female released courtship pheromone in the signal crayfish *Pacifastacus leniusculus*. J. Chem. Ecol. 29, 465–475.

Steel, C.G.H., Vafopoulou, X., 1998. Ecdysteroid titres in haemolymph and other tissues during moulting and reproduction in the terrestrial isopod, *Oniscus asellus* (L.). Invertebr. Reprod. Dev. 34, 187–194.

Steele, D.H., 1967. The life cycle of the marine amphipod *Stegocephalus inflatus* Kroyer in the Northwest Atlantic. Can. J. Zool. 45, 623–628.

Stefani, R., 1960. L'Artemia salina pathenogenetica a Cagliari. Rivista di Biologia 53, 463–490.

Stefani, R., 1963. La digametia femminile in *Artemia salina* Leach e la costituzione del corredo cromosomico nei biotipi diploide anfigonico e diploide partenogenetico. Caryologia 16 (3), 625–636.

Stefani, R., 1967. La maturazione dell'uovo nell' *Artemia salina* di Se'te. Rivista di Biologia 60, 599–615.

Stephens, G.C., 1952. The control of cement gland development in the crayfish, *Cambarus*. Biol. Bull. 103, 242–258.

Stewart, M.J., Stewart, P., Soonklang, N., Linthong, V., Hanna, P.J., Duan, W., Sobhon, P., 2010. Spermatogenesis in the blue swimming crab, *Portunus pelagicus*, and evidence for histones in mature sperm nuclei. Tissue Cell 42 (3), 137–150.

Stoner, A.W., 1986. Cohabitation on algal habitat islands by two hermaphroditic Tanaidacea (Crustacea: Peracarida). J. Crust. Biol. 6, 719–728.

Strong Jr., D.R., 1973. Amphipod amplexus, the significance of ecotypic variation. Ecology 1383–1388.

Subrahmanyam, C.B., 1963. A note on annual reproductive cycle of prawn *Penaeus indicus* of Madras Coast. Curr. Sci. 32, 165.

Subramoniam, T., Gunamalai, V., 2003. Breeding biology of the intertidal sand crab, *Emerita* (Decapoda: Anomura). Adv. Mar. Biol. 46, 91–182.

Subramoniam, T., Kirubagaran, K., 2010. Endocrine regulation of vitellogenesis in lobsters. J. Mar. Biol. Assoc. India 52, 229–236.

Subramoniam, T., Panneerselvam, M., 1985. Semi-annual breeding pattern in the burrowing sand crab *Albunea symmysta* (L.) (= symnista) of Madras coast. Indian J. Mar. Sci. 14, 226–227.

Subramoniam, T., Tirumalai, R., Gunamalai, V., Hoffmann, K.H., 1999. Embryonic ecdysteroids in a mole crab, *Emerita asiatica* Milne Edwards. J. Biosci. 24, 91–96.

Subramoniam, T., 1977a. Continuous breeding in the tropical anomuran crab, *Emerita asiatica* Milne Edwards from the Madras coast. In: Adiyodi, K.G., Adiyodi, R.G. (Eds.), Advances in Invertebrate Reproduction. vol. 1. Peralam-Kenoth, Karivellur, Kerala, pp. 166–174.

Subramoniam, T., 1977b. Aspects of sexual biology of the anomuran crab *Emerita asiatica*. Mar. Biol. 43 (4), 369–377.

Subramoniam, T., 1979a. Some aspects of reproductive ecology of a mole crab *Emerita asiatica* Milne Edwards. J. Exp. Mar. Biol. Ecol. 36, 259–268.

Subramoniam, T., 1979b. Sex reversal in a mole crab *Emerita asiatica* (Abstract). In: Inter. Symp. Invest. Reprod. Univ. Calif. Davis, Aug. 27–31.

Subramoniam, T., 1981. Protandric hermaphroditiism in a mole crab *Emerita asiatica* (Decapoda: Anomura). Biol. Bull. 160, 161–174.

Subramoniam, T., 1984. Spermatophore formation in two intertidal crabs *Albunea symnista* and *Emerita asiatica* (Decapoda: Anomura). Biol. Bull. 166 (1), 78–95.

Subramoniam, T., 1991a. Yolk utilization and esterase activity in the mole crab *Emerita asiatica* (Milne Edwards). Crustacean Issue. 7, 19–29.

Subramoniam, T., 1991b. Chemical composition of spermatophores in decapods Crustaceans. In: Bauer, R.T., Martin, J.W. (Eds.), Crustacean Sexual Biology. Columbia University Press, New York, pp. 308–321.

Subramoniam, T., 1993. Spermatophores and sperm transfer in marine crustaceans. Adv. Mar. Biol. 29, 129–214.

Subramoniam, T., 1995. Light and electron microscopic studies on the seminal secretions and the vas deferens of the penaeiodean shrimp, *Sicyonia ingentis*. J. Biosci. 20 (5), 691–706.

Subramoniam, T., 1999. Endocrine regulation of egg production in economically important crustaceans. Curr. Sci. 76, 350–360.

Subramoniam, T., 2000. Crustacean ecdysteroids in reproduction and embryogenesis. Comp. Biochem. Physiol. 125C, 135–156.

Subramoniam, T., 2011. Mechanisms and control of vitellogenesis in crustaceans. Fisher. Sci. 77 (1), 1–21.

Subramoniam, T., 2013. Origin and occurrence of sexual and mating systems in Crustacea: a progression towards communal living and eusociality. J. Biosci. 38, 951–969.

Sudha, K., Anilkumar, G., 2007. Elevated ecdysteroid titer and precocious molt and vitellogenesis induced by eyestalk ablation in the estuarine crab, *Metopograpsus messo (*Brachyura: Decapoda). J. Crust. Biol. 27, 304–308.

Suganthi, A.S., Anilkumar, G., 1998. Ultrastructural studies on the male accessory glands of two estuarine crabs, *Metopograpsus messor* (Forskal) and *Sesarma quadratum* (Fabricious) (Grapsidae: Brachyura, Decapoda). Cytobios 93, 7–21.

Sukumaran, M., 1985. Certain Aspects of Reproductive Biology of the Ghost Crab, Ocypode Platytarsis (Ph.D. thesis). University of Calicut, Kerala, India.

Sundermeyer, K., Hendricks, J.K., Prasad, S.V., Wells, M.A., 1996. The precursor protein of the structural apolipoproteins of lipophorin: cDNA and deduced amino acid sequence. Insect Biochem. Mol. Biol. 26 (8), 735–738.

Sunil, K.K.M., Diwan, A.D., 1994. Vitellogenesis in the Indian white prawn *Penaeus indicus* (Crustacea: Decapoda: Penaeidae. J. Aquact. Trop. 9, 157–172.

Sutcliffe, D.W., 1992. Reproduction in *Gammarus* (Crustacea, Amphipoda): basic processes. Freshwater Forum 2, 102–129.

Suzuki, S., Yamasaki, K., 1997. Sexual Bipotentiality of developing ovaries in the terrestrial Isopod *Armadillidium vulgare* (Malacostraca, Crustacea). Gen. Comp. Endocrinol. 107, 136–146.

Suzuki, S., 1987. Vitellins and vitellogenins of the terrestrial isopod, *Armadillidium vulgare*. Biol. Bull. 173 (2), 345–354.

Swartz, R.C., 1976a. Agonistic and sexual behavior of the xanthid crab, *Neopanope sayi*. Chesapeake Sci. 17, 24–34.

Swartz, R.C., 1976b. Sex ratio as a function of size in the xanthid crab, *Neopanope sayi*. Am. Nat. 110, 898–900.

Swartz, R.C., 1978. Reproductive and molt cycles in the xanthid crab, *Neopanope sayi* (Smith, 1869). Crustaceana 34, 1532.

Takahashi, T., Araki, A., Nomura, Y., Koga, M., Arizono, K., 2000. The occurrence of dual-gender imposex in Japanese freshwater crab. J. Health Sci. 46, 376–379.

Takayanagi, H., Yamamoto, Y., Takeda, N., 1986. An ovary-stimulating factor in the shrimp, *Paratya compressa*. J. Exp. Zool. 240, 203–209.

Talbot, P., Chanmanon, P., 1980a. The structure of sperm from the lobster, *Homarus americanus*. J. Ultrastruc. Res. 70, 275–286.

Talbot, P., Chanmanon, P., 1980b. Morphological features of the acrosome reaction of lobster (Homarus) sperm and the role of the reaction in generating forward sperm movement. J. Ultrastruc. Res. 70, 287–297.

Talbot, P., Demers, D., 1993. Tegumental Glands of Crustacea. The Crustacean Integument: Morphology and Biochemistry. CRC Press, New York, pp. 151–191.

Talbot, P., Summers, R.G., 1978. The structure of sperm from *Panulirus*, the spiny lobster, with special regard to the acrosome. J. Ultrastruc. Res. 64, 341–351.

Talbot, P., Zao, P., 1991. Secretion at molting by the pleopod tegumental glands of the lobster *Homarus americanus* (Milne Edwards). J. Crust. Biol. 11, 1–9.

Talbot, P., 1981. The ovary of the lobster, *Homarus americanus*: II. Structure of the mature follicle and origin of the chorion. J. Ultrastruc. Res. 76 (3), 249–262.

Tan-Fermin, J.D., Pudadera, R.A., 1989. Ovarian maturation stages of the wild giant tiger prawn, *Penaeus monodon* Fabricius. Aquaculture 77 (2), 229–242.

Tan-Fermin, J.D., 1991. Effects of unilateral eyestalk ablation on ovarian histology and oocyte size frequency of wild and pond-reared *Penaeus monodon* (Fabricius) broodstock. Aquaculture 93, 77–86.

Tatarazako, N., Oda, S., Watanabe, H., Morita, M., Iguchi, T., 2003. Juvenile hormone agonists affect the occurrence of male *Daphnia*. Chemosphere 53, 827–833.

Tata, J.R., Smith, D.F., 1979. Vitellogenesis: a versatile model for hormonal regulation of gene expression. Rec. Progr. Horm. Res. 35, 47–95.

Taylor, D.M., 1986. A Bilateral gynandromorphy of the snow crab *Chianocetes opilio*, from NL, Canada. Crustaceana 51, 309–312.

Terossi, M., Tudge, C., Lopez Greco, L.S., Mantelatto, F.L., 2011. A novel spermatozoon ultrastructure in the shrimp *Hippolyte obliquimanus* Dana, 1852 (Decapoda: Caridea: Hippolytidae). Invertebr. Reprod. Dev.. http://dx.doi.org/10.1080/07924259.2011.631040.

Terwilliger, N.B., 1991. Hemocyanins in crustacean oocytes and embryos. In: Wenner, A.M., Kuris, A.M. (Eds.), Crustacean Issues: Crustacean Egg Production. Balkema press, Rotterdam, pp. 31–39.

Teshima, S.I., Kanazawa, A., Horinouchi, K., Koshio, S., 1988. Lipid metabolism in destalked prawn *Penaeus japonicus*: induced maturation and transfer of lipid reserves to the ovaries. Nippon Suisan Gakkaishi 54 (7), 1123–1129.

Teshima, S.I., Kanazawa, A., Koshio, S., Horinouchi, K., 1989. Lipid metabolism of the prawn *Penaeus japonicus* during maturation: variation in lipid profiles of the ovary and hepatopancreas. Comp. Biochem. Physiol. Part B 92 (1), 45–49.

Thampy, D.M., John, P.A., 1974. Sex-reversal and androgenic gland in the fish parasite Irona far (Cymothoidae: Isopoda: Crustacea). Int. J. Parasitol. 4 (6), 575–583.

Thiel, M., Baeza, J.A., 2001. Factors affecting the social behaviour of crustaceans living symbiotically with other marine invertebrates: a modelling approach. Symbiosis 30, 163–190.

Thiel, M., Correa, C., 2004. Female rock shrimp *Rhynchocinetes typus* mate in rapid succession up a male dominance hierarchy. Behav. Ecol. Sociobiol. 57, 62–68.

Thiel, M., Hinojosa, I.A., 2003. Mating behavior of female rock shrimp *Rhynchocinetes typus* (Decapoda: Caridea) indication for convenience polyandry and cryptic female choice. Behav. Ecol. Sociobiol. 55, 113–121.

Thiel, M., Chak, S.T., Dumont, C.P., 2010. Male morphotypes and mating behavior of the dancing shrimp *Rhynchocinetes brucei* (Decapoda: Caridea). J. Crust. Biol. 30, 580–588.

Thiel, M., 2002. Reproductive biology of a small isopod symbiont living on a large isopod host: from the maternal marsupium to the protective grip of guarding males. Mar. Biol. 141, 175–183.

Thomas, J.D., Barnard, J.L., 1983. Transformation of the Leucothoides morph to the *Anamixis morph* (Amphipoda). J. Crust. Biol. 3, 154–158.

Thompson, G.A., McLay, C.L., 2005. Mating behaviour of *Heterozius rotundifrons* (Crustacea: Brachyura: Belliidae): is it a hard or soft shell mater? Mar. Freshwater Res. 56, 1107–1116.

Thorne, B.L., 1997. Evolution of eusociality in termites. Annu. Rev. Ecol. Syst. 27–54.

Ting, J.H., Kelly, L.S., Snell, T.W., 2000. Identification of sex, age and species-specific proteins on the surface of the harpacticoid copepod *Tigriopus japonicus*. Mar. Biol. 137, 31–37.

Tinikul, Y., Mercier, A.J., Soonklang, N., Sobhon, P., 2008. Changes in the levels of serotonin and dopamine in the central nervous system and ovary, and their possible roles in the ovarian development in the giant freshwater prawn, *Macrobrachium rosenbergii*. Gen. Comp. Endocrinol. 158, 250–258.

Tinikul, Y., Soonthornsumrith, B., Phoungpetchara, I., Meeratana, P., Poljaroen, J., Duangsuwan, P., Soonklang, N., Mercier, A.J., Sobhon, P., 2009. Effects of serotonin, dopamine, octopamine, and spiperone on ovarian maturation and embryonic development in the giant freshwater prawn, *Macrobrachium rosenbergii* (De Man, 1879). Crustaceana 82, 1007–1022.

Tinikul, Y., Poljaroen, J., Kornthong, N., Chotwiwatthanakun, C., Anuracpreeda, P., Poomtong, T., Sobhon, P., 2011. Distribution and changes of serotonin and dopamine levels in the central nervous system and ovary of the Pacific white shrimp, *Litopenaeus vannamei*, during ovarian maturation cycle. Cell Tis. Res. 345 (1), 103–124.

Tinturier-Hamelin, E., 1963. Polycromatisme et de¨termination ge¨ne¨tique du sexe chez l'espe©ce polytypique *Idotea baltica* (Pallas) (isopode valvifere©re). Cahiers Biol. Mar. 4, 473–591.

Tirmizi, N.M., Kazmi, M.A., 1971. Sexual dimorphism in *Saron marmoratus* (Olivier) (Decapoda, Hippolytidae). Crustaceana 21, 283–293.

Tirumalai, R., Subramoniam, T., 1992. Purification and characterization of vitellogenin and lipovitellins of the sand crab *Emerita asiatica*: molecular aspects of crab yolk proteins. Mol. Reprod. Dev. 58, 54–62.

Tirumalai, R., Subramoniam, T., 2001. Carbohydrate components of lipovitellin of the sand crab *Emerita asiatica*. Mol. Reprod. Dev. 58, 54–62.

Tirumalai, R., 1996. Some Molecular Aspects of Vitellogenesis in Decapod Crustaceans (Ph.D. thesis). University of Madras.

Tiu, S.H.K., Hui, J.H.L., He, J.G., Tobe, S.S., Chan, S.M., 2006. Characterization of vitellogenin in the shrimp *Metapenaeus ensis*: expression studies and hormonal regulation of MeVg1 transcription in vitro. Mol. Reprod. Dev. 73, 424–436.

Tiu, S.H.K., Benzie, J., Chan, S.M., 2008. From hepatopancreas to ovary: molecular characterization of a shrimp vitellogenin receptor involved in the processing of vitellogenin. Biol. Reprod. 79 (1), 66–74.

Tiu, S.H., Hui, H.L., Tsukimura, B., Tobe, S.S., He, J.G., Chan, S.M., 2009. Cloning and expression study of the lobster (*Homarus americanus*) vitellogenin: conservation in gene structure among decapods. Gen. Comp. Endocrinol. 160, 36–46.

Tokishita, S.I., Kato, Y., Kobayashi, T., Nakamura, S., Ohta, T., Yamagata, H., 2006. Organization and repression by juvenile hormone of a vitellogenin gene cluster in the crustacean, *Daphnia magna*. Biochem. Biophys. Res. Comm. 345 (1), 362–370.

Tom, M., Goren, M., Ovadia, M., 1987. Localization of the vitellin and its possible precursors in various organs of *Parapenaeus longirostris* (Crustacea, Decapoda, Penaeidae). Int. J. Invertebr. Reprod. Dev. 12 (1), 1–12.

Tomy, S., Saikrithi, P., James, N., Balasubramanian, C.P., Panigrahi, A., Otta, S.K., Subramoniam, T., Ponniah, A.G., 2016. Serotonin induced changes in the expression of ovarian gene network in the Indian white shrimp, *Penaeus indicus*. Aquaculture 452, 239–246.

Tóth, E., Bauer, R.T., 2007. Gonopore sexing technique allows determination of sex ratios and helper composition in eusocial shrimps. Mar. Biol. 151, 1875–1886.

Toyota, K., Kato, Y., Sato, M., Sugiura, N., Miyagawa, S., Miyakawa, H., Mizutani, T., 2013. Molecular cloningof doublesex genes of four cladocera (water flea) species. BMC Genomics 14 (1), 239.

Tracey, E., Pereira, A., Hughes, M., Korey, C.A., 2013. The embryonic development of the snapping shrimp, *Alpheus angulosus* McClure, 2002 (Decapoda, Caridea). Crustaceana 86 (11), 1367–1381.

Treeattrakool, S., Panyim, S., Udomkit, A., 2011. Induction of ovarian maturation and spawning in *Penaeus monodon* broodstock by double-stranded RNA. Mar. Biotech. 13, 163–169.

Treerattrakool, S., Panyim, S., Chan, S.M., Withyachumnarnkul, B., Udomkit, A., 2008. Molecular characterization of gonad-inhibiting hormone of *Penaeus monodon* and elucidation of its inhibitory role in vitellogenin expression by RNA interference. FEBS 275, 970–980.

Trentini, M., Scanabissi, F.S., 1978. Ultrastructural observations on the oogenesis of *Triops cancriformis* (Crustacea, Notostraca), I. Origin and differentiation of nurse cells. Cell Tissue Res. 194, 71–77.

Trivers, R., 1972. Parental Investment and Sexual Selection, vol. 136. Biological Laboratories, Harvard University, p. 179.

Trujillo, L.R., Primavera, J.H., 1986. Induced maturation in ablated *Penaeus notialis* and *Penaeus schmitti*. In: Maclean, J.L., Dizon, L.B., Hosillos, L.V. (Eds.), The First Asian Fisheries Forum. Proceedings of the First Asian Fisheries Forum, 26-31 May 1986, Manila, Philippines. Asian Fisheries Society, Manila, Philippines, pp. 697–700.

Tsang, W.S., Quackenbush, L.S., Chow, B.K., Tiu, S.H., He, J.G., Chan, S.M., 2003. Organization of the shrimp vitellogenin gene: evidence of multiple genes and tissue specific expression by the ovary and hepatopancreas. Gene 303, 99–109.

Tsukimura, B., Borst, D.W., 1992. Regulation of methyl farnesoate levels in the lobster, *Homarus americanus*. Gen. Comp. Endocrinol. 86, 287–303.

Tsukimura, B., Kamemoto, F.I., Borst, D.W., 1993. Cyclic nucleotide regulation of methyl farnesoate synthesis by the mandibular organ of the lobster, *Homarus americanus*. J. Exp. Zool. 265, 427–431.

Tsukimura, B., 2001. Crustacean vitellogenesis: its role in oocyte development. Am. Zool. 41 (3), 465–476.

Tsutsui, N., Kawazoe, I., Ohira, T., Jasmani, S., Yang, W.J., Wilder, M.N., Aida, K., 2000. Molecular characterization of a cDNA encoding vitellogenin and its expression in the hepatopancreas and ovary during vitellogenesis in the kuruma prawn, *Penaeus japonicus*. Zool. Sci. 17 (5), 651–660.

Tsutsui, N., Saido-Sakanaka, H., Yang, W.J., Jayasankar, V., Jasmani, S., Okuno, A., Ohira, T., Okumura, T., Aida, K., Wilder, M.N., 2004. Molecular characterization of a cDNA encoding vitellogenin in the coonstriped shrimp, *Pandalus hypsinotus* and site of vitellogenin mRNA expression. J. Exp. Zool. 301A, 802–814.

Tsutsui, N., Katayama, H., Ohira, T., Nagasawa, H., Wilder, M.N., Aida, K., 2005a. The effects of crustacean hyperglycemic hormone family peptides on vitellogenin gene expression in the kuruma prawn, *Marsupenaeus japonicus*. Gen. Comp. Endocrinol. 144, 232–239.

Tsutsui, N., Kim, Y.K., Jasmani, S., Ohira, T., Wilder, M.N., Aida, K., 2005b. The dynamics of vitellogenin gene expression differs between intact and eyestalk ablated kuruma prawn *Penaeus (Marsupenaeus) japonicus*. Fisher. Sci. 71, 249–256.

Tsutsui, N., Ohira, T., Kawazoe, I., Takahashi, A., Wilder, M.N., 2007. Purification of sinus gland peptides having vitellogenesis-inhibiting activity from the whiteleg shrimp *Litopenaeus vannamei*. Mar. Biotech. 9, 360–369.

Tudge, C.C., Jamieson, B.G.M., 1991. Ultrastructure of the mature spermatozoon of the coconut crab *Birgus latro* (Coenobitidae: Paguroidea: Decapoda). Mar. Biol. 108 (3), 395–402.

Tudge, C.C., Jamieson, B.G., 1996. Spermatophore and spermatozoal morphology in the Porcellanidae. II. The genera *Petrolisthes* and *Polyonyx* (Decapoda: Anomura). J. Crust. Biol. 16 (3), 535–546.

Tudge, C.C., Scheltinga, D.M., 2002. Spermatozoa morphology of the freshwater anomuran *Aegla longirostris* bond-Buckup & Buckup, 1994 (Crustacea: Decapoda: Aeglidae) from South America. Proc. Biol. Soc. Wash. 115, 118–128.

Tudge, C.C., Jamieson, B.G., Sandberg, L., Erséus, C., 1998. Ultrastructure of the mature spermatozoon of the king crab *Lithodes maja* (Lithodidae, Anomura, Decapoda): further confirmation of a lithodid-pagurid relationship. Invertebr. Biol. 57–66.

Tudge, C.C., Scheltinga, D.M., Jamieson, B.G.M., 1999. Spermatozoal ultrastructure in the Hippoidea (Anomura, Decapoda). J. Submicro. Cytol. Pathol. 31, 1–13.

Tudge, C.C., Scheltinga, D.M., Jamieson, B.G.M., 2001. Spermatozoal morphology in the symmetrical hermit crab *Pylocheles* (Bathycheles) sp. (Crustacea, Decapoda, Anomura, Paguroidea, Pylochelidae). Zoosystema 23, 117–130.

Tudge, C.C., 1991. Spermatophore diversity within and among the hermit crab families, Coenobitidae, Diogenidae, and Paguridae (Paguroidea, Anomura, Decapoda). Biol. Bull. 181 (2), 238–247.

Tudge, C.C., 1992. Comparative ultrastructure of hermit crab spermatozoa (Decapoda: Anomura: Paguroidea). J. Crust. Biol. 12 (3), 397–409.

Tudge, C.C., 1995. Ultrastructure and Phylogeny of Anomuran Crab Spermatozoa, PhD, Department of Zoology. The University of Queensland, Australia.

Tudge, C.C., 1997. Phylogeny of the Anomura (Decapoda, Crustacea): spermatozoa and spermatophore morphological evidence. Contrib. Zool. 67, 125–142.

Tudge, C.C., 1999. Spermatophore morphology in the hermit crab families Paguridae and Parapaguridae (Paguroidea, Anomura, Decapoda). Inv. Rep. Dev. 35 (3), 203–214.

Tudge, C.C., 2003. Endemic and enigmatic: the reproductive biology of *Aegla* (Crustacea: Anomura: Aeglidae) with observations on sperm structure. Mem. Mus. Vic. 60 (1), 63–70.

Tudge, C., 2009. Spermatozoal morphology and its bearing on decapod phylogeny. Decapod crustacean phylogenetics. Crustaceana 18, 101–119.

Turra, A., 2004. Intersexuality in hermit crabs: reproductive role and fate of gonopores in intersex individuals. J. Mar. Biol. Assoc. U.K. 84, 757–759.

Tweedie, M.W.F., 1950. Notes on the grapsoid crabs from the Raffles Museum. Bull. Rafels Mus. 310–324.

Uma, K., Subramoniam, T., 1979. Histochemical characteristics of spermatophore layers of *Scylla serrata* (Forskal) (Decapoda: Portunidae). Int. J. Invertebr. Reprod. Dev. 1, 31–40.

Uma, K., Subramoniam, T., 1984. A comparative study of the spermatophore in *Scylla serrata* (Forskal) (Decapoda, Brachyura) and *Clibanarius longitarsus* (de Haan) (Decapoda, Anomura). J. Mar. Biol. Assoc. India 26, 103–108.

Vaca, A.A., Alfaro, J., 2000. Ovarian maturation and spawning in the white shrimp, *Penaeus vannamei*, by serotonin injection. Aquaculture 182 (3), 373–385.

Vacquier, V.D., Moy, G.W., 1977. Isolation of bindin: the protein responsible for adhesion of sperm to sea urchin eggs. Proc. Natl. Acad. Sci. U.S.A. 74, 2456–2460.

Vacquier, V.D., 1979. The interaction of sea urchin gametes during fertilization. Am. Zool. 19, 839–849.

Van Beek, E., De Loof, A., 1988. Radio immunological determinations of concentrations in C21, C19 and C18 steroids during the reproductive cycle of female *Artemia* sp. Crustacea. Comp. Biochem. Physiol. A89, 595.

Van Beek, E., Van Brussel, M., Crielt, G., De Loof, A., 1987. A possible extra-ovarian site for synthesis of lipovitellin during vitellogenesis in *Artemia* sp. (Crustacea; Anostraca). Int. J. Invertebr. Rep. Dev. 12 (3), 227–238.

Van Berkel, T.J.C., Kruijt, J.K., Kempen, H.J.M., 1985. Specific targeting of high density lipoproteins to liver hepatocytes by incorporation of a trisgalactoside- terminated cholesterol derivative. J. Biol. Chem. 260, 12203–12207.

Vandel, A., 1941. Recherches sur la génétique et la sexualité des isopodes terrestres. VI: Les phénomènes de monogénie chez les Oniscoïdes. Bull. Biol. Fr. Belg. 75, 316–363.

Vandenberg, G.F., Adriaens, D., Verlycke, T., Janssen, C.R., 2003. Effects of 17 alpha-ethinylestradiol on sexual development of the amphipod *Hyalella azteca*. Ecotoxicol. Environ. Saf. 54, 216–222.

Vanichviriyakit, R., Kruevaisayawan, H., Weerachatyanukul, W., Pongtippatee-Tawipreeda, P., Withyachumnarnkul, B., Pratoomchat, B., Chavadej, J., Sobhon, P., 2004. Molecular modification of *Penaeus monodon* sperm in female thelycum and its consequent responses. Mol. Rep. Dev. 69, 356–363.

Vannini, M., 1980a. Burrows and digging behavior in Ocypode and other crabs (Crustacea: Brachyura). Monitore Zool. Ital. (N.S.) (Suppl. 13), 11–44.

Vannini, M., 1980b. Notes on the behavior of *Ocypode ryderi* Kinglsey (Crustacea, Brachyura). Mar. Behav. Physiol. 7, 171–183.

Varadarajan, S., Subramoniam, T., 1980. Histological changes during vitellogenesis in the ovary of the hermit crab, *Clibanarius clibanarius*. Proc. Indian Nat. Sci. Acad. 46B, 645–651.

Varadarajan, S., Subramoniam, T., 1982. Reproduction of the continuously breeding tropical hermit crab *Clibanarius clibanarius*. Mar. Ecol. Prog. Ser. 8, 197–201.

Varadarajan, M., 1982. Studies on the Reproductive Tract and Accessory Sex Organs of the Ghost Crab, *Ocypoda platytasis* Milne Edwards (Crustacea: Brachyura) (MPhil Dissertation). University of Madras.

Varaskin, A.A., Reunova, O.V., 1993. Effects of indomethacin, arachidonic acid and some metabolites on proliferation of spermatogonia in the scallop *Mizuhopecten yessoensis*. Z. Evol. Biokhim. Fiziol 29, 330–333.

Vazquez, F.J., Tropea, C., López Greco, L.S., 2008. Development of the female reproductive system in the freshwater crayfish *Cherax quadricarinatus* (Decapoda, Parastacidae). Invertebr. Biol. 127 (4), 433–443.

Veillet, A., 1945. Recherches sur le parasitisme de crabs et de Galathees par le Rhizocephals et le Epicarides. Ann. Inst. Ocean Monaco 22, 194–341.

Ventura, T., Manor, R., Aflalo, E.D., Weil, S., Raviv, S., Glazer, L., Sagi, A., 2009. Temporal silencing of an androgenic gland-specific insulin-like gene affecting phenotypical gender differences and spermatogenesis. Endocrinology 150, 1278–1286.

Ventura, T., Rosen, O., Sagi, A., 2011. From the discovery of the crustacean androgenic gland to the insulin-like hormone in six decades. Gen. Comp. Endocrinol. 173 (3), 381–388.

Vershinin, A., Lukyanova, O.N., 1993. Carotenoids in the developing embryos of sea urchin *Strongylocentrotus intermedius*. Comp. Biochem. Physiol. 104 (2), 371–373.

Vincent, S.G., Keller, R., Subramoniam, T., 2001. Development of vitellogenin-ELISA, an in vivo bioassay, and identification of two vitellogenesis-inhibiting hormones of the tiger shrimp *Penaeus monodon*. Mar. Biotechnol. 3, 561–571.

Visconti, P.E., Moore, G.D., Bailey, J.L., Leclerc, P., Connors, S.A., Pan, D., Olds-Clarke, P., Kopf, G.S., 1995. Capacitation of mouse spermatozoa. II. Protein tyrosine phosphorylation and capacitation are regulated by a cAMP-dependent pathway. Development 121 (4), 1139–1150.

Visudtiphole, V., Klinbunga, S., Kirtikara, K., 2009. Molecular characterization and expression profiles of cyclin A and cyclin B during ovarian development of the giant tiger shrimp *Penaeus monodon*. Comp. Biochem. Physiol. A 152 (4), 535–543.

Vitagliano, E., Marchetti, E., Vitagliano, G., 1996. Skewed sex-ratio, monogeny, and maternal sex determination in two geographical populations of *Asellus aquaticus* (L., 1758) (Isopoda). Crustaceana 69, 455–475.

Vogt, G., Holdich, D.M., 2002. Functional anatomy. Biol. Freshwater Crayfish 53–151.

Volff, J.N., Zarkower, D., Bardwell, V.J., Schartl, M., 2003. Evolutionary dynamics of the DM domain gene family in metazoans. J. Mol. Evol. 57, S241–S249.

Voordouw, M.J., Anholt, B.R., 2002. Environmental sex determination in a splash pool copepod. Biol. J. Linnean Soc. 76, 511–520.

Wada, S., Ashidate, M., Goshima, S., 1997. Observations on the reproductive behavior of the spiny king crab *Paralithodes brevipes* (Anomura: Lithodidae). Crust. Res. 26, 56–61.

Wada, K., 1981. Growth, breeding and recruitment in *Scopimera globosa* and *Hyoplax pusillus* (Crustacea: Ocypodidae) in the estuary of Waka River, middle Japan. Publ. Seto Mar. Biol. Lab. 26, 447–454.

Wada, K., 1983. Movement of burrow locations in *Scopimera globosa* and *Ilyopax pusillus* (Decapoda: Ocypodae). Physiol. Ecol. Jpn. 20, 1–21.

Wada, K., 1984. Pair formation in the two forms of *Macrophthalamus japonicus* de Haan (Crustacea: Brachyura) at a Co-occurring Area. J. Ethol. 2, 7–10.

Waddy, S.L., Aiken, D.E., 1991. Mating and insemination in the American lobster, *Homarus americanus*. In: Bauer, R.T., Martin, J.W. (Eds.), Crustacean Sexual Biology. Columbia University Press, New York, pp. 126–144.

Waddy, S.L., Aiken, D.E., 1995. Temperature regulation of reproduction in female American lobsters (*Homarus americanus*). In: ICES Marine Science Symposia, vol. 199. International Council for the Exploration of the Sea, Copenhagen, Denmark, pp. 54–60. 1991.

Waddy, S.L., Aiken, D.E., De Klejin, D.P.V., 1995. Control of growth and reproduction. Biol. Am. Lobster 217–266.

Wainwright, G., Prescott, M.C., Webster, S.G., Rees, H.H., 1996. Mass spectrometric determination of methyl farnesoate profiles and correlation with ovarian development in the edible crab, *Cancer pagurus*. J. Mass Spectrom. 31, 1338–1344.

Wainwright, G., Websters, S.G., Rees, H.H., 1998. Neuropeptide regulation of biosynthesis of the juvenoid, methyl farnesoate, in the edible crab, *Cancer pagurus*. Biochem. J. 334, 651–657.

Walker, W.F., 1980. Sperm utilization strategies in nonsocial insects. Am. Nat. 115, 780–799.

Walker, G., 1992. Cirripedia. Microsc. Anat. Invertebr. 9, 249–311.

Wall, D.A., Patel, S., 1987. Multivesicular bodies play a key role in vitellogenin endocytosis by *Xenopus oocytes*. Dev. Biol. 119 (1), 275–289.

Wallace, R.A., Walker, S.L., Hausehka, P.V., 1967. Crustacean lipovitellin. Isolation and characterization of the major high density lipoprotein from the eggs of decopods. Biochemistry 9, 1582–1590.

Walley, L.J., White, F., Brander, K.M., 1971. Sperm activation and fertilization in *Balanus balanoides*. J. Mar. Biol. Assoc. U.K. 51, 489–494.

Walley, L.J., 1965. The development and function of the oviducal gland in *Balanus balanoides*. J. Mar. Biol. Assoc. U.K. 45, 115–128.

Wang, Q., Fang, D.A., Sun, J.L., Wang, Y., Wang, J., Liu, L.H., 2012. Characterization of the vasa gene in the Chinese mitten crab *Eriocheir sinensis*: a germ line molecular marker. J. Insect Physiol. 58, 960–965.

Wang, J., Fang, D.A., Wang, Y., Wang, Y.L., Cheng, L., He, L., Wang, Q., 2013. Cathepsin A protein from the accessory sex gland of the Chinese mitten crab (*Eriocheir sinensis*) plays a key role in spermatophore digestion. J. Insect Physiol. 59, 953–960.

Wang, L.H., Lin, C., Huang, C.Y., Tsai, S., 2015. Studies on lipid content and composition in banded coral shrimp (*Stenopus hispidus*) embryos. J. Crust. Biol. 35 (5), 622–626.

Warner, G.F., 1967. The life history of the mangrove tree crab, *Aratus pisoni*. J. Zool.. Lond. 153, 321–335.

Warner, G.F., 1970. Behaviour of two species of graspid crab during intrasepcific encounter. Behaviour 36, 6–19.

Warner, R.R., 1988. Sex change and size-advantage model. Trends Ecol. Evol. 3, 133–136.

Warrier, S., Subramoniam, T., 2002. Receptor mediated yolk protein uptake in the crab *Scylla serrata*: crustacean vitellogenin receptor recognizes related mammalian serum lipoproteins. Mol. Reprod. Dev. 61 (4), 536–548.

Warrier, S.R., Subramoniam, T., 2003. Instability of crab vitellogenin and its immunological relatedness with mammalian atherogenic lipoproteins. Mol. Reprod. Dev. 64 (3), 329–340.

Warrier, S., Tirumalai, R., Subramoniam, T., 2001. Occurrence of vertebrate steroids, estradiol 17-β and progesterone in the reproducing females of the mud crab *Scylla serrata*. Comp. Biochem. Physiol. 130A, 283–294.

Watts, M.M., Pascoe, D., Carroll, K., 2002. Population responses of the freshwater amphipod *Gammarus pulex* (L.) to an environmental estrogen, 17α-ethinylestradiol. Environ. Toxicol. Chem. 21, 445–450.

Webster, S.G., Keller, R., Dircksen, H., 2012. The CHH-superfamily of multifunctional peptide hormones controlling crustacean metabolism, osmoregulation, moulting, and reproduction. Gen. Comp. Endocrinol. 175, 217–233.

Webster, S.G., 1991. Amino acid sequence of putative moult-inhibiting hormone from the crab *Carcinus maenas*. Proc. R. Soc. Lond. B Biol. Sci. 244, 247–252.

Weeks, S.C., Benvenuto, C., Reed, S.K., 2006. When Males and Hermaphrodites Coexist: Review of Androdioecy in Animals, Program in Evolution, Ecology, and Organismal Biology, Department of Biology. The University of Akron, Akron, OH, USA, pp. 44325–53908.

Weeks, S.C., Crosser, B.R., Bennett, R., Gray, M., Zucker, N., 2000. Maintenance of androdioecy in the freshwater shrimp, *Eulimnadia texana*: estimates of inbreeding depression in two populations. Evolution 54, 878–887.

Weeks, S.C., Crosser, B.R., Gray, M.M., 2001. Relative fitness of two hermaphroditic mating types in the androdioecious clam shrimp, *Eulimnadia texana*. J. Evol. Biol. 14 (1), 83–94.

Weeks, S.C., Posgai, R.T., Cesari, M., Scanabissi, F., 2005. Androdioecy inferred in the clam shrimp *Eulimnadia agassizii* (Spinicaudata: Limnadiidae). J. Crust. Biol. 25, 323–328.

Weeks, S.C., Benvenuto, C., Reed, S.K., 2006a. When males and hermaphrodites coexist: a review of androdioecy in animals. Integr. Comp. Biol. 46, 449–464.

Weeks, S.C., Sanderson, T.F., Reed, S.K., Zofkova, M., Knott, B., Balaraman, U., Hoeh, W.R., 2006b. Ancient androdioecy in the freshwater crustacean *Eulimnadia*. Proc. Royal Soc. London B 273, 725–734.

Weghofer, A., Schnepf, S., Barad, D., 2007. The impact of luteinizing hormone in assisted reproduction: a review. Curr. Opin. Obstet. Gynecol. 19 (3), 253–257 17.

Wehrtmann, I.S., Graeve, M., 1998. Lipid composition and utilization in developing eggs of two tropical marine caridean shrimps (Decapoda: Caridea: Alpheidae, Palaemonidae). Comp. Biochem. Physiol. Part B 121 (4), 457–463.

Weinheimer, A.J., Spraggins, 1969. The occurrence of two new prostaglandins (15-epi-PGA2 and its acetate, methyl ester) in the gorgonian *Plexaura homomalla*. Tetrahedron Lett. 59, 5185–5188.

Weitzman, M.C., 1966. Oogenesis in the tropical land crab, *Gecarcinus lateralis* (Freminville). Z. für Zellforschung mikroskopische Anatomie 75 (1), 109–119.

Wenner, A.M., Haley, S.R., 1981. On the question of sex reversal in mole crabs (Crustacea, Hippidae). J. Crust. Biol. 1, 506–517.

Wenner, A.M., 1972. Sex ratio as a function of size in marine Crustacea. Ann. Nat. 10, 321–330.

Werren, J.H., 1987. The coevolution of autosomal and cytoplasmic sex ratio factors. J. Theor. Biol. 124, 317–334.

Wharton, G.W., 1942. A typical beach animal, the mole crab *Emerita talpoidea* (Say). In: Ecology of Sand Beaches at Beaufort Ecol. Monogr., vol. 12. , pp. 137–181.

Whitaker, M.J., Irvine, R.F., 1984. Inositol 1, 4, 5-trisphosphate microinjection activates sea urchin eggs. Nature 312, 636–639.

Wickler, W., Seibt, U., 1981. Monogamy in Crustacea and man. Z. für Tierpsychologie 57, 215–234.

Wikramanayake, A.H., Uhlinger, K.R., Griffin, F.J., Clark, W.H., 1992. Sperm of the shrimp *Sicyonia ingentis* undergo a bi-phasic capacitation accompanied by morphological changes. Dev. Grow. Differ. 34 (3), 347–355.

Wilber, D.H., 1989. The influence of sexual selection and predation on the mating and postcopulatory guarding behavior of stone crabs (Xanthidae, *Menippe*). Behav. Ecol. Sociobiol. 24, 445–451.

Wilder, N.M., Okumura, T., Aida, K., 1991. Accumulation of ovarian ecdysteroids in synchronization with gonadal development in the giant freshwater prawn *Macrobrachium rosenbergii*. Zool. Sci. 8, 919–927.

Wilder, M.N., Okumura, T., Suzuki, Y., Fusetani, N., Aida, K., 1994. Vitellogenin production induced by eyestalk ablation in juvenile giant freshwater prawn *Macrobrachium rosenbergii* and trial methyl farnesoate administration. Zool. Sci. 11, 45–53.

Wilder, N.M., Fusetani, N., Aidu, K., 1995. The presence of 20-hydroxyscdysonic acid and ecdysonic acid in the eggs of the giant freshwater prawn *Macrobrachium rosenbergii*. Fish. Sci. 61, 101–106.

Wilder, M.N., Subramoniam, T., Aida, K., 2002. Yolk proteins of Crustacea. In: Adiyodi, K.G., Adiyodi, R.G., Raikhel, A.S., Sappington, T.W. (Eds.), Reproductive Biology of Invertebrates. Recent Progress in Vitellogenesis, vol. XII. Science Publishers Inc, Enfield, NH, USA, pp. 131–174.

Wilder, M.N., Okumura, T., Tsutsui, N., 2010. Reproductive mechanisms in Crustacea focusing on selected prawn species: vitellogenin structure, processing and synthetic control. Aqua BioSci. Monogr. 3 (3), 73–110.

Wilson, E.O., 1971. The Insect Societies. Belknap Press, Cambridge, MA.

Wilson, G.D.E., 1991. Functional morphology and evolution of isopod genitalia. In: Bauer, R.T., Martin, J.W. (Eds.), Crustacean Sexual Biology. Columbia University Press, New York, USA, pp. 228–245.

Wilson, G.D.F., 2009. The phylogenetic position of the Isopoda in the Peracarida (Crustacea: Malacostraca). Arthropod Syst. Phylogeny 67, 159–198.

Wingstrand, K.G., 1978. Comparative spermatology of the Crustacea Entomostraca 1. Subclass Branchiopoda. Biol. Skr. K. Dan. Vidensk. Selsk. 22, 1–66.

Wirtz, P., 1997. Crustacean symbionts of the sea anemone *Telmatactis cricoides* at Madeira and the Canary Islands. J. Zool. 242, 799–811.

Wittmann, K.J., 1982. Untersuchungen zur Sexualbiologie einer mediterranen Mysidacee (Crustacea), *Leptomysis lingvura* GO Sars. Zool. Anzeiger 209 (5–6), 362–375.

Wong, J.L., Wessel, G.M., 2006. Defending the zygote: search for the ancestral animal block to polyspermy. Curr. Top. Dev. Biol. 72, 1–151.

Wongprasert, K., Asuvapongpatana, S., Poltana, P., Tiensuwan, M., Withyachumnarnkul, B., 2006. Serotonin stimulates ovarian maturation and spawning in the black tiger shrimp *Penaeus monodon*. Aquaculture 261, 1447–1454.

Worley, E.R., 1939. A study on the sperm forming components in three species of decapods (*Pagurus pollicaris*, Say, *Homarus americanus* Miln. Edwards and *Libinia emarginata* Leach). Cellule 48, 149–176.

Wourms, J.P., 1987. Oogenesis. In: Giese, A.C., Pearse, J.S., Pearse, V.B. (Eds.), Reproduction of Marine Invertebrates, vol. 9. Blackwell, California, pp. 50–178.

Wu, L.T., Chu, K.H., 2008. Characterization of heat shock protein 90 in the shrimp *Metapenaeus ensis*: evidence for its role in the regulation of vitellogenin synthesis. Mol. Reprod. Dev. 75 (5), 952–959.

Wu, X., Chen, H., Liu, Z., Cheng, Y., 2014. Immunorecognition and distribution of progesterone receptors in the Chinese mitten crab *Eriocheir sinensis* during ovarian development. J. Shel. Res. 33 (1), 35–43.

Yager, J., 1981. Remipedia, a new class of Crustacea from a marine cave in Bahamas. J. Crust. Biol. 1, 328–333.

Yager, J., 1989. The male reproductive system, sperm, and spermatophores of the primitive, hermaphroditic, remipede crustacean *Speleonectus benjamini*. Invertebr. Reprod. Dev. 15, 75–81.

Yager, J., 1991. The reproductive biology of two species of Remipedia. In: Bauer, R.T., Martin, J.W. (Eds.), Crustacean Sexual Biology. Colombia University Press, New York, pp. 271–289.

Yaldwyn, J.C., 1966. Protandrous hermaphroditism in decapod prawns of the families Hippolytidae and Campylonotidae. Nature 209, 1366.

Yamano, K., Qiu, G.F., Unuma, T., 2004. Molecular cloning and ovarian expression profiles of thrombospondin, a major component of cortical rods in mature oocytes of penaeid shrimp, *Marsupenaeus japonicus*. Biol. Reprod. 70 (6), 1670–1678.

Yamashita, Y., Okumura, T., Yamada, H., 2001. Intersexuality in Acanthomysis mitskurii (Mysidacea) in Sendai Bay, Northeastern Japan. Plankton. Ecol. 48, 128–132.

Yanagimachi, R., 1994. Mammalian fertilization. In: Knobil, Neil (Eds.), The Physiology of Reproduction. Raven press, New York, pp. 189–317.

Yanagisawa, Y., Hamaishi, A., 1986. Mate acquisition by a solitary crab *Zebrida adamsii*, a symbiont of the sea urchin. J. Ethol. 4, 153–162.

Yang, W.J., Ohira, T., Tsutsui, N., Subramoniam, T., Huong, D.T.T., Aida, K., Wilder, M.N., 2000. Determination of amino acid sequence and site of mRNA expression of four vitellins in the giant freshwater prawn, *Macrobrachium rosenbergii*. J. Exp. Zool. 287 (6), 413–422.

Yano, I., Chinzei, Y., 1987. Ovary is the site of vitellogenin synthesis in kuruma prawn, *Penaeus japonicus*. Comp. Biochem. Physiol. Part B 86 (2), 213–218.

Yano, I., Hoshino, R., 2006. Effects of 17 β-estradiol on the vitellogenin synthesis and oocyte development in the ovary of kuruma prawn (*Marsupenaeus japonicus*). Comp. Biochem. Physiol. Part A 144, 18–23.

Yano, I., Wyban, J.A., 1992. Induced ovarian maturation of *Penaeus vannamei* by injection of lobster brain extract. Bull. Natl. Res. Inst. Aquacult. 21, 1–7.

Yano, I., Kanna, R.A., Oyama, R.N., Wyban, J.A., 1988. Mating behaviour in the penaeid shrimp *Penaeus vannamei*. Mar. Biol. 97, 171–175.

Yano, I., Fingerman, M., Nagabhushanam, R., 2000. Endocrine control of reproductive maturation in penaeid shrimp. In: Fingerman, M., Nagabhushanam, R. (Eds.), Recent Advances in Marine Biotechnology: Aquaculture, Part A: Seaweeds and Invertebrates. Science Publishers, Enfield, NH, pp. 161–176.

Yano, I., 1988. Oocyte development in the kuruma prawn *Penaeus japonicus*. Mar. Biol. 99, 547–553.

Yano, I., 1993. Ultra intensive culture and maturation in captivity of penaeid shrimp. In: McVey, J.P. (Ed.), CRC Handbook of Mariculture, Crustacean Aquaculture, vol. 1. second ed. CRC, Florida, pp. 289–313.

Yano, I., 1998. Hormonal control of vitellogenesis in penaeid shrimp. In: Flegel, T.W. (Ed.), Advances in Shrimp Biotechnology. National Center for Genetic Engineering and Biotechnology, Bangkok, pp. 29–31.

Yasuzumi, G., 1960. Spermatogenesis in animals as revealed by electron microscopy VII. Spermatid differentiation in the crab *Eriocheir japonicus*. J. Biophys. Biochem. Cytol. 7, 73–77.

Ye, H.H., Ma, J., Lin, Q.W., 2011. Occurrence of follicle-stimulating hormone-like substance in the Kuruma prawn, *Marsupenaeus japonicus*, during ovarian maturation. Mar. Biol. 7 (3), 304–309.

Young, N.J., Webster, S.G., Jones, D.A., Rees, H.H., 1991. Profile of embryonic ecdysteroids in the decapod crustacean *Macrobrachium rosenbergii*. Invertebr. Reprod. Dev. 20, 201–212.

Young, N.J., Webster, S.G., Ress, H.H., 1993. Ovarian and haemolymph ecdysteroid titers during vitellogenesis in *M. rosenbergii*. Gen. Comp. Endocrinol. 90 (2), 183–191.

Yue, G.H., Wang, G.L., Zhu, B.Q., Wang, C.M., Zhu, Z.Y., Lo, L.C., 2008. Discovery of four natural clones in a crayfish species *Procambarus clarkii*. Int. J. Biol. Sci. 4, 279–282.

Zadorozhny, P.A., Borisovets, E.E., Yakush, E.V., Davidyuk, T.S., 2008. Change of carotenoid composition in crabs during embryogenesis. J. Evol. Biochem. Physiol. 44 (4), 450–461.

Zaffagnini, F., 1968. Contributo alla conoscenza della biologia riproduttiva dei Fillopodi Concostraci. 2. Osservazioni sull'apparato riproduttore e sull'accrescimento ovocitario di *Limnadia lenticularis* (L.). In: Mem. 1st. Ital. Idrobiol. Dott. Marco di Marchi, vol. 23, pp. 129–140.

Zagalsky, P.F., Gilchrist, B.M., 1976. Isolation of a blue canthaxanthin-lipovitellin from the yolk platelets of *Branchipus stagnalis* (L.) (Crustacea: Anostraca). Comp. Biochem. Physiol. Part B 55 (2), 195–200.

Zagalsky, P.F., Cheesman, D.F., Ceccaldi, H.J., 1967. Studies on carotenoid –containing lipoproteins isolated from the eggs and ovaries of certain marine invertebrates. Comp. Biochem. Physiol. 22, 851–871.

Zagalsky, P.F., 1985. A study of the astaxanthin-lipovitellin, ovoverdin, isolated from the ovaries of the lobster, *Homarus gammarus* (L.). Comp. Biochem. Physiol. Part B 80 (3), 589–597.

Zara, F.J., Toyama, M.H., Caetano, F.H., Lopez-Greco, L.S., 2012. Spermatogenesis, spermatophore, and seminal fluid production in the adult blue crab *Callinestes danae* (Portunidae). J. Crust. Biol. 32 (2), 249–262.

Zara, F.J., Pereira, G.R.R., Sant'anna, B.S., 2014. Morphological changes in the seminal receptacle during ovarian development in the speckled swimming crab *Arenaeus cribrarius*. Biol. Bull. 227, 19–32.

Zardus, J.D., Hadfield, M.G., 2004. Larval development and complemental males in *Chelonibia testudinaria*, a barnacle commensal with sea turtles. J. Crust. Biol. 24, 409–421.

Zarkower, D., 2001. Establishing sexual dimorphism: conservation amidst diversity. Nat. Rev. Gen. 2, 175–185.

Zerbib, C., 1964. Evolution post-embryonnaire de la voie deferente chez le male et chez la femelle normale et masculinisee *d'Orchestia gammarella* Pallas (Crustace Amphipode). Bull. Biol. Fr. Belg. 98, 391–408.

Zerbib, C., 1976. Nature chimique des enclaves vitellines de l'ovocyte du Crustacea Amphipoda *Orchestia gammarellus* (Pallas). Ann. Histochim. 21, 279–285.

Zhang, D., Lin, J., 2004. Mating without anterior pleopods in a simultaneous hermaphroditic shrimp, *Lysmata wurdemanni*. (Decapoda, Caridea). Crustaceana 77, 1203–1212.

Zhang, D., Lin, J., Harley, M., Hardege, J.D., 2010a. Characterization of a sex pheromone in a simultaneously hermaphroditic shrimp *Lysmata wurdemanni*. Mar. Biol. 157, 1–6.

Zhang, D., Zhu, J., Lin, J., Hardege, J.D., 2010b. Surface glycoproteins are not the contact pheromones in the *Lysmata* shrimp. Mar. Biol. 157, 171–176.

Zhang, D., Terschak, J.A., Harley, M.A., Lin, J., Hardege, J.D., 2011. Simultaneously hermaphroditic shrimp use lipophilic cuticular hydrocarbons as contact sex pheromones. PLoS ONE 6, 717–720.

Zhang, Y., Fu, H., Qiao, H., Jin, S., Jiang, S., Xiong, Y., Gong, Y., Zhang, X., 2013. Molecular cloning and expression analysis of transformer-2 gene during development in *Macrobrachium nipponense* (de Haan 1849). J. World Aquacul. Soc. 44 (3), 338–349.

Zhao, G.F., Li, G.L., Zhu, C.H., 2009. Preliminary study on sex differentiation of *Litopenaeus vannamei*. J. Guangdong Ocean Univ. 3, 19–23.

Zhu, L.S., Miura, K., Chen, L., Raikhel, A.S., 2003. Cyclicity of mosquito vitellogenic ecdysteroid-mediated ignalling is modulated by alternative dimerization of the RXR homologue Ultraspiracle. Proc. Natl. Acad. Sci. U.S.A. 100, 544–549.

Zhu, J., Zhang, D., Lin, J., Grace, M.S., 2012. Aesthetascs in *Lysmata* spp. Shrimp: sexual dimorphism and relationship with social environments. Mar. Biol. 159, 507–517.

Zimmer, M., 2001. Why do male terrestrial isopods (Isopoda: Oniscidea) not guard females. Anim. Behav. 62, 815–821.

Zmora, N., Trant, J., Chan, S.M., Chung, J.S., 2007. Vitellogenin and its messenger RNA during ovarian development in the female blue crab, *Callinectes sapidus*: gene expression, synthesis, transport, and cleavage. Biol. Reprod. 77 (1), 138–146.

Zmora, N., Trant, J., Zohar, Y., Chung, J.-S., 2009. Molt-inhibiting hormone stimulates vitellogenesis at advanced ovarian developmental stages in the female blue crab, *Callinectes sapidus* 1: an ovarian stage dependent involvement. Saline Syst. 5, 7. http://dx.doi.org/10.1186/1746-1448-5-7.

Zou, E., Fingerman, M., 1999. Effects of estrogenic agents on chitobiase activity in the epidermis and hepatopancreas of the fiddler crab, *Uca pugilator*. Ecotoxicol. Environ. Saf. 42, 185–190.

Zou, E., Fingerman, M., 2000. External features of an intersex fiddler crab, *Uca pugilator*. Crustaceana 73, 417–423.

Zucker, N., Cunningham, M., Adams, H.P., 1997. Anatomical evidence for androdioecy in the clam shrimp *Eulimnadia texana*. Hydrobiologia 359, 171–175.

Zucker, N., Stafki, B., Weeks, S.C., 2001. Maintenance of androdioecy in the freshwater shrimp, *Eulimnadia texana*: relative longevity of males to hermaphrodites. Can. J. Zool. 79, 393–401.

Zucker, N., Aguilar, G.A., Weeks, S.C., McCandless, L.G., 2002. Impact of males on variation in reproductive cycle in an androdioecious desert shrimp. Invertebr. Biol. 121, 66–72.

Zucker, N., 1983. Courtship variation in the neo-tropical fiddler crab *Uca deichmanni*: another example of female incitation to male competition. Mar. Behav. Physiol. 10, 57–79.

Zucker, N., 1984. Delayed courtship in the fiddler crab *Uca musica terpsichores*. Anim. Behav. 32, 735–742.

Żukowska-Arendarczyk, M., 1981. Effect of hypophyseal gonadotropins (FSH and LH) on the ovaries of the sand shrimp *Crangon crangon* (Crustacea: Decapoda). Mar. Biol. 63 (3), 241–247.

Zupo, V., Messina, P., 2007. How do dietary diatoms cause the sex reversal of the shrimp *Hippolyte inermis* Leach (Crustacea, Decapoda). Mar. Biol. 151, 907–917.

Zupo, V., 1994. Strategies of sexual inversion in *Hippolyte inermis* Leach (Crustacea, Decapoda) from a Mediterranean seagrass meadow. J. Exp. Mar. Biol. Ecol. 178, 131–145.

Index

'*Note*: Page numbers followed by "f" indicate figures and "t" indicate tables.'

N

O

P

Kolibri 360 GmbH
Ferdinand-Jühlke-Straße 7
99095 Erfurt, Deutschland
produktsicherheit@kolibri360.de